CHEMISTRY
FOR PHARMACY
STUDENTS

CHEMISTRY
FOR PHARMACY
STUDENTS

General, Organic and
Natural Product Chemistry

Second Edition

LUTFUN NAHAR
Liverpool John Moores University
UK

SATYAJIT SARKER
Liverpool John Moores University
UK

Edition History
1e published 2007, ISBN 9780470017807

Registered Offices
John Wiley & Sons, Inc., 111 River Street, Hoboken, NJ 07030, USA
John Wiley & Sons Ltd, The Atrium, Southern Gate, Chichester, West Sussex, PO19 8SQ, UK

Editorial Office
The Atrium, Southern Gate, Chichester, West Sussex, PO19 8SQ, UK

For details of our global editorial offices, customer services, and more information about Wiley products visit us at www.wiley.com.

Wiley also publishes its books in a variety of electronic formats and by print-on-demand. Some content that appears in standard print versions of this book may not be available in other formats.

Library of Congress Cataloging-in-Publication Data
Names: Nahar, Lutfun, author. | Sarker, Satyajit, author.
Title: Chemistry for pharmacy students : general, organic and natural
   product chemistry / Lutfun Nahar (Liverpool John Moores University, UK),
   Satyajit Sarker (Liverpool John Moores University, UK).
Description: Second edition. | Hoboken, NJ : Wiley, 2019. | Includes index. |
   Identifiers: LCCN 2019009751 (print) | LCCN 2019016343 (ebook) | ISBN
   9781119394464 (Adobe PDF) | ISBN 9781119394488 (ePub) | ISBN 9781119394433
   (pbk.)
Subjects: LCSH: Chemistry–Textbooks. | Pharmaceutical chemistry–Textbooks.
Classification: LCC QD31.3 (ebook) | LCC QD31.3 .S377 2020 (print) | DDC
   540–dc23
LC record available at https://lccn.loc.gov/2019009751

Cover Design: Wiley
Cover Images: © fotohunter /iStock/Getty Images Plus, © Elena Elisseeva/Getty Images, © Thomas Northcut/Getty Images, © REB Images/Getty Images

Set in 9/13pts Ubuntu by SPi Global, Chennai, India
Printed and bound in Singapore by Markono Print Media Pte Ltd

10 9 8 7 6 5 4 3 2 1

Dedicated to pharmacy students, from home and abroad

Contents

Chapter 5: Organic Reactions   215

Chapter 6: Heterocyclic Compounds 327

Preface to the second edition

The first edition of *Chemistry for Pharmacy Students: General, Organic and Natural Product Chemistry* was written to address the need for the right level and appropriate coverage of chemistry in any modern Pharmacy curricula. The first edition reflected on the changing face of Pharmacy profession and the evolving role of pharmacists in the modern healthcare systems, and was aimed at placing chemistry more in the context of medicines and patients. Since the publication in 2007, in subsequent years, the first edition has been translated into the Greek, Japanese and Portuguese languages, and has acclaimed huge acceptance and popularity among Pharmacy students, as well as among academics who teach chemistry in Pharmacy curricula all over the world.

It has been over a decade since the publication of the first edition. We feel that it has now become necessary to compile a second edition, which should be a thoroughly revised and enhanced version of the first. The second edition will also cater for the chemistry requirements in any 'Integrated Pharmacy Curricula', where science in general is meant to be taught 'not in isolation', but together with, and as a part of, other practice and clinical elements of Pharmacy curricula. Whatever may be the structure and content of any Pharmacy curriculum, there will always be two fundamental aspects in it – medicines (drugs) and patients.

Pharmacy began its journey as a medicine (drug)-focused science subject but, over the years, it has evolved as a more patient-focused subject. Irrespective of the focus, the need for chemistry knowledge and understanding in any Pharmacy curricula cannot be over-emphasized. We know that all drugs are chemicals. The ways any drug exerts its pharmacological actions and also toxicity in a patient are governed by a series of biochemical reactions. Therefore, chemistry knowledge and understanding are fundamental to any Pharmacy programme, which is essentially the study of various aspects of drugs, their applications in patients, patient care and overall treatment outcome.

Like the first edition, this revised, reorganized and significantly enhanced second edition covers all core topics related to general, organic and natural product chemistry currently taught in Pharmacy undergraduate curricula in the UK, USA and various other developed countries, and relates these topics to drug molecules, their development and their fate once given to patients. While the second edition still provides a concise coverage of the essentials of general, organic and natural product chemistry into a manageable, affordable and student-friendly text, by concentrating purely on the basics of various topics without going into exhaustive detail or repetitive examples, the first chapter, which deals with various properties of drug molecules, has been significantly 'beefed up' in this second edition. Generally, the contents of the second edition are organized and dealt with in a similar way, to the first to ensure that the contents are suitable for year 1 (level 4) and year 2 (level 5) levels of most of the Pharmacy curricula. Theoretical aspects have been covered in the context of applications of these theories in relation to drug molecules, their discovery and developments.

Chapter 1 presents an account of general aspects of chemistry and their contributions to modern life, with particular emphasis on modern medicine and discussions on various important properties of drug molecules, for example, pH, polarity and solubility; it also covers some related fundamental concepts like electrolytes, zwitterion, osmosis, tonicity and so on. Chapter 2 incorporates the fundamentals of atomic structure and bonding and discusses the relevance of chemical bonding in drug molecules and drug–receptor interactions, while Chapter 3 covers key aspects of stereochemistry with particular focus given on the significance of stereoisomerism in determining drug action and toxicity. Chapter 4 deals with organic functional groups, their preparations, reactions and applications. All major types of organic reactions and their importance in drug discovery, development, delivery and metabolism in patient's body are outlined in Chapter 5. Chapter 6 is about heterocyclic compounds; their preparations, reactions and applications. While nucleic acids are covered in Chapter 7, various aspects of natural products including the origins, chemistry, biosynthesis and pharmaceutical importance of alkaloids, carbohydrates, glycosides, iridoids and secoiridoids, phenolics, steroids and terpenoids are presented in Chapter 8.

Although the primary readership of the second edition still remains to be the Pharmacy undergraduate students (BPharm/MPharm), especially in their first and second years of study, further readership can come from the students of various other subject areas within Biomedical Science and the Food Sciences, Life Sciences and Health Sciences, where the basic chemistry knowledge is essential for their programmes.

Dr Lutfun Nahar

Professor Satyajit Sarker

Preface to the first edition

The pharmacy profession and the role of pharmacists in the modern healthcare systems have evolved quite rapidly over the last couple of decades. The services that pharmacists provide are expanding with the introduction of supplementary prescribing, provision of health checks, patient counselling and many others. The main ethos of pharmacy profession is now as much about keeping people healthy as treating them when they are not well. Modern pharmacy profession is shifting away from a product-focus and towards a patient-focus. To cope with these changes, and to meet the demand of the modern pharmacy profession, pharmacy curriculum, especially in the developed world, has evolved significantly. In the western countries, almost all registered pharmacists are employed by the community and hospital pharmacies. As a consequence, the practice, law, management, care, prescribing science and clinical aspects of pharmacy have become the main components of pharmacy curriculum. In order to incorporate all these changes, naturally, the fundamental science components, e.g. chemistry, statistics, pharmaceutical biology, microbiology, pharmacognosy, and a few other topics, have been reduced remarkably. The impact of these recent changes is more innocuous in the area of pharmaceutical chemistry.

As all drugs are chemicals, and pharmacy is mainly about the study of various aspects of drugs, including manufacture, storage, actions and toxicities, metabolisms and managements, chemistry still plays a vital role in pharmacy education. However, the extent at which chemistry used to be taught a couple of decades ago has certainly changed remarkably. It has been recognised that, while pharmacy students need a solid foundation in chemistry knowledge, the extent cannot be the same as the chemistry students may need.

There are several books on general, organic and natural product chemistry available today, but all of them are written in a manner that the level is only suitable for undergraduate Chemistry students, not for Pharmacy undergraduates. Moreover, in most modern pharmacy curricula, general, organic and natural products chemistry is taught at the first and second year undergraduate levels only. There are also a limited number of Pharmaceutical Chemistry books available to the students, but none of them can meet the demand of the recent changes in Pharmacy courses in the developed countries. Therefore, there has been a pressing need for a chemistry text covering the fundamentals of general, organic and natural products chemistry written at a correct level for the Pharmacy undergraduates. Physical (Preformulation) and Analytical Chemistry (Pharmaceutical Analysis) are generally taught separately at year 2 and year 3 levels of any modern MPharm course, and there are a number of excellent and up-to-date texts available in these areas.

During our teaching careers, we have always struggled to find an appropriate book that can offer general, organic and natural products chemistry at the right level for pharmacy undergraduate students, and address the current changes in Pharmacy curricula all over the world, at least in the UK. We have always ended up recommending several books and also writing notes for the students. Therefore, we have decided to address this issue by compiling a chemistry book for Pharmacy students, which will cover general, organic and natural product chemistry in relation to drug molecules. Thus, the aims of our book are to provide the fundamental knowledge and overview of all core topics related to general, organic and natural product chemistry currently taught in pharmacy undergraduate courses in the UK, USA and various other developed countries, relate these topics to the better understanding of drug molecules and their development, and meet the demand of the recent changes in pharmacy curricula. This book attempts to condense the essentials of general, organic and natural product chemistry into a manageable, affordable and student-friendly text, by concentrating purely on the basics of various topics without going into exhaustive detail or repetitive examples.

In Pharmacy undergraduate courses, especially in the UK, we get students of heterogeneous educational backgrounds; while some of them have very good chemistry background, the others have the bare minimum or not at all. From our experience in teaching Pharmacy undergraduate students, we have been able to identify the appropriate level that is required for all these students to learn properly. While we recognise that learning styles and levels vary from student to student, we can still try to strike the balance in terms of the level and standard at a point, which is not too difficult or not too easy for any students, but will certainly be student-friendly. Bearing this in mind, the contents of this book are organised and dealt with in a way that they are suitable for year 1 and year 2 levels of pharmacy curriculum. While the theoretical aspects of various topics are covered adequately, much focus has been given to the applications of these theories in relation to drug molecules, their discovery and developments. Chapter 1 provides an overview of some general aspects of chemistry and their importance in modern life, with particular emphasis on medicinal applications, and brief discussions on various physical characteristics of drug molecules, e.g. pH, polarity, and solubility. While Chapter 2 deals with the fundamentals of atomic structure and bonding, Chapter 3 covers various aspects of stereochemistry. Chapter 4 incorporates organic functional groups, and various aspects of aliphatic, aromatic and heterocyclic chemistry, amino acids, nucleic acids and their pharmaceutical importance. Major organic reactions are covered adequately in Chapter 5, and various types of pharmaceutically important natural products are discussed in Chapter 6.

While the primary readership of this book is the pharmacy undergraduate students (BPharm/MPharm), especially in their first and second year of study, the readership could also extend to the students of various other subject areas within Food Sciences, Life Sciences and Health Sciences who are not becoming chemists, yet they need to know the fundamentals of chemistry for their courses.

Dr Satyajit Sarker

Dr Lutfun Nahar

Chapter 1
Introduction

Learning Objectives

After completing this chapter, students should be able to

- describe the role of chemistry in modern life;
- define some of the physical properties of drugs, for example, melting point, boiling point, polarity, solubility and acid-base properties;
- explain the terms pH, pK_a, buffer and neutralization.

1.1 ROLE OF CHEMISTRY IN MODERN LIFE

Chemistry is the science of the composition, structure, properties and reactions of matters, especially of atomic and molecular systems.

Life itself is full of chemistry, that is, life is the reflection of a series of continuous biochemical processes. Right from the composition of the cell to the whole organism, the presence of chemistry is conspicuous. Human beings are physically constructed of chemicals, live in a plethora of chemicals and are dependent on chemicals for their quality of modern life. All living organisms are composed of numerous organic substances. Evolution of life begins from one single organic compound called a *nucleotide*. Nucleotides join together to form the building blocks of life. Our identities, heredities and continuation of generations, all are governed by chemistry.

In our everyday life, whatever we see, use or consume have been the gifts of research in chemistry for thousands of years. In fact, chemistry is applied everywhere in modern life. From the colour of our clothes to the shapes of our PCs,

Chemistry for Pharmacy Students: General, Organic and Natural Product Chemistry, Second Edition. Lutfun Nahar and Satyajit Sarker.
© 2019 John Wiley & Sons Ltd. Published 2019 by John Wiley & Sons Ltd.

all are possible due to chemistry. It has played a major role in pharmaceutical advances, forensic science and modern agriculture. Diseases and their remedies have also been a part of human lives. Chemistry plays an important role in understanding diseases and their remedies; that is, drugs.

Medicines or drugs that we take for the treatment of various ailments are chemicals, either organic or inorganic molecules. However, most drugs are organic molecules. These molecules are either obtained from natural sources or synthesized in chemistry laboratories. Some important drug molecules are discussed here.

Aspirin, an organic molecule, is chemically known as acetyl salicylic acid and is an analgesic (relieves pain), antipyretic (reduces fever) and anti-inflammatory (reduces swelling) drug. Studies suggest that aspirin can also reduce the risk of heart attack. It is probably the most popular and widely used analgesic drug because of its structural simplicity and low cost. Salicin is the precursor of aspirin. It is found in the willow tree bark, whose medicinal properties have been known since 1763. Aspirin was developed and synthesized in order to avoid the irritation in the stomach caused by salicylic acid, which is also a powerful analgesic, derived from salicin. In fact, salicin is hydrolysed in the gastrointestinal tract to produce D-glucose and salicyl alcohol (see Section 8.4). Salicyl alcohol, on absorption, is oxidized to salicylic acid and other salicylates. However, aspirin can easily be synthesized from phenol using the *Kolbe reaction* (see Section 4.7.10.6).

Salicin
(A precursor of aspirin)

Salicyl alcohol

Salicylic acid

Aspirin
Acetyl salicylic acid

Paracetamol (acetaminophen), an *N*-acylated aromatic amine having an acyl group (R—CO—) substituted on nitrogen, is an important over-the-counter headache remedy. It is a mild analgesic and antipyretic medicine. The synthesis of paracetamol involves the reaction of *p*-aminophenol and acetic anhydride (see Section 4.7.10.6).

4-Aminophenol

Paracetamol
Acetaminophen

L-Dopa (L-3,4-dihydroxyphenylalanine), an amino acid, is a precursor of the neurotransmitters dopamine, norepinephrine (noradrenaline) and epinephrine (adrenaline), collectively known as catecholamines, and found in humans as well as in some animals and plants. It has long been used as a treatment for Parkinson's disease and other neurological disorders. L-Dopa was first isolated from the seedlings of *Vicia faba* (broad bean) by Marcus Guggenheim in 1913, and later it was synthesized in the lab for pharmaceutical uses.

(L)-Dopa
(The precursor of dopamine)

Morphine

Morphine is a naturally occurring opiate analgesic found in opium and is a strong pain reliever, classified as a narcotic analgesic (habit-forming) (see Section 8.2.2.5). Opium is the dried latex obtained from the immature poppy (*Papaver somniferum*) seeds. Morphine is widely used in clinical pain management, especially for pain associated with terminal cancers and post-surgery pain.

Penicillin V (phenoxymethylpenicillin), an analogue of the naturally occurring penicillin G (see Section 7.3.2), is a semisynthetic narrow-spectrum antibiotic useful for the treatment of bacterial infections. Penicillin V is quite stable even in high humidity and strong acidic medium (e.g. gastric juice). However, it is not active against beta-lactamase-producing bacteria. As we progress through various chapters of this book, we will come across a series of other examples of drug molecules and their properties.

Penicillin G
(The first penicillin of the penicillin group of antibiotics)

Penicillin V
Phenoxymethylpenicillin

In order to have proper understanding and knowledge about these drugs and their behaviour, there is no other alternative but to learn chemistry. Everywhere, from discovery to development, from production and storage to administration, and from desired actions to adverse effects of drugs, chemistry is directly involved.

In the drug discovery stage, suitable sources of potential drug molecules are explored. Sources of drug molecules can be natural, such as a narcotic analgesic, morphine, from *P. somniferum* (poppy plant), synthetic, such as a popular analgesic and antipyretic, paracetamol, and semisynthetic, such as penicillin V. Whatever the source is, chemistry is involved in all processes in the discovery phase. For example, if a drug molecule has to be purified from a natural source, for example, plant, the processes like extraction, isolation and identification are used, and all these processes involve chemistry (see Section 8.1.3.1).

Similarly, in the drug development steps, especially in pre-formulation and formulation studies, the structures and the physical properties (e.g. solubility and pH), of the drug molecules are exploited. Chemistry, particularly physical properties of drugs, is also important to determine storage conditions. Drugs having an ester functionality, for example, aspirin, could be quite unstable in the presence of moisture and should be kept in a dry and cool place. The chemistry of drug molecules dictates the choice of the appropriate route of administration. Efficient delivery of drug molecules to the target sites requires manipulation of various chemical properties and processes; for example, microencapsulation, nanoparticle-aided delivery and so on. When administered, the action of a drug inside our body depends on its binding to the appropriate receptor and its subsequent metabolic processes, all of which involve complex enzyme-driven biochemical reactions.

All drugs are chemicals, and pharmacy is a subject that deals with the study of various aspects of drugs. Therefore, it is needless to say that to become a good pharmacist the knowledge of the chemistry of drugs is essential. Before moving on to the other chapters, let us try to understand some of the fundamental chemical concepts in relation to the physical properties of drug molecules (see Section 1.6).

1.2 SOLUTIONS AND CONCENTRATIONS

A *solution* is a mixture where a solute is uniformly distributed within a solvent. A *solute* is the substance that is present in smaller quantities and a *solvent* usually the component that is present in greater quantity. Simply, a solution is a special type of homogenous mixture composed of two or more substances. For example, sugar (solute) is added to water (solvent) to prepare sugar solution. Similarly, saline (solution) is a mixture of sodium chloride (NaCl) (solute) and water (solvent). Solutions are extremely important in life as most chemical reactions, either in laboratories or in living organisms, take place in solutions.

Ideally, solutions are transparent and light can pass through the solutions. If the solute absorbs visible light, the solution will have a colour. We are familiar with liquid solutions, but a solution can also be in any state, such as solid, liquid or gas. For example, air is a solution of oxygen, nitrogen and a variety of other gases all in the gas state; steel is also a solid-state solution of carbon and iron. Solutes may be crystalline solids, such as sugars and salts that dissolve readily into solutions, or colloids, such as large protein molecules, which do not readily dissolve into solutions (see Section 1.3).

In Chemistry, especially in relation to drug molecules, their dosing, therapeutic efficacy, adverse reactions and toxicity, we often come across with the term *concentration*, which can simply be defined as the amount of solute per unit of solvent. Concentration is always the ratio of solute to solvent and it can be expressed in many ways. The most common method of expressing the concentration is based on the amount of solute in a fixed amount of solution where the quantities can be expressed in weight (w/w), in volume (v/v) or both (w/v). For example, a solution containing 10 g of NaCl and 90 g of water is a 10% (w/w) aqueous solution of NaCl.

Weight measure (w/w) is often used to express concentration and is commonly known as *percent concentration* (parts per 100), as shown in the previous example of 10% NaCl aqueous solution. It is the ratio of one part of solute to one hundred parts of solution. To calculate percent concentration, simply divide the mass of the solute by the total mass of the solution, and then multiply by 100. Percent concentration also can be displayed, albeit not so common, as *parts per thousand* (ppt) for expressing concentrations in grams of solute per kilogram of solution. For more diluted solutions, *parts per million* (ppm), which is the ratio of parts of solute to one million parts of solution, is often used. To calculate ppm, divide the mass of the solute by the total mass of the solution, and then multiply by 10^6. Grams per litre is the mass of solute divided by the volume of solution in litres. The ppt and ppm can be either w/w or w/v.

Molality of a solution is the number of moles of a solute per kilogram of solvent, while *molarity* of a solution is the number of moles of solute per litre of solution. Molarity (M) is the most widely used unit for concentration. The unit of molarity is mol/l or M. One mole is equal to the molecular weight (MW) of the solute in grams. For example, the MW of glucose is 180. To prepare a 1 M solution of glucose, one should add 180 g of glucose in a 1.0 l volumetric flask and then fill the flask with distilled water to a total volume of 1.0 l. Note that molarity is defined in terms of the volume of the solution, not the volume of the solvent. Sometimes, the term *normality* (N), which can be defined as the number of mole equivalents per litre of solution, is also used, especially for various acids and bases, to express the concentration of a solution. Like molarity, normality relates the amount of solute to the total volume of solution. The mole equivalents of an acid or base are calculated by determining the number of H^+ or HO^- ions per molecule: $N = n \times M$ (where n is an integer). For an acid solution, n is the number of H^+ ions provided by a formula unit of acid. For example, a 3 M H_2SO_4 solution is the same as a 6 N H_2SO_4 solution. For a basic solution, n is the number of HO^- ions provided by a formula unit of base. For example, a 1 M $Ca(OH)_2$ solution is the same as a 2 N $Ca(OH)_2$ solution. Note that the normality (N) of a solution is never less than its molarity.

A *concentrated solution* has a lot of solute per solvent, a *diluted solution* has a lot of solvent, a *saturated solution* has maximum amount of solute, and a *supersaturated solution* has more solute than it can hold. Supersaturated solutions are relatively unstable, and solute tends to precipitate out of the mixture to form crystals, resulting in a saturated solution. The equilibrium of a solution depends on the temperature.

A *stock solution* is prepared with a known concentration, from which a diluted solution can be made. The process of adding more solvent to a solution or removing some of the solute is called *dilution*. In other words, dilution is the process of reducing the concentration of a solute in solution, usually simply by mixing with more solvent. Any unit can be used for both volume and concentration as long as they are the same on both sides of the equation. The concentration of the diluted solution can easily be calculated from the following equation:

$$C_1V_1 = C_2V_2$$

Where, C_1 and C_2 are the initial and final concentrations and V_1 and V_2 are the initial and final volumes of the solution.

A *serial dilution*, often used in various *in vitro* assays, is simply a series of simple dilutions. Serial dilutions are made in increments of 1000 (10^3), 100 (10^2), 10 (10-fold) or 2 (twofold), but 10-fold and twofold serial dilutions are commonly used. Serial dilutions are an accurate method of making solutions of low molar concentrations. The first step in making a 10-fold serial dilution is to take stock solution (1 ml) in a tube and then to add distilled water (9 ml) or other suitable solvents. For making a twofold serial dilution one should take stock solution (1 ml) in a tube and then add distilled water (1 ml) or other suitable solvents.

1.3 SUSPENSION, COLLOID AND EMULSION

A *suspension* is a heterogeneous mixture between two substances one of which is finely dispersed into the other. Note that in a suspension, the solute particles do not dissolve, but are suspended throughout the bulk of the solvent. Most common suspensions include sand in water, dust in air and droplets of oil in air. The size of the particles is large enough (more than 1 µm) to be visible to the naked eye. In suspension, particles are so large that they settle out of the solvent if not constantly stirred. Therefore, it is possible to separate particles in any suspension through filtration. A suspension of liquid droplets or fine solid particles in a gas is called an *aerosol*. In relation to the atmosphere, the suspended particles, for example, fine dust and soot particles, sea salt, biogenic and volcanogenic sulphates, nitrates and cloud droplets, are called *particulates*.

A *colloid* is a mixture, where microscopically dispersed insoluble particles (10–1000 nm) of one substance are evenly suspended throughout another substance indefinitely. Note that to quality as a colloid, the mixture must not settle. Like a suspension, a colloid consists of two separate phases, a dispersed phase (solute) and a dispersing medium (continuous phase or solvent). *Colloidal particles* consist of small particles of one substance dispersed in a continuous phase of a different composition, known as *colloidal dispersions*. The properties of colloids and solutions are different due to their particle size. A colloidal dispersion, for example, milk, is not a true solution but it is not a suspension either, because it does not settle out on standing over time like a suspension.

Colloidal particles can be studied by various methods, for example, diffusion, electrophoresis and scattering of visible light and X-rays. There are several types of colloids, and the most popular one is called *colloidal solution*, where the solid forms the dispersed phase and the liquid forms the dispersion medium. The particles of the dispersed phase in a colloidal solution are known as *colloidal particles* or *micelles*. A gas may be dispersed in a liquid to form a foam (e.g. shaving lather) or in a solid to form a solid foam (e.g. marshmallow); a liquid may be dispersed in a gas to form an aerosol (e.g. aerosol spray), in another liquid to form an emulsion (e.g. mayonnaise) or in a solid to form a gel (e.g. cheese); a solid may be dispersed in a gas to form a solid aerosol (e.g. smoke in air), in a liquid to form a sol (e.g. ink) or in a solid to form a solid sol (e.g. certain alloys). Colloids are often purified by dialysis, which is a slow process.

Colloids are important in drug delivery, as colloidal carriers (e.g. nanoparticles) are used in controlled or sustained release and site-specific delivery of drugs. *Nanoparticles* are solid, colloidal particles consisting of macromolecular substances that vary in size from 10–1000 nm; they are natural or synthetic polymers. Depending on the interactions between the dispersed phase and the dispersing medium, colloidal solutions are classified as *lyophilic* (solvent loving) and *lyophobic* (solvent hating). The colloidal particles are strongly solvated in the dispersing medium of a lyophilic colloidal solution, for example, emulsion. When water is the dispersing medium, it is known as *hydrophilic*. The colloidal particles are not solvated in the dispersing medium of a lyophobic colloidal solution, such as a suspension. When water is the dispersing medium, it is called *hydrophobic*.

An *emulsion* is an integrated mixture of two immiscible liquids such as oil and water, stabilized by an emulsifying agent (emulsifier or surfactant). Simply, an emulsion is a fine dispersion of minute droplets of one liquid in another in which it is not soluble or miscible. For example, a type of paint used for walls, consisting of pigment bound in a synthetic resin, which forms an emulsion with water. An *emulsifying agent* (emulsifier) is a substance that keeps the parts of an emulsion mixed together. Water soluble emulsifiers form oil in water (o/w) emulsion, while oil soluble emulsifiers usually give water in oil (w/o) emulsion. Emulsions are usually prepared by vigorously shaking the two components together, often with the addition of an emulsifying agent, in order to stabilize the product formed.

1.4 ELECTROLYTES, NONELECTROLYTES AND ZWITTERIONS

Electrolytes are species that form ions, when dissolved in water and commonly exist as solutions of acids, bases or salts. They are essential minerals in the body, they control osmosis of water between body compartments, and help maintain the acid-base balance required for normal cellular activities. Many salts dissociate in water and break up into electrically charged ions. The salt NaCl breaks up into one ion of sodium (Na^+) and one ion of chloride (Cl^-). These charged particles can conduct electricity. The number of ions that carry a positive charge (cations) and ions that carry a negative charge (anions) should be equal.

$$NaCl \xrightarrow{H_2O} Na^+_{(aq)} + Cl^-_{(aq)}$$

The sweat that evaporates from the skin contains a variety of electrolytes, for example, cations such as sodium (Na^+), potassium (K^+) calcium (Ca^{2+}) and magnesium (Mg^{2+}), and anions such as chloride (Cl^-), bicarbonate (HCO_3^-), phosphate (HPO_4^{2-}) and sulphate (SO_4^{2-}).

Nonelectrolytes are species that do not form ions when dissolved in water. Thus, aqueous solutions of nonelectrolyte do not conduct electricity, for example, aqueous glucose ($C_6H_{12}O_6$). Glucose does not dissociate when dissolved in water. Most organic molecules are nonelectrolytes as they have covalent bonds and they do not form ions when dissolved in water.

$$C_6H_{12}O_6 + H_2O \longrightarrow C_6H_{12}O_{6(aq)}$$

Zwitterions (ion pair) can bear both a positive and a negative charge, for example, amino acids. Amino acids are the building blocks of proteins (see Section 7.2). They contain functional groups, amino groups ($-NH_2$) that can accept protons, and carboxyl groups ($-COOH$) that can lose protons. Under certain conditions, both of these events can occur, and the resulting molecule becomes a zwitterion. The simplest of the 20 amino acids that occur in proteins is glycine, H_2NCH_2COOH, whose solutions are distributed between the acidic-, zwitterion- and basic–species as shown next.

$$\overset{+}{N}H_3CH_2COOH \longleftrightarrow \overset{+}{N}H_3CH_2COO^- \longleftrightarrow NH_2CH_2COO^-$$

1.5 OSMOSIS AND TONICITY

Living cells have the potential of gaining or losing water through semipermeable membranes by osmosis. *Osmosis* is the process by which molecules of a solvent tend to pass through a semipermeable membrane from a less concentrated solution into a more concentrated one. Generally, osmosis occurs when the concentration of solutes on one side of the cell membrane is higher than the other. Molecules can move across the cell membranes from a low concentrated solution (dilute solution/

pure solvent) to a high concentrated one (concentrated solution) by diffusion as shown next. Eventually, the concentrations of the two solutions become equal.

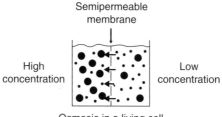

Osmosis in a living cell

In the body, water is the solvent, and the solutes include electrolytes, O_2, CO_2, glucose, urea, amino acids and proteins. *Osmole* is the measure of the total number of particles in a solution. Number of particles can be either molecules (e.g. sugar) or ions (e.g. NaCl). For example, 1 g mole of non-ionizable sugar is 1 Osm, whereas 0.5 g mol of NaCl ionizes into two ions (Na^+ and Cl^-) is also 1 Osm.

The concentration of solutes in body fluids is usually expressed as the *osmolality*, which is a measure of the osmoles (Osm) of solute per kilogram of solvent (Osm/kg). The ability of a semipermeable membrane solution to make water move into or out of a cell by osmosis is known as its *tonicity*. In general, a solution's tonicity can be defined by its osmolarity, which is defined as the number of osmoles of solute per litre of solution (Osm/l). A solution with *low* osmolarity has *fewer* solute particles per litre of solution, while a solution with *high* osmolarity has *more* solute particles per litre of solution.

A *hypertonic solution* has a higher concentration of solutes than the surrounding semipermeable membrane (lower concentration) and water will move out of the cells. This can cause cell to shrink. So, a hypertonic solution has higher osmolarity than blood plasma and red blood cells. A *hypotonic solution* has a lower concentration of solutes than the surrounding semipermeable membrane (higher concentration) and the net flow of water will be into the cells. This can result in cell to swell and eventually burst. So, a hypotonic solution has lower osmolarity than blood plasma and red blood cells. An *isotonic solution* has same concentration of solutes as the surrounding semipermeable membrane and there will be no net movement of water into or out of the cell. Therefore, an isotonic solution has same osmolarity as blood plasma and red blood cells.

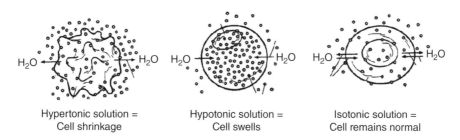

Hypertonic solution = Cell shrinkage

Hypotonic solution = Cell swells

Isotonic solution = Cell remains normal

1.6 PHYSICAL PROPERTIES OF DRUG MOLECULES

1.6.1 Physical State

Drug molecules exist in various physical states, for example, amorphous solid, crystalline solid, hygroscopic solid, liquid or gas. Physical state of drug molecules is an important factor in the formulation and delivery of drugs.

1.6.2 Melting Point and Boiling Point

Melting point (mp) is the temperature at which a solid becomes a liquid, and *boiling point (bp)* is the temperature at which the vapour pressure of the liquid is equal to the atmospheric pressure. Boiling point of a substance can also be defined as the temperature at which it can change its state from a liquid to a gas throughout the bulk of the liquid at a given pressure. For example, the melting point of water at 1 atm of pressure is 0 °C (32 °F, 273.15 K); this is also known as the *ice point*, and the boiling point of H_2O is 100 °C.

Melting point is used to characterize organic compounds and to confirm the purity. The melting point of a pure compound is always higher than the melting point of that compound mixed with a small amount of an impurity. The more impurity is present, the lower the melting point is. Finally, a minimum melting point is reached. The mixing ratio that results in the lowest possible melting point is known as the *eutectic point*.

The melting point increases as the molecular weight increases, and the boiling point increases as the molecular size increases. The increase in melting point is less regular than the increase in boiling point, because packing influences the melting point of a compound.

Packing of the solid is a property that determines how well the individual molecules in a solid fit together in a crystal lattice. The tighter the crystal lattice, the more energy is required to break it and eventually melt the compound. Alkanes with an odd number of carbon atoms pack less tightly, which decreases their melting points. Thus, alkanes with an even number of carbon atoms have higher melting points than the alkanes with an odd number of carbon atoms.

$CH_3CH_2CH_2CH_3$	$CH_3CH_2CH_2CH_2CH_3$	$CH_3CH_2CH_2CH_2CH_2CH_3$
Butane	Pentane	Hexane
mp: −138.4°C	mp: −129.7°C	mp: −93.5°C
	bp: 36.1°C	

On the contrary, between two alkanes having same molecular weights, the more highly branched alkane has a lower boiling point.

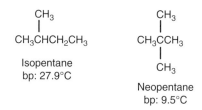

Isopentane
bp: 27.9°C

Neopentane
bp: 9.5°C

Ionic compounds are held together by extremely strong interactions of positive and negative charges, and they tend to have high boiling and melting points. For example, the ionic solid NaCl (salt) melts at >800°C.

1.6.3 Polarity and Solubility

Polarity is a physical property of a compound, which relates to other physical properties, for example, melting and boiling points, solubility and intermolecular interactions between molecules. Generally, there is a direct correlation between the polarity of a molecule and the number and types of polar and nonpolar covalent bonds (see Section 2.3.4.2). In a few cases, a molecule having polar bonds, but in a symmetrical arrangement, may give rise to a nonpolar molecule, for example, carbon dioxide (CO_2).

$$\delta^- \quad \delta^+ \quad \delta^-$$
$$:\!\ddot{O}\!=\!C\!=\!\ddot{O}:$$

Carbon dioxide
(A nonpolar molecule)

The term *bond polarity* is used to describe the sharing of electrons between atoms (see Section 2.4). In a nonpolar covalent bond, the electrons are shared equally between two atoms. A polar covalent bond is one in which one atom has a greater attraction for the electrons than the other atom (see Section 2.3.4.2). When this relative attraction is strong, the bond is an ionic bond (see Section 2.3.4.1).

The polarity in a bond arises from the different electronegativities of the two atoms that take part in bond formation (see Section 2.3.3). The greater the difference in electronegativity between the bonded atoms, the greater the bond polarity. Thus, electronegativity of an atom is related to bond polarity (see Section 2.4). For example, water is a polar molecule, whereas cyclohexane is nonpolar.

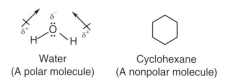

Water
(A polar molecule)

Cyclohexane
(A nonpolar molecule)

More examples of polar and nonpolar molecules are shown in the following Table. The bond polarity and electronegativity are discussed in Chapter 2.

Polar	Nonpolar
Water (H_2O)	Toluene (Ph—Me)
Methanol (MeOH)	n-Hexane (C_6H_{12})
Ethanol (EtOH)	Benzene (Ph—H)
Acetic acid (AcOH)	Toluene (Ph—Me)

Life occurs exclusively in water. Solutions in which water is the dissolving medium are called *aqueous solutions*. In aqueous solutions, the polar parts are hydrated and the nonpolar parts are excluded. Hydrogen bonding is a consequence of the basic molecular structure of water. Water has very high boiling point compared with small organic molecules due to the hydrogen bonding. The hydrogen bonding and other nonbonding interactions between molecules are described in Chapter 2. Examples of some common solvents and their boiling points are compared with the boiling point of water in the following Table.

Solvent	Formula	Molecular weight	bp (°C)
Acetone	C_3H_6O	58.08	56.05
Benzene	C_6H_6	78.11	80.10
Chloroform	$CHCl_3$	119.38	61.2
Cyclohexane	C_6H_{12}	84.16	80.70
Ethanol	C_2H_6O	46.07	78.50
Ethyl acetate	$C_4H_8O_2$	88.11	77.00
Methanol	CH_4O	32.04	64.60
Water	H_2O	18.02	100.00

The concept of solution has already been outlined earlier (see Section 1.2). Let's now delve into the concept of solubility. *Solubility* is the amount of a solute that can be dissolved in a specific solvent under given conditions. Therefore, solubility is a measure of how much of the solute can be dissolved into the solvent at a specific temperature. The process of dissolving solute in solvent is called *solvation*, or *hydration* when the solvent is water. In fact, the interaction between a dissolved species and the molecules of a solvent is solvation. The process of mixing solute (s) and solvent to form a solution is called *dissolution*. The stronger the intermolecular attractions (interactions) between solute and solvent, the more likely the solute will dissolve in a solvent.

The *rate of solution* is a measure of how fast a solute is dissolved in water or a particular solvent. It also depends on size of the particle, stirring, temperature and the amount of solid already dissolved. For example, glucose (which has hydrogen bonding) is highly soluble in water, but cyclohexane (which only has dispersion forces) is insoluble in water. Solubility largely depends on temperature, polarity, molecular size and stirring. Temperature always affects solubility and an increasing temperature usually increases the solubility of most solids in a liquid solvent. The solubility of gases decreases with increase in temperature. The polarity of the solute and solvent also affects the solubility. The stronger the attractions between solute and solvent molecules, the greater the solubility. Thus the solubility of molecules can also be explained on the basis of the polarity of molecules. In general, like dissolves like; that is, materials with similar polarity are soluble in each other. Thus, polar solvent, for example, water (H_2O), and nonpolar solvent, for example, benzene (C_6H_6), do not mix.

The term *miscible* is used to describe two substances (usually liquids) that are soluble in each other. If they do not mix, as oil and water, they are said to be *immiscible*. For example, ethyl alcohol and water are miscible liquids as both are polar molecules, *n*-hexane and dodecane are also miscible in one another as both are nonpolar molecules, whereas chloroform (nonpolar) and water (polar) are immiscible. A polar solvent, such as H_2O, has partial charges that can interact with the partial charges on a polar compound, such as sodium chloride (NaCl). As nonpolar compounds have no net charge, polar solvents are not attracted to them. For example, alkanes are nonpolar molecules and are insoluble in polar solvents such as H_2O, but are soluble in nonpolar solvents such as chloroform.

CH_3CH_2OH	H_2O	$CHCl_3$	C_6H_{12}
Ethanol (polar)	Water (polar)	Chloroform (nonpolar)	Dodecane (nonpolar)

Remember, size matters. Organic molecules with a branching carbon increases the solubility than a long-chain carbon, because branching reduces the size of the molecule and makes it easier to solvate. For example, isobutanol is more soluble in water than butanol.

$$CH_3CH_2CH_2CH_2OH \qquad \overset{\overset{\displaystyle OH}{|}}{CH_3CHCH_2CH_3}$$

Butanol   Isobutanol

1.7 ACID–BASE PROPERTIES AND pH

Drug molecules contain various types of functional groups, and these functional groups contribute to the overall acidity or basicity of drug molecules. One of the adverse effects of aspirin is stomach bleeding, which is partly due to its acidic nature. In the stomach, aspirin is hydrolysed to salicylic acid and acetic acid (see Section 4.9). The carboxylic acid group (—COOH) and a phenolic hydroxyl group (—OH) present in salicylic acid, make this molecule acidic. Moreover, acetic acid is formed and that is also moderately acidic. Thus, intake of aspirin increases the acidity of stomach significantly, and if this increased acidic condition stays in the stomach for a long period, it may cause stomach bleeding. Like aspirin, there are a number of other drug molecules that are acidic in nature. Similarly, there are basic and neutral drugs as well. Now, let us see what these terms *acid*, *base* and *neutral* compounds really mean, and how these parameters are measured. Most drugs are organic molecules and can be acidic, basic or neutral in nature.

Aspirin → (Hydrolysis in the stomach) → Salicylic acid + Acetic acid (CH_3CO_2H)

Simply, an electron-deficient species that accepts an electron pair is called an *acid*, for example, hydrochloric acid (HCl), and a species with electrons to donate is a *base*, for example, sodium hydroxide (NaOH). A neutral species does not do either of these. Most of the organic reactions are either *acid–base* reactions or involve catalysis by an acid or base at some point.

1.7.1 Acid–Base Definitions

Acids turn blue litmus red and have a sour taste, whereas bases turn red litmus to blue and have a bitter taste. Litmus is the oldest known pH indicator. Acid reacts with certain metals to produce hydrogen gas.

$$Na + HCl \longrightarrow NaCl + H_2 \uparrow \qquad Zn + H_2SO_4 \longrightarrow ZnSO_4 + H_2 \uparrow$$

Acids and bases are important classes of chemicals that control carbon dioxide (CO_2) transport in the blood. Carbon dioxide (CO_2) dissolves in the body fluid (H_2O) to form carbonic acid (H_2CO_3), and is excreted as a gas by the lungs.

$$CO_2 + H_2O \longrightarrow H_2CO_3$$

Stomach acid is hydrochloric acid (HCl), which is a strong acid. Acetic acid (CH_3COOH) is a weak organic acid that can be found in vinegar. Citrus fruits such as lemons, grapefruit, oranges and limes have citric acid ($C_6H_8O_7$) as well as ascorbic acid (vitamin C). Both these acids increase the acidity of foods and make it harder for bacteria to grow. Also, because of the antioxidant property, ascorbic acid prevents food items from oxidative spoilage. Sour milk, sour cream, yoghurt and cottage cheese have lactic acid from the fermentation of the sugar lactose. Certain bacteria break down the sugars in milk and make lactic acid, which reacts with milk proteins. This causes the milk to thicken and develop a creamy or curdy texture and sour flavour. Yoghurt is an example of a fermented dairy product whose texture and flavour both depend on the presence of lactic acid. Both citric acid and lactic acid are weak organic acids. They are used largely as food preservatives, curing agents and flavouring agents.

Several definitions have been used to describe the acid-base properties of aqueous solvents as well as other solvents. The Arrhenius definitions or the Brønsted–Lowry definitions adequately describe aqueous acids and bases.

1.7.1.1 Arrhenius Acids and Bases

According to Arrhenius' definition, an *acid* produces hydrogen ion (H^+), and a base produces hydroxide or hydroxyl ion (HO^-) in water. Salts are formed in the acid–base reactions, usually in neutralization reactions. Thus, a salt is an ionic compound that is made with the anion of an acid and the cation of a base. Arrhenius' definition only works for strong acids and strong bases and it is limited to aqueous solutions.

$$HCl \text{ (Acid)} + NaOH \text{ (Base)} \rightleftharpoons NaCl \text{ (Salt)} + H_2O \text{ (Water)}$$

1.7.1.2 Brønsted–Lowry Acids and Bases

Danish chemist Johannes Brønsted and the English chemist Thomas Lowry expanded the Arrhenius definition. They defined an *acid* as a proton (H^+) donor, and a *base* as a proton (H^+) acceptor. Brønsted–Lowry definitions work better for weak acids and weak bases.

$$HNO_2 \text{ (Acid)} + H_2O \text{ (Base)} \rightleftharpoons NO_2^- \text{ (A conjugate base)} + H_3O^+ \text{ (A conjugate acid)}$$

Each acid has a *conjugate base*, and each base has a *conjugate acid*. An acid reacts with a base to produce conjugate base and conjugate acid. These conjugate pairs only differ by a proton. In the example, HNO_2 is the acid, H_2O is the base, NO_2^- is the conjugated base, and H_3O^+ is the conjugated acid. Thus, a conjugate acid can lose a H^+ ion to form a base, and a conjugate base can gain a H^+ ion to form an acid. Water can be an acid or a base. It can gain a proton to become a hydronium ion (H_3O^+), its conjugate acid, or lose a proton to become the hydroxide ion (HO^-), its conjugate base.

When an acid transfers a proton to a base, it is converted to its conjugate base. By accepting a proton, the base is converted to its conjugate acid. In the following acid-base reaction, H_2O is converted to its conjugate base, hydroxide ion (HO^-), and NH_3 is converted to its conjugate acid, ammonium ion ($^+NH_4$). Therefore, the conjugate acid of any base always has an additional hydrogen atom and an increase in positive charge or a decrease in negative charge. On the other hand, the conjugate base of an acid has one hydrogen atom less and an increase in negative charge or lone pair of electrons, and also a decrease in positive charge. The stronger the acid, the weaker the conjugate base and vice versa.

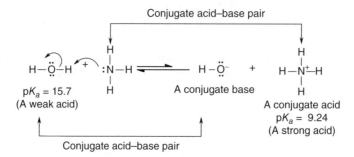

According to the Brønsted–Lowry definitions, any species that contains hydrogen can potentially act as an acid, and any compound that contains a lone pair of electrons can act as a base. Therefore, neutral molecules can also act as bases if they contain an oxygen, nitrogen or sulphur atom. Both an acid and a base must be present in a proton transfer reaction, because an acid cannot donate a proton unless a base is present to accept it. Thus, proton-transfer reactions are often called *acid–base reactions*. For example, in the following reaction between acetic acid (CH_3COOH) and ammonia (NH_3), a proton is transferred from CH_3COOH, an acid, to NH_3, a base.

Acid strength is related to base strength of its conjugate base. For an acid to be weak, its conjugate base must be strong. In general, in the reaction between an acid and base, the equilibrium favours the weaker acid or base. In the acid–base reaction that follows, NH_3 is a base because it accepts a proton, and CH_3COOH is an acid because it donates a proton. In the reverse reaction, ammonium ion ($^+NH_4$) is an acid because it donates a proton, and acetate ion (CH_3COO^-) is a base because it accepts a proton. The curly arrows show the flow of electrons in an acid-base reaction. Two half-headed arrows are used for the equilibrium reactions. A longer arrow indicates that the equilibrium favours the formation of acetate ion (CH_3COO^-) and ammonium ion ($^+NH_4$). Because acetic acid (CH_3COOH) is a stronger acid than ammonium ion ($^+NH_4$), the equilibrium lies towards the formation of weak acid and weak base.

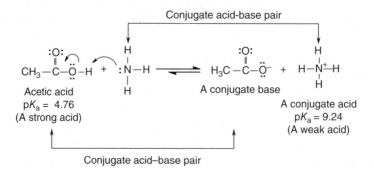

1.7.1.3 Lewis Theory of Acids and Bases

The Lewis definitions describe acids and bases for both organic and inorganic solvents. The advantage of Lewis definitions is that many more organic reactions can be considered as acid–base reactions because they do not have to occur in solutions.

The Lewis theory of acids and bases defines an acid as an electron-pair acceptor, and a base as an electron-pair donor to form a covalent bond. A *Lewis acid* is a species that accepts electrons and it is termed as *an electrophile*. A *Lewis base* is a species that donates electrons to a nucleus with an empty orbital, and is termed as *a nucleophile*. Thus, Lewis acids are electron-deficient species, whereas Lewis bases are electron-rich species. For example, the methyl cation (CH_3^+) may be regarded as a Lewis acid or an electrophile, because it accepts electrons from reagent such as chloride ion (Cl^-). In turn, because chloride ion (Cl^-) donates electrons to the methyl cation (CH_3^+), it is classified as a Lewis base or a nucleophile.

$$:\overset{..}{\underset{..}{Cl}}:^- \quad + \quad \overset{+}{CH_3} \quad \longrightarrow \quad CH_3Cl$$

Chloride ion	Methyl cation	Methyl chloride
(A Lewis base and a nucleophile)	(A Lewis acid and an electrophile)	

Lewis acids are known as *aprotic acids* and they react with Lewis bases by accepting pairs of electrons, not by donating protons. Since aprotic acids do not have any acidic hydrogens. Borane (BH_3), boron trichloride (BCl_3) and boron trifluoride (BF_3) are known as Lewis acids, because boron has a vacant d orbital that accepts a pair of electrons from a donor species. For example, diethyl ether ($C_2H_5OC_2H_5$) acts as a Lewis base towards BCl_3 and forms a complex of diethyl ether and boron trichloride (a salt).

$$C_2H_5 - \overset{..}{\underset{..}{O}} - C_2H_5 \quad + \quad BCl_3 \quad \longrightarrow \quad C_2H_5 - \overset{+}{\underset{\underset{C_2H_5}{|}}{O}} - \bar{B}Cl_3$$

Diethyl ether	Boron trichloride	A complex of diethyl ether
(A Lewis base)	(A Lewis acid)	and boron trichloride

Boron trifluoride (BF_3) reacts with a Lewis bases, such as tertiary methyl amine and generates a complex of complex of trimethyl amine and boron trifluoride (a salt).

$$H_3C - \overset{..}{\underset{\underset{CH_3}{|}}{N}} - CH_3 \quad + \quad BF_3 \quad \longrightarrow \quad H_3C - \overset{\overset{CH_3}{\overset{+|}{|}}}{\underset{\underset{CH_3}{|}}{N}} - \bar{B}F_3$$

Trimethyl amine	Boron trichloride	A complex trimethyl amine
(A Lewis base)	(A Lewis acid)	and boron trifluoride

1.7.2 Electronegativity and Acidity

Electronegativity is a measure of the tendency of an atom to attract a bonding pair of electrons (see Section 2.3.3). The relative acidity of HA within a period is determined by the stability of A⁻. The greater the electronegativity, the greater is the stability of A⁻. We know that carbon is less electronegative than nitrogen, which in turn is less electronegative than oxygen, and that oxygen is less electronegative than fluorine. Therefore, the strength of acidity increases from methane to hydrogen fluoride as shown next.

	H_3C-H	H_2N-H	$HO-H$	$F-H$
pK_a	51	38	15.7	3.5
Electronegativity of A in A—H	2.5	3.0	3.5	4.0

A molecule is said to have resonance when its structure cannot be adequately described by a single Lewis structure. Resonance may delocalize the electron pair that A⁻ needs to form the new bond with a proton. Delocalization increases the stability of A⁻ that also decreases the reactivity. A base that has resonance delocalization of the electron pair that is shared with the proton will therefore be less basic than a base without this feature. Since a weaker base has a stronger conjugate acid, a compound whose conjugate base has resonance stabilization will be more acidic.

Both carboxylic acids and alcohols contain an —OH group, but a carboxylic acid is a stronger acid than an alcohol. As we can see, that deprotonation of ethanol (CH_3CH_2OH) affords the ethoxide ion ($CH_3CH_2O^-$), which has no resonance (only one Lewis structure can be drawn), but deprotonation of acetic acid ($CH_3CH_2CO_2H$) affords an acetate ion ($CH_3CH_2CO_2^-$) that has resonance (two contributing Lewis structures can be drawn).

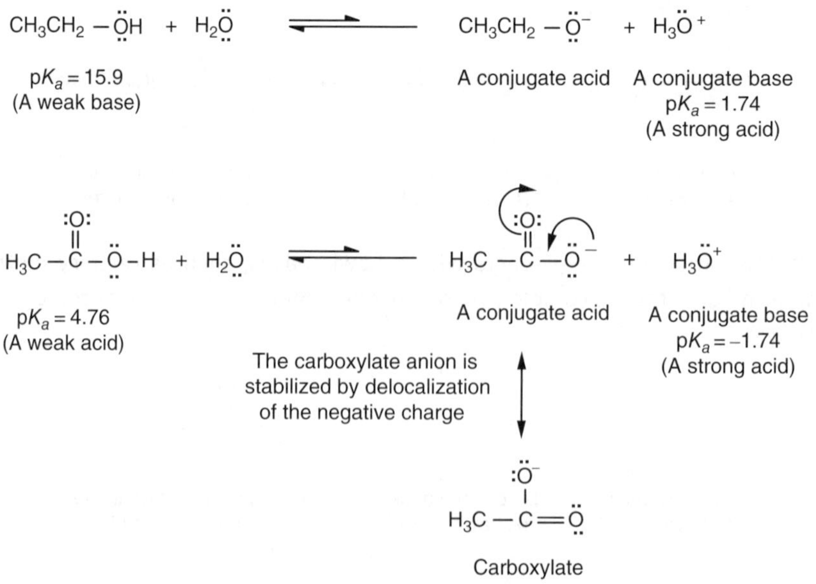

$$CH_3CH_2-\overset{..}{\underset{..}{O}}H + H_2\overset{..}{\underset{..}{O}} \rightleftharpoons CH_3CH_2-\overset{..}{\underset{..}{O}}{}^- + H_3\overset{..}{O}{}^+$$

$pK_a = 15.9$ A conjugate acid   A conjugate base
(A weak base) $pK_a = 1.74$
                                                 (A strong acid)

$pK_a = 4.76$ A conjugate acid   A conjugate base
(A weak acid) $pK_a = -1.74$
                The carboxylate anion is   (A strong acid)
                stabilized by delocalization
                of the negative charge

Carboxylate

This is because the acetate (CH_3COO^-) ion has resonance that delocalizes the electron pair to be shared with a proton, and the ethoxide (CH_3O^-) ion does not. Therefore, the acetate (CH_3COO^-) ion is a weaker base than the ethoxide (CH_3O^-) ion, which makes acetic acid (CH_3COOH) a stronger acid than ethanol (CH_3OH). Recalling that weaker bases have stronger conjugate acids, acetic acid has a $pK_a = 4.76$ (the lower the pK_a, the stronger the acid), whereas ethanol has a $pK_a = 15.9$ (the higher the pK_a, the weaker the acid).

1.7.3 Acid–Base Properties of Organic Functional Groups

Let us see the acid–base properties of some molecules with different functional groups. Organic functional groups have been discussed in Chapter 4. Most organic acids and bases are weak acids and weak bases. The most common examples are carboxylic acids, amines, alcohols, amides, ethers and ketones. Drug molecules also contain various types of functional groups, and these functional groups contribute to the overall acidity or basicity of drug molecules. Organic compounds with nonbonding electrons on nitrogen, oxygen, sulphur or phosphorus can act as Lewis bases.

The most common organic acids are carboxylic acids. They are moderately strong acids having pK_a values ranging from about 3 to 5. Acetic acid ($pK_a = 4.76$) can behave as an acid and donate a proton, or as a base and accept a proton. A protonated acetic acid ($pK_a = -6.1$) is a stronger acid than H_2SO_4 ($pK_a = -5.2$). Equilibrium always favours a reaction of the stronger acid and stronger base to give the weaker acid and weaker base.

Amines are the most important organic bases as well as weak acids. Thus, an amine can behave as an acid and donate a proton, or as a base and accept a proton. The most common organic bases are alkyl amines. They are moderately strong bases having pK_a values ranging from about 30 to 40. Methylamine ($pK_a = 38.0$) can behave as an acid and donate a proton, or as a base and accept a proton. A protonated methylamine ($pK_a = 10.64$) is a much weaker acid than H_2SO_4 ($pK_a = -5.2$), but a much weaker base than methylamide ion. Equilibrium always favours a reaction of the stronger acid and stronger base to give the weaker acid and weaker base.

$$H_3C - \ddot{N}H_2 \ + \ H - SO_3OH \ \rightleftharpoons \ H_3C - \overset{+}{N}H_3 \ + \ HSO_4^-$$

$pK_a = -5.2$
(A strong acid)

A conjugate base   A conjugate acid
pKa = 10.64
(A weak acid)

$$H_3C - \ddot{N}H_2 \ + \ H\ddot{O}^- \ \rightleftharpoons \ H_3C - \overset{..}{\underset{..}{N}}H \ + \ H\ddot{O} - H$$

$pK_a = 38.0$
(A weak acid)

A conjugate acid   A conjugate base
pKa = 15.7
(A strong acid)

An alcohol can behave like an acid and donate a proton. However, alcohols are much weaker organic acids with pK_a values close to 16. Alcohol may also behave as a base, for example, ethanol is protonated by H_2SO_4 and gives ethyloxonium ion $(CH_3CH_2O^+H_2)$. A protonated ethanol $(pK_a = -2.4)$ is a stronger acid than ethanol $(pK_a = 15.9)$.

$$H_5C_2 - \ddot{O} - H \ + \ H\ddot{O}^- \ \rightleftharpoons \ H_5C_2 - \ddot{O}^- \ + \ H\ddot{O} - H$$

$pK_a = 15.9$
(A weak acid)

A conjugate acid   A conjugate base
$pK_a = 15.7$
(A strong acid)

$$H_5C_2 - \ddot{O}H \ + \ H - SO_3OH \ \rightleftharpoons \ H_5C_2 - \overset{+}{\underset{H}{O}} - H \ + \ HSO_4^-$$

$pK_a = -5.2$
(A strong acid)

A conjugate base
$pK_a = -2.4$
(A weak acid)

A conjugate acid

Some organic compounds have more than one atom with nonbonding electrons, thus more than one site in such a molecule can react with acids. For example, acetamide (CH_3CONH_2) has nonbonding electrons on both nitrogen and oxygen atom, and either may be protonated. However, generally the reaction stops when one proton is added to the molecule.

Both acetamide (CH_3CONH_2) and acetic acid (CH_3COOH) are more readily protonated at the carbonyl oxygen rather than the basic site. The protonation of the nonbonding electrons on the oxygen atom of a carbonyl $(-C=O)$ or hydroxyl $(-OH)$ group is an important first step in the reactions under acidic conditions of compounds like acetamide, acetic acid, diethyl ether and alcohol. The conjugate acids of these compounds are more reactive towards Lewis bases than the unprotonated forms. Therefore, acids are used as catalysts to enhance reactions of organic compounds.

$$H_3C-\overset{\overset{\displaystyle :O:}{\|}}{C}-NH_2 \ + \ H-SO_3OH \ \longrightarrow \ H_3C-\overset{\overset{\displaystyle :\overset{+}{O}H}{\|}}{C}-NH_2 \ + \ HS\overset{-}{O_4}$$

Acetamide $pK_a = -5.2$ Protonated acetamide   A conjugate acid
(A strong acid)   (A weak acid)

$$H_3C-\overset{\overset{\displaystyle :O:}{\|}}{C}-OH \ + \ H-Cl \ \longrightarrow \ H_3C-\overset{\overset{\displaystyle :\overset{+}{O}H}{\|}}{C}-OH \ + \ Cl^-$$

Acetic acid $pK_a = -7.0$ Protonated acetic acid   A conjugate acid
(A strong acid) $pK_a = -6.1$
           (A weak acid)

The reaction of diethyl ether with concentrated hydrogen chloride (HCl) is typical of that of an oxygen base with a protic acid. *Protic acids* have hydrogens that can form hydrogen bonds with water, alcohols and ammonia. Therefore, just like water, organic oxygenated compounds are protonated to give oxonium ions, for example, protonated ether.

$$C_2H_5-\overset{\overset{\displaystyle ..}{\underset{\displaystyle ..}{O}}}{}-C_2H_5 \ + \ H-Cl \ \longrightarrow \ C_2H_5-\overset{\overset{\displaystyle +}{\underset{\displaystyle |}{\overset{..}{O}}}}{\underset{\displaystyle H}{}}-C_2H_5 \ + \ Cl^-$$

Diethyl ether $pK_a = -7.0$ A conjugate acid
(A strong acid)   Protonated ether
           $pK_a = -3.6$
           (A weak acid)

Ketones can behave as a base. Acetone donates electrons to boron trichloride, a Lewis acid and forms a complex of acetone and boron trichloride.

$$H_3C-\overset{\overset{\displaystyle :O:}{\|}}{C}-CH_3 \ + \ BCl_3 \ \longrightarrow \ H_3C-\overset{\overset{\displaystyle :O:}{\|}}{\underset{\displaystyle CH_3}{\overset{+}{C}}}-\overset{-}{B}Cl_3$$

Acetone   Boron trichloride   A complex of acetone
(A Lewis base)   (A Lewis acid)   and boron trichloride

Most organic reactions are either acid–base reactions or catalysed by an acid or a base. The reaction of an organic compound as an acid depends on how easily it can lose a proton to a base. Electronegativity is the ability of an atom, which is bonded to another atom or atoms, to attract electrons strongly towards it. The acidity of the hydrogen atom depends on the electronegativity of the bonded central atom. The more electronegative the bonded central atom, the more acidic are the protons. Carbon is less electronegative than nitrogen and oxygen. Thus, carbon attracts and holds electrons less strongly than nitrogen and oxygen do. For example, ethane is a very weak acid in which the hydrogen atoms are bonded to carbon atoms.

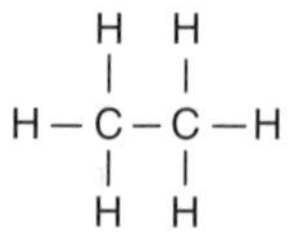

$$
\begin{array}{cc}
& H \quad H \\
& | \quad\; | \\
H-&C-C-H \\
& | \quad\; | \\
& H \quad H
\end{array}
$$

Ethane

Nitrogen is less electronegative than oxygen. Thus, nitrogen attracts and holds the electrons less strongly than oxygen does. For example, in methylamine, the hydrogen atoms on nitrogen are acidic, but the hydrogen atom bonded to the oxygen atom in methanol is even more acidic.

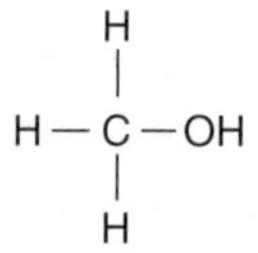

$$
\begin{array}{cc}
H & \qquad\qquad H \\
| & \qquad\qquad | \\
H-C-OH & \qquad H-C-NH_2 \\
| & \qquad\qquad | \\
H & \qquad\qquad H
\end{array}
$$

Methanol 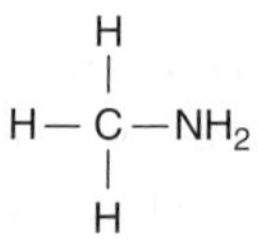　　　　　　Methyl amine

Weak acids produce strong conjugate bases. Thus, ethane gives stronger conjugate base than methylamine and methanol. The conjugate bases of ethane, methylamine and methanol are shown next.

CH_3CH_3(Ethane) $\rightarrow CH_3NH_2$(Methylamine) $\rightarrow CH_3OH$(Methanol)

(Increasing acidity of hydrogen bonded to carbon, nitrogen, and oxygen)

CH_3O^-(Methoxide anion) $\rightarrow CH_3NH^-$(Methylamide anion) $\rightarrow CH_3CH_2^-$(Ethyl anion)

(Increasing basicity of the conjugate base)

1.7.4 pH, pOH and pK_a Values

The pH and pOH logarithmic scales are used to keep track of a large concentration of acids and bases. When an acid is added to a solution, the pH gets lower (acidic), but when base is added, the pH gets higher (basic). The pH value is defined as the negative of the logarithm to base 10 of the concentration of the hydrogen ion. The acidity or basicity of a substance is defined most typically by the pH value. We write log as $\log_{10}$ for convenience.

$$
pH = -\log_{10}\left[H_3O^+\right] = -\log\left[H_3O^+\right]
$$

The acidity of an aqueous solution is determined by the concentration of H_3O^+ ions. Thus, the pH of a solution indicates the concentration of hydrogen ions in the solution. The pH is a measure of the concentration of hydrogen ions in a solution. The pH equation also can be written as the logarithm of the reciprocal of the hydrogen ion concentration.

$$pH = -\log\left[H^+\right] = \log\frac{1}{\left[H^+\right]}$$

A substance that can act as an acid or a base is known as *amphiprotic* or *ampho-teric*. Water is amphiprotic, as it acts both as an acid or a base. Pure water ionizes partially or undergoes auto-ionization (autoprotolysis). Two molecules of water react with each other to form hydronium and hydroxide ions.

H_2O (Acid) + H_2O (Base) $\rightleftharpoons$ H_3O^+ (A conjugate acid) + HO^- (A conjugate base)

The acidity of an aqueous solution is determined by the concentration of hydrogen (H_3O^+, hydronium) ions. The pH of a solution indicates the concentration of hydronium (H_3O^+) ions in the solution. The acidity or basicity of a substance is defined most typically by the pH value. Because the $[H_3O^+]$ in an aqueous solution is typically quite small, chemists have found an equivalent way to express $[H_3O^+]$ as a positive number, whose value normally lies between 0 and 14. The lower the pH, the more acidic is the substance or the solution. The pH of a solution can be changed simply by adding either acid or base to the solution.

When water molecules react with one another they form hydronium (H_3O^+) and hydroxide (HO^-) ions. The ratio of the molar concentrations of reactants and products is a constant at certain temperature and is known as equilibrium constant, K.

H_2O + H_2O $\rightleftharpoons$ H_3O^+ + HO^-

$pK_a = 7.0$ A conjugate base   A conjugate acid
(A weak acid and a $pK_a = -1.74$
weak base)   (A strong acid)

Equilibrium constant, K can be written as:

$$K = \frac{\left[H_3O^+\right]\left[HO^-\right]}{[H_2O]}$$

Where, $[H_3O^+]$, $[HO^-]$ and $[H_2O]$ are molar concentrations of hydronium ion, hydroxide ion and water, respectively. Remember, molar concentration (mol dm^{-3} or mol l^{-1}) is also called *molarity* (M). Only a very few water molecules are ionized, so $[H_2O]$ can be regarded as constant. Therefore, for pure water the reaction equilibrium can be expressed as:

$$K[H_2O] = \left[H^+\right]\left[HO^-\right] = K_w$$

Here, K_w is the ionic product of water. We write $[H^+]$ instead of $[H_3O^+]$ for convenience.

The hydrogen [H⁺] and hydroxide [HO⁻] ions are always present in aqueous solutions. In acid or basic solutions, the concentrations of H^+ and HO^- ions are not equal. For a neutral solution H^+ and HO^- ions are at the same concentration and the pH of the neutral solution is exactly 7.

In an acidic solution, there is an excess of hydrogen ions over hydroxide ions, therefore the pH of the acidic solution is below 7. In a basic solution, there is an excess of hydroxide ions over hydrogen ions, so the pH of the basic solution is above 7.

The pK_w is the negative logarithm (to base 10) of the K_w. The relationship between K_w and pK_w is exactly the same as that between [H⁺] and pH.

$$pK_w = -\log K_w$$

The pOH is the negative logarithm (to base 10) of the hydroxide (HO⁻) ion concentration.

$$pOH = -\log\left[HO^-\right]$$

The pOH gives us another way to measure the acidity of a solution. It is just the opposite of pH. The higher pOH means the solution is acidic and the lower pOH means the solution is basic. Whereas, the higher pH indicates basic solution and lower pH means acidic solution. Do not confuse pH with pK_a. The pH scale is used to describe the acidity of a solution. The pK_a is characteristic of a particular compound, and it tells how readily the compound gives up a proton. At equilibrium, the concentration of H^+ is 10^{-7}, so we can calculate the pH of water at equilibrium as: pH = $-\log$ [H+] = $-\log$ [10^{-7}] = 7. Solutions with a pH of 7 are said to be *neutral*, while those with pH values below 7 are defined as acidic, and those are above pH of 7 as being basic.

The pH of blood plasma is around 7.4 (slightly basic), whereas the gastric juice is around 1.6 (highly acidic). The pH scales of acids and bases are shown next.

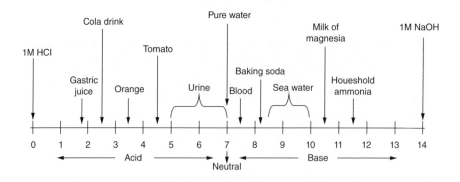

Acids and bases that are *strong electrolytes* are called strong acids and strong bases. Those that are weak electrolytes are called weak acids and weak bases. *Strong acids,* for example, HCl, HBr, HI, H_2SO_4, HNO_3, $HClO_3$ and $HClO_4$, completely ionize in solution, and are always represented in chemical equations in their ionized form. Similarly, *strong bases,* for example, LiOH, NaOH, KOH, RbOH, $Ca(OH)_2$, $Sr(OH)_2$ and $Ba(OH)_2$, completely ionize in solution and are always represented in their ionized form in chemical equations. A *salt* is formed when an acid and a base are mixed and the acid releases H^+ ions, while the base releases HO^- ions. This process is called *hydrolysis.* The conjugate base of a strong acid is very weak and cannot undergo hydrolysis. Similarly, the conjugate acid of a strong base is very weak and likewise does not undergo hydrolysis.

Acidity and *basicity* are described in terms of equilibria. Acidity is the measure of how easily a compound gives up a proton, and basicity is a measure of how well a compound shares its electrons with a proton. A strong acid is one that gives up its proton easily. This means that its conjugate base must be weak, because it has little affinity for a proton. A weak acid gives up its proton with difficulty, indicating that its conjugate base is strong because it has a high affinity for a proton. Thus, the stronger the acid, the weaker its conjugate base. When acids or bases are ionized, the amount of ionization will depend on the strength of acids and bases. Strong acids and bases are almost completely ionized in water. A stronger acid produces a weaker conjugate base, likewise a stronger base produces a weaker conjugate acid. For strong acids and bases, we can directly calculate the pH, if we know the molar concentration (M) of the solution.

An acid–base reaction is favoured in the direction from the stronger member to the weaker member of each conjugate acid–base pair. Thus, in an acid–base reaction, the equilibrium will always favour the reaction that moves the proton to the stronger base. For example, when a strong acid, such as hydrochloric acid (an inorganic or mineral acid), is dissolved in water, it dissociates almost completely, which means that the products are favoured at equilibrium. When a much weaker acid, such as acetic acid (an organic acid), is dissolved in water, it dissociates only to a small extent, so the reactants are favoured at equilibrium.

$$H-Cl \ + \ H-\ddot{\underset{\cdot\cdot}{O}}-H \ \rightleftharpoons \ H-\overset{+}{\underset{|}{\overset{\cdot\cdot}{O}}}-H \ + \ Cl^-$$

$pK_a = -7.0$	$\quad\quad\quad\quad$ H $\quad\quad\quad$ A conjugate acid
(A strong acid)	A conjugate base
	$pK_a = -1.74$
	(A weak acid)

The H_3O^+ ($pK_a = -1.74$) is the conjugate acid of water. The pK_a of H_3O^+ is higher than HCl ($pK_a = -7.0$). This means that HCl will give up its protons to water completely to form the H_3O^+ ion. The lower the value of pK_a stronger is the acid and the higher the value of pK_a stronger is the base.

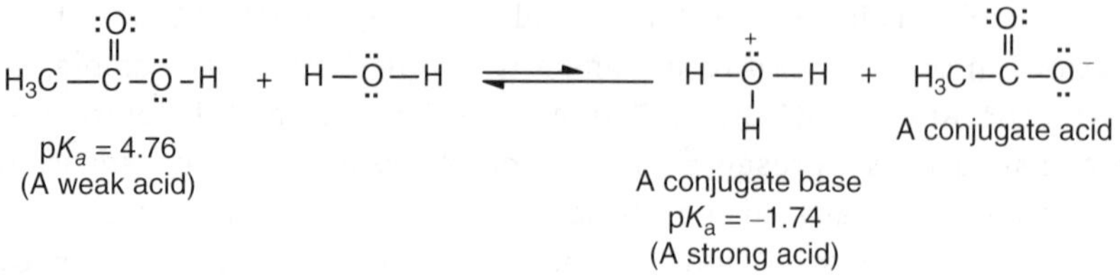

The pK_a is the negative logarithm (to base 10) of the K_a. The relationship between K_a and pK_a is exactly the same as that between K_w and pK_w.

$$pK_a = -\log K_a$$

We know that the pK_a is characteristic of a particular compound, and it tells how readily the compound gives up a proton. K_a values are very small and hard to remember. Therefore, the strength of an acid is generally indicated by its pK_a values rather than its K_a value. The lower the value of pK_a, the stronger is the acid. For example, hydrochloric acid is a strong acid and has $pK_a = -7.0$, and acetic acid is a weak acid and has $pK_a = 4.76$.

Very strong acids	$pK_a < 1$	Moderately strong acids	$pK_a = 1–5$
Weak acids	$pK_a = 5–15$	Extremely weak acids	$pK_a > 15$

The strength of an acid and a base is described in terms of the acid–base reaction equilibrium. Acid and base dissociation constants (K_a and K_b) are the measure of the strength of acids and bases. Usually, K_a and K_b values are used to compare, respectively, the strength of weak acids and weak bases. Acid dissociation constant (K_a) is a measure of the strength of an acid. For example, ethanol is a very weak acid ($pK_a = 15.9$), acetic acid ($pK_a = 4.76$) is somewhat stronger than ethanol, but trichloroacetic acid ($pK_a = 0.66$) is much stronger acid than acetic acid.

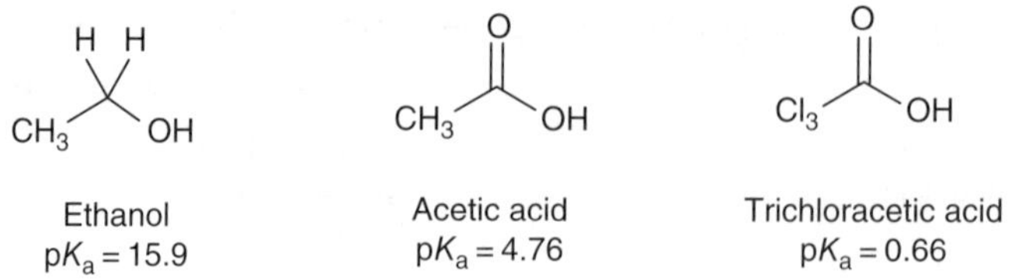

The strength of an acid depends on its degree of dissociation in water. The acid dissociation constant, K_a, is the equilibrium constant for a dissociation reaction. Let us consider first a simple reaction and its equilibrium features as shown next.

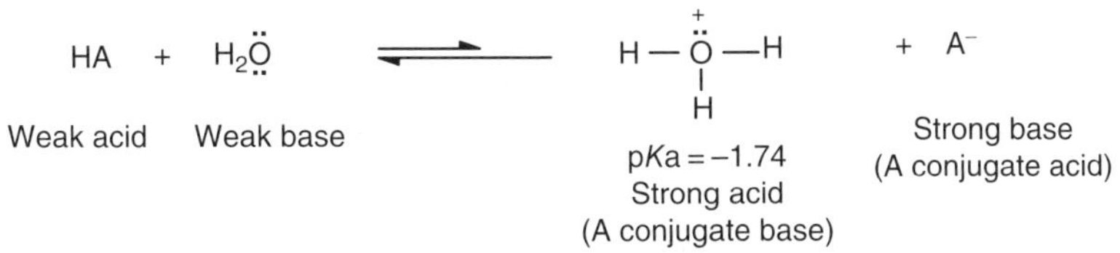

$$HA \quad + \quad H_2\ddot{O} \quad \rightleftharpoons \quad H - \overset{+}{\underset{|}{\overset{\displaystyle \ddot{O}}{H}}} - H \quad + \quad A^-$$

Weak acid	Weak base		Strong base
		pKa = −1.74	(A conjugate acid)
		Strong acid	
		(A conjugate base)	

Weak acid is only partially ionized in water and the reaction equilibrium lies on the left. Whether a reversible reaction favours reactants or products at equilibrium is indicated by the equilibrium constant of the reaction (K_{eq}). Remember that brackets are used to indicate concentration in moles per litre = molarity (M). The degree to which an acid (HA) dissociates is described by its acid dissociation constant (K_a). The acid dissociation constant is obtained by multiplying the equilibrium constant (K_{eq}) by the concentration of the solvent in which the reaction takes place. Thus at equilibrium, the strength of an acid dissociation constant, K_a can be expressed as:

$$K_a = K_{eq}[H_2O] = \frac{[H_3O^+][A^-]}{[HA]}$$

Here, [H_2O] is constant and is included in the acid dissociation constant, K_a. The K_a values can be used to describe the relative strength of acids. It is a constant at constant temperature. Strong acids have K_a larger than 1 and weak acids have K_a smaller than 1. Therefore, the larger the acid dissociation constant, the stronger is the acid. Hydrochloric acid (strong acid) has an acid dissociation constant of 10^7, whereas acetic acid (weak acid) has an acid dissociation constant of 1.74×10^{-5}.

A weak base is only partially ionized and the reaction equilibrium lies on the left.

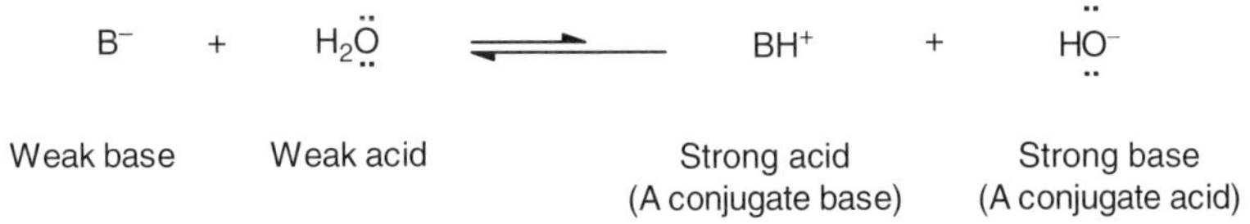

$$B^- \quad + \quad H_2\ddot{O} \quad \rightleftharpoons \quad BH^+ \quad + \quad H\ddot{O}^-$$

| Weak base | Weak acid | Strong acid | Strong base |
| | | (A conjugate base) | (A conjugate acid) |

At equilibrium, the strength of a base dissociation constant, K_b, can be expressed as:

$$K_b = K[H_2O] = \frac{[BH^+][HO^-]}{[B]}$$

Here, $[H_2O]$ is a constant and is included in the base dissociation constant, K_b. The K_b is the negative logarithm (to base 10) of the K_b. The relationship between K_b and pK_b is exactly the same as that between K_a and pK_a.

$$pK_b = -\log K_b$$

The pK_b is also characteristic of a particular compound, and it tells how readily the compound gives up a hydroxide (hydroxyl) ion. The K_b values are very small and hard to remember. The lower the value for pK_b or higher the value of K_b, the stronger the base. The higher the value for pK_b or lower the value of K_b, the weaker the base.

Weak acids and bases are only partially ionized. The dissociation constant can be used to calculate the amount ionized and the pH of the acids or bases. For weak acids and bases k_a and k_b always have values that are smaller than one. Acids with a larger K_a are stronger acids, and bases with smaller K_a are stronger bases. Bases with a larger K_b are stronger bases, and acids with smaller K_b are stronger acids.

Weak acids, for example, HF, HCN, H_2S, H_2O, NH_4, HNO_2 (nitrous acid), HCO_2H (formic acid), CH_3CO_2H (acetic acid), $C_6H_5CO_2H$ (benzoic acid) and *weak bases*, for example, H_2O, NH_3, NH_4OH (ammonium hydroxide), N_2H_4 (hydrazine), CH_3NH_2 (methylamine), $CH_3CH_2NH_2$ (ethylamine), are only partially ionized in solution and are always represented in their ionized form in chemical equations. When acids or bases are ionized, the amount of ionization will depend on the strength of acids and bases. An acid–base reaction is favoured in the direction from the stronger member to the weaker member of each conjugate acid–base pair. When a weak acid, such as acetic acid (ethanoic acid or acetyl hydroxide), is dissolved in water, it dissociates only a small amount so the reactants are favoured at equilibrium.

$pK_a = 4.76$	Weak base		Strong base	$pK_a = -1.74$
Weak acid			(A conjugate acid)	Strong acid
				(A conjugate base)

The H_3O^+ ($pK_a = -1.74$) is the conjugate acid of water. The pK_a of H_3O^+ is lower than CH_3CO_2H ($pK_a = 4.76$). This means that H_3O^+ (strong acid) will give up its protons to $CH_3CO_2^-$ (strong base) completely to form the CH_3CO_2H.

A salt is an ionic compound formed by the reaction of an acid and a base. This is called a *neutralization* reaction. The reverse of the neutralization reaction is called *hydrolysis*. In a hydrolysis reaction, a salt reacts with water to reform the acid and base. The conjugate base of a strong acid is very weak and cannot undergo hydrolysis. Similarly, the conjugate acid of a strong base is very weak and likewise does not undergo hydrolysis.

$$H_2SO_4 \text{ (Strong acid)} + Zn(OH)_2 \text{ (Strong base)} \longrightarrow ZnSO_4\text{(Salt)} + 2H_2O \text{ (water)}$$

There are four combinations of strong and weak acids and bases. Salt hydrolysis usually affects the pH of a solution. The pH of the salt depends on the strengths of the original acids and bases as shown here.

Acid	Base	Salt pH
Strong	Strong	7.0
Weak	Strong	>7.0
Strong	Weak	<7.0
Weak	Weak	Depends on which one is stronger

Strong acids and strong bases react with each other almost completely, and salt and water are produced. The strong acid and strong base neutralizes each other and forms a neutral salt.

$$HCl \text{ (Strong acid)} + NaOH \text{ (Strong base)} \longrightarrow NaCl \text{ (Salt)} + H_2O \text{ (water)}$$

Hydrolysis of NaCl produces Na^+ and Cl^- ions and they do not react with water. This is called a neutralization reaction.

$$NaCl \text{ (Salt)} \xrightarrow{H_2O} Na^+ + Cl^-$$

Salts of strong acids and weak bases form acidic solutions. The reaction between a strong acid and a weak base only yields a salt, and water is not formed since a weak base tends not to be a hydroxide.

$$HCl \text{ (Strong acid)} + NH_3 \text{ (Weak base)} \longrightarrow NH_4Cl \text{ (Salt)}$$

Hydrolysis of ammonium chloride produces NH_4^+ and Cl^- ions when water is added.

$$NH_4Cl \text{ (Salt)} \xrightarrow{H_2O} NH_4^+ + Cl^-$$

The Cl^- ion does not react with water but the NH_4^+ ion reacts with water and produces H_3O^+ and NH_3. Since the H_3O^+ ($pK_a = -1.74$) ion is present in the solution, it is acidic.

$$NH_4^+ + H_2O \longrightarrow H_3O^+ + NH_3$$

Salts of weak acids and strong bases form basic solutions. A weak acid reacts with a strong base and produces a basic salt.

$$CH_3CO_2H \text{ (Weak acid)} + NaOH \text{ (Strong base)} \longrightarrow CH_3CO_2Na \text{ (Salt)} + H_2O$$

Hydrolysis of CH_3CO_2Na forms $CH_3CO_2^-$ and Na^+ ions. The Na^+ ion does not react with water but $CH_3CO_2^-$ ion reacts with water and produces CH_3CO_2H and HO^-. Since the HO^- ($pK_a = 15.7$) ion is present in the solution, it is basic. The higher the value of pK_a, the stronger the base.

$$CH_3CO_2Na \xrightarrow{H_2O} CH_3CO_2^- + Na^+$$

$$CH_3CO_2^- + H_2O \longrightarrow CH_3CO_2H + HO^-$$

When a weak acid reacts with a weak base, a salt is formed and the resulting pH of the solution depends on the relative strength of the acid and base. The solution formed in the reaction may be slightly acidic, basic or neutral depending on the relative concentrations of H^+ and HO^- ions. If K_a for the cation is greater than K_b for the anion, the solution is acidic. If K_b for the anion is greater than K_a for the cation, the solution is basic. If K_a and K_b are similar, the solution is close to neutral. For example, if the acid HClO has a $K_a = 3.4 \times 10^{-8}$ and the base NH_3 has a $K_b = 1.6 \times 10^{-5}$, then the aqueous solution of HClO and NH_3 will be basic, because the K_b of NH_3 is higher than the K_a of HClO. Acid with higher K_a (acid dissociation constant) is a strong acid, and base with higher K_b (base dissociation constant) is a strong base.

$$HClO \text{ (Weak acid)} + NH_3 \text{ (Weak base)} \rightleftharpoons NH_4ClO \text{ (Basic salt)}$$

Most organic acids and bases are weak acids and weak bases. The pK_a and pK_b values of weak acids and bases at 25 °C are shown here.

Acid	pK_a	pK_b
Acetic acid	4.76	9.24
Ammonium ion	9.25	4.75
Benzoic acid	4.19	9.81
Formic acid	3.74	10.26
Lactic acid	3.86	10.14
Phenol	9.89	4.11

1.7.5 Acid–Base Titration: Neutralization

The process of obtaining quantitative information of a sample using a fast chemical reaction by reacting with a certain volume of reactant whose concentration is known is called *titration*. The goal is to determine the pH during the course of titration. In a titration, a solution of accurately known concentration is added gradually to another solution of unknown concentration until the chemical reaction between the two solutions is complete. Titration is also called *volumetric analysis*, which is a type of quantitative chemical analysis. Generally, the *titrant* (the known solution) is added from a burette to a known quantity of the analyte (the unknown solution) until the reaction is complete.

The titration curve is the pH change against the amount of titrant added. From the added volume of the titrant, it is possible to determine the concentration of the unknown. Often, an indicator is used to detect the end of the reaction, known as the *endpoint* at which the indicator changes colour. The endpoint is determined by the sudden change of pH.

An *acid–base titration* is a method that allows quantitative analysis of the concentration of an unknown acid or base solution. In an acid–base titration (neutralization), the base will react with the weak acid and form a solution that contains the weak acid and its conjugate base until the acid is completely neutralized. The following equation is used frequently, when trying to find the pH of buffer solutions.

$$pH = pK_a + \log \frac{[Base]}{[Acid]}$$

where, pH is the log of the molar concentration of the hydrogen,

pK_a is the equilibrium dissociation constant for an acid, [Base] is the molar concentration of a basic solution and [Acid] is the molar concentration of an acidic solution.

The point at which the acid and base are present in equal stoichiometric amounts is called the *equivalence point*. In other words, an equivalence point is reached, when the reaction is completed. Generally, for the titration of a strong base with a weak acid, the equivalence point is reached when the pH is greater than 7. The half equivalence point is when half of the total amount of base needed to neutralize the acid has been added. It is at this point, where the pH = pK_a of the weak acid.

An *indicator* is a substance that changes colour at the equivalence point over a specific pH range. An indicator can exist in either its acid or base form and it changes colour when proton gain or loss occurs. In acid–base titrations, a suitable acid–base indicator is used to detect the endpoint from the change of colour of the indicator used. An acid–base indicator is a weak acid or a weak base. For example, when a weak acid is titrated with a strong base containing phenolphthalein indicator, it produces a conjugate base of weak acid at the equivalence point pH > 7. The names and the pH range of some common acid-base indicators are shown in the following Table.

Indicator	pH range	Colour in acid	Colour in base
Thymol blue	1.2–2.8	Red	Yellow
Bromophenol blue	3.0–4.6	Yellow	Blue-violet
Methyl orange	3.1–4.4	Orange	Yellow
Methyl red	4.2–6.3	Red	Yellow
Chlorophenol blue	4.8–6.4	Yellow	Red
Bromothymol blue	6.0–7.6	Yellow	Blue
Cresol red	7.2–8.8	Yellow	Red
Phenolphthalein	8.0–10.0	Colourless	Red

1.8 BUFFER AND ITS USE

A *buffer* is a solution containing a weak acid and its conjugate base (i.e. CH_3COOH and CH_3COO^-) or a weak base and its conjugate acid (i.e. NH_3 and $^+NH_4$). Buffers are commonly used, when pH must be maintained at a relatively constant value in many biological systems as well as useful for monitoring pH for reaction at an optimum value. A buffer solution has the ability to resist changes in pH on addition of small amounts of acid or base. For example, when a strong acid is added, the base present in the buffer solution neutralizes the hydronium (H_3O^+) ions. Similarly, when a strong base is added, the acid present in the buffer solution neutralizes the hydroxide (HO^-) ions.

By choosing the appropriate components, a solution can be buffered virtually at any pH. The pH of a buffered solution depends on the ratio of the concentrations of buffering components. When the ratio is least affected by adding acids or bases, the solution is most resistant to a change in pH. It is more effective when the acid-base ratio is equal to 1. Therefore, the pK_a of the weak acid selected for the buffer should be as close as possible to the desired pH, because it follows the following equation.

$$pH = pK_a$$

An acidic buffer solution is simply the one, which has a pH less than 7. Therefore, acidic buffer solutions are commonly made from a weak acid and one of its salts, often a sodium salt. A basic buffer solution has a pH greater than 7. Thus, basic buffer solutions are commonly made from a weak base and one of its salts. An acidic buffer solution consists of a mixture of a weak acid and its salt at pre-determined concentration. For a weak acid HA that ionizes to H^+ and its salt A^-.

$$HA \rightleftharpoons [H^+] + [A^-]$$

The acid equilibrium can be written as:

$$K_a = \frac{[H^+][A^-]}{[HA]}$$

The hydrogen ion concentration can be written as:

$$[H^+] = K_a \frac{[HA]}{A^-}$$

Taking the –log on both sides of the equation gives:

$$-\log[H^+] = -\log K_a - \log\frac{[HA]}{[A^-]}$$

$$pH = pK_a - \log\frac{[HA]}{[A^-]}$$

On converting the last log term, it becomes positive.

$$pH = pK_a + \log\frac{[A^-]}{[HA]}$$

$$pH = pK_a + \log\frac{[Salt]}{[Acid]}$$

Therefore, pH = pK_a + log [Salt] – log [Acid]. This equation is known as *Henderson–Hasselbalch equation*, or simply *buffer equation*. It is useful for calculating the pH of a weak acid solution containing its salt.

Buffers play an important role in cellular processes, because they maintain the pH at an optimal level for biological processes. Strong acids, such as HCl, are poor buffers, while weaker acids, such as acetic acid, are good buffers in the pH ranges found in biological environments. Thus, the role of a buffer system in the body is important as it tends to resist any pH changes as a result of metabolic processes. The most important practical example of a buffered solution is blood, which can absorb the acids and bases produced by biological reactions without changing its pH. The normal pH of blood is 7.4. Blood can absorb the acids and bases produced in biological reactions without changing its pH. Thus, blood is a buffer solution. A constant pH for blood is vital, because cells can survive only this narrow pH range around 7.4.

One common buffer system in the body is the carbonic acid buffer system. It helps keep the blood at a fairly constant pH at around 7.4. If it goes below 7.35, someone may experience a condition known as *acidosis* and triggers *hyperventilation,* which causes to breathe at an abnormally rapid rate. If it goes above 7.45, someone may experience a condition known as *alkalosis* and trigger *hypoventilation,* which causes to breathe at an abnormally slow rate. Therefore, if blood pH goes below 6.8 or above 7.8, cells of the body can stop functioning and the person eventually can die.

Water is the medium for metabolic reactions within cells. Carbon dioxide from metabolism combines with water in blood plasma to produce carbonic acid (H_2CO_3), which is weakly ionic and can dissociate to from hydronium (H_3O^+) ions and the bicarbonate (HCO_3^-) ions as shown next. The amount of H_2CO_3 depends on the amount of CO_2 present.

$$CO_2 + H_2O \longrightarrow H_2CO_3$$

$$H_2CO_3 + H_2O \longrightarrow H_3O^+ + HCO_3^-$$

The pH of blood is maintained at about 7.4 by the carbonic acid-bicarbonate ion buffering system. When any acidic substance enters the bloodstream, the bicarbonate (HCO_3^-) ions neutralize the hydronium (H_3O^+) ions forming carbonic acid and water. Carbonic acid (H_2CO_3) is already a component of the buffering system of blood. Thus hydronium (H_3O^+) ions are removed, preventing the pH of blood from becoming acidic.

$$HCO_3^- + H_3O^+ \longrightarrow H_2CO_3 + H_2O$$

On the other hand, when a basic substance enters the bloodstream, carbonic acid (H_2CO_3) reacts with the hydroxide (HO^-) ions producing bicarbonate (HCO_3^-) ions and water. Bicarbonate ions (HCO_3^-) are already a component of the buffer. In this manner, the hydroxide (HO^-) ions are removed from blood, preventing the pH of blood from becoming basic.

$$H_2CO_3 + HO^- \longrightarrow HCO_3^- + H_2O$$

Therefore, by far the most important buffer for maintaining acid-base balance in the blood is the carbonic acid–bicarbonate buffer.

1.8.1 Common Ion Effects and Buffer Capacity

The most important application of acid–base solutions containing a *common (same) ion* is buffering. The pH of a buffer does not depend on the absolute amount of the conjugate acid–base pair. It is based on the ratio of the two. In other words, it depends on the *common ion effect.* The solution containing both the weak acid and its conjugate base has a pH much higher than the solution containing only the weak base. The conjugate base is referred to as a common ion, because it is found in both the weak acid and the anion. The common ion effect can also be applied to a solution containing a weak base and its conjugate acid. The common ion effect is the suppression of the ionization of a weak acid or a weak base by the presence of a common ion from a strong electrolyte.

The *buffer capacity* is a measure of the effectiveness of a buffer. The amount of acid and base that can be added without causing a large change in pH is governed by the buffering capacity of the solution. This is determined by the concentrations of acid and its conjugate base. The higher the concentrations of the acid and its conjugate base, the addition of more acid or base can tolerate the solution. In other words, the more concentrated the acid and its conjugate base in a solution, the more added acid or base the solution can neutralize. The buffering capacity is governed by the ratio of acid and its conjugate base. Therefore, buffer capacity is a measure of how well a solution resists changes in pH when strong acid or based is added. A buffer is most effective if the concentrations of the buffer acid and its salt are equal.

$$pH = pK_a + \log(1) = pK_a + 0$$

$$\text{Thus, } pH = pK_a$$

Chapter 2
Atomic Structure and Bonding

Learning objectives

After completing this chapter students should be able to

- describe the fundamental concepts of atomic structure;
- explain various aspects of chemical bonding;
- discuss the relevance of chemical bonding in drug molecules and drug-receptor interactions, protein-protein and protein-DNA interactions.

2.1 ATOMS, ELEMENTS AND COMPOUNDS

The basic building block of all matter is called an *atom*. The term 'atom' comes from the Greek word for indivisible, because it was once thought that atoms were the smallest things in the universe and could not be divided! However, we now know that atoms are a collection of various subatomic particles containing negatively charged *electrons*, positively charged *protons* and neutral particles called *neutrons*. A *neutral atom* has an equal number of protons and electrons. Atoms can lose electrons and become positively charged, known as *cations*, or they can gain electrons and thereby become negatively charged, called *anions*. However, there won't be any change in the number of protons.

Each element has its own unique number of protons, neutrons and electrons. Both protons and neutrons have mass, whereas the mass of electrons is negligible.

Chemistry for Pharmacy Students: General, Organic and Natural Product Chemistry, Second Edition. Lutfun Nahar and Satyajit Sarker.

Protons and neutrons exist at the centre of the atom in the *nucleus*. Electrons move around the nucleus, and are arranged in shells at increasing distances from the nucleus. These shells represent different energy levels, the outermost shell being the highest energy level.

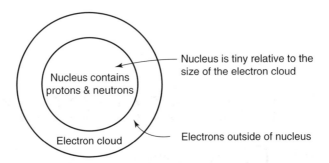

The number of protons that an atom has in its nucleus is called *atomic number*. The total number of protons and neutrons in the nucleus of an atom is known as the *mass number*. For example, a carbon atom containing six protons and six neutrons has a mass number of 12. The neutron is used as a comparison to find the relative mass of protons and electrons, and has a physical mass of 1.6749×10^{-27} kg. The number of neutrons of an atom can be varied.

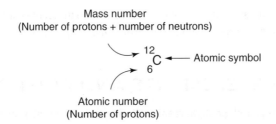

Isotopes are atoms with the same number of electrons and protons, but different numbers of neutrons. Thus, they have the same atomic number but different mass numbers. Isotopes are different forms of a single element. For example, the most common isotope of carbon is carbon-12 (^{12}C), which has six neutrons and six protons, whereas the rare isotope of carbon-13 (^{13}C) contains seven neutrons and six protons. Both ^{12}C and ^{13}C are *stable isotopes*, as they do not undergo *radioactive decay*, the process by which an unstable atomic nucleus loses energy by emitting radiation. Isotopes are extensively used in nuclear medicine to diagnose diseases, such as blood disorders and hyperthyroidism. For example, sodium-24 (^{24}Na) is used to trace blood flow in the body so that obstructions can be detected, and iodine-131 (^{131}I) is used to test the activity of the thyroid gland.

Elements are substances containing atoms of only one type, for example, O_2, N_2 and Cl_2. There are 118 elements listed in the periodic table, staring with the element 1, hydrogen, and ends with the element 118, oganesson. Some of these elements are extremely rare (see Section 2.2). Of them, 98 elements occur naturally on the earth, and the elements 99–118 have been synthesized in laboratories or nuclear reactors.

Compounds are substances formed when atoms of two or more different elements are joined together, for example, NaCl, H_2O, and HCl. *Molecules* are made up of atoms stuck together. Note that all compounds are molecules, but not all molecules are compounds. For example, molecular hydrogen (H_2), molecular oxygen (O_2) and molecular nitrogen (N_2) are not compounds as they are composed of a single element, whereas a single molecule of H_2O comprises two atoms of H and one atom of O, and is also a compound. Some molecules, such as proteins, contain hundreds or even thousands of atoms that held together in chains, known as *macromolecules*.

Molecular weight is in atomic mass units because it is the mass of one molecule, whereas *molar mass* is in grams per mole because it is the mass of one mole of molecule. In fact, there is no difference between molecular weight (MW) and molar mass (MM), except the units. For example, the MW of H_2O = 18 g; and the mass of one molecule of $H_2O = 3.0 \times 10^{-23}$ g. A mole is simply an Avogadro's number, which is 6.022×10^{23}, regardless of it is H_2O or NaCl or anything else. The term molar mass in chemistry refers to the mass of one mole of molecule. Thus, the molar mass of $H_2O = 6.022 \times 10^{23} \times 3.0 \times 10^{-23}$ g mol^{-1} = 18.0 g mol^{-1}.

2.2 ATOMIC STRUCTURE: ORBITALS AND ELECTRONIC CONFIGURATIONS

The electrons, which belong to the first generation of the lepton particle family, and are generally thought to be elementary particles, because they have no known components or substructure. Note that leptons are the basic building blocks of matter; that is, they are considered as the elementary particles. Electrons are extremely important in organic chemistry. To understand the fundamentals of chemistry, one must understand the location of electrons, as it is the arrangement of the electrons that creates the bonds between the atoms.

Electrons are said to be occupied *orbitals* in an atom. An *orbital* is a region of space that can hold two electrons. In fact, often an orbital is described as a three-dimensional region within which there is a 95% probability of finding the electrons. Electrons do not move freely in the space around the nucleus, but are confined to regions of space called *shells*. Each shell can contain up to $2n^2$ electrons, where *n* is the number of the *shell*. Each shell contains subshells known as *atomic orbitals* as shown here.

Shell	Number of orbitals contained each shell
4	$4s, 4p_x, 4p_y, 4p_z$, five $4d$, seven $4f$
3	$3s, 3p_x, 3p_y, 3p_z$, five $3d$
2	$2s, 2p_x, 2p_y, 2p_z$
1	$1s$

Atomic orbitals are commonly designated by a combination of numerals and letters (as shown) that represent specific properties of the electrons associated with the orbitals, for example, $1s$, $2p$, $3d$, $4f$. The numerals, called *principal quantum numbers*, indicate energy levels as well as relative distance from the nucleus.

The first shell contains a single orbital known as $1s$ orbital. The second shell contains one $2s$ and three $2p$ orbitals. These three $2p$ orbitals are designated as $2p_x$, $2p_y$, and $2p_z$. The third shell contains one $3s$ orbital, three $3p$ orbitals, and five $3d$ orbitals. Thus, the first shell can hold only two electrons, the second shells eight electrons, third shells up to 18 electrons and so on. The electron shells are identified by the principal quantum number, $n = 1, 2, 3, 4$ and so on. As the number of electrons goes up, the shell numbers as well as the relative energies of the shell electrons also increase.

Shell	Total number of shell electrons	Relative energies of shell electrons
4	32	
3	16	↑
2	8	
1	2	

The electronic configuration of an atom describes the number of electrons that an atom possesses, and the orbitals in which those electrons are placed in. Simply, the arrangements of electrons in orbitals, subshells and shells are called *electronic configurations*. Electronic configurations can be represented by using noble gas symbols to show some of the inner electrons, or by using Lewis structures in which the valence electrons are represented by dots. In chemistry, *valence* describes how easily an atom can combine with other chemical species to form a compound or a molecule.

Valence is the number of electrons an atom must lose or gain to attain the nearest noble gas or inert gas electronic configuration. The *ground-state electronic configuration* known as the lowest energy, and the *excited-state electronic configuration* is the highest energy orbital. If energy is applied to an atom in the ground state, one or more electrons can jump into a higher energy orbital. Thus, it takes a greater energy to remove an electron from the first shell of an atom than from any other shells. For example, sodium atom has electronic configuration as 2, 8 and 1. Therefore, to attain the stable configuration, Na atom must lose one electron from its outermost shell and become the nearest noble gas configuration, that is, the configuration of neon that has the electronic configuration as 2 and 8. Thus, sodium has a valence of 1. Since all other elements of Group I in the periodic table have one electron in their outermost shell, it can be said that Group I elements have a valence of 1.

At the far end on the left hand side of the periodic table, let us take another example, chlorine, which has the electronic configuration as 2, 8 and 7, and the nearest noble gas is argon, which has electronic configuration as 2, 8 and 8. To attain the argon electronic configuration, chlorine must gain one electron. Therefore, chlorine has a valence of 1. Since all other elements of Group 7A in the periodic table have seven electrons in their outermost shell and they can gain one electron, we can say that the Group 7A elements have a valence of 1.

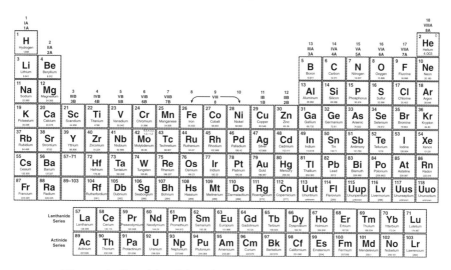

The periodic table of the elements with valence electrons

Each atom has an infinite number of possible electronic configurations, but the main concern is the ground-state electronic configuration, which is the lowest energy. The ground-state electronic configuration of an atom can be determined by the following *three* principles.

- The *Aufbau principle* (introduced by Niels Bohr and Wolfgang Pauli in the early 1920s) states that the orbitals fill in order of increasing energy, from lowest to highest. Because a 1s orbital is closer to the nucleus, it is lower in energy than a 2s orbital, which is lower in energy than the 3s orbital. While comparing atomic orbitals in the same shell, s atomic orbital is lower in energy than p atomic orbital, and p atomic orbital is lower in energy than d atomic orbital.

- The *Pauli exclusion principle* (introduced by the Austrian physicist, Wolfgang Pauli, in 1925) states that no more than two electrons can occupy each orbital, and if two electrons are present, their spins must be paired. For example, the two electrons of a helium atom must occupy the $1s$ orbital in opposite spins.

- The *Hund's rule* (introduced by the German physicist Friedrich Hund in 1927) explains that when *degenerative orbitals* (orbitals that have same energy) are present, but not enough electrons are available to fill all the shells completely, then single electron will occupy an empty orbital first before it will pair up with another electron. It is understandable, as it takes energy to pair up electrons. Therefore, the six electrons in the carbon atom are filled as follows: the first four electrons will go to the $1s$ and $2s$ orbitals, a fifth electron goes to the $2p_x$, the sixth electron to the $2p_y$ orbital and the $2p_z$ orbital will remain empty.

The ground-state electronic configurations for elements 1–18 are listed next (electrons are listed by symbol, atomic number and ground-state electronic configuration)

First period			Second period			Third period		
H	1	$1s^1$	Li	3	[He] $2s^1$	Na	11	[Ne] $3s^1$
He	2	$1s^2$	Be	4	[He] $2s^2$	Mg	12	[Ne] $3s^2$
			B	5	[He] $2s^2\,2p^1$	Al	13	[Ne] $3s^2\,3p^1$
			C	6	[He] $2s^2\,2p^2$	Si	14	[Ne] $3s^2\,3p^2$
			N	7	[He] $2s^2\,2p^3$	P	15	[Ne] $3s^2\,3p^3$
			O	8	[He] $2s^2\,2p^4$	S	16	[Ne] $3s^2\,3p^4$
			F	9	[He] $2s^2\,2p^5$	Cl	17	[Ne] $3s^2\,3p^5$
			Ne	10	[He] $2s^2\,2p^6$	Ar	18	[Ne] $3s^2\,3p^6$

Let us see how we can write the ground-state electronic configuration for carbon, nitrogen, oxygen, sulphur and chlorine showing the occupancy of each p orbital. Carbon has the atomic number 6, and the ground-state electronic configuration for carbon can be written as: $1s^2\,2s^2\,2p_x^1 2p_y^1 2p_z^0$. Similarly, we can write the others as follows:

Nitrogen (atomic number 7):	$1s^2\,2s^2\,2p_x^1\,2p_y^1\,2p_z^1$
Oxygen (atomic number 8):	$1s^2\,2s^2\,2p_x^2\,2p_y^1\,2p_z^1$
Sulphur (atomic number 16):	$1s^2\,2s^2\,2p_x^2\,2p_y^2\,2p_z^2\,3s^2\,3p_x^2\,3p_y^1\,3p_z^1$
Chlorine (atomic number 17):	$1s^2\,2s^2\,2p_x^2\,2p_y^2\,2p_z^2\,3s^2\,3p_x^2\,3p_y^2\,3p_z^1$

2.3 CHEMICAL BONDING THEORIES: FORMATION OF CHEMICAL BONDS

In a chemical reaction, atom forms a bond to obtain a stable electronic configuration; that is, the electronic configuration of the nearest noble gas. All noble gases are inert, because their atoms have stable electronic configuration in which they have eight electrons in the outermost shell except helium (two electrons). Therefore, they cannot donate or gain electrons.

One of the driving forces behind the bonding in an atom is to obtain a stable valence electron configuration. A filled shell is also known as a noble gas configuration. Electrons in filled shells are called *core electrons*. The core electrons do not participate in chemical bonding. Electrons in shells that are not completely filled are called *valence electrons*, also known as *outermost shell electrons*, and the outermost shell is called the *valence shell*. Carbon, for example, has two core electrons and four valence electrons.

Elements tends to gain or lose electrons, so they will have the same number of electrons as a noble gas to become more stable. The valence electrons are the number of electrons in the outermost shell of an atom and Lewis structure represents the outermost electrons of an atom. The chemical behaviour of an element depends on its electronic configuration. In the periodic table (see Section 2.2), elements in the same column have the same number of valence electrons, and similar chemical properties. Likewise, elements in the same row are similar in size.

2.3.1 Lewis Structures

Molecular structures that use the notation for the electron pair bond are called *Lewis structures*, named after Gilbert N. Lewis, an American physical chemist, who introduced this in 1916. Lewis structures, also known as Lewis dot diagrams, Lewis dot formulas, Lewis dot structures, electron dot structures, or Lewis electron dot structures (LEDSs), offer information about how atoms are bonded to each other, and the total electron pairs involved. According to the *Lewis theory*, an atom will give up, accept or share electrons in order to achieve a filled outermost shell that contains eight electrons. The Lewis structure of a covalent molecule shows all the electrons in the valence shell of each atom; the bonds between atoms are shown as shared pairs of electrons. Atoms are most stable if they have a filled valence shell of electrons. Atoms transfer or share electrons in a way that they can attain a filled shell of electrons, this stable configuration of electrons is called *octet*. Except for hydrogen and helium, a filled valence shell contains eight electrons.

Lewis structures help to track the valence electrons and predict the types of bonds. The number of valence electrons present in each of the elements is to be considered first. The number of valence electrons determines the number of electrons needed to complete the octet of eight electrons. Simple ions are atoms that have gained or lost electrons to satisfy the octet rule. However, not all compounds follow the octet rule.

Sodium (Na) loses a single electron from its $3s$ orbital to attain a more stable neon gas configuration ($1s^2\ 2s^2\ 2p^6$) with no electron in the outermost shell. An atom having a filled valence shell is said to have a *closed shell configuration*. The total number of electrons in the valence shell of each atom can be determined from its group number in the periodic table. The shared electrons are called the *bonding electrons* and may be represented by a line or lines between two atoms. The valence electrons that are not being shared are the *nonbonding electron pairs* or simply *lone pairs*, and they are shown in the Lewis structure by dots around the symbol of the atom. A nonbonding electron pair often dictates the reactivity of a molecule. Any species that have an unpaired electron are called *radicals*. Radical species can be electrically neutral and they are often called free radicals (see Section 5.3). Usually they are highly reactive, and are believed to play significant roles in ageing, cancer and many other ailments.

Lewis structure shows the connectivity between atoms in a molecule by a number of dots equal to the number of electrons in the outermost shell of an atom of that molecule. A pair of electrons is represented by two dots, or a dash. When drawing Lewis structures, it is essential to keep track of the number of electrons available to form bonds and the location of the electrons. The number of valence electrons of an atom can be obtained from the periodic table because it is equal to the group number of the atom. For example, hydrogen (H) in Group 1A has one valence electron, carbon (C) in Group 4A has four valence electrons and fluorine (F) in Group 7A has seven valence electrons.

To write the Lewis formula of CH_3F, first of all we have to find the total number of valence electrons of all the atoms involved in this structure, for example, C, H and F having valence electrons 4, 1 and 7, respectively.

$$4+3(1)+7=14$$

$$C3 \times HF$$

The carbon atom bonds with three hydrogen atoms and one fluorine atom, and it requires four pairs of electrons. The remaining six valence electrons are with the fluorine atom in the three nonbonding pairs.

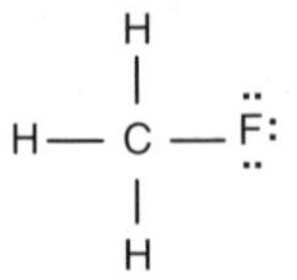

$$
\begin{array}{c}
\quad\ \text{H} \\
\quad\ | \\
\text{H} - \text{C} - \ddot{\underset{\cdot\cdot}{\text{F}}}{:} \\
\quad\ | \\
\quad\ \text{H}
\end{array}
$$

In the periodic table, the period 2 elements C, N, O, and F have the valence electrons that belong to the second shell ($2s$ and three $2p$). The shell can be completely filled with eight electrons. In the period 3 elements Si, P, S, and Cl have the valence electrons that belong to the third shell ($3s$, three $3p$, and five $3d$). The shell is only partially filled with eight electrons in $3s$ and three $3p$, but the five $3d$ orbitals can accommodate an additional 10 electrons. For these differences in valence shell orbitals available to elements of the second and third periods, we see significant differences in the covalent bonding of oxygen and sulphur, and of nitrogen and phosphorus. Although oxygen and nitrogen can accommodate no more than eight electrons in their valence shells, many phosphorus-containing compounds have 10 electrons in the valence shell of phosphorus, and many sulphur-containing compounds have 10 and even 12 electrons in the valence shell of sulphur.

So, to derive Lewis structures for most molecules one should follow this sequence.

i. Draw a tentative structure. The element with the least number of atoms is usually the central element.

ii. Calculate the number of valence electrons for all atoms in the compound.

iii. Put a pair of electrons between each symbol.

iv. Place pairs of electrons around atoms beginning with the outer atom until each has eight electrons, except for hydrogen. If an atom other than hydrogen has less than eight electrons, then move unshared pairs to form multiple bonds.

If the structure is an ion, electrons are added or subtracted to give the proper charge. Lewis structures show the connectivity or bonding sequence of the atoms, indicating single, double, or triple bonds. They also show what atoms are bonded together, and whether any atoms possess lone pair of electrons or have a formal charge. In polar covalent compounds some valence electrons may remain unshared (nonbonding electrons) and they are found in the outermost shell. For example, water (H_2O) has six valence electrons. Two of these combine with hydrogens to make two O—H covalent bonds and four of the valence electrons remain unshared, therefore oxygen has two nonbonding electron pairs in the outermost shell.

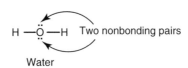

Water

A *formal charge* is the difference between the number of valence electrons an atom actually has, when it is not bonded to any other atoms, and the number of nonbonding electrons and half of its bonding electrons. Thus, a positive or negative charge assigned to an atom is called a formal charge. The sum of the formal charges on the individual atoms must be equal to the total charge on the ion. The decision as to where to put the charge is made by calculating the formal charge for each atom in an ion or a molecule. For example, the hydronium ion (H_3O^+), is positively charged and the oxygen atom has a formal charge of +1.

$$\text{Formal charge} = [(\text{group number}) - (\text{nonbonding pairs}) - \tfrac{1}{2}\,(\text{bonding electrons})]$$
$$= [6 - 2 - \tfrac{1}{2}\,(6)] = 1$$

An uncharged oxygen atom must have six electrons in its valence shell. In the hydronium ion, oxygen bonds with three hydrogen atoms. So, only five electrons effectively belong to oxygen, which is one less than the valence electrons. Thus, oxygen bears a formal charge of +1. Elements of the second period, including carbon, nitrogen, oxygen and fluorine, cannot accommodate more than eight electrons as they have only four orbitals ($2s$, $2p_x$, $2p_y$ and $2p_z$) of their valence shells.

In neutral organic compounds, a carbon atom can have four single bonds or two double bonds or a double and two single bonds or a triple and a single bond. Nitrogen forms three single bonds and a lone pair or one double bond, a single bond and a lone pair or a triple and a lone pair. Oxygen can have two single bonds and two lone pairs or one double bond and two lone pairs, and hydrogen only forms one single bond. In Lewis structures, these bonds are represented by electron dot or line bond formulas. A pair of shared electrons are shown as a line between two atoms. The hydrogen atom of molecular hydrogen has one covalent bond and no lone pair. The nitrogen atom of ammonia has three covalent bonds and one lone pair. The oxygen atom of water has two covalent bonds and two lone pairs. The chlorine atom has one covalent bond and three lone pairs. Each atom has a complete octet, except hydrogen that has a completely full outermost shell.

Note that hydrogen is exceptional as it has one and only valence electron. Thus, a hydrogen atom can achieve a completely empty shell by losing an electron and become a positively charged hydrogen ion, known as a proton. A hydrogen also atom can achieve a filled outermost shell by gaining an electron, thereby forming a negatively charged hydrogen ion, called a hydride ion.

$$H\cdot \longrightarrow H^+ + e^-$$

Hydrogen atom   Proton   Electron

$$H\cdot + e^- \longrightarrow H\colon^-$$

Hydrogen atom   Electron   Hydride ion

2.3.2 Resonance and Resonance Structures

The structures of some compounds cannot be sufficiently described by a single Lewis structure. Resonance is possible whenever a Lewis structure has a multiple bond and an adjacent atom with at least one lone pair of electrons. *Resonance* means the use of two or more Lewis structures to represent a particular molecule or ion. *Resonance structures* are used to depict delocalized electrons within a molecule or ionic compound that cannot be described fully with only one Lewis structure.

Resonance structures only differ in the position of the electron pairs, never the atom positions. Therefore, the location of the lone pairs and bonding pairs differs in resonance structures. Resonance structures are not real. An individual resonance structure does not accurately represent the structure of a molecule or ion, only the hybrid does. Resonance structures are not in equilibrium with each other. There is no movement of electrons from one form to another. Thus, resonance structures are not isomers. Two isomers differ in the arrangement of both atoms and electrons, whereas resonance structures differ only in the arrangement of electrons. The most important example is benzene (C_6H_6).

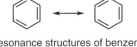

Benzene

Benzene has six carbon atoms linked to each other in a six-membered ring and its Lewis structure is usually drawn with alternative double and single bonds. Chemists often simplify it further as a hexagon containing an inscribed circle. The resonance structures of benzene can be drawn as shown. A double-headed arrow is used to separate the two resonance structures.

Resonance structures of benzene
Kekulé structures

These different structures are known as resonance structures and often called *Kekulé structures* (see Section 4.7.6). The two Kekulé structures for benzene have the same arrangement of atoms, but differ only in the placement of electrons.

2.3.3 Electronegativity and Chemical Bonding

Electronegativity is defined as the attraction of an atom for its outermost shell electrons. It is a measure of the tendency of an atom to attract or willing to accept a bonding pair of electrons. Thus, electronegativity is the ability of an atom, which is bonded to another atom or atoms, to attract electrons strongly towards it. This competition for electron density is scaled by electronegativity values. Elements with higher electronegativity values have greater attraction for bonding electrons. Thus, electronegativity of an atom is related to bond polarity (see Section 2.4). The difference in electronegativity between two bonded atoms can be used to measure the polarity of the bonding between them.

The greater the difference in electronegativity between the bonded atoms, the greater is the polarity of the bond. If the difference is great enough, electrons are transferred from the less electronegative atom to the more electronegative one, hence ionic bond is formed. Only if the two atoms have exactly the same electronegativity is a nonpolar bond is formed. Electronegativity increases from left to right, and down to up in the periodic table as shown here.

H																
2.1																

Li	Be											B	C	N	O	F
1.0	1.5											1.5	2.5	3.0	3.5	4.0

Na	Mg											Al	Si	P	S	Cl
0.9	1.2											1.5	1.8	2.1	3.5	3.0

K	Ca	Sc	Ti	V	Cr	Mn	Fe	Co	Ni	Cu	Zn	Ga	Ga	AS	Se	Br
0.8	1.0	1.3	1.5	1.6	1.6	1.5	1.8	1.9	1.8	1.9	1.6	1.6	1.8	2.0	2.4	2.8

Rb	Sr	Y	Zr	Nb	Mo	Tc	Ru	Rh	Pd	Ag	Cd	In	Sn	Sb	Te	I
0.8	1.0	1.2	1.4	1.6	1.8	1.9	2.2	2.2	2.2	1.9	1.7	1.7	1.8	1.9	2.1	2.5

Periodic table with electronegativity values
of some important elements

Electronegative elements have a strong affinity for electrons, and can easily gain electrons to form anions in chemical reactions, whereas electropositive elements have relatively weak attractions for electrons and can easily lose electrons to form cations. Fluorine (F) is the most electronegative element on the periodic table. Elements are less electronegative the further away they are from fluorine, except hydrogen (H) that has an electronegativity between boron (B) and carbon (C). This trend mainly applies to elements within the same group or period. Different bonds and respective difference in electronegativity are shown here.

Bond	Difference in electronegativity	Types of bond
C—Cl	3.0–2.5 = 0.5	Polar covalent
P—H	2.1–2.1 = 0	Nonpolar covalent
C—F	4.0–2.5 = 1.5	Polar covalent
S—H	2.5–2.1 = 0.4	Nonpolar covalent
O—H	3.5–2.1 = 1.4	Polar covalent

Elements with higher electronegativity values have a greater attraction for bonding electrons. In general, if the electronegativities are less than 0.5, the bond is nonpolar covalent, and if the electronegativity difference between bonded atoms are 0.5–1.9, the bond is polar covalent. If the difference in electronegativities between the two atoms is 2.0 or greater, the bond is ionic.

Electrons in a polar covalent bond are unequally shared between the two bonded atoms, which result in partial positive and negative charges. The separation of the partial charges creates a *dipole*. The word dipole means two poles, the separated partial positive and negative charges. A polar molecule results when a molecule contains polar bonds in an unsymmetrical arrangement. Nonpolar molecules, whose atoms have equal or nearly equal electronegativities, have zero or very small dipole moments. Molecules, where atoms have polar bonds with symmetrical molecular geometry, also do not have dipole moments as they cancel out each other (see Section 2.4).

2.3.4 Various Types of Chemical Bonding

A *chemical bond* is the attractive force that holds two atoms together. Valence electrons take part in bonding. An atom that gains electrons becomes an *anion*, a negatively charged ion, and an atom that loses electrons becomes a *cation*, a positively charged ion. *Metals* tend to lose electrons and become electropositive and *non-metals* tend to gain electrons and become electronegative. While cations are smaller than atoms, anions are larger. Atoms get smaller in size as they go across a period, and get larger in size as they go down a group and increase the number of shells to hold electrons.

The energy required for removing an electron from an atom or ion is called *ionization energy*. Atoms can have a series of ionization energies, since more than one electron can always be removed, except for hydrogen. In general, the first ionization energies increase across a period and decrease down the group. Adding electrons is easier than removing electrons. It requires a vast amount of energy to remove electrons. Atoms can form either ionic or covalent bonds to attain a complete outer shell electronic configuration.

2.3.4.1 Ionic Bonds

Ionic bonds result from the transfer of one or more electrons between atoms. The more electronegative atom always gains one or more valence electrons, and hence becomes an anion. The less electronegative atom always loses one or more valence electrons, and becomes a cation. A single-headed arrow indicates a single electron transfer from a less electronegative atom to the more electronegative atom. Ionic compounds are held together by the attraction of opposite charges. Therefore, ionic bonds consist of the electrostatic attraction between positively and negatively charged ions.

Ionic bonds involve the complete transfer of electrons between two atoms of widely different electronegativities, and both atoms obtain a stable octet outermost shell of electrons. Anions have usually gained sufficient electrons to complete their outermost shells. Cations have usually lost their outermost shells electrons, so that the next inner shells become the stable octet outermost shells. Thus, ionic bonds are commonly formed between reactive metals, electropositive elements on the left side of the periodic table, and non-metals, electronegative elements on the right side of the periodic table (see Section 2.2). For example, Na (electronegativity: 0.9) easily gives up an electron, and Cl (electronegativity: 3.0) readily accepts an electron to form an ionic bond. In the formation of ionic compound Na^+Cl^-, the single $3s$ valence electron of Na is transferred to the partially filled valence shell of chlorine.

$$Na\ ([Ne]\ 3s^1) + Cl\ ([Ne]\ 3s^23p^5) \rightarrow Na^+([Ne]\ 3s^0) + Cl^-([Ne]\ 3s^23p^6) \rightarrow Na^+Cl^-$$

2.3.4.2 Covalent Bonds

Covalent bonds are formed from the sharing of electron pairs between bonded atoms instead of giving up or gaining electrons. In this case, an atom can obtain a filled valence shell by sharing electrons. For example, two chlorine atoms can achieve a filled valence shell and form a chlorine molecule (Cl_2). Similarly, hydrogen and fluorine can form a covalent bond by sharing electrons and form hydrogen fluoride (**HF**) molecule.

$$:\overset{..}{\underset{..}{Cl}}:\overset{..}{\underset{..}{Cl}}: \longrightarrow Cl-Cl \qquad H:\overset{..}{\underset{..}{F}}: \longrightarrow H-F$$

2.3.4.2.1 Nonpolar and Polar Covalent Bonds

In general, most bonds within organic molecules, including various drug molecules, are covalent. The exceptions are compounds that possess metal atoms where the metal atoms should be treated as ions. If a bond is covalent, it is possible to identify whether it is a polar or nonpolar bond. In a *nonpolar covalent bond*, the electrons are shared equally between two atoms. An ideal example is the bonding between two identical atoms, for example, H_2, O_2, N_2, Cl_2 and F_2. Nonpolar covalent bonds can also occur between different atoms that have identical electronegativity values, for example, CH_4. In many covalent bonds, the electrons are not shared equally between two bonded atoms. Bonds between different atoms usually result in the electrons being attracted to one atom more strongly than the other. Such an unequal sharing of the pair of bonding electrons results in a polar covalent bond, for example, HF, HCl and H_2O.

Nonpolar covalent bonds		Polar covalent bonds	
H:H	F:F	$\overset{\delta^+}{H}:\overset{\delta^-}{\underset{..}{\overset{..}{F}}}:$	$\overset{\delta^+}{H}:\overset{\delta^-}{\underset{..}{\overset{..}{Cl}}}:$
H_2	F_2	HF	HCl

A *polar covalent bond* is composed of atoms with slightly different electronegativities. Therefore, one atom has a greater attraction for the electrons than the other atom. In other words, in a polar covalent bond, more electronegative atom is partially electronegative and the other one is partially electropositive. For example, in chloromethane (CH_3Cl), where the more electronegative chlorine atom (electronegativity: 3.0) is bonded to a less electronegative carbon atom (electronegativity: 2.5), the bonding electrons are attracted more strongly towards chlorine. This results in a partial positive charge on the carbon and a partial negative charge on the chlorine.

$$H \overset{H}{\underset{H}{\overset{|}{\underset{|}{\overset{\delta^+}{C}}}}} \overset{\delta^-}{Cl}$$

$\mu = 1.87$ D

Chloromethane
(A polar molecule)

The greater the difference in electronegativity between the bonded atoms, the more polar the bond is (see Section 2.3.3). The direction of bond polarity can be indicated with an arrow. The head of the arrow is at the negative end of the bond; a short perpendicular line near the tail of the arrow marks the positive end of the bond. Bond polarity (see Section 2.4) is usually measured by bond dipole moment (μ) and the unit is called the *debye* (D).

Chloromethane has a bond dipole moment, $\mu = 1.87$ D. The C—H bond is considered nonpolar, as the carbon atom (electronegativity: 2.5) and the hydrogen atom (electronegativity: 2.1) have very similar electronegativity values. Thus, methane (CH$_4$) has no bond dipole moment, hence it is a nonpolar molecule.

$$\begin{array}{c} \text{H} \\ | \\ \text{H}-\text{C}-\text{H} \\ | \\ \text{H} \end{array}$$

Methane
(A nonpolar molecule)

2.3.4.2.2 Covalent Bond Formation

Covalent bonds are formed when atomic orbitals overlap. The overlap of atomic orbitals is called *hybridization*, and the resulting atomic orbitals are called *hybrid orbitals*. There are two types of orbital overlap, which form sigma (σ) and pi (π) bonds. Pi (π) bonds never occur alone without the bonded atoms being joined by a σ bond. Therefore, a double bond consists of a σ bond and a π bond, whereas a triple bond consists of a σ bond and two π bonds. A sigma overlap occurs when there is one bonding interaction that results from the overlap of two *s* orbitals or an *s* orbital overlaps a *p* orbital or two *p* orbitals overlap head to head. A pi overlap occurs only when two bonding interactions result from the sideways overlap of two parallel *p* orbitals. The *s* orbital is spherical in shape and *p* orbitals are dumb-bell shapes.

Sigma overlap of a *s* orbital with a *p* orbital

Pi overlap of two parallel *p* orbitals

Let us consider the formation of the σ overlap in hydrogen molecule (H$_2$), from two hydrogen atoms. Each hydrogen atom has one electron, which occupies the 1*s* orbital. The overlap of two *s* orbitals, one from each of two hydrogen atoms, forms a σ bond. The electron density of a σ bond is greatest along the axis of the bond. Since *s* orbitals are spherical in shape, two hydrogen atoms can approach one another from any direction resulting in a strong σ bond.

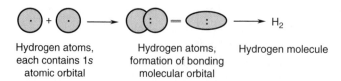

Hydrogen atoms, each contains 1*s* atomic orbital

Hydrogen atoms, formation of bonding molecular orbital

Hydrogen molecule

2.3.4.2.3 Covalent Bond Cleavage

Covalent bonds are broken either homolytically or heterolytically. A *homolytic cleavage* is the breaking of a covalent bond, where each fragment gets one of the shared electrons. Thus, homolytic cleavage produces free radicals (see Section 5.3). In a homolysis reaction, a single-headed arrow is used to show the movement of a single electron.

$$A \overset{\frown}{-} B \xrightarrow[\text{homolysis}]{\text{Bond}} A \cdot + B \cdot$$

Radicals

A *heterolytic bond cleavage* is the breaking of a covalent bond where one atom gets both of the shared electrons. Typically, heterolytic bond cleavage is most likely to occur in polar bonds, and the electrons will move towards the more electronegative atom. Whereas homolytic bond cleavage produces electrically charged ions, as cation and anion. In a heterolysis reaction, a double-headed arrow is used to show the movement of an electron pair.

$$A \overset{\frown}{-} B \xrightarrow[\text{heterolysis}]{\text{Bond}} A^+ \quad + \quad B \overset{..}{:}$$

A cation   An anion

Elements in organic compounds are joined by covalent bonds, where electrons are shared and each element contributes one electron to the bond. The number of electrons necessary to complete the octet determines the number of electrons that must be contributed and shared by a different element in a bond. This analysis finally determines the number of bonds that each element may enter into with other elements.

In a *single bond,* two atoms share one pair of electrons and form a σ bond. In a *double bond,* they share two pairs of electrons and form a σ bond and a π bond. In a *triple bond,* two atoms share three pairs of electrons and form a σ bond and two π bonds. For example, alkanes are saturated hydrocarbons in which all the bonds are single bonds, for example, ethane (see Section 4.3), whereas alkenes are unsaturated hydrocarbons that contain carbon-carbon double bonds, for example, ethylene (see Section 4.5), and alkynes are also unsaturated hydrocarbons that have carbon-carbon triple bonds, for example, acetylene (see Section 4.6).

Ethane
(Alkane)   Ethene or ethylene
                  (Alkene)   Ethyne or acetylene
                                    (Alkyne)

2.4 BOND POLARITY AND INTERMOLECULAR FORCES

Bond polarity is a useful concept that depicts the sharing of electrons between atoms. The shared electron pairs between two atoms are not necessarily shared equally and this leads to a *bond polarity*. The higher the difference between the electronegativities of the bonded atoms, the more polar their bond gets. Even though there is difference in the electronegativity of bonded atoms, the resulting molecule may not be polar if the entire molecule is symmetric or linear. The polarity of individual bonds is necessary for a molecule to be polar, but it is not a sufficient reason. It also depends on the geometry of the molecule. For example, carbon dioxide (CO_2) is a linear molecule, albeit it has two polar bonds, because of its geometry CO_2 is a nonpolar molecule. Whereas H_2O is a bent molecule, which also has two polar bonds, because of its geometry.

Carbon dioxide molecule is not a polar molecule, even though it has polar bonds. Carbon dioxide is linear, and the C—O bond dipoles are oriented in opposite directions. So they cancel out each other and the dipole moment is zero. Water is a bent molecule with no symmetry. Thus, the dipole moments do not cancel out, causing the molecule to have a net dipole moment, making H_2O a polar molecule.

Carbon dioxide
(A nonpolar molecule)

Water
(A polar molecule)

Atoms, such as nitrogen, oxygen and halogen that are more electronegative than carbon have a tendency to have partial negative charges. Atoms, such as carbon and hydrogen, have a tendency to be more neutral or have partial positive charges. Thus, bond polarity arises from the difference in electronegativities of two atoms participating in the bond formation. This also depends on the attraction forces between molecules, and these interactions are called *intermolecular interactions or forces*. The physical properties, for example, boiling points, melting points and solubilities of the molecules are determined, to a large extent, by intermolecular non-bonding interactions. There are three types of non-bonding intermolecular interactions: *dipole-dipole interactions, van der Waals forces* and *hydrogen bonding*. These interactions increase significantly as the molecular weights increase, and also increase with increasing polarity of the molecules.

2.4.1 Dipole–Dipole Interactions

The interactions between the positive end of one dipole and the negative end of another dipole are called *dipole–dipole interactions*. As a result of dipole–dipole interactions, polar molecules are held together more strongly than nonpolar molecules. Dipole–dipole interactions arise when electrons are not equally shared in the covalent bonds because of the difference in electronegativity. For example, hydrogen fluoride has a dipole moment of 1.98 D, which lies along the H—F bond. As the fluorine atom (electronegativity: 4.0) has greater electronegativity than the hydrogen atom (electronegativity: 2.1), electrons are therefore pulled strongly towards fluorine as shown.

$$\overset{\delta^+}{H} \!-\! \overset{\delta^-}{F}$$
$$\mu = 1.98 \text{ D}$$

The arrow indicates that the electrons are towards the more electronegative atom fluorine. The δ^+ and δ^- symbols indicate partial positive and negative charges. Dipole–dipole interactions are stronger than van der Waals forces, but not as strong as ionic or covalent bonds.

2.4.2 van der Waals Forces

Relatively weak forces of attraction that exist between nonpolar molecules are called *van der Waals forces* or *London dispersion forces*, named after Dutch scientist Johannes Diderik van der Waals. These forces are distance-dependent interactions between atoms or molecules. Dispersion forces between molecules are much weaker than the covalent bonds within molecules. Electrons move continuously within bonds and molecules, so at any time, one side of the molecule can have more electron density than the other side, which gives rise to a temporary dipole. Because the dipoles in the molecules are induced, the interactions between the molecules are also called *induced dipole–induced dipole interactions*.

van der Waals forces are the weakest of all the intermolecular interactions. Alkenes are nonpolar molecules, because the electronegativities of carbon and hydrogen are similar. Consequently, there are no significant partial charges on any of the atoms in an alkane. Therefore, the size of the van der Waals forces that hold alkane molecules together depends on the area of contact between the molecules. The greater the area of contact, the stronger are the van der Waals forces, and the greater the amount of energy required to overcome these forces. For example, the isobutane (bp: $-10.2\,^\circ$C) and butane (bp: $-0.6\,^\circ$C), both with the molecular formula C_4H_{10}, have different boiling points. Isobutane is a more compact molecule than butane. Thus, butane has a greater surface area for interaction with each other than isobutane. The stronger interactions that are possible for *n*-butane are reflected in its boiling point, which is higher than the boiling point of isobutane.

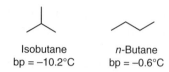

Isobutane
bp = −10.2°C

n-Butane
bp = −0.6°C

2.4.3 Hydrogen Bonding

Hydrogen bonding is the attractive force between the hydrogen attached to an electronegative atom of one molecule and an electronegative atom of the same (*intramolecular*) or a different molecule (*intermolecular*). It is an unusually strong force of attraction between highly polar molecules in which hydrogen is covalently bonded to nitrogen, oxygen or fluorine. Therefore, a hydrogen bond is a special type of interaction between atoms.

A hydrogen bond is formed whenever a polar covalent bond involving a hydrogen atom is in close proximity to an electronegative atom such as O or N. The attractive forces of hydrogen bonding are usually indicated by a dashed line rather than the solid line used for a covalent bond. For example, water molecules form intermolecular hydrogen bonding, and the hydrogen bonding in H_2O graphic shows a cluster of water molecules in the liquid state.

Hydrogen bond

Donor

Acceptor

Hydrogen bonding in H_2O

Water is a polar molecule due to the electronegativity difference between hydrogen (electronegativity: 2.1) and oxygen (electronegativity: 3.5) atoms. The polarity of the water molecule with the attraction of the positive and negative partial charges is the basis for the hydrogen bonding. Hydrogen bonding is responsible for certain characteristics of water, for example, surface tension, viscosity and vapour pressure.

Hydrogen bonding occurs with hydrogen atoms covalently bonded to oxygen, fluorine or nitrogen, but not with chlorine that has larger atom size. The strength of a hydrogen bond involving an oxygen, a fluorine, or a nitrogen atom ranges from 3 to 10 kcal mol⁻¹, making hydrogen bonds the strongest known type of intermolecular interactions. Due to the intermolecular hydrogen bonding in water, it is responsible for the unexpectedly high boiling point (bp) of water (bp: 100 °C). Hydrogen bonds are interactions between molecules and should not be confused with covalent bonds to hydrogen within a molecule. Hydrogen bonding is usually stronger than normal dipole forces between molecules, but not as strong as ionic or covalent bonds. The strongest hydrogen bonds are linear (180°), when the two electronegative atoms and the hydrogen between them lie in a straight line.

Hydrogen bond is of fundamental importance in biology. Hydrogen bond is said to be the 'bond of life'. The double helix structure of DNA is formed and held together with the hydrogen bonds (see Section 7.1.2.2). The nature of the hydrogen bonds in proteins dictates their properties and behaviour. *Intramolecular hydrogen bonds* (within the molecule) in proteins result in the formation of globular proteins, for example, enzymes or hormones. On the other hand, *intermolecular hydrogen bonds* (between different molecules) tend to give insoluble proteins such as fibrous protein. In cellulose, a polysaccharide, molecules are held together through hydrogen bonding that provides plants with rigidity and protection (see Section 8.3.6.3). In drug-receptor binding, hydrogen bonding often plays an important role.

2.5 HYDROPHILICITY AND LIPOPHILICITY

The high polarity of water makes it an excellent solvent for polar and ionic molecules and a poor solvent for nonpolar molecules. Hydrophilic substances tend to be polar. Water is a polar solvent and will dissolve molecules that are polar or ionic, also known as *hydrophilic or lipophobic molecules*. For example, vitamin C (ascorbic acid), which is a natural antioxidant having polar functional groups (hydroxyl groups), is soluble in water (see Sections 5.3.4 and 8.3.7.3).

Ascorbic acid
Vitamin C
(A natural antioxidant)

Lipophilic substances appear to be nonpolar molecules, are not very soluble in water and they are known as *hydrophobic or lipophilic molecules*. For example, vitamin E (α-tocopherol) is a natural antioxidant (see Section 5.3.4) having mainly nonpolar functional groups (long chain alkane and methyl groups) and it is insoluble in water but soluble in *n*-octanol.

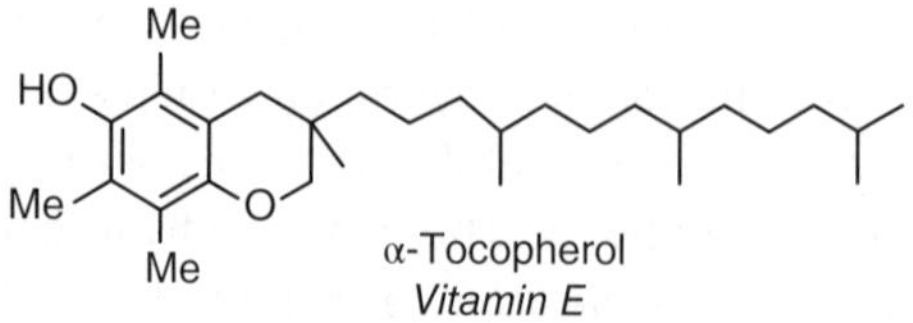

α-Tocopherol
Vitamin E
(A natural antioxidant)

Hydrophilicity and lipophilicity refer to the solubility of something in water or lipids, respectively. *Hydrophilicity* ('hydro' means water and 'philicity' means loving or liking) is the ability of a chemical compound or drug to go into solution in water and polar solvents (e.g. MeOH). This is also known as *lipophobicity* (lipo means fat and phobicity means hating). For example, atenolol (a β-blocker to treat hypertension) is a hydrophilic drug that is soluble in water. Thus, hydrophilicity represents the affinity of a compound or a moiety for a hydrophilic environment. Hydrophilic compounds are fat insoluble. They cannot cross the cell membrane. Thus, they bind to receptor molecules on the outer surface of target cells, initiating reactions within the cell that ultimately modifies the functions of the cells.

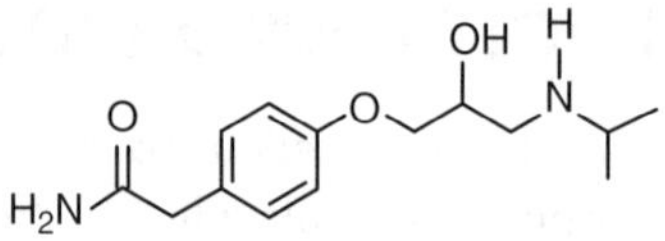

Atenolol, a hydrophilic drug
(A β-blocker to treat hypertension)

Lipophilicity (lipo means fat and philicity means loving or liking) is the ability of a chemical compound or drug to go into solution in lipids (fat and oil) and nonpolar solvents (e.g. *n*-octanol). This is also known as *hydrophobicity* (hydro means water and phobicity means hating). For example, clotrimazole (an antifungal drug to treat thrush) is a lipophilic drug that is soluble in *n*-octanol but insoluble in water.

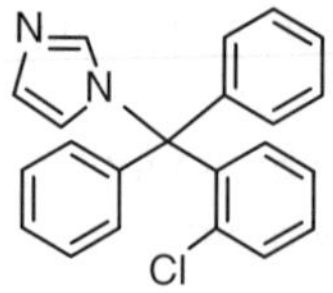

Clotrimazole, a lipophilic drug
(An antifungal drug to treat trush)

Lipophilicity represents the affinity of a compound or a moiety for a lipophilic environment. Lipophilic compounds are water insoluble. They can easily cross the cell membrane. Thus, they can enter target cells and bind to intracellular receptors to carry out their action. The concept of hydrophilicity and lipophilicity is important in relation to drug's interactions with biological membranes. Since biological membranes are lipophilic in nature, the rate of drug transfer for passively absorbed drugs is directly related to the lipophilicity of the molecule.

The cell membrane (plasma membrane) is a permeable lipid bilayer (phospholipid bilayer) found in all cells. Lipid or phospholipid bilayer is a fatty, nonpolar barrier that creates separate aqueous compartments within the body. Some of the important chemical and physiological functions occur in or on the lipid bilayer.

A *phospholipid* has both polar part (hydrophilic head) and nonpolar part (hydrophobic tail) ends known as amphipathic (amphiphilic), where one end tends to dissolve in polar solvent (such as water) and the other end in nonpolar (oil). Thus, the phospholipids form a bilayer that acts like a barrier between the cell and the environment.

Fatty acid ester (hyrophobic tail)

Phosphate ester (hydrophilic head)

Cholesterol (a C_{27} sterol) belongs to a group of compounds known as steroids (see Sections 8.6.1 and 8.6.5). Mammalian cell membranes, especially human cell membranes, have high concentrations of cholesterol. The structure of cholesterol is quite similar to that of phospholipid having polar and nonpolar parts.

Cholesterol
(A C_{27} sterol)

The fluid mosaic model (first devised by S. J. Singer and G. L. Nicolson in 1972), which explains various observations regarding the structure of functional cell membranes, shows the lipid bilayer is composed of a number of proteins that are similar to shifting tiles. The spaces between the tiles are filled with phospholipids, which also contain cholesterol that makes the lipid bilayer stronger, more flexible and more permeable.

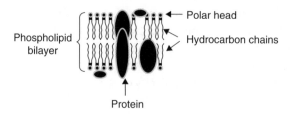

The fluid mosaic model of lipid bilayer

The plasma membrane is selectively permeable. Some substances can move across the plasma membrane but some cannot. Macromolecules are unable to cross the membrane as they are too large. Polar or charged compounds do not cross cell membranes (lipid) very easily because they are hydrophilic (polar) and unable to enter the hydrophobic (lipid) phase of the lipid bilayer. Only small polar molecules such as CO_2 and H_2O can pass through membrane easily. Non-charged molecules such as alcohols and oxygen are also able to cross the membrane with ease, as they are able to slip through the hydrophilic heads of the phospholipids and pass between the hydrophobic tails of the membrane.

2.6 SIGNIFICANCE OF CHEMICAL BONDING IN DRUG–RECEPTOR INTERACTIONS

Most drugs interact with receptor sites localized in macromolecules that have protein-like properties and specific three-dimensional shapes. Proteins are made up of amino acids that determine the types of bonding and interaction they can take part. They are capable of binding other molecules. Any molecule that is bound to a protein is known as a *ligand*. Ligand has strong affinity for receptor. A *receptor* is the specific chemical constituents (protein or protein complex) of the cell with which a drug interacts to produce its pharmacological effects. Often receptors are transmembrane proteins that relay signals by binding a ligand on one side of the membrane, altering their conformation on the other side to transmit a signal for example, activating a kinase internal domain. One may consider that every protein that acts as the molecular target for a certain drug should be called a receptor. However, this term mainly incorporates those proteins that play an important role in the intercellular communication via chemical messengers. As such, enzymes, ion channels and carriers are usually not classified as receptors. The term receptor is mostly reserved for those protein structures that serve as intracellular antennas for chemical messengers. On recognition of the appropriate chemical signal (ligand), the receptor proteins transmit the signal into a biochemical change in the target cell via a wide variety of possible pathways.

A minimum three-point attachment of a drug to a receptor site is essential for desired effect. In the most cases, a specific chemical structure is required for the receptor site and a complementary drug structure. Slight changes in the molecular structure of the drug may drastically change specificity and thus the efficacy. However, there are some drugs that act exclusively by physical means outside of cells and do not involve any binding to the receptors. These sites include external surfaces of skin and gastrointestinal tract. Drugs also act outside of cell membranes by chemical interactions, for example, neutralization of stomach acid by antacids.

$$\text{Drug} + \text{Receptor} \rightarrow \text{Drug} - \text{Receptor Complex} \rightarrow \text{Pharmacological Response}$$

The drug-receptor interaction, that is, the binding of a drug molecule to its receptor, is governed by various types of chemical bonding that have been discussed earlier. A variety of chemical forces may result in a temporary binding of the drug to its receptor. Interaction takes place by using the same bonding forces involved when simple molecules interact, for example, covalent (40–140 kcal mol^{-1}), ionic (10 kcal mol^{-1}), ion-dipole (1–7 kcal mol^{-1}), dipole-dipole (1–7 kcal mol^{-1}), van der Waals (0.5–1 kcal mol^{-1}), hydrogen bonding (1–7 kcal mol^{-1}), and hydrophobic interactions (1 kcal mol^{-1}). However, most useful drugs bind through the use of multiple weak bonds (ionic and weaker).

Covalent bonds are strong and practically irreversible. Since the drug-receptor interaction is a reversible process, covalent bond formation is rather rare except in a few situations. Drugs that interfere with DNA function by chemically modifying specific nucleotides are mitomycin C, cisplatin and anthramycin. Mitomycin C is a well characterized antitumour agent, which forms a covalent interaction with DNA after reductive activation forming a cross-linking structure between guanine bases on adjacent strands of DNA thereby inhibiting single strand formation.

Cisplatin
(An anticancer drug)

Mitomycin C
(An antitumour agent)

Similarly, anthramycin is another antitumour drug that binds covalently to N-2 of guanine located in the minor groove of DNA. It has a preference of purine-G-purine sequences (purines are adenine and guanine) with bonding to the middle G. Cisplatin, an anticancer drug, is a transition metal complex *cis-diamine-dichloro-platinum*. The effect of this drug is due to the ability to palatinate the N-7 of guanine on the major groove site of DNA double helix. This chemical modification of platinum atom cross-links two adjacent guanines on the same DNA strand interfering with the mobility of DNA polymerases (see Section 7.1.3.1).

Anthramycin
(An antitumour agent)

Many drugs are acids or amines, easily ionized at physiological pH and able to form ionic bonds by the attraction of opposite charges in the receptor site. For example, the ionic interaction between protonated amino group on salbutamol or quaternary ammonium on acetylcholine and the dissociated carboxylic acid group of its receptor site.

Protonated salbutamol

Dissociated carboxylic acid group
on a receptor site

Similarly, the dissociated carboxylic group on the drug can bind with amino groups on the receptor. Ion–dipole and dipole–dipole bonds have similar interactions, but are more complicated and are weaker than ionic bonds. Polar–polar interaction, for example, hydrogen bonding, is also an important binding force in drug-receptor interaction because the drug-receptor interaction is basically an exchange of the hydrogen bond between a drug molecule, surrounding water and the receptor site. Formation of hydrophobic bonds between nonpolar hydrocarbon groups on the drug and those in the receptor site is also common. Although these bonds are not very specific, the interactions take place to exclude water molecules. Repulsive forces that decrease the stability of the drug-receptor interaction include repulsion of like charges and *steric hindrance* (see Section 5.6.1.3).

Drugs that target enzymes usually act by reversible enzyme *inhibition. Enzymes* are generally complex proteins produced by living cells. They catalyse specific biochemical reactions without undergoing any change in themselves. Enzymes speed up chemical reactions in the body, but do not get used up in the process. Almost all biochemical reactions in living beings need enzymes. The drugs upon which an enzyme may act are called *substrates*. Enzymes convert the substrates into different molecules known as *products*. An enzyme must have an active site, which is specific to a specific substrate. The correct binding of the substrate to enzyme provides both the specificity of the reaction and the catalytic power of the enzyme. The binding forces responsible for enzyme-substrate recognition are predominantly the same nonpolar interactions. Inhibitors bind to an enzyme and prevent either the formation of the enzyme-substrate complex or catalysis to enzyme and product.

Competitive inhibitors bind to the active site of the enzyme, because they resemble the substrate in shape and chemical composition. *Non-competitive inhibitors* bind to a site other than the active site. The binding of this type of inhibitor affects the shape of the enzyme active site so that the substrate can still bind, but it will not be converted into a product. Aspirin's binding to the enzyme cyclo-oxygenase (COX) is irreversible, unlike other reversible drug-enzyme binding, and results in complete inactivation of COX. This occurs via a covalent bond formation between the acetyl group of aspirin and a serine residue in the active site of the COX enzyme.

2.7 SIGNIFICANCE OF CHEMICAL BONDING IN PROTEIN–PROTEIN INTERACTIONS

Chemical bonding is of immense importance in protein–protein interactions/bindings, which take place for several reasons including inhibition, signalling, catalysis and producing macrostructures, for example, collagen filaments. Protein–protein interaction is the basis of our immune system; *antibodies* are Y shaped proteins of the immune system that recognize and bind to other proteins, called *antigens*. The structure of all antibodies is essentially the same, with only the *antigen-binding site* varying between them. The antigen-binding site comprises several loops of polypeptide chain that protrude from each arm of the antibody and makes one antibody target a specific antigen. The diversity of antigen-binding sites is generated by changing the amino acid sequence of these loops. The presence of these different amino acids govern formation of different bonds between the antibody and the antigen.

2.8 SIGNIFICANCE OF CHEMICAL BONDING IN PROTEIN–DNA INTERACTIONS

Chemical bonding dictates the binding/interaction between proteins and DNA either specifically or non-specifically. For example, *histones*, found in eukaryotic cell nuclei and responsible for the primary level of chromatin packing in eukaryotic cells, are highly alkaline structural proteins that bind DNA in a non-specific way. Histones are the main protein components of *chromatin*, acting as spools around which DNA winds, and playing a pivotal role in gene regulation. They form a complex called a *nucleosome*, and the DNA is wound around it. These non-specific interactions depend on the formation of ionic bonds between the basic amino acid residues in the histones and the acidic sugar phosphate backbone of the DNA. Chemical alterations, for example, methylation and phosphorylation, to the basic residues in the histones, can alter the strength of these interactions, making the DNA more or less tightly packed and affecting transcription, which is the first step of gene expression (where a particular segment of DNA is copied into RNA, especially mRNA, by the enzyme RNA polymerase). *Transcription factors* are proteins that help turn specific genes 'on' or 'off' by binding to nearby DNA. They regulate transcription of genes by binding to DNA in a specific way. Each transcription factor only binds to a particular DNA sequence, which is in or close to the gene promoter. Once assembled at the promoter, the transcription factors position the RNA polymerase and pull the double helix apart to expose the template strand of the required gene to allow transcription to take place. These specific interactions depend on the different bases of the DNA strand as they govern the binding of specific transcription factors.

Chapter 3
Stereochemistry

Learning objectives

After completing this chapter, students should be able to

- define stereochemistry;

- outline different types of isomerisms;

- distinguish between conformational isomers and configurational isomers;

- discuss conformational isomerism in alkanes;

- explain the terms torsional energy, torsional strain, angle strain, enantiomers, chirality, specific rotation, optical activity, diastereomers, meso compounds and racemic mixture;

- designate configuration of enantiomers using D and L system, and (R) and (S) system;

- explain geometrical isomerism in alkenes and cyclic compounds;

- outline the synthesis of chiral molecules;

- explain resolution of racemic mixtures;

- discuss the significance of stereoisomerism in determining drug action and toxicity.

Chemistry for Pharmacy Students: General, Organic and Natural Product Chemistry, Second Edition. Lutfun Nahar and Satyajit Sarker.
© 2019 John Wiley & Sons Ltd. Published 2019 by John Wiley & Sons Ltd.

3.1 STEREOCHEMISTRY: DEFINITION

Stereochemistry is the chemistry of molecules in *three dimensions*. This is an important branch of chemistry that deals with various chiral molecules and stereoisomers.
In organic chemistry, subtle difference in spatial arrangements in organic molecules
renders prominent effects on their chemical reactivity and pharmacological actions.
Thus, a clear understanding of stereochemistry is crucial for the study of complex
molecules that are biologically important, for example, proteins, carbohydrates and
nucleic acids, and also drug molecules, especially in relation to their behaviour and
pharmacological actions including toxicities.

Louis Pasteur, a well-known French biologist, chemist, and microbiologist, is the
first stereochemist, who observed that salts of tartaric acid collected from wine
production vessels could rotate plane polarized light, but that salts from other
sources did not; this observation was made in 1849. Before we go into further
details, let us have a look at different types of isomerism that may exist in organic
molecules.

3.2 ISOMERISM

Compounds with the same molecular formula but different structures are called *isomers*. For example, 1-butene and 2-butene have the same molecular formula, C_4H_8,
but structurally they are different because of the different position of the double
bond. There are two types of isomers: *constitutional isomers* and *stereoisomers*.

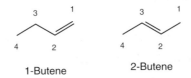

1-Butene   2-Butene

3.2.1 Constitutional Isomers

When two different compounds have the same formula but different connections
between atoms, they are called *constitutional isomers*. To determine whether two
molecules are constitutional isomers, simply count the number of each atom in both
molecules and see how the atoms are arranged. If both molecules possess the same
count for all of the different atoms, and the atoms are arranged in different ways,
the molecules will be considered as constitutional isomers. For example, ethanol
and dimethylether have the same molecular formula, C_2H_6O, but they differ in the
sequence of bonding. Similarly, butane and isobutane are two constitutional isomers.
Constitutional isomers generally have different physical and chemical properties.

Ethanol   Dimethylether   Butane   Isobutane

3.2.2 Stereoisomers

Stereoisomers are compounds that have the same molecular formula and the same connection between atoms, but differ in the arrangement of atoms in the space, that is, in three dimensions (3D). For example, in α-glucose and β-glucose, the atoms are connected in the same order, but the 3D orientation of the hydroxyl group at C-1 is different in each case. Similarly, *cis*- and *trans*-cinnamic acid only differ in the three dimensional orientation of the atoms or groups.

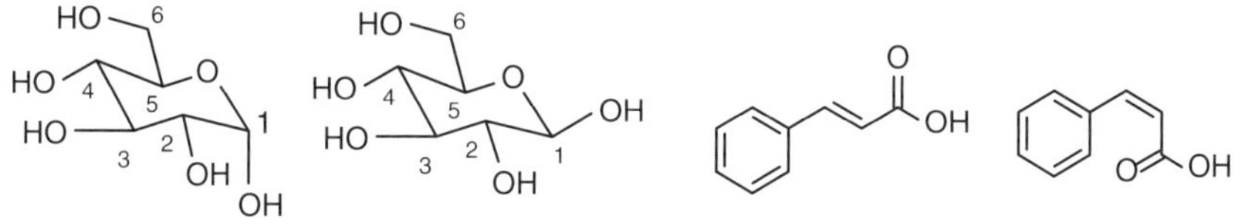

α-Glucose   β-Glucose   *trans*-Cinnamic acid   *cis*-Cinnamic acid

There are two major types of stereoisomers: *conformational isomers* and *configurational isomers*. Configurational isomers include optical isomers, geometrical isomers, enantiomers and diastereomers.

3.2.2.1 Conformational Isomers

Atoms within a molecule move relative to one another by the rotation around covalent single bonds (σ bonds) and the 3D tetrahedral shape of the sp^3-hybridized centres. Such rotation of covalent bonds gives rise to different conformations of a compound. Conformers may also result from restricted rotations in rings. The number of different conformers depends on the number of single bonds and on the number and size of the flexible rings. Each structure is called a *conformer* or *conformational isomer*.

Generally, conformers rapidly interconvert at room temperature.

3.2.2.1.1 Conformational Isomers of Alkanes

Conformational isomerism can be presented with the simplest example, ethane (C_2H_6), which can exist as an infinite number of conformers by the rotation of C—C σ bond. Ethane has two sp^3-hybridized carbon atoms and the tetrahedral angle about each is 109.5°. The most significant conformers of ethane are the *staggered* and *eclipsed* conformers. The staggered conformation is the most stable as it is of the lowest energy.

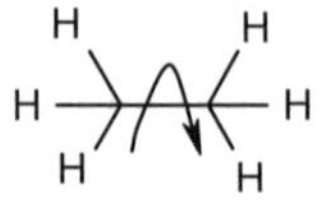

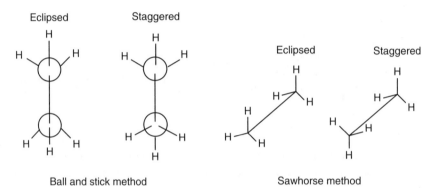

Rotation about the C—C bond in ethane

3.2.2.1.2 Visualization of Conformers

There are four conventional methods for visualization of 3D structures on paper, that is, the ball and stick method, sawhorse method, wedge and broken line method and Newman projection method. Using these methods, the staggered and eclipsed conformers of ethane can be drawn as follows.

Eclipsed   Staggered

Eclipsed   Staggered

Ball and stick method   Sawhorse method

3.2.2.1.3 Eclipsed and Staggered Conformers

In the staggered conformation, the H atoms are as far apart as possible. This reduces the repulsive forces between them. This is why staggered conformers are stable. In the eclipsed conformation, H atoms are closest together. This gives rise to higher repulsive forces between them. As a result, eclipsed conformers are unstable. At any moment, more molecules will be in staggered form than any other conformations. The angles in the eclipsed conformation are 0°, whereas the angles in the staggered conformation are 60°.

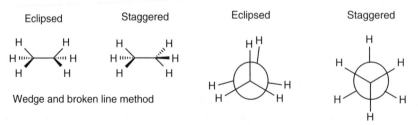

Eclipsed   Staggered   Eclipsed   Staggered

Wedge and broken line method

Newman projection method

3.2.2.1.4 Torsional Energy and Torsional Strain

Torsional energy is the energy required for rotating about the C—C σ bond. In ethane, this is very low (only 3 kcal). *Torsional strain* is the strain observed when a conformer rotates away from the most stable conformation, that is, staggered form. Torsional strain is due to the slight repulsion between electron-clouds in the C—H bonds as they pass close by each other in the eclipsed conformer. In ethane, this is also low.

3.2.2.1.5 Conformational Isomerism in Propane

Propane is a three carbon (*sp³*-hybridized) atoms containing linear alkane. All are tetrahedrally arranged. When a hydrogen atom of ethane is replaced by a methyl (CH_3) group, we get propane. There is rotation about two C—C σ bonds.

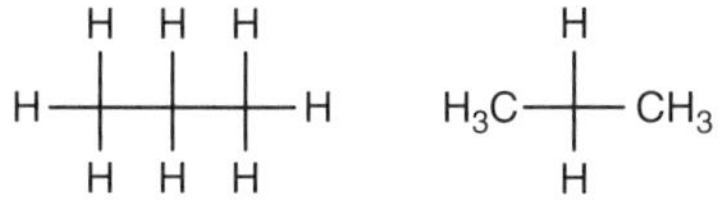

Propane

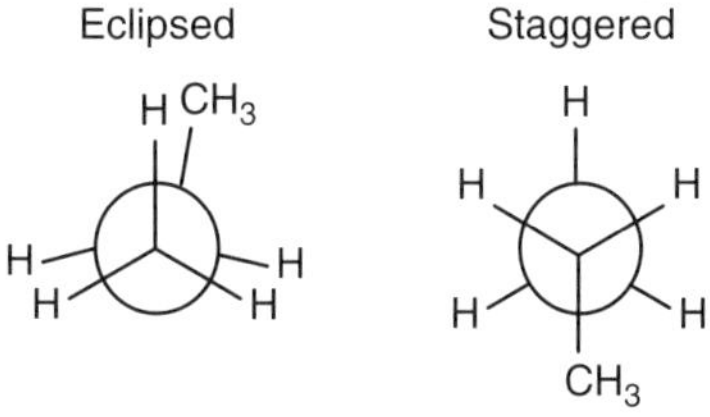

Newman projection of propane conformers

In the eclipsed conformation of propane, we now have a larger CH_3 close to H atom. This results in an increased repulsive force or increased steric strain. The energy difference between the eclipsed and staggered forms of propane is greater than that of ethane.

3.2.2.1.6 Conformational Isomerism in Butane

Butane is a four carbon (*sp³*-hybridized) atoms containing linear alkane. All are tetrahedrally arranged. When a hydrogen atom of propane is replaced by a methyl (CH_3) group, we get butane. There is rotation about two C—C σ bonds, but the rotation about C_2—C_3 is the most important of all.

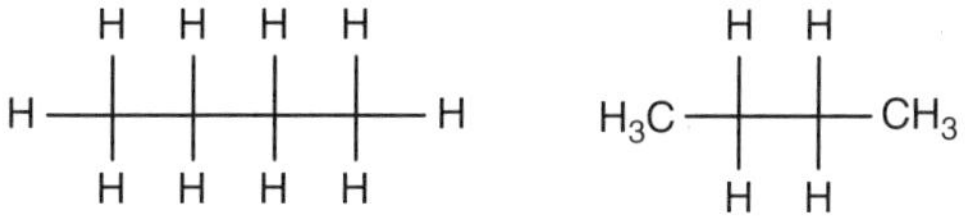

Butane

Among the conformers, the least stable is the first eclipsed structure where two CH_3 groups are totally eclipsed, and the most stable is the first staggered conformer, where two CH_3 groups are staggered, and far apart from each other. When two bulky groups are staggered, we get the *anti-conformation*, and when they are 60° to each other, we have the *gauche* conformer. In butane, the torsional energy is even higher than propane. Thus, there is slightly restricted rotation about C_2-C_3 bond in butane. The order of stability (from the highest to the lowest) among the following conformers is: anti → gauche → another eclipsed → eclipsed. The most stable conformer has the lowest steric strain and torsional strain.

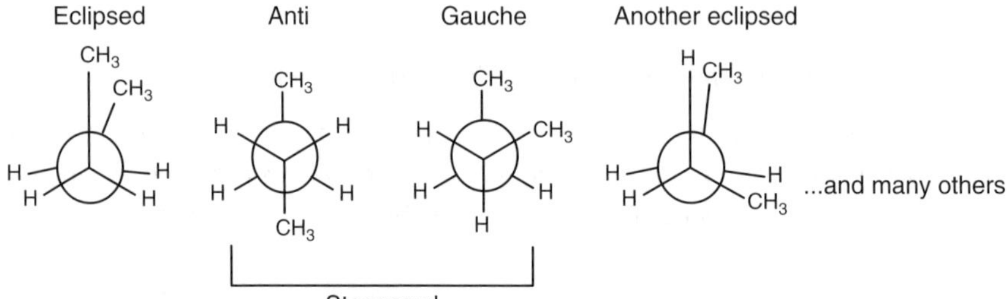

Staggered
Newman projection of butane conformers

3.2.2.1.7 Conformational Isomerism in Cyclopropane

Cyclopropane is the first member of the cycloalkane series, and composed of three carbons and six hydrogen atoms (C_3H_6). The rotation about C—C bonds is quite restricted in cycloalkanes, especially in smaller rings; for example, cyclopropane.

Cyclopropane

In cyclopropane, each C atom is still *sp³*-hybridized so we should have a bond angle of 109.5°, but each C atom is at the corner of an equilateral triangle, which has angles of 60°! As a result, there is considerable *angle strain*. The *sp³* hybrids still overlap but only just. This gives a very unstable and weak structure. The *angle strain* can be defined as the strain induced in a molecule when bond angle deviates from the ideal tetrahedral value. For example, this deviation in cyclopropane is from 109.5° to 60°.

3.2.2.1.8 Conformational Isomerism in Cyclobutane

Cyclobutane comprises four carbons and eight hydrogen atoms (C_4H_8). If we consider cyclobutane to have a flat or planar structure, the bond angles will be 90°. So the angle strain (109.5°) will be much less than that of cyclopropane. However, cyclobutane, in its planar form, will give rise to torsional strain since all H atoms are eclipsed.

Cyclobutane as a planar molecule
All H atoms are eclipsed
Bond angle = 90°

Most stable folded conformation
of cyclobutane
H atoms are not eclipsed

Cyclobutane, in fact, is not a planar molecule. To reduce torsional strain, this compound attains this nonplanar folded conformation. Hydrogen atoms are not eclipsed in this conformation and torsional strain is much lower than its planar structure. However, in this form, angles are <90°, which means a slight increase in angle strain.

3.2.2.1.9 Conformational Isomerism in Cyclopentane

Cyclopentane is a five-carbon cyclic alkane. If we consider cyclopropane as a planar and regular pentagon, the angles are 108°. So, there is very little angle strain or almost strain-free (109.5° for sp^3 hybrids). However, in this form the torsional strain is quite large, because most of its hydrogen atoms are eclipsed. Thus, to reduce torsional strain, cyclopentane twists to adopt a puckered or envelope-shaped, nonplanar conformation that strikes a balance between increased angle strain and decreased torsional strain. In this conformation, most of the hydrogen atoms are almost staggered.

Cyclopentane as a planar molecule
Bond angle = 108°

Most stable puckered conformation
of cyclopentane
Most H atoms are nearly staggered

3.2.2.1.10 Conformational Isomerism in Cyclohexane

Cyclohexane (C_6H_{12}) is a six-carbon cyclic alkane that occurs extensively in nature. Many pharmaceutically important compounds possess cyclohexane rings, for example, steroidal molecules (see Section 8.6). If we consider cyclohexane as a planar and regular hexagon, the angles are 120° (109.5° for sp^3 hybrids).

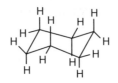

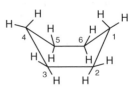

| Cyclohexane as a planar molecule (bond angle 120°) | Most stable chair conformation of cyclohexane All neighbouring C—H bonds are staggered | Boat conformation of cyclohexane |

Again, in reality, cyclohexane is not a planar molecule. To strike a balance between torsional strain and angle strain, and to achieve more stability, cyclohexane attains various conformations, among which the *chair* and *boat conformations* are most significant. At any one moment 99.9% cyclohexane molecules will have chair conformation.

The chair conformation of cyclohexane is the most stable conformer. In chair conformation, the C—C—C angles can reach the strain free tetrahedral value (109.5°) and all neighbouring C—H bonds are staggered. Therefore, this conformation does not have any angle strain or torsional strain. Another conformation of cyclohexane is the *boat conformation*. Here, the H atoms on C_2—C_3 and C_5—C_6 are eclipsed, which results in an increased torsional strain. Also, the H atoms on C_1 and C_4 are close enough to produce steric strain.

In the chair conformation of cyclohexane, there are two types of positions for the substituents on the ring, *axial* (perpendicular to the ring, i.e. parallel to the ring axis) and *equatorial* (in the plane of the ring, i.e. around the ring equator) positions. Six hydrogen atoms are in the axial position and six other in the equatorial position. Each carbon atom in cyclohexane chair conformation has an axial hydrogen and an equatorial hydrogen atom, and each side of the ring has three axial and three equatorial hydrogen atoms.

| Chair conformation of cyclohexane Six axial (a) and six equatorial (e) hydrogen atoms | Chair conformation of cyclohexane Diaxial interaction |

When all 12 substituents are hydrogen atoms, there is no steric strain. The presence of any groups larger than H changes the stability by increasing the steric strain, especially, if these groups are present in axial positions. When axial, *diaxial interaction* can cause steric strain. In the case of equatorial, there is more room and less steric strain. Bulky groups always preferably occupy equatorial position. Because of axial and equatorial positions in the chair conformation of cyclohexane, one might expect to see two isomeric forms of a monosubstituted cyclohexane. However, in reality, only one monosubstituted form exists, because cyclohexane rings are conformationally mobile at room temperature. Different chair conformations interconvert resulting in the exchange of axial and equatorial positions. This interconversion of chair conformations is known as a *ring-flip*. During ring-flip, the middle four carbon atoms remain in place, while the two ends are folded in opposite directions. As a result, an axial substituent in one chair form of cyclohexane becomes an equatorial substituent in the ring-flipped chair form, and vice versa.

R = Any substituent group or atom other than H

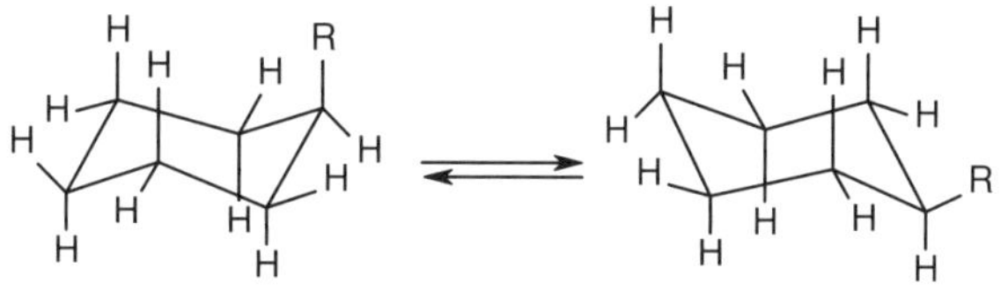

R in axial position   R in equatorial position

Ring-flip in chair conformation
of monosubstituted cyclohexane

3.2.2.2 Configurational Isomers

Configurational isomers differ from each other only in the arrangement of their atoms in the space (3D), and cannot be converted from one into another by rotations about single bonds within the molecules. Before we look into the details of various configurational isomers, we need to understand the concept of *chirality*. Chirality is extremely important in biological systems. Enzyme active sites are capable of chiral discrimination and can distinguish between enantiomers. For example, the enantiomers of *S*-(+)-carvone, smells like caraway seeds and *R*-(−)-carvone smells like spearmint. But not all enantiomers have distinguishable odours.

S-(+)-Carvone
(Smells like caraway seeds)

R-(−)-Carvone
(Smells like spearmint)

3.2.2.2.1 Chirality and Chiral Molecules

Many objects around us are handed. For example, our left and right hands are mirror images of each other, and cannot be superimposed on each other as shown next. Other chiral objects include shoes, gloves and printed pages. Many molecules are also handed, that is, they cannot be superimposed on their mirror images. Such molecules are called *chiral* molecules. Many compounds that occur in living organisms, for example, carbohydrates and proteins, are chiral.

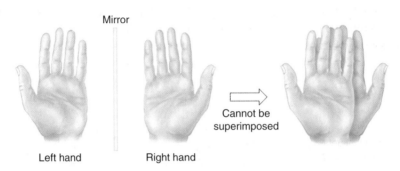

Mirror

Cannot be superimposed

Left hand   Right hand

The most common feature in *chiral molecules* is a tetrahedral, that is, sp^3-hybridized, carbon atom with four different atoms or groups attached. Such a carbon atom is called a *chiral carbon* or an *asymmetric carbon* or *asymmetric centre*. Chiral molecules do not have a plane of symmetry.

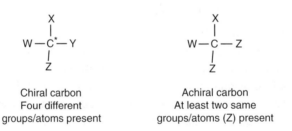

Chiral carbon
Four different
groups/atoms present

Achiral carbon
At least two same
groups/atoms (Z) present

When there are two or more same atoms/groups, the carbon is called *achiral*. For example, the carbon atom in bromo-chloro-methane has a bromine, a chlorine, but two hydrogen atoms linked to it as shown next. Thus, the carbon atom in this compound is achiral. Achiral molecules often have a *plane of symmetry*. A plane of symmetry is an imaginary plane that divides a molecule into two equal halves that are mirror image of each other. With achiral molecules, the compound and its mirror image are the same, which means that they can be superimposed.

If you rotate the mirror image through 180°, it is identical to the original structure.

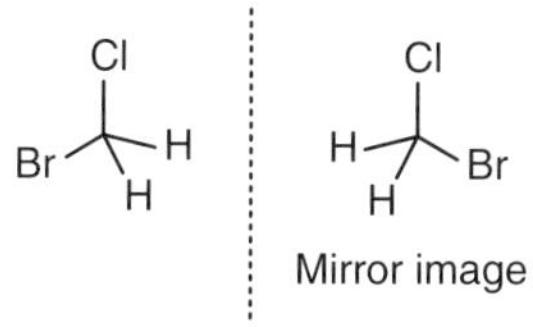

Achiral molecule
Superimposable mirror image

3.2.2.2.2 Enantiomers

Every chiral molecule has enantiomers, but achiral molecules do not have enantiomers. Greek word *enantio* means 'opposite'. A chiral molecule and its mirror image are called *enantiomers* or *enantiomeric pairs*. They are non-superimposable. The actual arrangement or orientation (in space) of atoms or groups attached to the chiral carbon (asymmetric centre, stereogenic centre, or stereocentre) is called the *configuration* of a compound.

The arrangement of W, X, Y and Z is the configuration

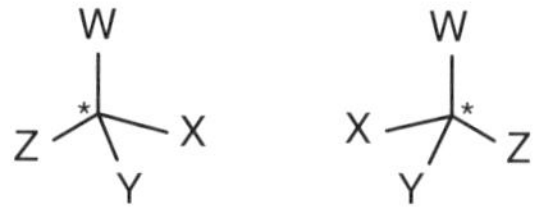

Mirror image

Enantiomers
(Not superimposable)

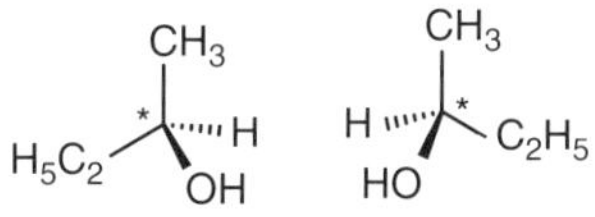

(−)-2-Butanol   (+)-2-Butanol

(Enantiomers of 2-butanol)

Enantiomers share same physical properties, for example, melting points, boiling points and solubilities. They also have same chemical properties. However, they differ in their activities with *plane polarized light* that gives rise to *optical isomerism*, and also in their pharmacological actions.

3.2.2.2.3 Drawing a Chiral Molecule: Enantiomer

On a plane paper, chiral molecules can be drawn using *wedge bonds*. There are also a few other methods that use horizontal bonds representing bonds pointing out of the paper and vertical bonds pointing into the paper. Some examples are given next.

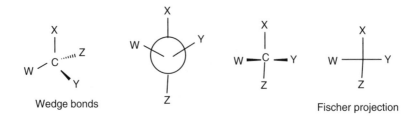

Wedge bonds   Fischer projection

3.2.2.2.4 Optical Isomers

Light consists of waves that are composed of electrical and magnetic vectors (at right angles). If we looked 'end-on' at light as it travels, we would see it oscillates in all directions. When a beam of ordinary light is passed through a *polarizer*, the polarizer interacts with the electrical vector in a way that the light emerging from it oscillates in one direction or plane. This is called *plane polarized light.*

Light oscillates in all direction   The plane of oscillation in
                                           a plane-polarized light

When plane-polarized light passes through a solution of an enantiomer, the plane of light rotates. Any compounds that rotate plane-polarized light are called *optically active*. If the rotation is in a clockwise direction, the enantiomer is said to be *dextrorotatory* and is given the (+) sign in front of its name. Anticlockwise rotation gives an enantiomer, called *levorotatory*, and is given the sign (–) in front of its name.

            or

Before passing through a   Rotates clockwise (right)   Rotates anti-clockwise (left)
solution of an enantiomer   Dextrorotatory (+)   Levorotatory (−)

                        After passing through a solution of an
                                       enantiomer

The amount of rotation can be measured with an instrument called *polarimeter*. A solution of optically active molecule (enantiomer) is placed in a sample tube, plane-polarized light is passed through the tube, and a rotation of the polarization plane takes place. The light then goes through a second polarizer called analyser. By rotating the analyser until the light passes through it, the new plane of polarization can be found, and the extent of rotation that has taken place can be measured.

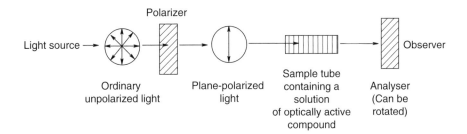

Enantiomers are optically active, and are called *optical isomers*, with one being (+) and the other (–).

When we have a pair of enantiomers, each rotates the plane-polarized light by the same amount, but in the opposite direction. A mixture of enantiomers with the same amount of each is called a *racemic mixture*. The first known racemic mixture was racemic acid, which Louis Pasteur found to be a mixture of the two enantiomeric isomers of tartaric acid. Racemic mixtures are optically inactive, that is, they cancel each other out and are denoted by (±).

There are several drugs available as racemic mixtures, that is, 50:50 mixture of two enantiomers. For example, the drugs atenolol (used to treat high blood pressure and angina, and to reduce heart attack risk), atropine (used to treat certain types of nerve agent and pesticide poisoning, and some types of slow heart rate), disopyramide (an antiarrhythmic medication used in the treatment of ventricular tachycardia) and warfarin (a blood thinning medicine: an oral anticoagulant) are available as racemic mixtures. However, over the last couple of decades, various regulatory authorities have gradually made it a requirement that when drugs at the developmental stage are made up of a racemic mixture, the individual efficacy and any toxicity of the both enantiomers must be assessed carefully. Thus, it can be assumed that there will be gradual reduction in the marketing of new drugs in racemic forms.

3.2.2.2.5 Specific Rotations

The more molecules (optically active) the light beam encounters, the greater the observed rotation. Thus, the amount of rotation depends on both sample concentration and sample path length.

If we keep the concentration constant but double the length of sample tube, the observed rotation doubles. The amount of rotation also depends on the temperature and the wavelength of the light used.

Therefore, to obtain a meaningful optical rotation data, we have to choose standard conditions, and here comes the concept of *specific rotation*.

The *specific rotation* of a compound, designated as $[\alpha]_D$, is defined as the observed rotation, α, when the sample path length l is 1 dm, the sample concentration C is 1 g ml^{-1}, and the light of 599.6 nm wavelength (the D line of a sodium lamp, which is the yellow light emitted from common sodium lamps) is used.

$$[\alpha]_D = \frac{\text{Observed rotation } \alpha \text{ in degrees}}{\text{Path length, } l \text{ (dm)} \times \text{Concentration, } C \text{ (g ml}^{-1})}$$

$$= \frac{\alpha}{l \times C}$$

As the specific rotation also depends on temperature, the temperature at which the rotation is measured is often shown in the equation. A specific rotation measured at 25 °C is denoted more precisely as:

$$[\alpha]_D^{25}$$

When optical rotation data are expressed in this standard way, the specific rotation, $[\alpha]_D$, is a physical constant, characteristic of a given optically active compound. For example, the specific rotation of morphine is −132°, that is,

$$[\alpha]_D^{25} = -132°$$

This means, the D line of a sodium lamp (l = 599.6 nm) was used for light, that a temperature of 25 °C was maintained, and that a sample containing 1.00 g ml^{-1} of the optically active morphine, in a 1 dm tube, produced a rotation of 132° in an anti-clockwise direction.

3.2.2.2.6 *How to Designate the Configuration of Enantiomers*

We have already seen that (+) or (–) sign indicates optical activity of an enantiomer. However, the optical activity does not tell us the actual configuration of an enantiomer. It only gives us the information whether an enantiomer rotates the plane-polarized light clockwise or anti-clockwise.

Let us look at the example of glyceraldehyde, an optically active molecule. Glyceraldehyde can exist as enantiomers, that is, (+) and (–) forms, but the sign does not describe the exact configuration.

Glyceraldehyde

There are two systems to designate configuration of enantiomers: D and L-system, and (R) and (S)-system, also known as Cahn–Ingold–Prelog system.

3.2.2.2.7 D and L system

Emil Fischer used glyceraldehyde as a standard for the D and L system of designating configuration. He arbitrarily took the (+)-glyceraldehyde enantiomer and assigned this D-glyceraldehyde. The other enantiomer is the (–)-glyceraldehyde and was assigned as L-glyceraldehyde. We can easily identify the only difference in the following structures, which is the orientation of the hydroxyl group at the chiral centre. In the case of D-glyceraldehyde, the —OH group on the chiral carbon is in on the right-hand side, whereas in L-glyceraldehyde, it is on the left. In the D and L system, structures that are similar to glyceraldehyde (at chiral carbon) are compared. For example, 2,3-dihydroxypropanoic acid.

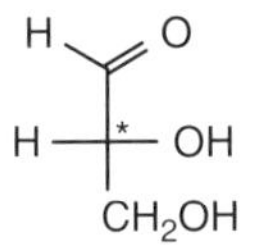

$$
\begin{array}{cc}
\text{H}\diagup\!\!\diagdown\text{O} & \text{H}\diagup\!\!\diagdown\text{O} \\
\text{H}\!\!-\!\!\overset{*}{|}\!\!-\!\!\text{OH} & \text{HO}\!\!-\!\!\overset{*}{|}\!\!-\!\!\text{H} \\
\text{CH}_2\text{OH} & \text{CH}_2\text{OH} \\
\text{(+)-D-Glyceraldehyde} & \text{(–)-L-Glyceraldehyde}
\end{array}
$$

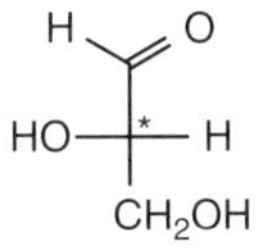

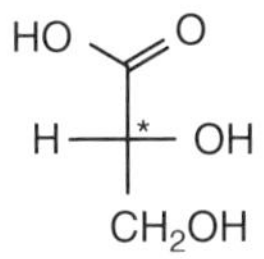

$$
\begin{array}{cc}
\text{HO}\diagup\!\!\diagdown\text{O} & \text{HO}\diagup\!\!\diagdown\text{O} \\
\text{H}\!\!-\!\!\overset{*}{|}\!\!-\!\!\text{OH} & \text{HO}\!\!-\!\!\overset{*}{|}\!\!-\!\!\text{H} \\
\text{CH}_2\text{OH} & \text{CH}_2\text{OH}
\end{array}
$$

D-2,3-dihydroxypropanoic acid The —OH on the chiral carbon (*) is on the right	D-2,3-dihydroxypropanoic acid The —OH on the chiral carbon (*) is on the left

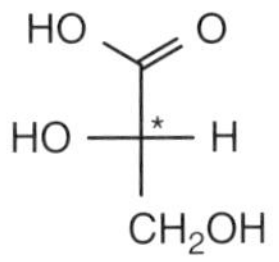

One must remember that there is no correlation between D and L configurations, and (+) and (–) rotations. D-isomer does not have to have a (+) rotation, and similarly, L-isomer does not have to have a (–) rotation.

For some compounds, the D-isomer is (+) and for others the L-isomer may be (+). Similarly, for some, we may have L (–) and others may have D (–).

This D and L system is common in biology/biochemistry, especially with sugars and amino acids (with amino acids, the —NH$_2$ configuration is compared with —OH of glyceraldehyde). This system is particularly used to designate various carbohydrate or sugar molecules, for example, D-glucose, L-rhamnose and L-alanine.

3.2.2.2.8 Cahn–Ingold–Prelog system: (R) and (S) System

Three chemists, R. S. Cahn (England), C. K. Ingold (England) and V. Prelog (Switzerland), devised a system of nomenclature that can describe the configuration of enantiomers more precisely. This system is called the *(R) and (S)* system, or the *Cahn–Ingold–Prelog system.*

According to this system, one enantiomer of 2-hexanol should be designated *(R)*-2-hexanol, and the other *(S)*-2-hexanol. *R* and *S* came from the Latin words *rectus* and *sinister,* meaning right and left, respectively.

$$
\begin{array}{cc}
\underset{\substack{\text{CH}_2\text{CH}_2\text{CH}_3 \\ 3 \quad 4 \quad 5}}{\overset{\substack{1 \\ \text{CH}_3}}{\text{HO} \underset{*|2}{\cdots} \text{H}}} &
\underset{\substack{\text{CH}_2\text{CH}_2\text{CH}_3 \\ 3 \quad 4 \quad 5}}{\overset{\substack{1 \\ \text{CH}_3}}{\text{H} \underset{*|2}{\cdots} \text{OH}}}
\end{array}
$$

Enantiomers of 2-hexanol

The following rules or steps are applied for designating any enantiomer to *R* or *S*.

i. Each of the four groups attached to the chiral carbon is assigned **1–4 (or a–d)** in terms of order of priority or preference, **1** being the highest and **4** being the lowest priority. Priority is first assigned on the basis of the atomic number of the atom that is directly attached to the chiral carbon.

Higher atomic number gets higher priority. This can be shown by the structure of 2-hexanol.

$$
\underset{\substack{\text{CH}_2\text{CH}_2\text{CH}_3 \\ 2}}{\overset{\substack{3 \\ \text{CH}_3}}{\underset{1}{\text{HO}} \underset{*}{\cdots} \overset{}{\underset{4}{\text{H}}}}}
$$

Priority determination
in 2-hexanol

ii. When a priority cannot be assigned on the basis of the atomic numbers of the atoms that are directly attached to the chiral carbon, the next set of atoms in the unassigned groups is examined. This process is continued to the first point of difference.

In 2-hexanol, there are two carbon atoms directly attached to the chiral carbon; one is of the methyl group, and the other is of the propyl group. In this case, only on the basis of the atomic number of the carbon atom, one cannot assign the priority of these two carbon atoms, and must consider the next set of atoms attached to these carbon atoms. When we examine the methyl group of the enantiomer, we find that the next set of atoms consists of three H atoms. On the other hand, in the propyl group, the next set of atoms consists of one C and two H atoms. Carbon has a higher atomic number than H, so the $CH_3CH_2CH_2-$ group gets the priority over CH_3.

iii. Groups containing double or triple bonds (π bonds) are assigned priorities as if both atoms were duplicated or triplicated. For example,

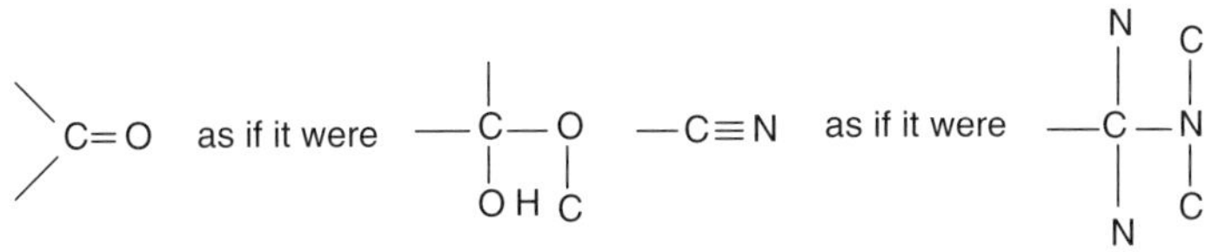

iv. Having decided on priority of the four groups, one has to arrange (rotate) the molecule in a way that group 4, that is, the lowest priority, is pointing away from the viewer.

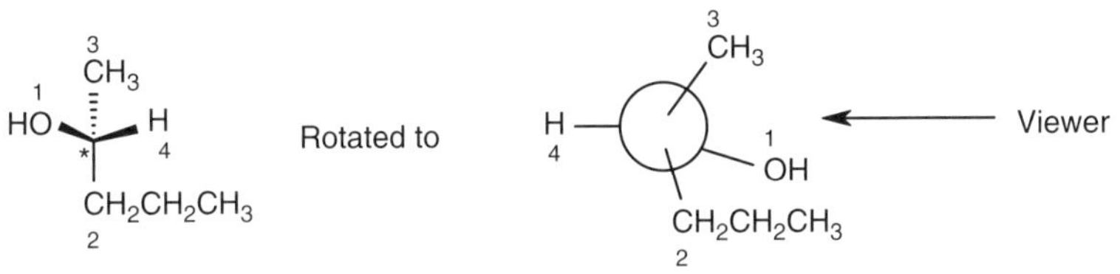

The least priority H atom (4) is away from the viewer

Then an arrow from group **1 → 2 → 3** is to be drawn. If the direction is clockwise, it is called an (R)-isomer. If it is anti-clockwise, it is called an (S)-isomer. With enantiomers, one will be the (R)-isomer and the other, the (S)-isomer. Again, there is no correlation between (R) and (S), and (+) and (−).

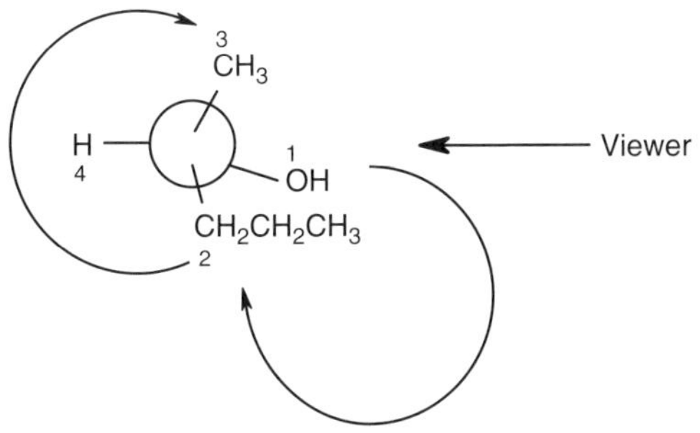

(R)-2-Hexanol
Priority order 1 to 4 is in
clockwise direction

Following the *Cahn–Ingold–Prelog system*, it is now possible to draw the structures of (R)- and (S)-enantiomers of various chiral molecules. For example,

2,3-dihydroxypropanoic acid where the priorities, 1 = OH, 2 = COOH, 3 = CH$_2$OH, and 4 = H.

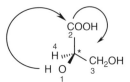

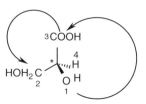

(R)-2,3-Dihydroxypropanoic acid
Priority order 1 to 4 is in
clockwise direction

(S)-2,3-Dihydroxypropanoic acid
Priority order 1 to 4 is in
clockwise direction

When there are more than one stereocentre (chiral carbon) present in a molecule, it is possible to have more than two stereoisomers. It is then necessary to designate all these stereoisomers using (R) and (S) system. In 2,3,4-trihydroxybutanal, there are two chiral carbons. The chiral centres are at C-2 and C-3. Using the (R) and (S) system, one can designate these isomers as follows.

C-2	C-3	Designation of stereoisomer
R	R	(2R, 3R)
R	S	(2R, 3S)
S	R	(2S, 3R)
S	S	(2S, 3S)

2,3,4-Trihydroxybutanal
A molecule with two chiral centres (*)

3.3 STEREOISOMERISM OF MOLECULES WITH MORE THAN ONE STEREOCENTRE

3.3.1 Diastereomers and Meso Structures

In compounds, whose stereoisomerism is due to tetrahedral stereocentres, the total number of stereoisomers will not exceed 2^n, where n is the number of tetrahedral stereocentres. For example, in 2,3,4-trihydroxybutanal, there are two chiral carbons. The chiral centres are at C-2 and C-3. Therefore, the maximum number of possible isomers will be $2^2 = 4$. All four stereoisomers of 2,3,4-trihydroxybutanal (A–D) are optically active, and among them, there are two enantiomeric pairs, A and B, and C and D, as shown in the structures next.

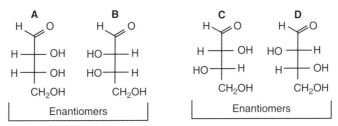

Four possible stereoisomers of 2,3,4-trihydroxybutanal

If we look at the structures **A** and **C** or **B** and **D**, we have stereoisomers, but not enantiomers. These are called *diastereomers*. Therefore, diastereomers are stereoisomers that are not mirror images. Other pairs of diastereomers among the stereoisomers of 2,3,4-trihydroxybutanal are **A** and **D**, and **B** and **C**. Diastereomers have different physical properties (boiling points, melting points and solubilities), they are often easy to separate by usual separation techniques such as distillation, recrystallization and chromatography. Enantiomers are much more difficult to separate.

A *meso compound* is an *achiral molecule* that has chiral atoms. Now, let us consider another similar molecule, tartaric acid, where there are two chiral carbons. In tartaric acid, four isomeric forms are theoretically expected ($2^2 = 4$).

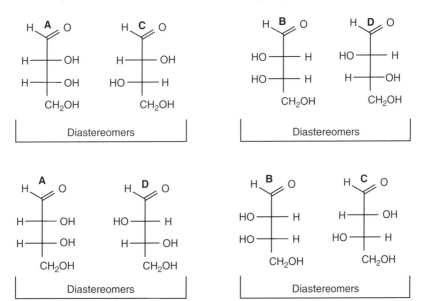

Diastereomers of 2,3,4-trihydroxybutanal

However, because one half of the tartaric acid molecule is a mirror image of the other half, we get a *meso* structure. A meso diastereomer is achiral since it has a mirror plane of symmetry. This means this compound and its mirror image are superimposable, that is, they are the same compound. Thus, instead of four, we get only three stereoisomers for tartaric acid.

Structures **1** and **2** are enantiomers, and both are optically active. In structures **3** and **4**, there is a plane of symmetry, that is, there is a mirror image within a single molecule.

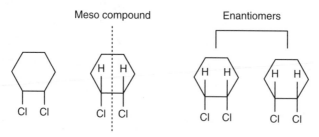

Stereoisomers of tartaric acid

The 'same compound' pair are called the *meso diastereomer.* Structures **3** and **4** are superimposable, and essentially are same compound. Hence, we have a *meso-*tartaric acid and it is achiral (since it has a plane of symmetry, and it is superimposable on its mirror image). Meso tartaric acid is optically inactive. Thus for tartaric acid, we have (+), (–) and *meso*-tartaric acid.

3.3.2 Cyclic Compounds

Depending on the type of substitution on a ring, the molecule can be chiral (optically active) or achiral (optically inactive).

For example, 1,2-dichlorocyclohexane can exists as *meso* compounds (optically inactive) and enantiomers (optically active).

Stereoisomerism in 1,2-dichlorocyclohexane

If the two groups attached to the ring are different, that is, no plane of symmetry, then there will be four isomers. For example, 1-bromo-2-chloro-cyclohexane.

1-Bromo-2-chloro-cyclohexane

3.3.3 Geometrical Isomers of Alkenes and Cyclic Compounds

Geometrical isomerism is found in alkenes and cyclic compounds. In alkenes, there is restricted rotation about the double bond. When there are substituent groups attached to the double bond, they can bond in different ways resulting in *trans* (opposite side) and *cis* (same side) isomers. These are called *geometrical isomers*. They have different chemical and physical properties. Each isomer can be converted to another when enough energy is supplied, for example, by absorption of UV radiation or being heated to temperatures around 300 °C. The conversion occurs because π bond breaks when energy is absorbed, and the two halves of the molecule can then rotate with respect to each other before the π bond forms again.

trans Isomer
(Substituent G is on opposite sides on the double-bonded carbons)

cis Isomer
(Substituent G is on same side on the double-bonded carbons)

When there is same substituent attached to the double-bonded carbons, as in the previous example, it is quite straightforward to designate *trans* or *cis*. However, if there are more than one different groups or atoms present, as in the following examples, the situation becomes a bit more complicated for assigning *cis* and *trans*.

Alkenes with different substituents on the double-bonded carbons

To simplify this situation, the *E/Z* system is used for naming geometrical isomers. (*Z*) stands for German *zusammen*, which means 'same side', and (*E*) for German *entgegen* meaning 'on the opposite side'.

In E and Z system, the following rules or steps are followed:

i. On each C atom of the double bond, we have to assign the priority of the atoms bonded. Priority should be on the same basis as (R)/(S)-system (i.e. on the basis of atomic number).

ii. If the two higher priority groups of the two C atoms are on the *same* side of the double bond, it is called (Z)-isomer.

iii. If the two higher priority groups of the two C atoms are on the *opposite* side of the double bond, it is called (E)-isomer.

Let us take a look at 1-bromo-1,2-dichloroethene as an example. In this molecule, atoms attached are: Cl and Br on C-1, and Cl and H on C-2.

Atomic numbers of these substituents are in the order of Br > Cl > H. So, once the priorities are assigned, we can easily draw the (E)- and (Z)-isomers of 1-bromo-1,2-dichloroethene in the following way.

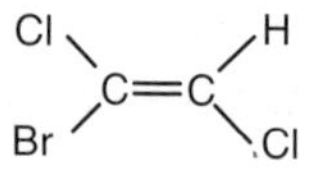

(Z)-1-Bromo-1,2-dichloroethene
(The two higher priority groups are
on the same side)

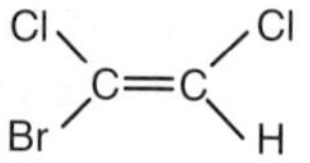

(E)-1-Bromo-1,2-dichloroethene
(The two higher priority groups are
on the opposite side)

Now, let us have a look at the cyclic compounds. We can use this (E)- and (Z)-system for a cyclic compound, when two or more groups attached to a ring. For example, if in the following substituted cyclopentane, A and B are different groups, each C atom attached to A and B is chiral carbons or stereocentres.

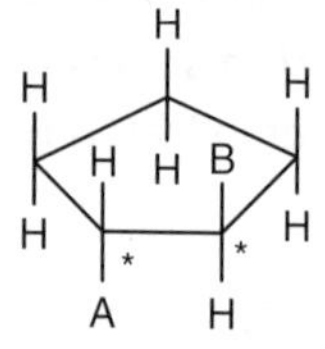

(E)-form
The two higher priority groups
(A or B > H)
are on the opposite side

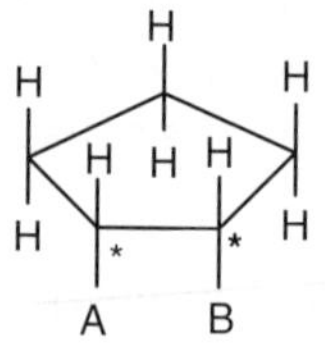

(Z)-form
The two higher priority groups
(A or B > H)
are on the same side

In 1-bromo-2-chlorocyclopentane, there are two chiral centres. Therefore, four possible stereoisomers can be expected ($2^2 = 4$).

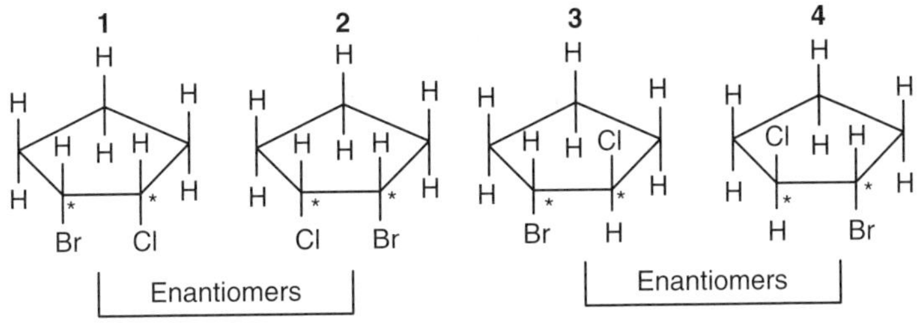

The figure shows four possible isomers of 1-bromo-2-chlorocyclopentane, labeled 1, 2, 3, 4, with enantiomer brackets.

Four possible isomers of 1-bromo-2-chlorocyclopentane

The isomers are: (+)-*cis*-2-bromo-1-chlorocyclopentane (**1**), (−)-*cis*-1-bromo-2-chlorocyclopentane (**2**), (+)-*trans*-2-bromo-1-chlorocyclopentane (**3**) and (−)-*trans*-1-bromo-2 chlorocyclopentane (**4**). However, when A = B, that is, two substituents are same, as in 1,2-dihydroxycyclopentane, only three isomers are possible, because of the presence of a plane of symmetry with this molecule. In this case, we have *meso* structure.

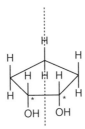

1,2-Dihydroxycyclopentane
(There is a plane of symmetry within the molecule)

In 1,2-dihydroxycyclohexane, a plane of symmetry exists within the molecule and instead of four, it produces three isomers as follows.

One is an optically inactive *meso* isomer *cis* or (Z)-isomer and two optically active *trans* or (E)-isomers. With cyclohexane, we can have equatorial and axial bonds. Thus, with *trans* structure, we get di-axial and di-equatorial bonds, and with *cis* structure we get axial-equatorial bonds.

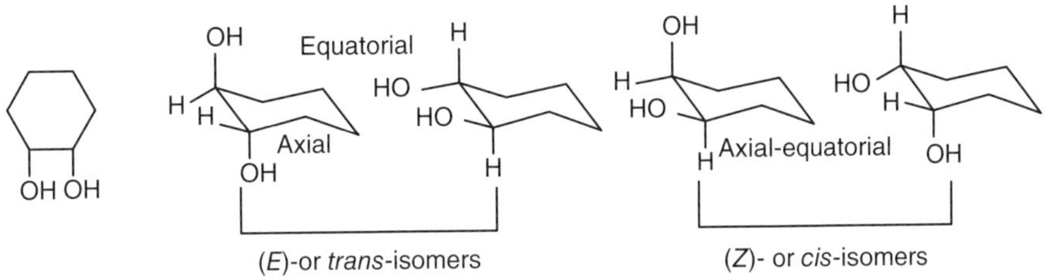

Equatorial

Axial

Axial-equatorial

(E)-or *trans*-isomers

(Z)- or *cis*-isomers

Four possible isomers of 1,2-dihydroxycyclohexane

3.4 SIGNIFICANCE OF STEREOISOMERISM IN DETERMINING DRUG ACTION AND TOXICITY

Pharmacy is a discipline of science that deals with various aspects of drugs including how they bind with receptors inside the body and exert pharmacological actions. All drugs are chemical entities and a great majority (30–50%) of them have stereocentres, show stereoisomerism and exist as enantiomers. Moreover, the current trend in drug markets has observed a rapid increase of the sales of chiral drugs at the expense of the achiral ones. Chiral drugs, whether enantiomerically pure or sold as a racemic mixture, are likely to continue to dominate drug markets. It is therefore important to understand how drug chirality affects its interaction with drug targets and to be able to use proper nomenclature in describing the drugs themselves and the nature of forces responsible for those interactions.

Most often, only one form shows correct physiological and pharmacological action. For example, only one enantiomer of morphine is active as an analgesic, only one enantiomer of glucose is metabolized in our body to give energy and only one enantiomeric form of adrenaline is a neurotransmitter.

One enantiomeric form of a drug may be active, and the other may be inactive, less active or even toxic. Not only drug molecules, but also various other molecules that are essential for living organisms also exist in stereoisomeric forms, and their biological properties often are specific to one stereoisomer. Most of the molecules that make up living organisms are chiral, that is, they show stereoisomerism. For example, all but one (glycine) of the 20 essential amino acids are chiral. Thus, it is important to understand stereochemistry for a better understanding of drug molecules, their action and toxicity.

$$H_2N \overset{\displaystyle R}{\underset{\displaystyle H}{\text{———}}} COOH$$

Glycine R = H, an achiral amino acid
Alanine R = Me, a chiral amino acid

Ibuprofen is a popular analgesic and anti-inflammatory drug, and belongs to the group called nonsteroidal anti-inflammatory drug (NSAID). There are two stereoisomeric forms of ibuprofen. This drug can exist as (S)- and (R)-stereoisomers (enantiomers). Only the (S)-form is active. The (R)-form is completely inactive, although it is slowly converted in the body to the active (S)-form. The drug marketed under the trade names, commercially known as Advil®, Anadin®, Arthrofen®, Brufen®, Nurofen®, Nuprin® and Motrin® is a racemic mixture of (R)- and (S)-ibuprofen.

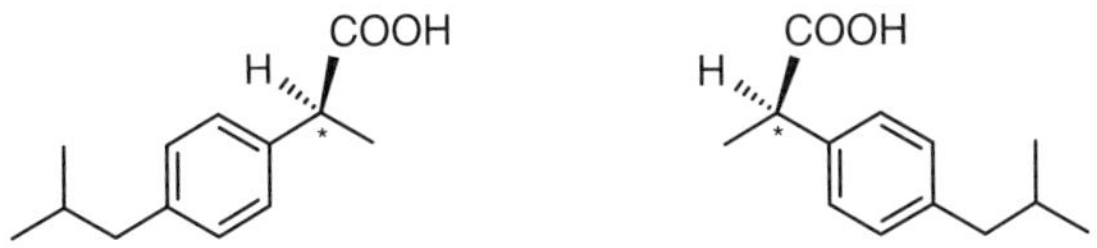

(S)-Ibuprofen
(An active stereoisomer)

(R)-Ibuprofen
(An inactive stereoisomer)

Similarly, another well-known NSAID, naproxen sodium is commonly used to reduce pain, fever and inflammation. It also has two stereoisomers, and only the S-enantiomer is an anti-inflammatory drug, but the R-enantiomer is a known liver toxin.

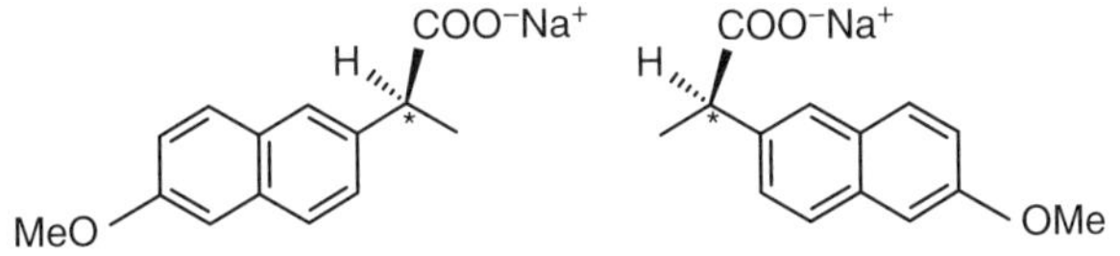

(S)-Naproxen sodium
(An anti-inflammatory drug)

(R)-Naproxen sodium
(A liver toxin)

In the early 1950s, Chemie Grunenthal, a German pharmaceutical company, developed a drug called thalidomide, which was marketed in 1957 under the name of contergan. It was prescribed to prevent nausea or morning sickness in pregnant women and soon it became an over-the-counter drug. The drug, however, caused severe adverse effects on thousands of babies, who were exposed to this drug while their mothers were pregnant. The drug caused 12 000 babies to be born with severe birth defects, including limb deformities such as missing or stunted limbs, and only 50% of them survived. Later, it was found that thalidomide molecule can exist in two stereoisomeric forms, one form is active as sedative but the other is responsible for its *teratogenic* activity (harmful effects on foetus).

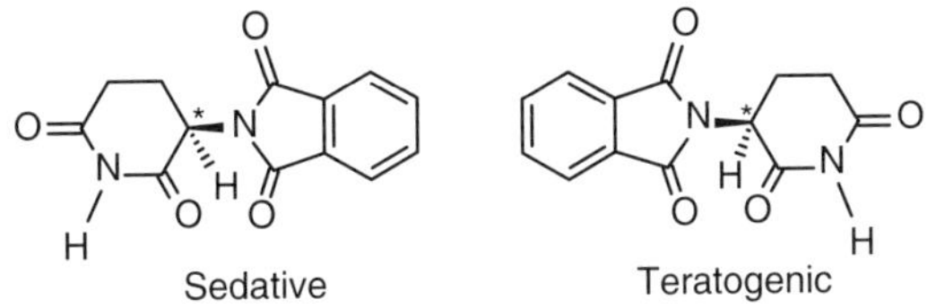

Sedative

Teratogenic

Thalidomide stereoisomers

Limonene is a monoterpene that occurs in citrus fruits. Two enantiomers of limonene produce two distinct flavours, (–)-limonene is responsible for the flavour of lemons and (+)-limonene for orange. Similarly, one enantiomeric form of carvone is the cause of caraway flavour, while the other enantiomer has the essence of spearmint.

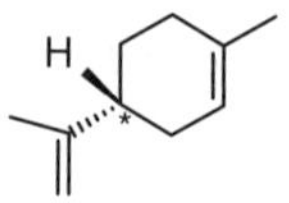

(+)-Limonene
(In orange)

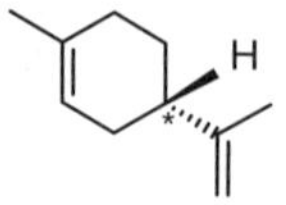

(−)-Limonene
(In lemon)

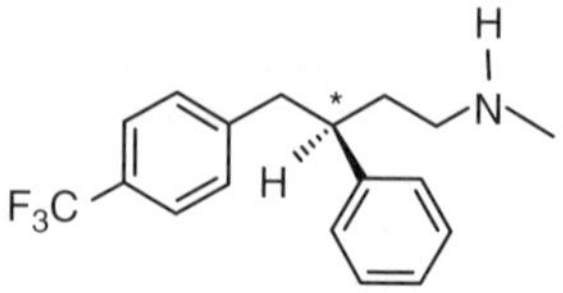

(S)-Fluoxetine
(Prevents migraine)

Fluoxetine, commonly known as Prozac®, as a racemic mixture is an antidepressant drug of the selective serotonin reuptake inhibitor (SSRI) class, but has no effect on migraine. The pure S-enantiomer works remarkably well in the prevention of migraine and is now under clinical evaluation. This drug was discovered by the company, Eli Lilly in 1972 and was marketed in 1986.

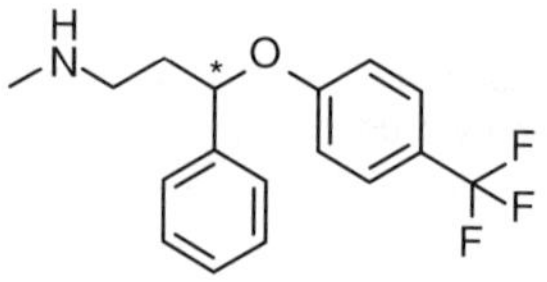

Fluxetine (commercial name: Prozac)

Salbutamol, salmeterol and terbutaline are sympathomimetic drugs that are selective β$_2$-adrenoreceptor agonists mainly used as bronchodilators in the treatment of asthma. They have long been marketed as a racemic mixture. However, only their (R)-(−)-isomer is effective and the other inactive (S)-(+)-isomer may be responsible for the occasional unpleasant side-effects associated with these drugs.

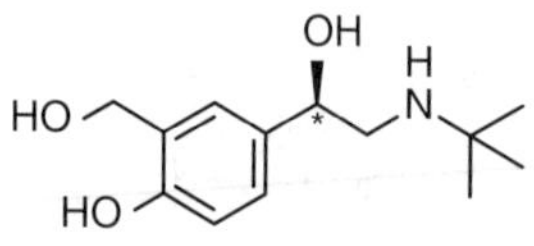

(R)-(−)-Salbutamol

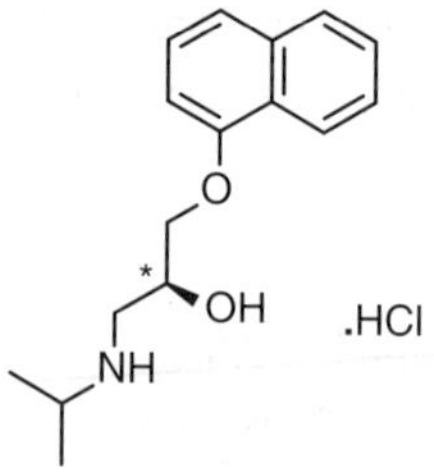

(S)-(−)-Propanolol hydrochloride

Levorotatory (L) isomer of all β-blockers is more potent in blocking β-adrenoreceptors than their dextrorotary (D) isomer; for example, (S)-(−)-propranolol is 100 times more active than its (R)-(+)-antipode. A number of β-blockers are still marketed as racemic forms such as acebutolol, atenolol, alprenolol, betaxolol, carvedilol, metoprolol, labetalol, pindolol and sotalol, except timolol and penbutolol are used as single L-isomers. D, L (racemic mixture) and D-propranolol can inhibit the conversion of thyroxin (T4) to triiodothyronin (T3), but not its L-form. Therefore, single D-propranolol might be used as a specific drug without β-blocking effects to reduce plasma concentrations of T3 particularly in patients suffering from hyperthyroidism in which racemic propranolol cannot be administered because of contraindications for β-blocking drugs.

Several calcium channel antagonists are used under racemic forms such as verapamil, nicardipine, nimodipine, nisoldipine, felodipine and mandipine; however, diltiazem is a diastereoisomer with two pairs of enantiomers. For example, the pharmacological potency of (S)-(–)-verapamil is 10–20 times greater than its (R)-(+)-isomer in terms of negative chromotropic effect on atrioventricular (AV) conduction and vasodilator. Methadone, a central-acting analgesic with high affinity for μ-opiod receptors, is prescribed for the treatment of opiate dependence and cancer pain. It is a chiral synthetic compound used in therapy as a racemic mixture. However, (R)-(–)-methadone is over 25-fold more potent as an analgesic than its (S)-(+) form.

(S)-(–)-Verapamil hydrochloride

(R)-(–)-Methadone or levamethadone

L-Dopa
L-3,4-dihydroxyphenylalanine

Several drugs are nowadays marketed as single enantiomeric forms solely because their other forms are toxic. For example, dopa (3,4-dihydroxyphenylalanine) is a precursor of dopamine that is effective in the treatment of Parkinson's disease. Dopa was used under racemic form D, L-dopa, but because of severe toxicity (agranulocytosis) of the D-isomer, only the L-form called L-Dopa (L-3,4-dihydroxyphenylalanine) is used in therapeutics.

3.5 SYNTHESIS OF CHIRAL MOLECULES

3.5.1 Racemic Forms

In many occasions, a reaction carried out with *achiral* reactants results in the formation of a chiral product. In the absence of any chiral influence, the outcome of such reactions is the formation of a racemic form. For example, hydrogenation of ethylmethylketone yields a racemic mixture of 2-hydroxybutane.

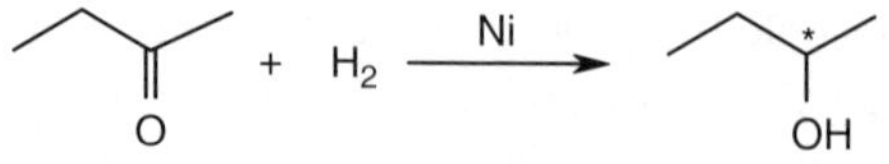

Ethylmethylketone $\quad\quad$ (±)-2-Hydroxybutane

Similarly, the addition of HBr to 1-butene produces a racemic mixture of 2-bromobutane.

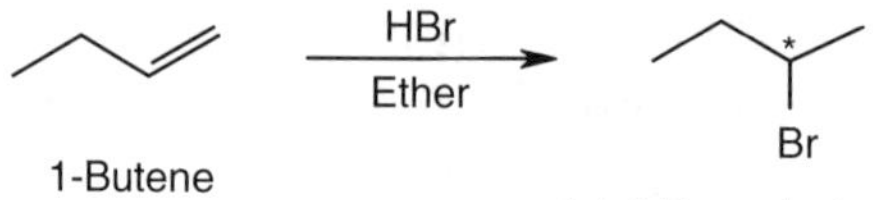

1-Butene

(±)-2-Bromobutane

3.5.2 Enantioselective Synthesis

A reaction that produces a predominance of one enantiomer over another is known as *enantioselective synthesis*.

To carry out an enantioselective reaction, a chiral reagent, solvent, or catalyst must assert an influence on the course of the reaction. In nature, most of the organic or bioorganic reactions are enantioselective, and the chiral influence generally comes from various *enzymes*.

Enzymes are chiral molecules and they possess an active site where the reactant molecules are bound momentarily during the reaction. The active site in any enzyme is chiral, and allows only one enantiomeric form of a chiral reactant to fit in properly. Enzymes are also used to carry out enantioselective reactions in the laboratories. *Lipase* is one such enzyme used frequently in labs. Lipase catalyses a reaction called *hydrolysis*, where an ester reacts with a molecule of water to produce a carboxylic acid and an alcohol. The use of lipase allows the hydrolysis to be used to prepare almost pure enantiomers.

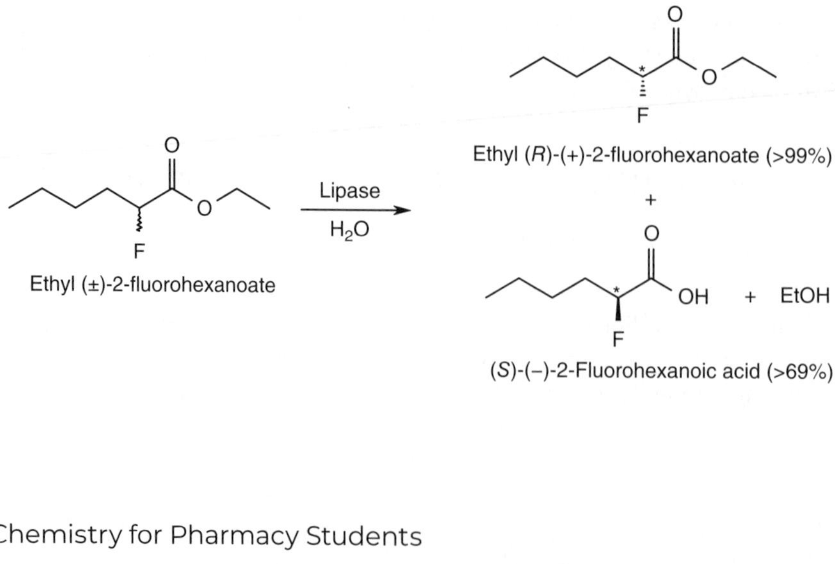

Ethyl (±)-2-fluorohexanoate

Ethyl (R)-(+)-2-fluorohexanoate (>99%)

(S)-(−)-2-Fluorohexanoic acid (>69%)

3.6 SEPARATION OF STEREOISOMERS: RESOLUTION OF RACEMIC MIXTURES

In nature, often only one enantiomer is produced. Living organisms such as plants and animals are the best sources of optically active compounds, but in organic synthesis it is different and often extremely difficult to obtain only one enantiomer. A number of compounds exist as racemic mixtures (±), that is, a mixture of equal amounts of two enantiomers, (–) and (+). Often, one enantiomer shows medicinal properties. Therefore, it is important to purify the racemic mixture so that active enantiomer can be obtained. The separation of a mixture of enantiomers is called the *resolution of a racemic mixture*. Enantiomers have the same physical properties (boiling points, melting points and solubilities), but they differ in chirality, so a chiral probe must be used for such a separation.

Through luck, in 1848, Louis Pasteur was able to separate or resolve racemic tartaric acid into its (+) and (–) forms by *crystallization*. Two enantiomers of the sodium ammonium salt of tartaric acid give rise to two distinctly different types of chiral crystals that can then be separated easily. However, only a very few organic compounds crystallize into separate crystals (of two enantiomeric forms) that are visibly chiral as the crystals of the sodium ammonium salt of tartaric acid.

Therefore, Pasteur's method of separation of enantiomers is not generally applicable to the separation of enantiomers.

One of the current methods for resolution of enantiomers is to react the racemic mixture with an enantiomerically pure compound. This reaction changes a racemic form into a mixture of diastereomers. As diastereomers have different boiling points, melting points and solubilities, they can be separated by conventional means, for example, recrystallization and chromatography. For example, alcohols react with the enantiomerically pure tartaric acid to give two diastereomeric esters. These esters are separated and then acid hydrolysis cleaves each ester back to an optically active alcohol and carboxylic acid. Later, the resolving agents, for example, tartaric acids, are recovered and recycled as they are expensive.

Resolution of a racemic mixture can also be achieved by using an enzyme. An enzyme selectively converts one enantiomer in a racemic mixture to another compound, after which the unreacted enantiomer and the new compound are separated. For example, lipase is used in the hydrolysis of chiral esters. Similar results can be achieved in the laboratory using chiral catalysts or reagents, although enantiomerically pure catalysts or reagents are expensive.

Among the recent instrumental methods, *chiral chromatography* can be used to separate enantiomers. The most commonly used chromatographic technique is chiral high performance liquid chromatography (HPLC).

Diastereomeric interaction between molecules of the racemic mixture, and the chiral chromatography medium causes enantiomers of the racemate to move through the stationary phase at different rates. Chiral HPLC columns have now become quite popular for the separation of chiral compounds.

3.7 COMPOUNDS WITH STEREOCENTRES OTHER THAN CARBON

Silicon (Si) and germanium (Ge) are in the same group of the periodic table as carbon and they form tetrahedral compounds as a carbon does.

When four different groups are situated around the central atom in silicon, germanium and nitrogen compounds, the molecules are chiral. Sulphoxides, where one of the four groups is a nonbonding electron pair, are also chiral.

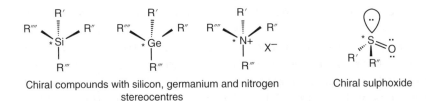

Chiral compounds with silicon, germanium and nitrogen stereocentres

Chiral sulphoxide

3.8 CHIRAL COMPOUNDS THAT DO NOT HAVE FOUR DIFFERENT GROUPS

A molecule is chiral if it is not superimposable on its mirror image. A tetrahedral atom with four different groups is just one of the factors that confer chirality on a molecule. There are a number of molecules where a tetrahedral atom with four different groups is not present, yet they are not superimposable, that is, chiral. For example, 1,3-dichloroallene is a chiral molecule, but it does not have a tetrahedral atom with four different groups.

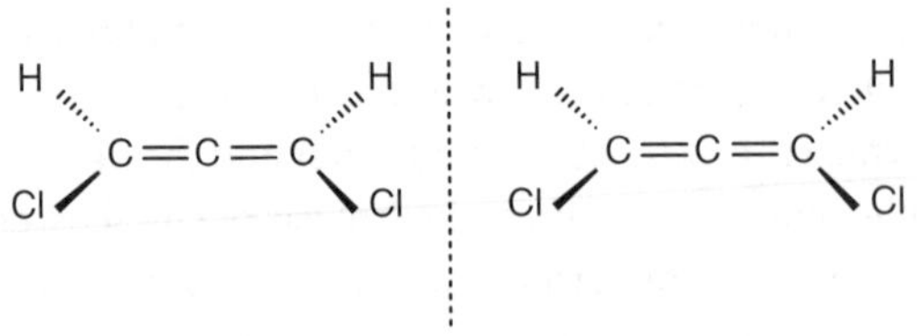

1,3-Dichloroallene

An allene is a hydrocarbon in which one atom of carbon is connected by double bonds with two other atoms of carbon. Allene also is the common name for the parent compound of this series, 1,2-propadiene. The planes of the π bonds of allenes are perpendicular to each other.

This geometry of the π bonds causes the groups attached to the end carbon atoms to lie in perpendicular planes. Because of this geometry, allenes with different substituents on the end carbon atoms are chiral. However, allenes do not show *cis–trans* isomerism.

Non-superimposability of the mirror images is a necessary and sufficient condition for chirality (optical activity), and this non-superimposability can be achieved by a compound in a number of ways, for example, restricted rotation along bonds (atropisomerism). Simply, if the substituents prevent the mirror image from being superimposable the compound attains chirality in spite of not having a defined chiral centre. Like allenes as shown before, some substituted biphenyl compounds do not have any chiral centres in them, yet they are chiral and optically active because of restricted rotation as shown next.

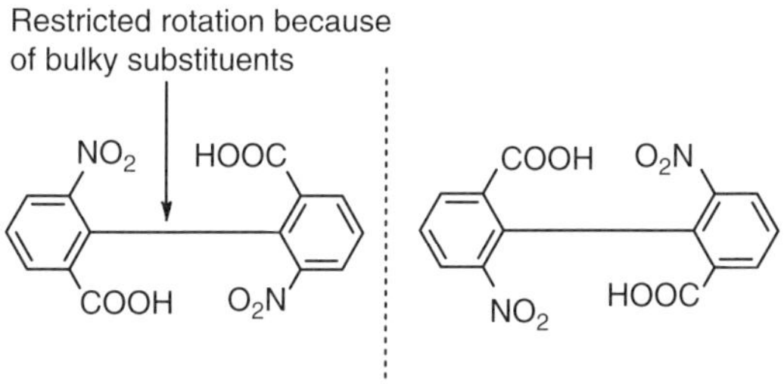

Optically active biphenyl compound
(Non-superimposable)

For biphenyl compounds to be optically active, there are two major conditions to be met. Firstly, the substituent in the *ortho* position must have a large size. If three bulky groups present on *ortho* position, they render restriction in rotation. The groups are large enough to interfere mechanically, that is, to behave as obstacles to restrict free rotation about the single bond. Thus, two benzene rings cannot be co-planner. Second, resolvable biphenyls must contain different *ortho* substitutions on each ring. If one or both rings contain two identical substituents, the molecule will not be chiral as shown in the following example. Plane of symmetry must be absent in biphenyls.

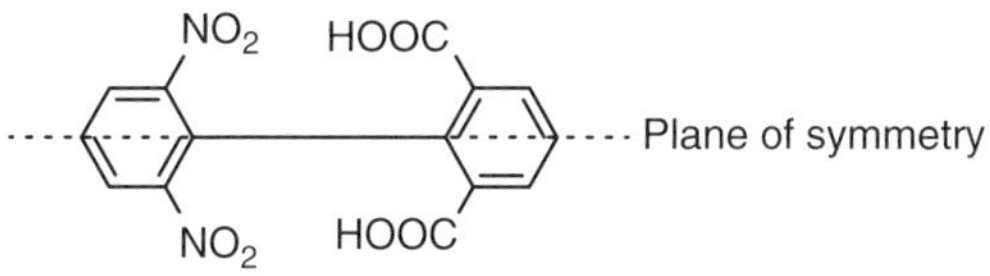

Optically inactive biphenyl compound

In this biphenyl compound, since two substituted groups are the same that is two nitro groups on the same phenyl ring and two carboxylic acid groups on the other phenyl ring, a plane of symmetry exists in this molecule and this compound cannot be optically active.

Chapter 4

Organic Functional Groups

Learning Objectives

After completing this chapter, students should be able to:

- recognize various organic functional groups;

- outline the properties, preparations, reactivity and reactions of various organic functional groups;

- discuss the importance of organic functional groups in determining drug action and toxicity;

- describe the significance of organic functional groups in determining stability of drug molecules.

4.1 ORGANIC FUNCTIONAL GROUPS: DEFINITION AND STRUCTURAL FEATURES

All organic compounds are grouped into classes based on characteristic features called *functional groups*. A functional group is an atom or a group of atoms within a molecule that serves as a site of chemical reactivity. Simply, functional groups are structural units in organic molecules that are portrayed by specific bonding arrangements between specific atoms. Carbon combines with other atoms such as H, N, O, S and halogens to form functional groups. The most important functional

Chemistry for Pharmacy Students: General, Organic and Natural Product Chemistry,
Second Edition. Lutfun Nahar and Satyajit Sarker.
© 2019 John Wiley & Sons Ltd. Published 2019 by John Wiley & Sons Ltd.

groups are shown in the following table, with the key structural elements and a simple example.

Name	General structure	Example
Alkane	R–H	$CH_3CH_2CH_3$ Propane
Alkene	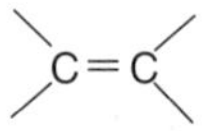 $\diagdown C = C \diagup$	$CH_3CH_2CH=CH_2$ 1-Butene
Alkyne	$—C≡C—$	$H—C≡C—H$ Ethyne
Aromatic	 $C_6H_5—= Ph = Ar$	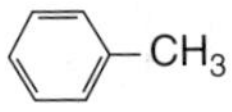 $\bigcirc—CH_3$ Toluene
Alkyl halide (haloalkane)	R-F, R-Cl, R-Br, R-I	$CH—Cl_3$ Chloroform
Alcohol	R–OH (R is never H)	$CH_3CH_2—OH$ Ethanol (ethyl alcohol)
Ether	R–OR (R is never H)	$CH_3CH_2—O—CH_2CH_3$ Diethyl ether
Thiol (mercaptan)	R–SH (R is never H)	$(CH_3)_3C—SH$ tert-Butyl thiol
Thioether (sulphide)	R–S–R (R is never H)	$CH_3CH_2—S—CH_3$ Methylthioethane
Amine	RNH_2, R_2NH, R_3N	$(CH_3)_2—NH$ Dimethyl amine
Aldehyde	$\overset{\text{O}}{\overset{\|}{R—C—H}}$ RCHO	$\overset{\text{O}}{\overset{\|}{H_3C—C—H}}$ Acetaldehyde
Ketone	$\overset{\text{O}}{\overset{\|}{R—C—R}}$ RCOR	$\overset{\text{O}}{\overset{\|}{H_3C—C—CH_3}}$ Acetone

Name	General structure	Example
Carboxylic acid	$\begin{matrix} O \\ \parallel \\ R-C-OH \end{matrix}$ RCO_2H	$\begin{matrix} O \\ \parallel \\ H_3C-C-OH \end{matrix}$ Acetic acid
Acid chloride	$\begin{matrix} O \\ \parallel \\ R-C-Cl \end{matrix}$ $RCOCl$	$\begin{matrix} O \\ \parallel \\ H_3C-C-Cl \end{matrix}$ Ethanoyl chloride *Acetyl chloride*
Acid anhydride	$\begin{matrix} O \quad\quad O \\ \parallel \quad\quad \parallel \\ R-C-O-C-R \end{matrix}$ $(RCO)_2O$	$\begin{matrix} O \quad\quad\quad O \\ \parallel \quad\quad\quad \parallel \\ H_3C-C-O-C-CH_3 \end{matrix}$ Acetic anhydride
Ester	$\begin{matrix} O \\ \parallel \\ R-C-OR \end{matrix}$ RCO_2R	$\begin{matrix} O \\ \parallel \\ H_3C-C-O-C_2H_5 \end{matrix}$ Ethyl acetate
Amide	$\begin{matrix} O \\ \parallel \\ R-C-NH_2 \end{matrix}$ $RCONH_2$	$\begin{matrix} O \\ \parallel \\ H_5C_2-C-NH_2 \end{matrix}$ Propanamide
Nitrile	$R-C\equiv N$ RCN	$CH_3-C\equiv N$ Acetonitrile

R = hydrocarbon group such as methyl or ethyl and can sometimes be H or phenyl. Where two R groups are shown in a single structure, they do not have to be the same, but they can be.

A *reaction* or *chemical reaction* is the chemical process by which one compound is transformed into a new compound. Thus, functional groups are important in chemical reactions. It is important to recognize these functional groups, because they dictate the physical, chemical and other properties of organic molecules including various drug molecules and drug interactions. Some examples of drug molecules containing various functional groups may include fluoxetine, commonly known as Prozac® (an antidepressant), aspirin (an analgesic, antipyretic and anti-inflammatory) and terbutaline (an inhaler that relieves asthma condition).

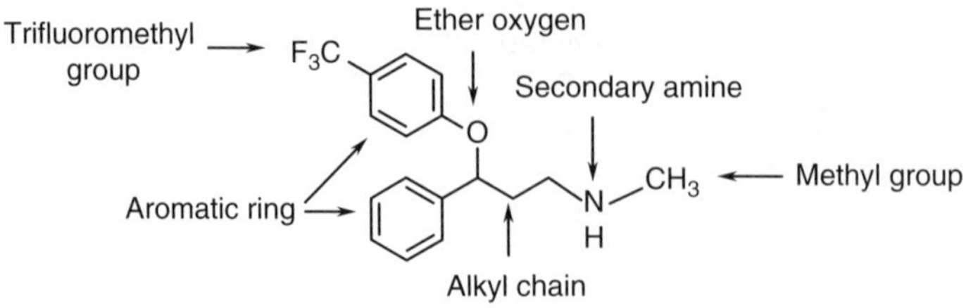

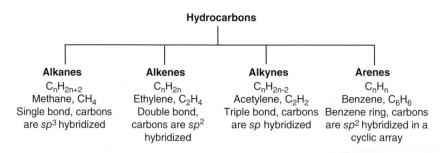

Trifluoromethyl group

Ether oxygen

Secondary amine

Methyl group

Aromatic ring

Alkyl chain

Fluoxetine
(Prozac®, an antidepressant)

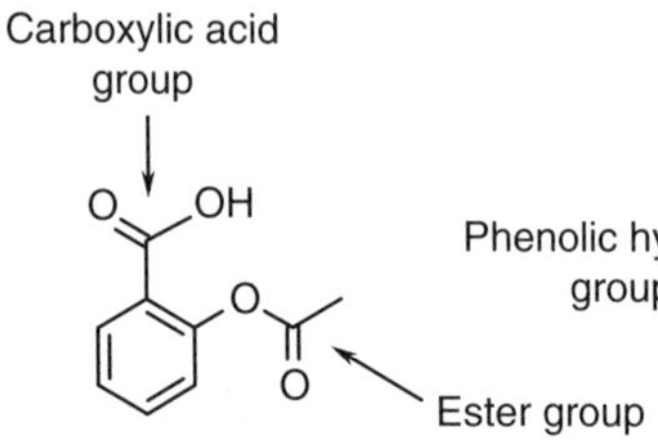

Carboxylic acid group

Alcoholic hydroxyl group

Phenolic hydroxyl group

Ester group

tert-Butyl group

Aspirin
(An analgesic, antipyretic and anti-inflammatory)

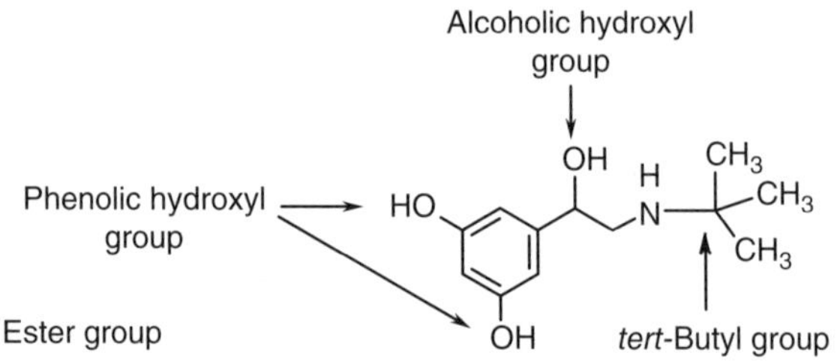

Terbutaline
(Relieves asthma condition)

4.2 HYDROCARBONS

Hydrocarbons are compounds that only contain carbon and hydrogen atoms, and they can be classified as follows depending on the bond types that are present within the molecules.

Hydrocarbons

Alkanes	Alkenes	Alkynes	Arenes
C_nH_{2n+2}	C_nH_{2n}	C_nH_{2n-2}	C_nH_n
Methane, CH_4	Ethylene, C_2H_4	Acetylene, C_2H_2	Benzene, C_6H_6
Single bond, carbons are sp^3 hybridized	Double bond, carbons are sp^2 hybridized	Triple bond, carbons are sp hybridized	Benzene ring, carbons are sp^2 hybridized in a cyclic array

4.3 ALKANES, CYCLOALKANES AND THEIR DERIVATIVES

4.3.1 Alkanes

Hydrocarbons that have no double or triple bond are classified as *alkanes*. The carbon atoms in alkanes can be arranged in chains (*alkanes or acyclic alkanes*) or in rings (*cycloalkanes*) (see Section 4.3.2). Acyclic alkanes have the general molecular formula C_nH_{2n+2}. All alkanes, made of carbon and hydrogen only, are also known as *saturated hydrocarbons* or *aliphatic hydrocarbons*. Thus, the alkane family is characterized by the presence of tetrahedral carbon (sp^3) atoms. Methane (CH_4) and ethane (C_2H_6) are the first two members of the alkane family.

A group derived from an alkane by removal of one of its hydrogen atoms is known as an *alkyl group*. For example, the methyl group (CH_3—) from methane (CH_4) and the ethyl group (CH_3CH_2—) from ethane (CH_3CH_3).

| Methane | Methyl group | Ethane | Ethyl group |

4.3.1.1 The IUPAC Nomenclature of Alkanes

Generally, organic compounds are given systematic names by using the order: prefix-parent-suffix, where prefix indicates how many branching groups present, parent indicates how many carbons in the longest chain, and suffix indicates the name of the family. Common names as well as systematic names are used for alkanes and their derivatives. However, it is advisable to use systematic names or the IUPAC (*International Union of Pure and Applied Chemistry*) nomenclature, which can be derived from a simple set of rules.

The IUPAC naming of the alkanes is based on a prefix indicating the number of carbon atoms in the chain (as shown) followed by the suffix -*ane*. For example, if a chain contains three carbons, the parent name is *propane*, if four carbons, the parent name is *butane* and so on. The remaining parts of the structure are treated as substituents on the chain. Numbers are used to indicate the positions of the substituents on the parent carbon chain as shown.

Prefix	Number of carbon atoms	Prefix	Number of carbon atoms
Meth-	1	Hept-	7
Eth-	2	Oct-	8
Prop-	3	Non-	9
But-	4	Dec-	10
Pent-	5	Undec-	11
Hex-	6	Dodec-	12

First, one has to identify and name the groups attached to the chain, and number the chain so that the substituent gets the lowest possible number. For example, one of the isomers of pentane is 2-methylbutane, where the parent chain is a four-carbon butane chain and is numbered starting from the end nearer the substituent group (methyl group). Therefore, the methyl group is indicated as being attached to carbon atom number 2.

$$CH_3$$
$$H_3C-\underset{1}{C}-\underset{2}{CH_2}-\underset{3}{CH_2}-\underset{4}{CH_3}$$
$$H$$

Systematic name: 2-Methylbutane

$$CH_3$$
$$H_3C-\underset{1}{C}-\underset{2}{CH_3}$$
$$H$$

Systematic name: 2-Methylpropane
Common name: Isobutane

Similarly, isobutane is the common name of one of the structural isomers of C_4H_{10} (butane). The longest continuous chain of carbons consists of three atoms in length, so the systematic name is based on propane. Finally, since a methyl group appears on the second carbon, the correct name is 2-methylpropane.

When multiple substituents are present, the location of each substituent should be designated by an appropriate name and number. The presence of two or more identical substituents is indicated by the prefixes di, tri, tetra and so on, and the position of each substituent is indicated by a number in the prefix.

A number and a word are separated by a hyphen, and numbers are separated by comma. For example, in 2,2-dimethylbutane, both methyl groups are attached to carbon atom 2 of a butane chain. The names of the substituents are arranged in alphabetical order, not numerical order; for example, 3-ethyl-2-methylhexane, *not* 2-methyl-3-ethylhexane.

$$CH_3$$
$$H_3C-\underset{2}{C}-\underset{3}{CH_2}-\underset{4}{CH_3}$$
$$CH_3$$

Systematic name: 2,2-Dimethylbutane

$$CH_3$$
$$CH_3\underset{1}{CH}-\underset{2}{CH}-\underset{3}{CH}-\underset{4}{CH_2}-\underset{5}{CH_2}\underset{6}{CH_3}$$
$$C_2H_5$$

Systematic name: 3-Ethyl-2-methylhexane

4.3.1.2 Isomerism and Physical Properties

Compounds that differ from each other in their molecular formulas by the unit $-CH_2-$ are called members of *homologous series*. Thus, methane and ethane belong to a homologous series of saturated hydrocarbons. Compounds, which have same molecular formula but different order of attachment of their atoms, are called *constitutional isomers* (see Section 3.2.1). For the molecular formulas CH_4, C_2H_6 and C_3H_8, only one order of attachment of atoms is possible.

The molecular formula C_4H_{10} gives rise to two different structural formulas in which four carbon atoms and 10 hydrogen atoms can be connected to each other in the following ways. These structures also can be drawn using line drawings, where *zigzag* lines represent carbon chains. Butane (C_4H_{10}) can exist in two different isomeric forms, for example, n-butane and isobutane (2-methylpropane), known as constitutional isomers (see Section 3.2.1). Their structures differ in connectivities and they are different compounds. They have different physical properties; for example, different boiling points.

n-Butane (bp: 0.6°C)　　　　　　　Isobutane (bp: 10.2°C)

Alkanes have similar chemical properties, but their physical properties vary with molecular weight and the shape of the molecule. Alkanes are *nonpolar hydro- carbon*, and are called *hydrophobic* (water hating), since they are not soluble in water (see Section 2.5).

The low polarity of all the bonds in alkanes means that the only intermolecular forces between molecules of alkanes are the weak *dipole–dipole forces* (see Section 2.4.1), which are easily overcome. Dipole–dipole forces are attractive forces between the positive end of one polar molecule and the negative end of another polar molecule, and are much weaker than ionic or covalent bonds. As a result, compared to other functional groups, alkanes have low melting and boiling points, and low solubility in polar solvents, for example, water, but high solubility in nonpolar solvents, for example, hexane and dichloromethane (DCM).

Name	Condensed structure	Molecular formula	Molecular weight	bp (°C)
Methane	CH_4	CH_4	16.04	−161.5
Ethane	CH_3CH_3	C_2H_6	30.07	−88.6
Propane	$CH_3CH_2CH_3$	C_3H_8	44.10	−42.1
Butane	$CH_3(CH_2)_2CH_3$	C_4H_{10}	58.12	−0.6
Pentane	$CH_3(CH_2)_3CH_3$	C_5H_{12}	72.15	36.1
Hexane	$CH_3(CH_2)_4CH_3$	C_6H_{14}	86.18	68.9
Heptane	$CH_3(CH_2)_5CH_3$	C_7H_{16}	100.20	98.4
Octane	$CH_3(CH_2)_6CH_3$	C_8H_{18}	114.23	125.7
Nonane	$CH_3(CH_2)_7CH_3$	C_9H_{20}	128.26	150.8
Decane	$CH_3(CH_2)_8CH_3$	$C_{10}H_{22}$	142.28	174.1

The more the molecular weight, the higher the boiling point of alkanes down the homologous series. Alkanes from methane to butane are colourless gases at room temperature and atmospheric pressure. Generally, the physical state changes from gas to liquid to solid as the molecular weight increases. Branching of alkanes also lowers the boiling points. Thus, among the isomers of pentane, neopentane (bp: 9.5°C) has the lowest boiling point.

$$CH_3CH_2CH_2CH_2CH_3$$

Pentane
bp: 36.1°C

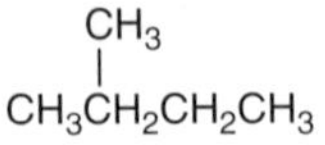

$$
\begin{array}{c}
CH_3 \\
| \\
CH_3CH_2CH_2CH_3
\end{array}
$$

Isopentane
bp: 27.9°C

$$
\begin{array}{c}
CH_3 \\
| \\
CH_3CH_2CH_3 \\
| \\
CH_3
\end{array}
$$

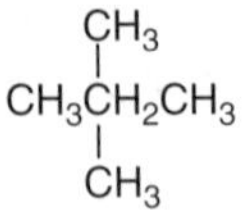

Neopentane
bp: 9.5°C

4.3.1.3 Structure and Conformation of Alkanes

Chemists use two parameters, bond lengths and bond angles, to describe the 3D-structures of covalently bonded compounds. A *bond length* is the average distance between the nuclei of the atoms that are covalently bonded together. A *bond angle* is the angle formed by the interaction of two covalent bonds at the atom common to both.

Alkanes have only sp^3-hybridized carbons. The conformation of alkanes is discussed in Chapter 3 (see Section 3.2.2.1). Methane (CH_4) is a nonpolar molecule, and has four covalent C—H bonds. In methane, all four C—H bonds have the same length (1.10 Å), and all the bond angles (109.5°) are the same. Therefore, all four covalent bonds in methane are identical. Three different ways to represent a methane molecule are shown next. In a perspective formula, bonds in a plane of the paper are drawn as solid lines, bonds sticking out of the plane of the paper towards you are drawn as solid wedges, and those on the back from the plane of the paper away from you are drawn as broken wedges.

CH_4

Condensed formula
Methane

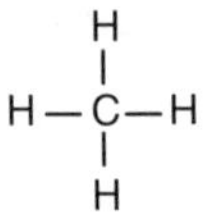

$$
\begin{array}{c}
H \\
| \\
H-C-H \\
| \\
H
\end{array}
$$

Lewis structure
Methane

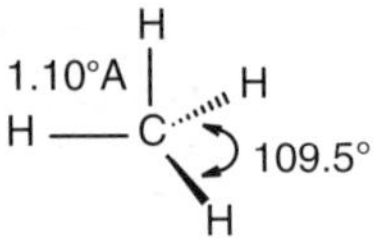

Perspective formula
Methane

One of the hydrogen atoms in CH_4 are replaced by another atom or group to afford a new derivative, such as alkyl halide or alcohol. Chloromethane (CH_3Cl) is a compound in which one of the hydrogen atoms in CH_4 is substituted by a Cl atom. Chloromethane (methyl chloride) is an alkyl halide, where the hydrocarbon part of the molecule is a methyl group (CH_3-). Similarly, in methanol (CH_3OH), one of the hydrogen atoms of CH_4 is replaced by an OH group.

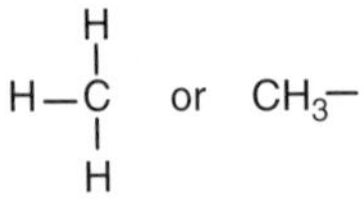

$$
\begin{array}{c}
H \\
| \\
H-C \\
| \\
H
\end{array}
\quad \text{or} \quad CH_3-
$$

Methyl group
(An alkyl group)

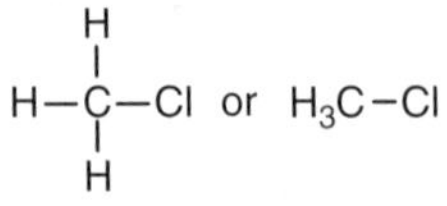

$$
\begin{array}{c}
H \\
| \\
H-C-Cl \\
| \\
H
\end{array}
\quad \text{or} \quad H_3C-Cl
$$

Chloromethane or methyl chloride
(An alkyl halide)

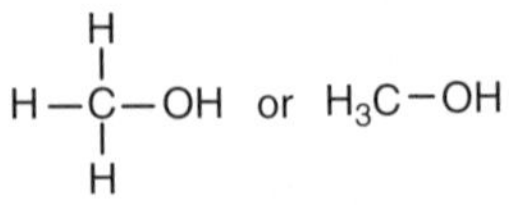

$$
\begin{array}{c}
H \\
| \\
H-C-OH \\
| \\
H
\end{array}
\quad \text{or} \quad H_3C-OH
$$

Methanol or methyl alcohol
(An alcohol)

In these examples, the first name given for each compound is its systematic name, and the second is the common name. The name of an alkyl group is obtained by changing the suffix -*ane* to -*yl*. Thus, the methyl group (CH_3—) is derived from methane (CH_4) and ethyl group (C_2H_5—) from ethane (C_2H_6) and so on. Sometimes, an alkane is represented by the symbol RH, the corresponding alkyl group is symbolized by R—.

4.3.1.4 Classification of Carbon Substitution

Carbon atoms in alkanes are sometimes labelled as primary (1°), secondary (2°), tertiary (3°) and quaternary (4°) carbon. This classification is based on the number of carbon atoms attached to the given carbon atom. A carbon atom bonded to only one carbon atom is known as 1°; when bonded to two carbon atoms, it is 2°; when bonded to three carbon atoms, it is 3° and when bonded to four carbon atoms, it is known as 4°. Different types of carbon atoms are shown in the following compound.

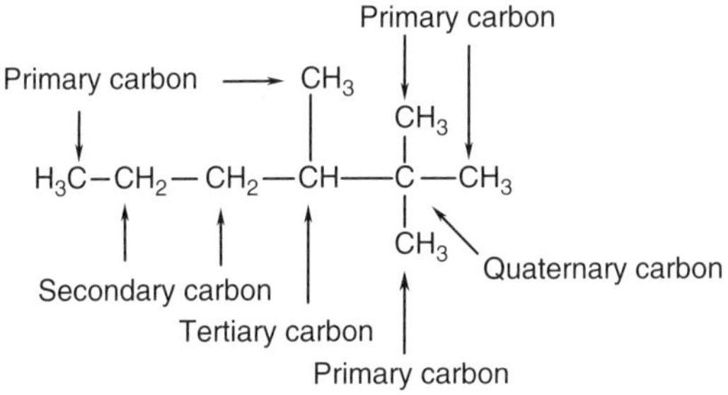

4.3.1.5 Natural Sources of Alkanes

The principal source of alkanes is naturally occurring petroleum and gas. Natural gas is rich in methane, it also contains ethane and propane with some other low molecular weight alkanes. Petroleum is a complex mixture of hydrocarbons with almost half of which are alkanes and cycloalkanes. Another fossil fuel, coal, is another rich source of alkanes. Usually alkanes are produced through refinement or hydrogenation of petroleum and coal.

4.3.1.6 Preparation of Alkanes

Alkanes are simply prepared by *catalytic hydrogenation* of alkenes or alkynes (see Sections 5.4.1.1 and 5.4.1.2). The most commonly used metal catalysts include Pt—C and Pd—C.

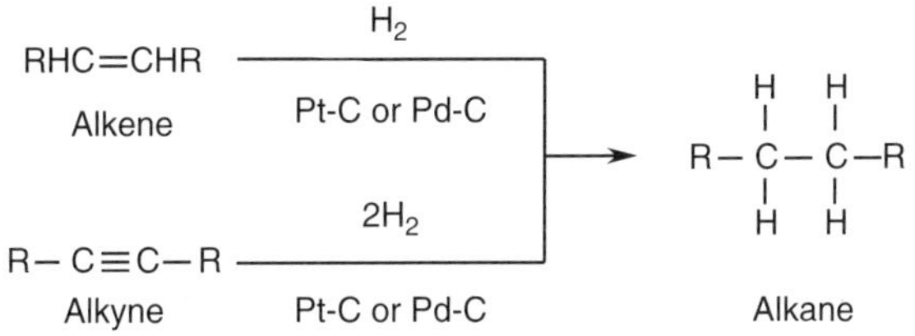

Alkanes are also prepared from alkyl halides by reduction, directly with lithium aluminium hydride (LiAlH$_4$) in dry tetrahydrofuran (THF) or zinc (Zn) and acetic acid (AcOH) (see Section 5.9.11) or via the Grignard reagent formation followed by hydrolytic work-up (see Section 5.9.12). A primary alkyl halide with Gilman reagent (R'$_2$CuLi, lithium organocuprate) also provides alkane, but the reaction is limited to 1° alkyl halide (see Section 5.6.2.1). This reaction is known as the *Corey–House reaction*.

$$
\begin{array}{ccccc}
\text{R—H} & \xleftarrow{\text{LiAlH}_4, \text{ THF or}} & \text{R—X} & \xrightarrow{\text{i. R}_2'\text{CuLi, ether}} & \text{R—R'} \\
\text{Alkane} & \text{Zn, AcOH} & \text{Alkyl halide} & \text{ii. H}_2\text{O} & \text{Alkane}
\end{array}
$$

i. Mg, dry ether
ii. H$_2$O

R—H
Alkane

Selective reduction of acyl benzenes as well as aldehydes or ketones, either by Clemmensen reduction (*see* Section 5.9.15) or Wolff–Kishner reduction (see Section 5.9.16), also yields alkanes.

$$
\begin{array}{ccc}
\overset{\displaystyle O}{\underset{\displaystyle \parallel}{\text{R—C}}}\text{—Y} & \xrightarrow[\text{NH}_2\text{NH}_2, \text{ NaOH}]{\text{Zn(Hg) in HCl or}} & \text{R—CH}_2\text{—Y} \\
\text{Y = H or R} & & \text{Alkane} \\
\text{Aldehyde or ketone} & &
\end{array}
$$

4.3.1.7 Reactivity of Alkanes

Alkanes are quite unreactive towards most reagents. In fact, the alkanes and cycloalkanes, with the only exception of cyclopropane, are perhaps the least chemically reactive class of organic compounds. Simply none of the C—H or C—C bonds in a typical saturated hydrocarbon are attacked by a strong acid, such as sulphuric acid (H$_2$SO$_4$), or by an oxidizing agent, such as bromine (in the dark), oxygen or potassium permanganate (KMnO$_4$) or by usual metal catalytic hydrogenation such as platinum (Pt), palladium (Pd) or nickel (Ni) at room temperatures.

4.3.1.8 Reactions of Alkanes

Alkanes contain only strong σ bonds (strong carbon-carbon single bonds and strong carbon-hydrogen bonds) and all the bonds (C—C and C—H) are nonpolar. As a result, alkanes are quite unreactive towards most reagents, although reactions do occur under severely forcing conditions. In fact, it is often convenient to regard the hydrocarbon framework of a molecule as an unreactive support for the more reactive functional groups. More branched alkanes are more stable and less reactive than linear alkanes. For example, isobutane is more stable than *n*-butane. Alkanes can be burned, destroying the entire molecule; they can react with some of the halogens, breaking carbon-hydrogen bonds and alkanes can crack by breaking the carbon-carbon bonds.

4.3.1.8.1 Combustion or Oxidation of Alkanes

Alkanes react with O_2 under certain conditions. For example, alkanes undergo combustion reaction with oxygen at high temperatures to give carbon dioxide (CO_2) and water. This is why alkanes are good fuels. Oxidation of saturated hydrocarbons is the basis for their use as energy sources for heat, for example, natural gas, liquefied petroleum gas (LPG) and fuel oil, and for power, for example, gasoline, diesel fuel and aviation fuel. However, incomplete combustion (where there is not enough oxygen) of alkanes forms carbon or carbon monoxide (CO), which is a colourless and tasteless poisonous gas.

After carbon monoxide is breathed in (CO poisoning), it enters the bloodstream and mixes with haemoglobin (the part of red blood cells that carry oxygen around the body), to form carboxyhaemoglobin, restricting or preventing altogether the blood's ability to carry oxygen. This lack of oxygen resulting from CO poisoning causes the living cells and tissues to fail with fatal consequences.

4.3.1.8.2 Halogenation of Alkanes: Free-Radical Chain Reaction

Halogenation of alkanes is a substitution reaction, where a hydrogen atom is replaced by a halogen atom. Alkanes react with halogen molecules (F_2, Cl_2, Br_2 or I_2) under UV light or at high temperatures to provide a mixture of alkyl halides (see Section 4.3.3.3). This reaction is called a *free-radical chain reaction* (see Section 5.3.1).

$$CH_4 + X_2 \xrightarrow{h\nu} CH_3X + CH_2X_2 + CHX_3 + CX_4$$
$$X = F, Cl, Br \text{ or } I$$

4.3.1.8.3 Catalytic Dehydrogenation of Alkanes

Catalytic dehydrogenation converts unreactive alkanes to electron-rich highly reactive alkenes. For example, *n*-pentane and isopentane can be converted to *n*-pentene and isopentene using chromium (III) oxide (Cr_2O_3) as a catalyst at 500 °C.

n-Pentane $\xrightarrow[500°C]{Cr_2O_3}$ *n*-Pentene   Isopentane $\xrightarrow[500°C]{Cr_2O_3}$ Isopentene

4.3.2 Cycloalkanes

Cycloalkanes are alkanes that are cyclic with the general formula C_nH_{2n}. The simplest members of this cyclic hydrocarbon class consist of a single, unsubstituted carbon ring and this forms a homologous series similar to the unbranched alkanes. The smallest cycloalkane is cyclopropane. There are also polycyclic alkanes, which are molecules that contain two or more cycloalkanes that are joined, forming multiple rings. The C_3 to C_6 simple cycloalkanes with their structural representations are shown.

Name	Molecular formula	Structural formula	Name	Molecular formula	Structural formula
Cyclopropane	C_3H_6		Cyclopentane	C_5H_{10}	
Cyclobutane	C_4H_8		Cyclohexane	C_6H_{12}	

4.3.2.1 Nomenclature of Cycloalkanes

The nomenclature of cycloalkanes is almost the same as that of alkanes, with the exception that the prefix *cyclo-* is to be added to the name of the alkane. When a substituent is present on the ring, the name of the substituent is added as a prefix to the name of the cycloalkane. No number is required for rings with only one substituent.

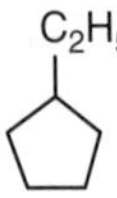

Ethylcyclopentane 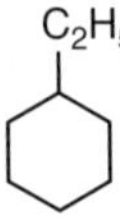   Ethylcyclohexane 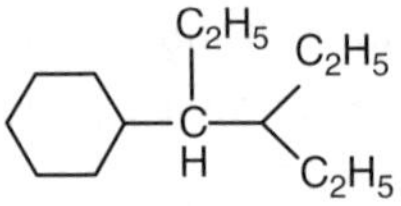   (1,1-Diethylbutyl)cyclohexane

However, if two or more substituents are present on the ring, numbering starts from the carbon that has the group of alphabetical priority, and proceeds around the ring so as to give the second substituent the lowest number.

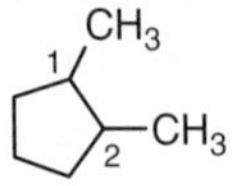

1,2-Dimethylcyclopentane
not 1,5-diethylcyclopentane

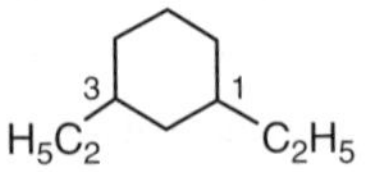

1,3-Diethylcyclohexane
not 1,5-diethylcyclohexane

When the number of carbons in the ring is greater than or equal to the number of carbons in the longest chain, the compound is named a cycloalkane. However, if an alkyl chain of the cycloalkane has a greater number of carbons, then the alkyl chain is used as the parent, and the cycloalkane is a cycloalkyl substituent.

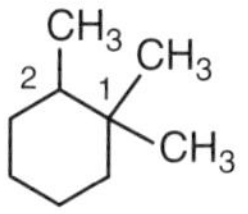

1,1,2-Trimethylcyclohexane
not 1,2,2-Trimethylcyclohexane

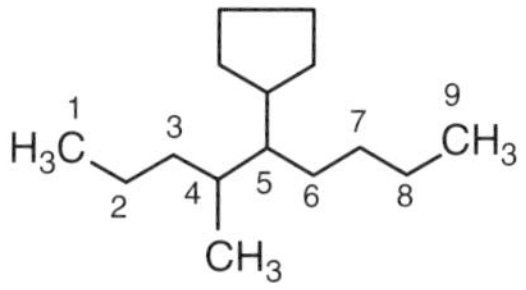

5-Cyclopentyl-4-methylnonane
not 5-Cyclopentyl-6-methylnonane

4.3.2.2 Geometrical Isomerism in Cycloalkanes

Open chain alkanes have free rotation in their C—C bonds, but cycloalkanes cannot undergo free rotation due to the cyclic structures. Thus, substituted cycloalkanes can give rise to *cis* and *trans* isomers, known as geometrical isomers (see Section 3.3.3). For example, 1,2-diethylcyclopentane can exist as geometrical isomers, *cis*-1,2-diethylcyclopentane and *trans*-1,2-diethylcyclopentane.

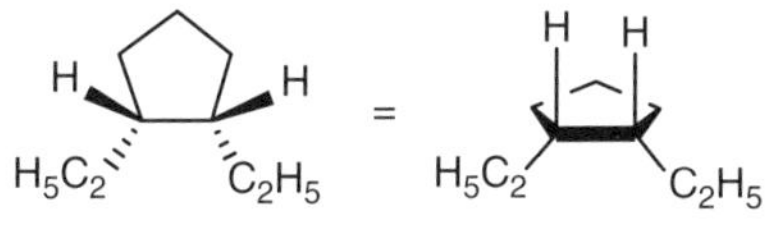

cis-1,2-Diethylcyclopentane

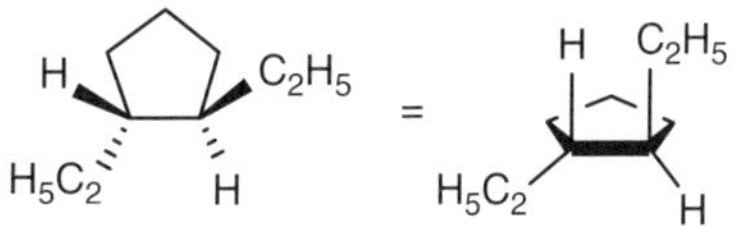

trans-1,2-Diethylcyclopentane

4.3.2.3 Physical Properties of Cycloalkanes

Cycloalkanes are nonpolar molecules like alkanes. As a result, they usually have low melting and boiling points compared to other functional groups. Cycloalkanes have different physical properties from acyclic alkanes due to the greater number of London dispersion forces that causes ring strain. Note that the *London dispersion* force is the weakest intermolecular force, and is a temporary attractive force resulting from the electrons in two adjacent atoms occupying positions that make the atoms to form temporary dipoles. This force is sometimes called an *induced dipole–induced dipole attraction*. The ring strain makes cycloalkane much harder to boil.

Name	Molecular formula	Molecular weight	bp (°C)
Cyclopropane	C_3H_6	42.08	−32.7
Cylcobutane	C_4H_8	56.11	−12.5
Cyclopentane	C_5H_{10}	70.13	49.3
Cyclohexane	C_6H_{12}	84.16	80.7
Cycloheptane	C_7H_{14}	98.19	118.5
Cyclooctane	C_8H_{16}	112.21	150.0

Cycloalkanes have higher boiling points than analogous acyclic alkanes (see Section 4.3.1.2). The smaller cycloalkanes such as cyclopropane and cyclobutane are gases and the rest are liquids and solids at room temperature.

4.3.2.4 Sources of Cycloalkanes

Cycloalkanes of ring sizes ranging from 3 to 30 are found abundantly in nature. While compounds containing five-membered rings (cyclopentane) and six-membered rings (cyclohexane) are especially common in natural products, cyclopentane and cyclohexane themselves are present in petroleum. However, unsubstituted cycloalkanes are rarely found in natural sources.

4.3.2.5 Preparations of Cycloalkanes

Cycloalkanes are mainly prepared simply by *catalytic hydrogenation* of cycloalkenes (see Section 5.4.1.1).

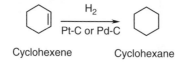

Cyclohexene   Cyclohexane

4.3.2.6 Reactions of Cycloalkanes

Cycloalkanes contain only strong σ bonds, and all the bonds (C—C and C—H) are nonpolar as mentioned with alkanes. As a result, cycloalkanes are also quite unreactive towards most reagents.

4.3.2.6.1 *Reduction of Smaller Cycloalkanes: Preparation of Smaller Alkanes*

Catalytic hydrogenation of smaller cycloalkanes yields open chain alkanes. Cyclopropane and cyclobutane are unstable due to their ring strain compared to the larger cycloalkanes; for example, cyclopentane and cyclohexane. The two smaller cycloalkanes react with hydrogen, even though they are not alkenes. In presence of a nickel (Ni) catalyst, the rings open up and form corresponding acyclic (open chain) alkanes. Cyclobutanes require a higher temperature than cyclopentane for ring opening.

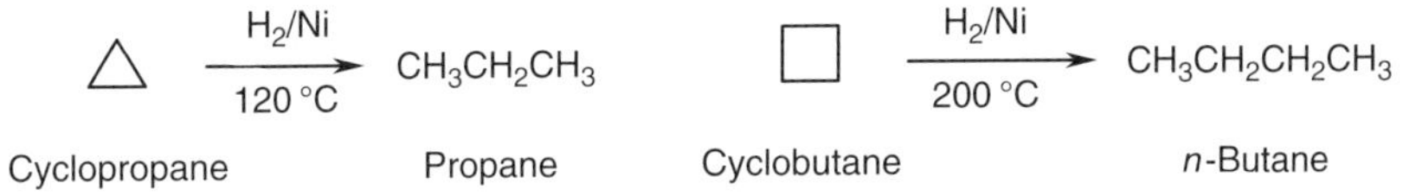

Cyclopropane   Propane   Cyclobutane   n-Butane

4.3.2.6.2 Chlorination of Cyclohexane: Preparation of Chlorocyclohexane

Chlorination of cyclohexane is a useful route for the synthesis of chlorocyclohexane and the reaction is carried out using radical chlorination of sulphuryl chloride (SO_2Cl_2). Note that sulphuryl chloride (SO_2Cl_2) is not thionyl chloride $(SOCl_2)$! Cyclohexane reacts with SO_2Cl_2 in presence of azobisisobutyronitrile (AIBN) to afford chlorocyclohexane.

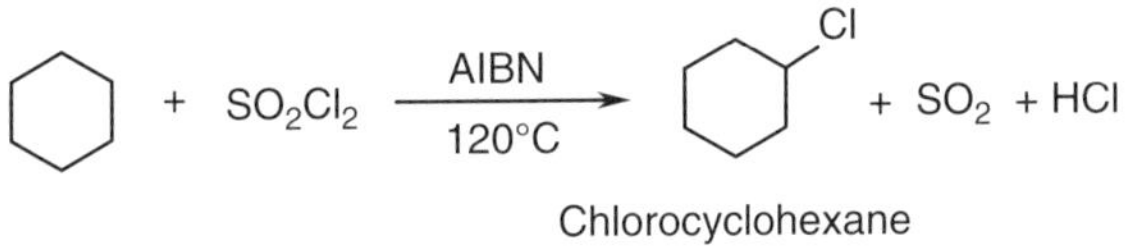

Chlorocyclohexane

4.3.3 Alkyl Halides

Alkyl halides contain one or more halogen atoms attached to a tetrahedral carbon (sp^3) atom. They resemble the alkanes in structure but contain one or more halogen atoms with carbon instead of hydrogen atoms. The functional group is -X, where X may be F, Cl, Br or I. The linear or acyclic alkyl halides having one halogen atom have formula $C_nH_{2n+1}X$, whereas the cyclohaloalkanes have general formula $C_nH_{2n-1}X$. Simple members of this class are methyl chloride (CH_3Cl), ethyl chloride (CH_3CH_2Cl) and cyclopentyl chloride (C_5H_9Cl).

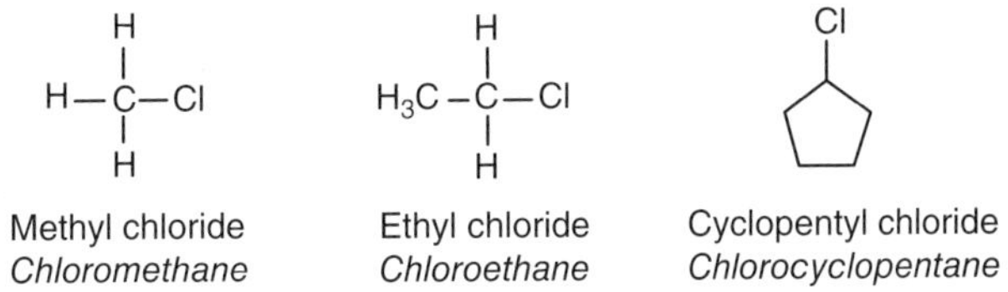

Methyl chloride   Ethyl chloride   Cyclopentyl chloride
Chloromethane   *Chloroethane*   *Chlorocyclopentane*

Based on the number of alkyl groups attached to the C—X unit, alkyl halides are classed as primary (1°), secondary (2°) or tertiary (3°).

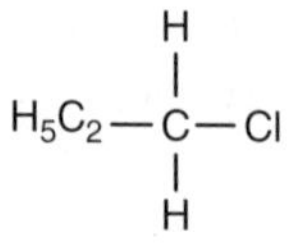

H$_5$C$_2$—C—Cl (with H above and H below)	H$_3$C—C—Cl (with H above and CH$_3$ below)	H$_3$C—C—Cl (with CH$_3$ above and CH$_3$ below)
Propyl chloride *Chloropropane* (1° halide)	Isopropyl chloride *2-Chloropropane* (2° halide)	*tert*-Butyl chloride *2-Chloro-2-methyl propane* (3° halide)

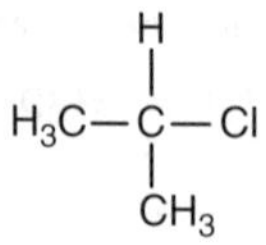

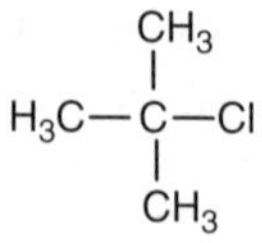

A *geminal* (*gem*) dihalide has two halogen atoms on the same carbon, and *vicinal* (*vic*) dihalide has halogen atoms on adjacent carbon atoms.

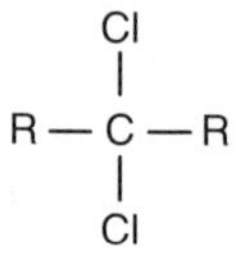

gem-Dichloride 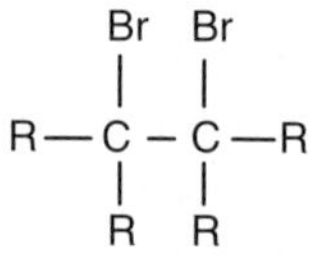   *vic*-Dibromide

Alkyl halides are commonly found in many drugs, such as anaesthetics, isoflurane and halothane, that are on the WHO's list of essential medicines.

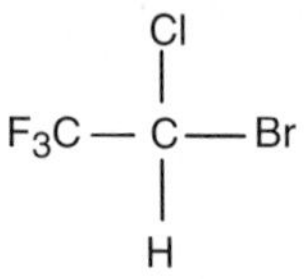

Halothane 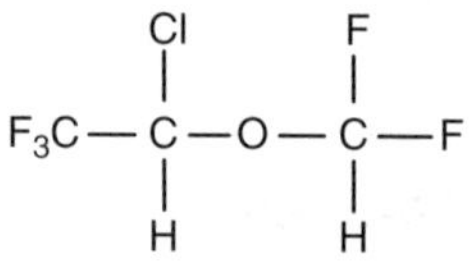   Isoflurane

4.3.3.1 Nomenclature of Alkyl Halides

According to the IUPAC system, alkyl halides are treated as alkanes with a halogen substituent. The halogen prefixes are fluoro-, chloro-, bromo- and iodo-. An alkyl halide is named as a haloalkane with an alkane as the parent structure.

CH$_3$CH$_2$CH$_2$CH$_2$—Cl	CH$_3$CH$_2$CH$_2$—C—CH$_3$ (with Br above and Br below)	(cyclopentane structure)
n-Butyl chloride *1-Chlorobutane*	2,2-Dibromopentane	*trans*-1-Chloro-3-ethylcyclopentane

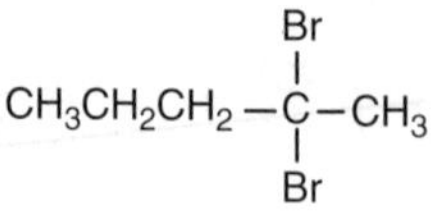

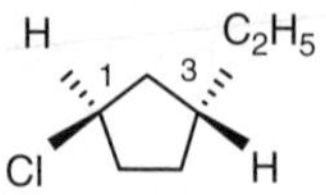

When the carbon chain bears both an alkyl substituent and a halogen, the two substituents are considered to have equal rank and the chain is numbered to give the lower number to the substituent nearer the end of the chain.

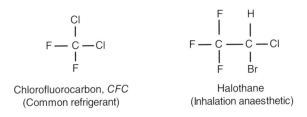

Br   CH₃
| |
CH₃CHCH₂ —CH — CH₂CH₃
1   2   3   4   5   6

2-Bromo-4-methylhexane

CH₃   Br
| |
CH₃CHCH₂—CH —CH₂CH₃
1   2   3   4   5   6

4-Bromo-2-methylhexane

Often, compounds of the CH_2X_2 type are called methylene halides, such as methylene chloride (CH_2Cl_2), CHX_3 type compounds are called *haloforms*, for example, chloroform ($CHCl_3$), and CX_4 type compounds are called carbon tetrahalides, for example, carbon tetrachloride (CCl_4). Methylene chloride (dichloromethane, DCM), chloroform and carbon tetrachloride are extensively used in organic synthesis as nonpolar solvents. Other widely used halogenated alkanes are chlorofluorocarbon (CF_2Cl_2, CFC), used as a refrigerant and halothane ($CF_3CHClBr$), used as an inhalation anaesthetic.

Cl
|
F — C — Cl
|
F

Chlorofluorocarbon, *CFC*
(Common refrigerant)

F   H
| |
F — C — C — Cl
| |
F   Br

Halothane
(Inhalation anaesthetic)

4.3.3.2 Physical Properties of Alkyl Halides

Halogen atoms (X = F, Cl, Br, or I) are more electronegative than carbon atoms. So, the C—X bond is highly polarized; a partial positive charge (δ^+) is on the carbon atom and a partial negative charge (δ^-) on the halogen atom. *Polarizability*, as can be observed in alkyl halides, is the ability of an atom to accommodate a change in electron density. Generally, the chemistry of alkyl halides is dominated by this polarizability effect and results in the C—X bond being broken either in an *elimination* or a *substitution* reaction (see Sections 5.5 and 5.6).

The polarizability of an atom decreases as its charge and electronegativity increases. Fluorine is a highly electronegative and nonpolarizable atom, whereas iodine is a less electronegative and highly polarizable atom. Because of the highly polarized C—X bond, alkyl halides possess considerably higher boiling points than analogous alkanes. The boiling point increases with the increase in atomic weight and polarizability of the halogen atom. Thus, alkyl fluoride has the lowest boiling point and alkyl iodide has the highest. Alkyl halides are usually soluble in nonpolar solvents such as ether and chloroform, but insoluble in water because of their inability to form hydrogen bonds.

Name	R = H	R = F	R = Cl	R = Br	R = I
CH₃—R	−161.7	−78.4	−24.2	3.4	42.4
CH₃CH₂—R	−88.6	−37.7	12.3	38.4	72.3
CH₃CH₂CH₂—R	−42.1	−2.5	46.6	71.0	102.5
CH₃CH₂CH₂CH₂—R	−0.6	32.5	78.4	101.6	130.5

4.3.3.3 Preparations of Alkyl halides

Alkyl halides can be prepared in many ways; free-radical halogenation of alkanes is one way (see Section 5.3.1). However, free-radical halogenation is an inefficient method that generates a mixture of mono-, di- and tri- halogenated compounds. Therefore, alkyl halides are almost always prepared from corresponding alcohols by the use of hydrogen halides (HX) or phosphorous halides (PX$_3$) in ether (*see* Sections 5.6.3.2 and 5.6.3.3). Specifically, alkyl chlorides are obtained by the reaction of alcohols with thionyl chloride (SOCl$_2$) in triethylamine (Et$_3$N) or pyridine (C$_5$H$_5$N) (see Section 5.6.3.4).

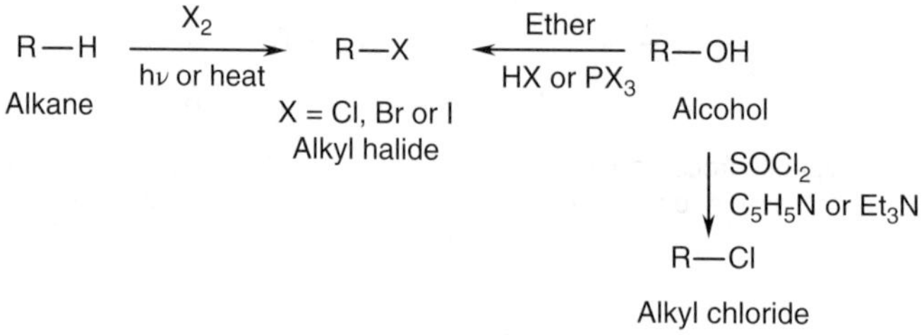

Other methods for the preparation of alkyl halides are electrophilic addition of hydrogen halides (HX) to alkenes (see Section 5.4.2.1), and the reaction follows *Markovnikov's rule* (see Section 5.4.2).

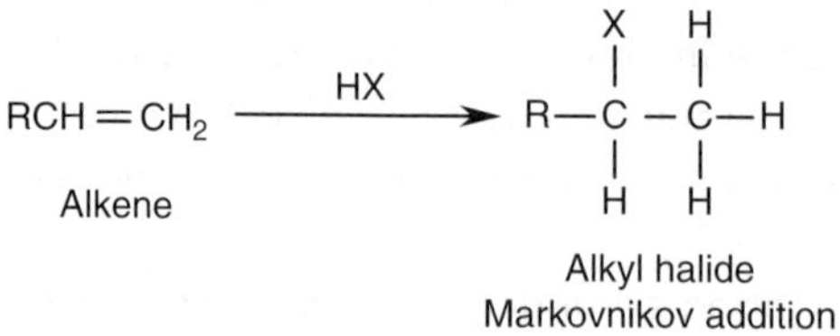

4.3.3.4 Reactivity of Alkyl Halides

The alkyl halide functional group consists of a *sp³*-hybridized carbon atom bonded to a halogen atom via a strong σ bond. The C—X bonds in alkyl halides are highly polar due to the electronegativity and polarizability of the halogen atoms, making the carbon atom electrophilic and the halogen nucleophilic. Generally, the electronegativity of halides decreases and the polarizability increases in the following order: $F > Cl > Br > I$. Typically, halogens (Cl, Br, and I) are good leaving groups in the nucleophilic substitution reactions, except fluorine that is highly electronegative and nonpolarizable. The 'leaving group ability' depends on the electronegativity and polarizability. So, fluorine is not a good leaving group because of its higher electronegativity and lower polarizability.

The bond dipole moment decreases in the following order: $C-Cl > C-F > C-Br > C-I$. Note that the bond dipole moment adopts the idea of electric dipole moment to measure the polarity of a chemical bond within a molecule, and it occurs whenever there is a separation of positive and negative charges. The bond dipole moment of CH_3Cl ($\mu = 1.87$ D) is quite similar to that of methanol (MeOH, $\mu = 1.70$ D) and water ($\mu = 1.85$ D). The chemistry of alkyl halides is predominantly governed by this effect, and results in the C—X bond being broken either in a *substitution* or *elimination* process. While primary alkyl halides are often reactive for displacement by nucleophiles, secondary and tertiary alkyl halides are significantly less reactive.

4.3.3.5 Reactions of Alkyl Halides

The alkyl halides undergo not only substitution but also elimination reactions. Both types of these reactions are carried out in basic reagents. An *elimination* reaction results when a proton and a leaving group are removed from adjacent carbon atoms, giving rise to a π bond between the two carbon atoms (see Section 5.5). In a *substitution* reaction, a nucleophile displaces the leaving group (very often a halide ion) from the carbon atom by using its lone pair to form a new σ bond to the carbon atom (see Section 5.6). Generally, halides are good leaving groups, except fluoride. Thus, in both reactions, the halogen atom leaves with its bonding pair of electrons to provide a halide ion (X:$^-$), which is a stable leaving group (see Sections 5.5.5 and 5.6.1.1).

$$RCH_2CH_2-X + Y\!:^- \quad \begin{cases} \xrightarrow{\text{Elimination}} RCH=CH_2 + H-Y + X\!:^- \\ \xrightarrow[\text{Substitution}]{} RCH_2CH_2-Y + X\!:^- \end{cases}$$

$X = Cl, Br \text{ or } I$

If the halide loss is along with the loss of another atom, the overall reaction is called an *elimination*. The elimination of H—X is common, and known as a dehydrohalogenation (see Section 5.5.5). If an atom replaces the halide atom, the overall reaction is a *substitution* and known as nucleophilic substitution of alkyl halides (see Section 5.6.2). Often, substitution and elimination reactions occur in competition with each other. Most nucleophiles can also act as bases. Therefore, the preference for elimination or substitution is determined by the reaction conditions and the alkyl halide used (see Sections 5.5.1, 5.5.2 and 5.6.2).

4.3.3.5.1 Organometallic Compounds or Organometallics

Electrophiles are usually neutral or charged reagents that accept electron pairs from nucleophiles and produce new covalent bonds. Carbon nucleophiles are extensively used in organic synthesis to build new C—C bonds when they react with electrophiles. The carbon atom must be bonded to a less electronegative atom to be a nucleophilic. In most cases, this less electronegative atom is a metal (for example, Li, Na, K and Mg), where the difference in electronegativity between the carbon and the metal atom dictates the degree of nucleophilicity or basicity of the carbon atom. Such species that contain carbon—metal bonds are known as organometallic compounds or simply organometallics.

When a compound has a covalent bond between a carbon and a metal, it is called an *organometallic compound*. Alkyl halides are most commonly converted to organometallic compounds that contain carbon-metal covalent bonds, usually Mg or Li.

$$\overset{\delta-}{H_3C}\!-\!\overset{\delta+}{Mg} \qquad \overset{\delta-}{H_3C}\!-\!\overset{\delta+}{Li}$$

Carbon—metal bonds vary widely in character from covalent to ionic depending on the metal. The greater the ionic character of the bond, the more reactive is the compound. The most common types of organometallic compounds are Grignard reagents, organolithium reagents and Gilman reagents (organocuprates, R_2CuLi). A carbon-metal bond is polarized with significant negative charge on the carbon, because metals are so electropositive. These compounds are powerful sources of nucleophilic carbons and, therefore, are strong bases.

Grignard Reagents

The most important organometallic reagent in organic synthesis are Grignard reagents, which were introduced by the French chemist François Auguste Victor Grignard in 1900. These reagents are prepared by the reaction of organic halides with magnesium turnings, usually in dry ether. An ether solvent is used because it forms a complex with the Grignard reagent, which stabilizes it. The halide reactivity order is as follows: I > Br > Cl > F. This reaction is versatile and primary (1°), secondary (2°) and tertiary (3°) alkyl halides can be used, as well as vinyl, allyl and aryl halides. Indeed, these halides do not form Grignard reagents in dry ether, instead dry tetrahydrofuran (THF) is used as the solvent for more vigorous reaction conditions.

$$R\!-\!X + Mg \xrightarrow{\text{Dry ether}} RMgX \qquad H_5C_2\!-\!Br + Mg \xrightarrow{\text{Dry ether}} C_2H_5MgBr$$

Organolithium Reagents

These reagents are prepared by the reaction of alkyl halides with lithium metals in ether. Unlike Grignard reagents, organolithiums (RLi) can also be prepared using a variety of hydrocarbon solvents, for example, dry hexane and dry pentane, as well as dry ether. The alkyl halide can be primary (1°), secondary (2°) or tertiary (3°), and the order of halide reactivity is: $I > Br > Cl > F$. Organolithium reagents are strong bases and react vigorously with water or any other weak proton sources. Thus, the reaction must be carried out under extremely dry conditions.

$$R-X + 2Li \xrightarrow[\text{Dry hexane}]{\text{Dry ether or}} R-Li + LiX$$

Gilman Reagents or Lithium Organocuprates

The most useful Gilman reagents are lithium organocuprates (R_2CuLi). They are easily prepared by the reaction of two equivalents of the organolithium reagent with copper(I) iodide in ether. Organocuprates are thermally labile, so they are prepared at low temperatures.

$$2R-X + Cu-I \xrightarrow{\text{Ether}} R_2CuLi + Li-I$$

Organometallics except Gilman reagents readily react with hydrogen atoms attached to oxygen, nitrogen or sulphur in addition to other acidic hydrogen atoms. They are versatile and useful reagents to prepare alcohols from aldehyde and ketones (see Section 4.4.1.5), acid chlorides (see Section 4.4.3.2), esters (see Section 4.4.5.3) and epoxides (see Sections 5.6.4.2 and 5.6.4.3). Organometallics react with terminal alkynes to form metal acetylides (alkynides) by acid–base reactions (see Section 4.6.3). Alkynides are good nucleophiles and extremely useful for the synthesis of a variety of other compounds (see Section 4.6.8). Organolithium reagents react similarly as Grignard reagents, but they are more reactive than Grignard reagents.

Gilman reagents (R_2CuLi) are weaker organometallics. They readily undergo Corey–House reactions with alkyl halides to yield alkanes (see Section 5.6.2.1), but the reaction is limited to 1° alkyl halides. Gilman reagents also react with acid chlorides, a highly reactive carbonyl compounds to give ketone (see Section 5.6.5.8). Gilman reagents do not react with relatively less reactive carbonyl compounds such as aldehydes, ketones, esters, amides, acid anhydrides and nitriles.

Alkanes can be prepared from alkyl halides by reduction directly with Zn and acetic acid (AcOH) or by hydride reduction with $LiAlH_4$ (see Section 5.9.11) or via the Grignard reagent formation followed by hydrolytic work-up (see Section 5.9.12). The Corey–House reactions of alkyl halides with Gilman reagents (R'_2CuLi, organocuprates) afford alkanes, but this reaction is limited to 1° alkyl halides (see Section 5.6.2.1). Base-catalysed dehydrohalogenation of alkyl halides is an important reaction for the preparation of alkenes (see Section 5.5.5).

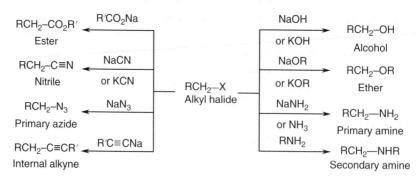

Alkyl halide reacts with triphenylphosphine to provide a phosphonium salt, which is an important intermediate for the preparation of phosphorus ylide (see Section 5.4.3.6).

$$RCH_2-X \quad + \quad (Ph)_3P: \quad \longrightarrow \quad RCH_2 -\overset{+}{P}(Ph)_3X^-$$

Alkyl halide Triphenylphosphine   Phosphonium salt

4.3.3.5.2 Nucleophilic Substitution Reactions of Alkul Halides

Alkyl halides undergo nucleophilic substitution reactions via S_N2 with a variety of nucleophiles, such as metal hydroxides (NaOH or KOH) and metal alkoxides (NaOR or KOR) to give alcohols (see Section 5.6.2.2) and ethers (see Section 5.6.2.3), respectively. They react with metal amides (NaNH$_2$) or NH$_3$, 1° amines or 2° amines to yield 1°, 2° or 3° amines, respectively (see Section 5.6.2.4). Alkyl halides react with metal carboxylate (R'CO$_2$Na), metal cyanides (NaCN or KCN), metal azides (NaN$_3$) and metal alkynides (R'C≡CM or R'C≡CMgX) to produce esters (see Section 5.6.2.5), nitriles (see Section 5.6.2.6), azides (see Section 5.6.2.7) and internal alkynes (see Section 5.6.2.8), respectively. Most of these transformations are limited to primary alkyl halides, simply because higher alkyl halides tend to react via elimination (see Section 5.5.5.1).

4.3.4 Alcohols

Alcohols are aliphatic organic compounds that contain hydroxyl (—OH) functional groups. Therefore, an alcohol has the general formula ROH. The simplest and most common alcohols are methyl alcohol (CH_3OH) and ethyl alcohol (CH_3CH_2OH).

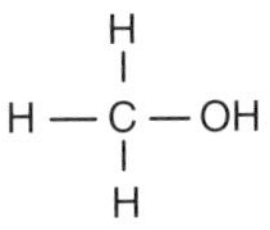

Methanol
Methyl alcohol

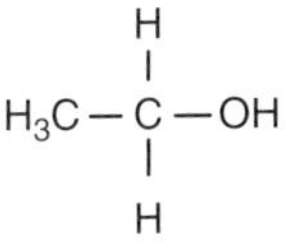

Ethanol
Ethyl alcohol

An alcohol may be acyclic or cyclic. It may contain a double bond, a halogen atom or additional hydroxyl groups. Alcohols are usually classified as primary (1°), secondary (2°) or tertiary (3°). When a hydroxyl group is linked directly to an aromatic ring, the compound is called a *phenol* (see Section 4.7.10), which differs distinctly from alcohols due to the aromatic ring.

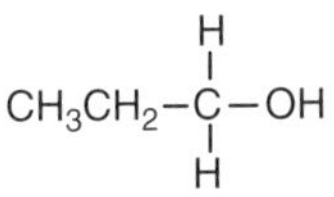

Propanol
Propyl alcohol (1°)

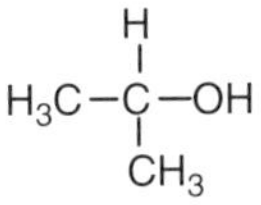

Isopropanol
2-Propanol (2°)

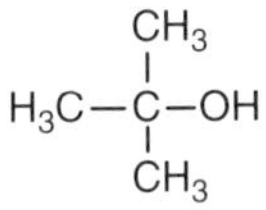

tert-Butanol
2-Methyl-2-propanol (3°)

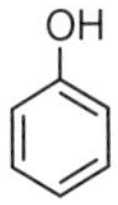

Phenol

4.3.4.1 Nomenclature of Alcohols

Generally, the name of an alcohol ends with *-ol*. An alcohol can be named as an alkyl alcohol, usually for small alkyl groups; for example, methyl alcohol and ethyl alcohol. The longest carbon chain bearing the —OH group is used as the parent, the last *-e* from this alkane is replaced by an *-ol* to obtain the root name. The longest chain is numbered starting from the end nearest to the —OH group and the position of the —OH group is numbered. Cyclic alcohols have the prefix *cyclo-*, and the —OH group is deemed to be on C-1.

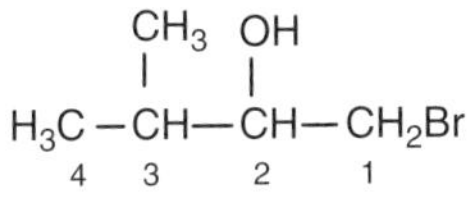

1-Bromo-3-methyl-2-butanol
(The—OH is at C-2 of butane)

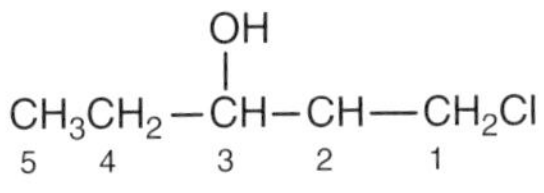

1-Chloropentanol
(The—OH is at C-3 of pentane)

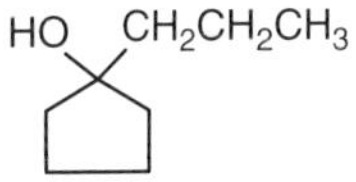

1-Propylcyclopentanol

Alcohols with double or triple bonds are named using the *-ol* suffix on the alkene or alkyne name. Numbering offers the hydroxyl group the lowest possible number. When numbers are also given for the multiple bond position, the position of the hydroxyl can be written immediately before the *-ol* prefix.

If the hydroxyl group is only a minor part of the structure, it may be named as a hydroxyl-substituent.

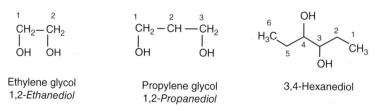

OH | CH₃CH₂—CH—CH₂CO₂H 5 4 3 2 1	OH | H₂C=CHCH₂CHCH₃ 5 4 3 2 1	
3-Hydroxypentanoic acid	Pent-4-en-2-ol	3-Bromo-5-chlorocyclohexanol

Diols are compounds with two hydroxyl groups on adjacent carbon. They are named as alcohols are, except that the suffix -*diol* is used and two numbers are required to locate the hydroxyls. 1,2-Diols are called glycols. The common names for glycols usually arise from the name of the alkene from which they are prepared.

Glycols are often called *vicinal* diols, when two hydroxyl groups are attached to adjacent carbon atoms. Two most commonly known *vicinal* diols are 1,2-ethanediol (ethylene glycol) and 1,2-propanediol (propylene glycol).

Ethylene glycol *1,2-Ethanediol*	Propylene glycol *1,2-Propanediol*	3,4-Hexanediol

4.3.4.2 Physical Properties of Alcohols

Alcohols can be considered as organic analogues of water. Both the C—O and O—H bonds are polarized due to the electronegativity of the oxygen atom. The highly polar nature of the O—H bond results in the formation of hydrogen bonds with other alcohol molecules or other hydrogen bonding systems; for example, water and amines. Thus, alcohols have considerably higher boiling points due to the hydrogen bonding between molecules (*intermolecular hydrogen bonding*). They are more polar than hydrocarbons and are better solvents for polar molecules.

Hydrogen bonds between
alcohol molecules

Hydrogen bonds to water
in aqueous solution

The hydroxyl group is *hydrophilic* (water loving), whereas the alkyl (hydrocarbon) part is *hydrophobic* (water repellent or fearing) (see Section 2.5). In general, small alcohols are miscible with water, but water solubility decreases as the size of the alkyl group increases. Thus, alcohols will also dissolve in non-polar solvents.

Most of the common alcohols of up to about 11 or 12 carbon atoms are liquids at room temperature. The boiling point of alcohols increases with the increase in the alkyl chain length as shown in the following table. Branching also lowers boiling point, so isopentanol has a lower boiling point than its isomer *n*-pentanol, and this pattern is observed in all isomeric alkanols (alkyl alcohols).

Name	Molecular formula	Molecular weight	bp (°C)	Solubility (g) in H_2O
Methanol	CH_3OH	32.04	64.5	Infinite
Ethanol	CH_3CH_2OH	46.07	78.3	Infinite
n-Propanol	$CH_3CH_2CH_2OH$	60.10	97.0	Infinite
Isopropanol	$CH_3CHOHCH_3$	60.10	82.5	Infinite
n-Butanol	$CH_3CH_2CH_2CH_2OH$	74.12	118.0	8.0
Isobutanol	$(CH_3)_2CHCH_2OH$	74.12	108.0	10.0

4.3.4.3 Acidity and Basicity of Alcohols

Alcohols resemble water in their acidity and basicity. They are stronger acids than terminal alkynes and primary or secondary amines.

But they are weaker acids than HCl, H_2SO_4 and even acetic acid. Alcohols dissociate in water and forms alkoxides (RO^-) and hydronium ion (H_3O^+).

$$R-\ddot{O}H + H_2\ddot{O} \longrightarrow R\ddot{O}^- + H-\overset{+}{\underset{H}{\ddot{O}}}-H$$

Alcohol   Alkoxide ion

Alcohols are considerably acidic to react with active metals to liberate hydrogen (H_2) gas. Thus, an *alkoxide* (RO^-) can be prepared easily by the reaction of an alcohol with Na or K metal.

$$R-\ddot{O}H + Na \longrightarrow R\ddot{O}^-Na^+ + 1/2\ H_2$$

Alcohol   Sodium alkoxide

$$R-\ddot{O}H + K \longrightarrow R\ddot{O}^-K^+ + 1/2\ H_2$$

Alcohol   Potassium alkoxide

Like hydroxide (HO⁻) ions, alkoxide (RO⁻) ions are strong bases and nucleo-philes. Halogens increase the acidity, but acidity decreases as the alkyl (R) group increases. The electron withdrawing groups on an alcohol increase the acidity by stabilizing the alkoxide (RO⁻) formed.

Name	Molecular formula	pK_a	Name	Molecular formula	pK_a
Methanol	CH_3OH	15.5	Cyclohexanol	$C_6H_{11}OH$	18.0
Ethanol	C_2H_5OH	15.9	Phenol	C_6H_5OH	10.0
2-Chloroethanol	ClC_2H_4OH	14.3	Water	H_2O	15.7
2,2,2-Trifluoroethanol	CF_3CH_2OH	12.4	Acetic acid	CH_3COOH	4.76
Tert-Butanol	$(CH_3)_3COH$	19.0	Hydrochloric acid	HCl	−7.0

The acidity of alcohols varies greatly, depending on the substituents. Note that lower the pK_a values (and higher the K_a values), the stronger the acids. Alkoxides can react with primary alkyl halides (see Section 5.6.2.3) or tosylate esters (see Section 5.6.3.6) to give ethers.

$$CH_3CH_2-\ddot{O}H + H_2O \longrightarrow CH_3CH_2\ddot{O}^- + H_3O^+$$

Ethyl alcohol

Ethoxide
$K_a = 1.3 \times 10^{-16}$

$$CF_3CH_2-\ddot{O}H + H_2O \longrightarrow CF_3CH_2\ddot{O}^- + H_3O^+$$

Trifluoroethyl alcohol

Trifluroethoxide
$K_a = 4.0 \times 10^{-13}$

Alcohols are basic enough to accept a proton from strong acids, for example, HCl and H_2SO_4, and able to dissociate completely in acidic solution. Sterically hindered alcohols, for example, tert-butyl alcohol, are strongly basic (higher pK_a values and lower K_a values), and react with strong acids to yield oxonium (RO⁺H₂) ions.

$$R-\ddot{O}H + H_2SO_4 \longrightarrow R-\overset{+}{\underset{H}{\ddot{O}}}-H + HSO_4^-$$

Alcohol   Oxonium ion

Note that *steric hindrance* results from steric effects, and can be described as the slowing of chemical reactions due to steric bulk. Remember that *steric effects* are nonbonding interactions that influence the shape and reactivity of ions and molecules; they complement electronic effects, which usually dictate shape and reactivity. Repulsive forces between overlapping electron clouds result in steric effects.

4.3.4.4 Preparations of Alcohols

Alcohols can be prepared from hydration of alkenes or oxymercuration–reduction of alkenes or hydroboration–oxidation of alkenes. Interestingly, the hydration and oxymercuration-reduction of alkenes (see Section 5.4.2.7), follow Markovnikov's rule (see Section 5.4.2), whereas hydroboration-oxidation of alkenes (see Section 5.4.2.8) follows anti-Markovnikov's rule (see Section 5.4.2).

$$
\underset{\substack{\text{Alcohol} \\ \text{anti-Markovnikov addition}}}{\overset{\displaystyle\underset{R\ \ R}{\overset{H\ \ OH}{R-\overset{|}{\underset{|}{C}}-\overset{|}{\underset{|}{C}}-H}}}{}}
\xleftarrow[\text{ii. } H_2O_2,\ NaOH]{\text{i. } BH_3.THF}
\underset{\text{Alkene}}{R_2C=CHR}
\xrightarrow[\substack{\text{i. } Hg(OAc)_2,\ THF,\ H_2O \\ \text{ii. } NaBH_4,\ NaOH}]{H_2O,\ H_2SO_4}
\underset{\substack{\text{Alcohol} \\ \text{Markovnikov addition}}}{\overset{\displaystyle\underset{R\ \ R}{\overset{OH\ H}{R-\overset{|}{\underset{|}{C}}-\overset{|}{\underset{|}{C}}-H}}}{}}
$$

However, the most important methods for preparing alcohols are catalytic hydrogenation (H_2/Pd–C), nucleophilic addition of organometallics (RMgX or RLi) to aldehydes or ketones (see Section 5.2.5), acid chlorides (see Section 5.4.2.3) or esters (see Section 5.4.4.3), or metal hydride ($NaBH_4$ or $LiAlH_4$) reduction of aldehydes or ketones (see Section 5.9.14) or acid chlorides (see Section 5.9.17) and metal hydride reduction ($LiAlH_4$) of esters (see Section 5.9.18) or carboxylic acids (see Section 5.9.19). Alcohols can also be synthesized from the reaction between alkyl halides (R–X) and metal hydroxides (NaOH or KOH) as shown earlier (see Section 4.3.3.5).

$$
\underset{\text{1° or 2°Alcohol}}{\overset{\displaystyle\overset{OH}{R-\overset{|}{C}H-Y}}{}}
\xleftarrow[NaBH_4 \text{ or } LiAlH_4]{H_2/Pd\text{-}C \text{ or}}
\underset{Y = H \text{ or } R}{\overset{\displaystyle\overset{O}{\overset{\|}{R-C-Y}}}{}}
\xrightarrow[\text{ii. } H_3O^+]{\text{i. } R'MgX \text{ or } R'Li}
\underset{\text{2° or 3° Alcohol}}{\overset{\displaystyle\overset{OH}{\underset{R'}{R-\overset{|}{\underset{|}{C}}-Y}}}{}}
$$

$$
\underset{\text{1°Alcohol}}{R-CH_2OH}
\xleftarrow[\text{ii. } H_3O^+]{\text{i. } LiAlH_4,\ dry\ THF}
\underset{Y = Cl \text{ or } OR}{\overset{\displaystyle\overset{O}{\overset{\|}{R-C-Y}}}{}}
\xrightarrow[\text{ii. } H_3O^+]{\text{i. } 2R'MgX \text{ or } 2R'Li}
\underset{\text{3° Alcohol}}{\overset{\displaystyle\overset{OH}{\underset{R'}{R-\overset{|}{\underset{|}{C}}-R'}}}{}}
$$

Primary alcohols are obtained from epoxides by acid-catalysed cleavage of weak nucleophiles, such as H_2O or ROH (see Section 5.6.4.2), and secondary alcohols are afforded by base-catalysed cleavage of strong nucleophiles, such as organometallics (RMgX or RLi), metal alkynides (R'C≡CM or R'C≡CMgX), metal hydroxides (KOH or NaOH), metal alkoxides (KOR or NaOR) or metal hydrides ($LiAlH_4$) (see Section 5.6.4.3).

2-substituted alcohol
(2° Alcohol)

Base-catalysed cleavage

(Nu:⁻ = R⁻, RC≡C⁻, C≡N⁻, N₃⁻, HO⁻, RO⁻ or H⁻)

Epoxide

Acid-catalysed cleavage

Nu: = H_2O or ROH

1-substituted alcohol
(1° Alcohol)

4.3.4.5 Preparations of Diols

Diols are generally prepared by *syn*-hydroxylation of alkenes using dilute basic $KMnO_4$ or OsO_4, H_2O_2 through *syn*-addition (see Section 5.9.4) or epoxidation of alkenes with peroxyacids followed by basic hydrolysis through *anti*-addition, most commonly known as *anti*-hydroxylation of alkenes (see Section 5.9.4).

$KMnO_4$, Na_2CO_3 or OsO_4, H_2O_2

Syn-addition

$R_2C=CR_2$
Alkene

i. RCOOH
ii. NaOH, H_2O

Anti-addition

4.3.4.6 Reactivity of Alcohols

The hydroxyl (O—H) group in alcohol is polarized due to the electronegativity difference between oxygen (electronegativity: 3.5) and hydrogen (electronegativity: 2.1) atoms. The oxygen of the —OH group can react as either a base or a nucleophile in the nucleophilic substitution reactions. Alcohol itself cannot undergo a nucleophilic substitution reaction because the hydroxyl group is strongly basic and a poor leaving group. Therefore, it needs to be converted to a better leaving group; for example, H_2O (weak base), a good leaving group.

One way to convert a —OH group into a weaker base (H_2O) is to protonate alcohol by adding acid to the reaction mixture. Only weakly basic nucleophiles such as halides (I⁻, Br⁻, Cl⁻) are used in the nucleophilic substitution reaction. Moderately basic nucleophiles for example, ammonia (NH_3), amines (RNH_2) and strongly basic nucleophiles, such as alkoxides (RO⁻), cyanides (CN⁻) would be protonated in the acidic solution resulting in the total loss or significant decrease in their nucleophilicity.

4.3.4.7 Reactions of Alcohols

Alcohols undergo a wide variety of transformations such as dehydration, esterification, substitution, oxidation and reduction via tosylate esters. Alkyl halides are almost always prepared from corresponding alcohols by the use of hydrogen halides (HX) or phosphorous halides (PX$_3$) in ether (see Sections 5.6.3.2 and 5.6.3.3). Alkyl chlorides are prepared by the reaction of alcohols with thionyl chloride (SOCl$_2$) in pyridine or Et$_3$N (see Section 5.6.3.4).

$$R-X \xrightarrow[\text{Ether}]{\text{HX or PX}_3} R-OH \xrightarrow[\text{Pyridine or Et}_3\text{N}]{\text{SOCl}_2} R-Cl$$

Alkyl halide   Alcohol   Alkyl chloride

Alkenes are usually prepared by dehydration of alcohols via elimination reactions (see Section 5.5.3), and esters are obtained by the acid-catalysed reaction of alcohols and carboxylic acids; this reaction is also known as *Fischer esterification* (see Section 5.6.5.1).

$$RCH_2CH_2-OR' \xleftarrow[\text{HCl}]{\text{R'CO}_2\text{H}} RCH_2CH_2\cdot OH \xrightarrow[\text{Heat}]{\text{H}_2\text{SO}_4} RCH=CH_2$$

Ester   Alcohol   Alkene

Symmetrical ethers are yielded from the dehydration of two molecules of alcohols with strong acids; for example, H$_2$SO$_4$ (see Section 5.6.3.1). Alcohols react with *p*-toluenesulphonyl chloride (tosyl chloride, TsCl), also commonly known as *sulphonyl chloride*, in pyridine or Et$_3$N to afford tosylate esters or alkyl tosylates (see Section 5.6.3.5), which are easily reduced to alkanes by metal hydrides (LiAlH$_4$) (see Section 5.9.13).

RCO$_2$H or RCOH $\xleftarrow{\text{Oxidation of } 1° \text{ alcohol}}$ Carboxylic acids or Aldehyde

RCOR $\xleftarrow{\text{Oxidation of } 2° \text{ alcohol}}$ Ketone

R–OH Alcohol

$\xrightarrow[140°C]{\text{ROH, H}_2\text{SO}_4}$ R–O–R Ether

$\xrightarrow[\text{Pyridine}]{\text{TsCl}}$ R–OTs Alkyl tosylate

i. LiAlH$_4$, dry THF
ii. H$_3$O$^+$

R–H Alkane

Carboxylic acids and aldehydes are prepared by the oxidation of primary (1°) alcohols, and ketones are obtained by oxidation of secondary (2°) alcohols by various oxidizing agents (see Section 5.9.8). Tertiary alcohols do undergo oxidation because they don't have hydrogen atoms attached to the oxygen bearing carbon atom (carbinol carbon).

4.3.5 Ethers

Ethers are organic relatives of water, where alkyl groups replace both hydrogen atoms. Ethers have two hydrocarbon bonded to an oxygen atom with a general formula R—O—R. The simplest and most common ethers are diethyl ether and tetrahydrofuran (THF), which is a cyclic ether.

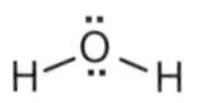

Water	Ether	Diethyl ether	Tetrahydrofuran
		Ether	THF

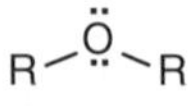

4.3.5.1 Nomenclature of Ethers

Ethers can be symmetrical, where two alkyl groups are the same, or unsymmetrical, where two alkyl groups are different. While diethyl ether is symmetrical, ethyl methyl ether is unsymmetrical. The common name of an unsymmetrical ether is obtained by quoting the names of the two alkyl groups in alphabetical order followed by the prefix ether.

$$H_3C-O-CH_3 \qquad\qquad H_3C-O-C_2H_5$$

Dimethylether (symmetrical)   Ethyl methyl ether (unsymmetrical)

In the nomenclature of ethers, either the suffix -*ether* or the prefix *alkoxy*- is used. For example, diethyl ether can be called ethoxyethane, and methyl *t*-butyl ether can be named 2-methyl-2-methoxypropane.

Diethylether
Ethoxyethane

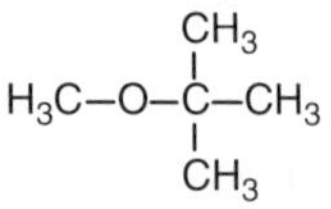

Methyl *t*-butyl ether
2-Methyl-2-methoxypropane

Three-membered cyclic ethers are known as *epoxides*. They are just a subclass of ethers containing a three-membered oxirane ring (C—O—C unit). Cyclic ethers have the prefix *epoxy*- and suffix -*alkene oxide*. Four-, five- and six-membered cyclic ethers are known as oxetane, oxolane and oxane, respectively. Naming of these heterocyclic compounds depends on the ring size and number of oxygen present in the ring.

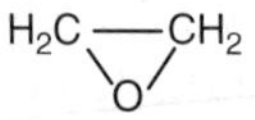

Ethylene oxide	Propylene oxide	Butylene oxide
Epoxy ethane	*1,2-Epoxy propane*	*1,2-Epoxy butane*

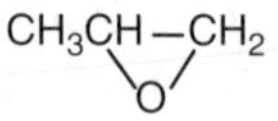

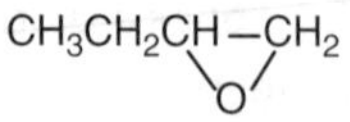

Oxirane ring	Oxetane	Oxolane	Oxane
Epoxide		THF	

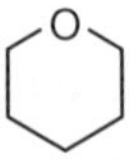

Crown ethers are macrocyclic ethers with ethylene bridges separated by oxygen atoms. They have exceptional ability to selectively coordinate with metal cations and thus allow inorganic salts to dissolve in organic solvents. Crown ethers are widely used in supramolecular chemistry and are important class of molecules in targeted drug delivery.

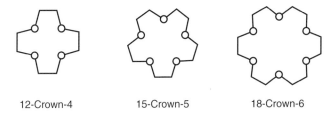

| 12-Crown-4 | 15-Crown-5 | 18-Crown-6 |

4.3.5.2 Physical Properties of Ethers

Ether cannot form hydrogen bonds with other ether molecules, since there is no H to be donated as there is no —OH group, but can be involved in hydrogen bonding with hydrogen bonding systems; for example, H_2O, alcohols and amines. Therefore, ethers have much lower melting points and boiling points and less water solubility than analogous alcohols.

Name	Molecular formula	Molecular weight	bp (°C)
Methyl ethyl ether	$CH_3OCH_2CH_3$	60.10	7.4
n-Propanol	$CH_3CH_2CH_2OH$	60.10	97.0
Diethyl ether	$CH_3CH_2OCH_2CH_3$	74.12	34.6
n-Butanol	$CH_3CH_2CH_2CH_2OH$	74.12	118.0

4.3.5.3 Preparations of Linear Ethers

Ethers are prepared from alkyl halides by the treatment of metal alkoxide, such as sodium ethoxide (EtO⁻Na⁺). This is known as *Williamson ether synthesis* (see Section 5.6.2.3). Williamson ether synthesis is an important laboratory method for the preparation of both symmetrical and unsymmetrical ethers.

$$R\ddot{O}-M^+ \quad + \quad R'-CH_2-X \quad \longrightarrow \quad RO-CH_2-R' + MX$$

Metal alkoxide	Primary halides	Ether
M = Na, K	or tosylates	Symmetrical or
	X = Cl or OTs	unsymmetrical

Symmetrical ethers also can be prepared by dehydration of two molecules of primary alcohols and strong acids (see Section 5.6.3.1), and from alkenes either by acid-catalysed addition of alcohols (see Section 5.4.2.10) or alkoxymercuration-reduction (see Section 5.4.2.11). The last two reactions follow Markovnikov's rules (see Section 5.4.2).

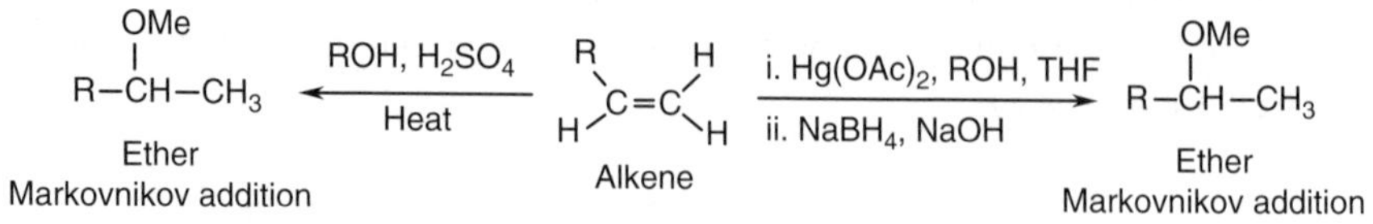

4.3.5.4 Preparations of Cyclic Ethers

The simplest epoxide (ethylene dioxide) is prepared by catalytic oxidation of ethylene, and alkenes are also oxidized to other epoxides by peracid or peroxy acid (see Section 5.9.2). For example, peroxyacetic acid oxidizes cyclopentene to cyclopentene oxide in high yield.

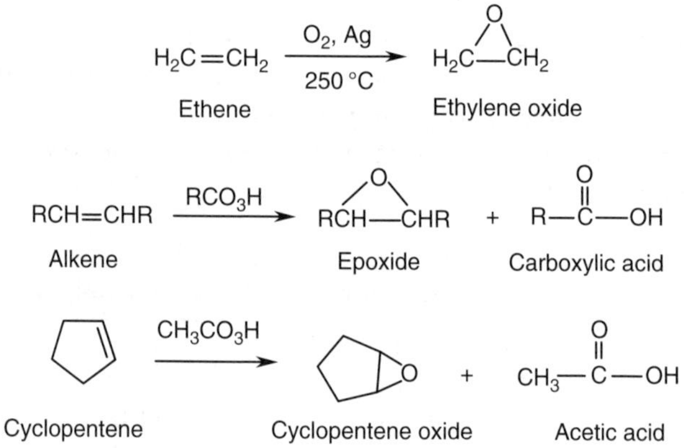

Alkenes are converted to halohydrins by the treatment of halides and water. When halohydrins are treated with a strong base (NaOH), an intramolecular cyclization occurs and epoxides are formed. For example, 1-butene can be converted to butylene oxide via butylene chlorohydrin.

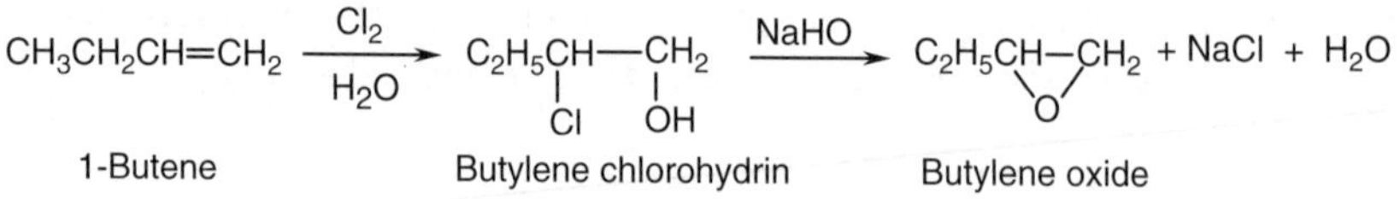

4.3.5.5 Reactivity of Ethers

Ethers are fairly unreactive making them useful as good polar protic solvents for many organic reactions. For example, dry diethyl ether and dry THF are the common solvents used in the Grignard reaction (see Section 4.3.3.5). Ethers often form complexes with molecules that have vacant orbitals; for example, diethyl ether and BCl_3 forms a complex of boron trichloride (see Section 1.7.1.3), and THF complexes with borane (BH$_3$.THF), which is used in the *hydroboration-oxidation* of alkenes and alkynes (see Sections 5.4.2.8 and 5.4.2.9). A few other common ether solvents are shown here.

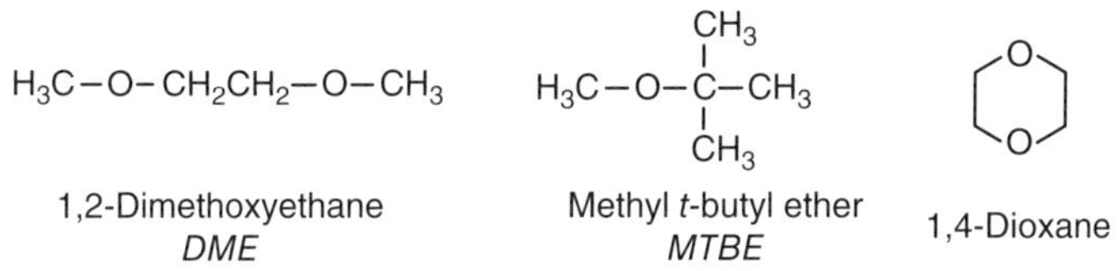

$$H_3C-O-CH_2CH_2-O-CH_3$$

1,2-Dimethoxyethane
DME

$$H_3C-O-\overset{\overset{\displaystyle CH_3}{|}}{\underset{\underset{\displaystyle CH_3}{|}}{C}}-CH_3$$

Methyl *t*-butyl ether
MTBE

1,4-Dioxane

4.3.5.6 Reactions of Ethers

Despite ethers being generally stable and chemically inert, they can undergo cleavage and auto-oxidation under certain conditions. Ethers may auto-oxidize to unstable ether peroxides if left in the presence of oxygen for a few months. These ether peroxides formed by auto-oxidation have caused many laboratory accidents. Thus, peroxide-forming chemicals should be stored in sealed, air-impermeable, light-resistant containers.

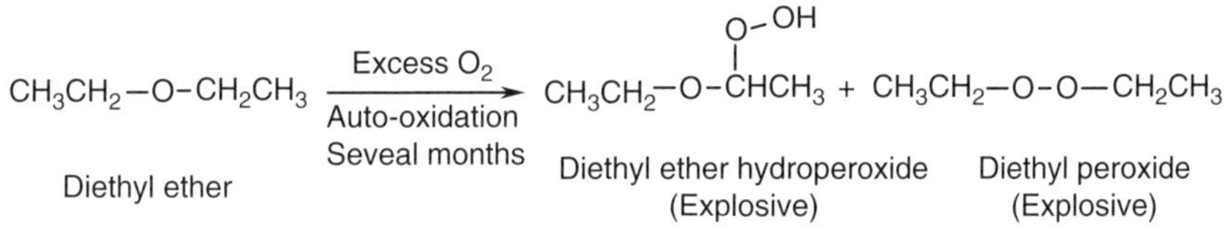

$$CH_3CH_2-O-CH_2CH_3 \xrightarrow[\substack{\text{Auto-oxidation} \\ \text{Seveal months}}]{\text{Excess O}_2} CH_3CH_2-O-\overset{\overset{\displaystyle O^{-OH}}{|}}{C}HCH_3 + CH_3CH_2-O-O-CH_2CH_3$$

Diethyl ether

Diethyl ether hydroperoxide
(Explosive)

Diethyl peroxide
(Explosive)

Simple ethers are relatively unreactive towards bases, oxidizing agents and reducing agents. They cannot undergo nucleophilic substitution reactions, except with excess haloacids (usually HBr or HI) at high temperatures, where cyclic ethers are protonated to undergo nucleophilic substitution reactions to form corresponding alkyl halides (see Section 5.6.4.1).

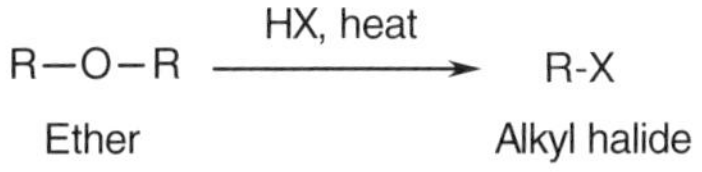

$$R-O-R \xrightarrow{\text{HX, heat}} R-X$$

Ether

Alkyl halide

On the contrary, epoxides (cyclic ethers) are much more reactive than simple ethers due to ring strain, and are useful intermediates because of their chemical versatility. They undergo nucleophilic substitution reactions with both acids and bases to give alcohols (see Sections 5.6.4.2 and 5.6.4.3).

4.3.6 Thiols

Thiols, with the general formula RSH, are the sulphur analogues of alcohols. They are also commonly called *mercaptans.* The functional group of a thiol is —SH. The simplest members of this class are methanethiol (CH$_3$SH), ethanethiol (C$_2$H$_5$SH) and propanethiol (C_3H_7SH). The amino acid cysteine contains a thiol functional group, and the thiol group plays an important role in biological systems (see Section 7.2).

H_3C-SH	H_5C_2-SH	$H_5C_2-CH_2-SH$	
Methanethiol	Ethanethiol	Propanethiol	
Methyl mercaptan	Ethyl mercaptan	Propyl mercaptan	Cysteine

Methanethiol, a natural substance found in the blood, brain and other tissues of humans and animals, is a colourless gas with a rotten cabbage smell. It also occurs naturally in certain foods, such as some nuts and cheese. As natural gas (methane) has no odour, ethanethiol, an odorant with a very distinctive smell, is added to make it easier to detect a leak.

4.3.6.1 Nomenclature of Thiols

The nomenclature of thiols is similar to alcohols, except that they are named using the suffix -*thiol*, and as a substituent as *mercapto-*.

Cyclohexanethiol   *p*-Mercaptobenzoic acid

4.3.6.2 Physical Properties of Thiols

The S—H bond in thiols is less polar than O—H bond in alcohols, since sulphur is less electronegative than oxygen atom, also the electronegativity difference between sulphur and hydrogen is negligible (see Section 2.3.3). Therefore, thiols form much weaker hydrogen bonding with water molecules or other thiol molecules compared to alcohols. So, thiols have much lower boiling points than alcohols with similar molecular weight. Because of the weaker hydrogen bonds, thiols are less soluble in H_2O and other polar solvents than analogous alcohols.

Name	Molecular formula	Molecular weight	bp (°C)
Ethanethiol	CH_3CH_2SH	62.13	35.0
n-Propanol	$CH_3CH_2CH_2OH$	60.10	97.0

4.3.6.3 Acidity and Basicity of Thiols

Thiols are much more acidic than analogous alcohols, for example, RSH ($pK_a = 10$) versus ROH ($pK_a = 16–19$), and more nucleophilic than analogous alcohols. In fact, RSH is almost as nucleophilic as alkoxide (RO^-) ion.

4.3.6.4 Preparations of Thiols

Thiols are prepared from alkyl halides and sodium hydrosulphide (Na$^+$SH$^-$) by S_N2 reaction. Usually, a large excess of Na$^+$SH$^-$ is used with unhindered alkyl halide to prevent dialkylation.

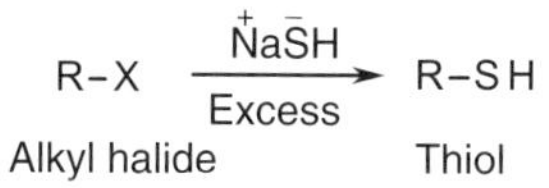

$$R-X \xrightarrow[\text{Excess}]{\overset{+}{\text{Na}}\overset{-}{\text{S}}\text{H}} R-SH$$

Alkyl halide   Thiol

Similarly, thiols are obtained from alkyl halides and thiourea by S_N2 reaction, followed by basic hydrolysis.

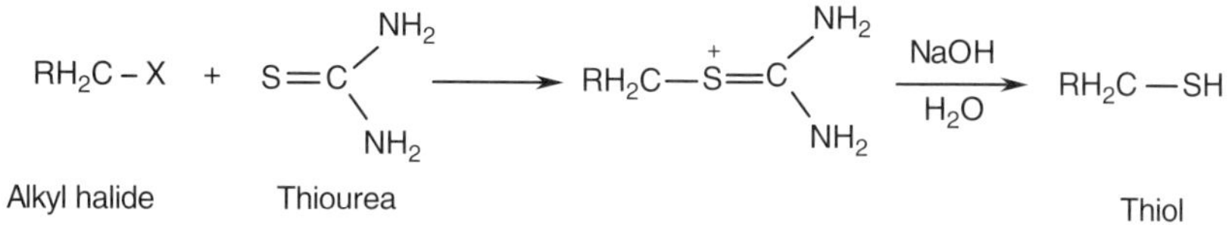

Alkyl halide   Thiourea   Thiol

4.3.6.5 Reactions of Thiols

The deprotonated form thiolate (RS$^-$) is more chemically reactive than the protonated thiol. In the presence of a base, a thiolate anion (RS$^-$) is formed, which is a powerful nucleophile. Thiols are easily oxidized to disulphides (R—S—S—R), which is an important feature of protein structure. They are also oxidized by Br$_2$ to produce disulphide.

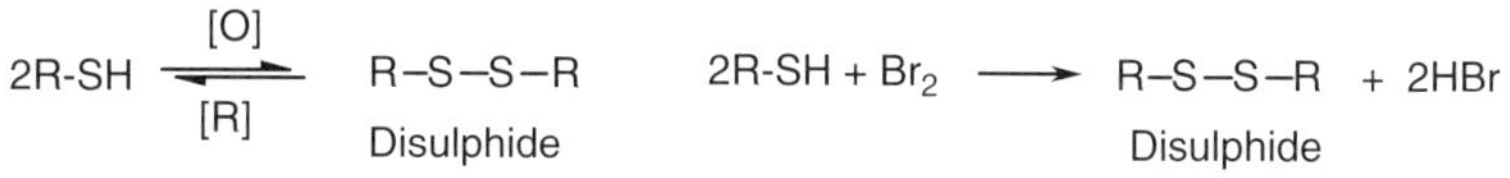

$$2R\text{-}SH \underset{[R]}{\overset{[O]}{\rightleftharpoons}} R\text{-}S\text{-}S\text{-}R \qquad 2R\text{-}SH + Br_2 \longrightarrow R\text{-}S\text{-}S\text{-}R + 2HBr$$

Disulphide   Disulphide

Vigorous oxidation of thiol with hydrogen peroxide (H$_2$O$_2$) or sodium hypochlorite (NaOCl), yields alkyl sulphonic acid (RSO$_3$H). For example, oxidation of ethanethiol using either H$_2$O$_2$ or NaOCl affords ethyl sulphonic acid.

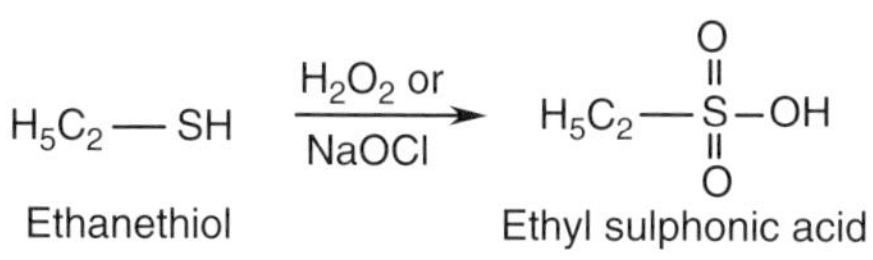

Ethanethiol   Ethyl sulphonic acid

4.3.7 Thioethers

Thioethers have a general formula R$_2$S. They are the sulphur analogues of ethers, and also widely known as *sulphides*. The functional group of a thioether is R—S—R′, where R and R' can be identical or different aliphatic or aromatic hydrocarbon. The simplest thioethers are methylthiomethane and ethylthioethane.

$$H_3C-S-CH_3 \qquad H_5C_2-S-C_2H_5$$

Methylthiomethane   Ethylthioethane
Dimethyl sulphide   *Diethyl sulphide*

Thioethers are widely distributed in nature. The biologically important natural products methionine and biotin are thioethers and they are often associated with foul odours, for example, allyl sulphide is found in garlic. Thioethers are also used as medicinal preparations, dyes and solvents.

4.3.7.1 Nomenclature of Thioethers

The nomenclature of thioethers is similar to ethers, except that they are named using the suffix -*sulphide*, and when a substituent as *thio-*.

$$H_5C_2-S-CH_3 \qquad HO-CH_2-CH_2-CH_2-S-CH_3$$

Ethyl methyl sulphide   3-Hydroxypropyl methyl sulphide
Methylthioethane   *3-Methylthiopropanol*

4.3.7.2 Physical properties of Thioethers

Thioethers cannot form hydrogen bonds with another thioether molecule, just like ethers. Similarly, they can be involved in hydrogen bonding with water, alcohols and amines. Interestingly, thioethers have higher boiling points than corresponding ethers. For example, dimethyl sulphide boils at 37.0 °C, whereas dimethylether $(CH_3)_2O$ boils at –23.6 °C.

4.3.7.3 Preparations of Thioethers

Thioethers are usually prepared from alkyl halides and sodium thiolates by S_N2 reaction. For example, sodium methanethiolate reacts with CH_3Br in dry THF to yield methylthiomethane.

$$CH_3-Br \;+\; CH_3-\overset{-}{S}\,\overset{+}{Na} \;\xrightarrow{\;THF\;}\; CH_3-S-CH_3 \;+\; NaBr$$

Methyl bromide   Methanethiolate   Methylthiomethane
                                              Dimethyl sulphide

Similarly, the reaction of disulphides with organolithium reagents produces thioethers by S_N2 reaction. For example, methyl lithium reacts with dimethyl disulphide in dry THF to give methylthiomethane.

$$\overset{-}{C}H_3-\overset{+}{Li} \quad + \quad CH_3-S-S-CH_3 \quad \xrightarrow{\text{THF}} \quad CH_3-S-CH_3 \quad + \quad CH_3SLi$$

| Methyl lithium | Dimethyl disulphide | Methylthiomethane
Dimethyl sulphide |

4.3.7.4 Reactions of Thioethers

Thioethers can be easily oxidized to dialkyl sulphoxide (R_2SO) by hydrogen peroxide (H_2O_2) at room temperature. The oxidation can be continued by reacting sulphoxide with concentrated nitric acid (HNO_3) or peroxyacid (RCO_3H) to afford dialkyl sulphone (R_2SO_2). For example, dimethyl sulphide reacts with hydrogen peroxide to produce dimethyl sulphoxide (DMSO), which then treated with concentrated HNO_3 or peracetic acid (CH_3CO_3H) to afford dimethyl sulphone.

$$H_3C-S-CH_3 \xrightarrow[\text{25 °C}]{H_2O_2} H_3C-\overset{\overset{O}{\|}}{S}-CH_3 \xrightarrow[\text{CH}_3\text{CO}_3\text{H}]{\text{HNO}_3 \text{ or}} H_3C-\overset{\overset{O}{\|}}{\underset{\underset{O}{\|}}{S}}-CH_3$$

| Dimethyl sulphide | Dimethyl sulphoxide
DMSO | Dimethyl sulphone |

Hydrogenation of thioethers in the presence of Raney Ni gives alkanes and hydrogen sulphide. For example, methylthioethane is easily hydrogenated to methane, ethane and H_2S.

$$H_3C-S-C_2H_5 + H_2 \xrightarrow{\text{Raney Ni}} H_3C-H + H_5C_2-H + H_2S$$

| Methylthioethane | | Methane | Ethane | Hydrogen sulphide |

Thioethers are readily alkylated to give stable sulphonium salts, when reacted with alkyl halides in dry THF. For example, dimethyl sulphide reacts with methyl iodide to give trimethyl sulphonium iodide.

$$H_3C-S-CH_3 \quad + \quad H_3C-I \quad \xrightarrow{\text{THF}} \quad H_3C-\overset{+}{\underset{\underset{CH_3}{|}}{S}}-CH_3 \quad I^-$$

| Dimethyl sulphide | Trimethyl sulphonium iodide |

4.3.8 Amines

Amines are nitrogen-containing compounds, where the functional group is an amino group (NH_2). They are organic relatives of ammonia, where one or more of the hydrogen atoms of ammonia are replaced by alkyl or aryl groups. Amine with one substituent called a primary amine. If it has two or three substituents, it is then called a secondary or a tertiary amine, respectively. Thus, an amine has the general formula RNH_2, R_2NH or R_3N. The simplest and most common amines are methylamine (CH_3NH_2) and ethylamine ($CH_3CH_2NH_2$).

$$H_3C-NH_2 \qquad H_5C_2-NH_2$$

Methylamine   Ethylamine

Amines are classified as primary (1°), secondary (2°), tertiary (3°) or quaternary (4°) depending on how many alkyl groups are attached to the N atom. Quaternary amines, $(CH_3)_4N^+$ are known as ammonium cations.

$CH_3CH_2CH_2-NH_2$	$H_3C-NH-CH_3$	$H_3C-\underset{CH_3}{\overset{CH_3}{N}}-CH_3$	$H_3C-\underset{CH_3}{\overset{CH_3}{\overset{+}{N}}}-CH_3$
n-Propylamine (1° amine)	Dimethylamine (2° amine)	Trimethylamine (3° amine)	Tetramethylamine (4° amine)

4.3.8.1 Nomenclature of Amines

Aliphatic amines are named by the alkyl group or groups attached to nitrogen with the suffix -*amine*. If there are more than one alkyl groups bonded to nitrogen atom, the prefixes *di*-and *tri*- are used to indicate the presence of two or three alkyl groups of the same kind. Often, the prefix *amino*- is used to the name of the parent chain for more complex amines. If other substituents are attached to the nitrogen atom, they are indicated by the prefix *N*- before the name of the substituents. The simplest aromatic amine is aniline ($C_6H_5NH_2$), where nitrogen is attached directly to a benzene ring.

$H_2N-CH_2CH_2-NH_2$	$H_5C_2-NH-CH_3$	$H_3C-NH-\underset{CH_3}{\overset{}{CH}}(CH_2)_3CH_3$
Ethylenediamine (1° amine)	Methylethylamine (2° amine)	2-(*N*-Methylamino)hexane (2° amine)

Aniline   *o*-Chloroaniline   *p*-Toluidine

Heterocyclic nitrogenous compounds are cyclic compounds containing nitrogen or some other atoms in addition to the carbon within the ring. They are quite abundant in both animals and plants, such as tryptophan (see Section 6.9.4), nicotine and quinine (see Sections 8.2.2.1 and 8.2.2.4).

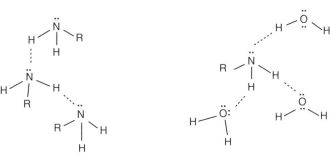

Tryptophan
(An amino acid)

Nicotine from tobacco leaf
(An active ingredient in cigarette smoke)

Quinine form the *Cinchona* tree
(An antimalarial drug)

4.3.8.2 Physical Properties of Amines

In amines, both the C—N and the N—H bonds are polar due to the electronegativity of the nitrogen atom compared to carbon and hydrogen atoms (see Section 2.3.3). The polar nature of the N—H bond results in the formation of intermolecular hydrogen bonds with other primary (1°) and secondary (2°) amines or other hydrogen-bonding systems; for example, water and alcohols. Primary and secondary amines can form hydrogen bonding but tertiary amines do not, because they have no N—H bonds.

Hydrogen bonds between
amine molecules

Hydrogen bonds to water
in aqueous solution

Because of hydrogen bonding, 1° and 2° amines have higher boiling points than analogous alkanes. While comparing alcohols, amines form weaker hydrogen bond as nitrogen is less electronegative than oxygen. Thus, they have lower boiling points than analogous alcohols. As tertiary amines (3°) cannot form hydrogen bonding as a result they are much lower boiling than analogous alcohols or 1° or 2° amines of similar molecular weight.

Name	Molecular formula	Molecular weight	bp (°C)
	1° amines		
Methylamine	CH_3NH_2	31.06	−6.0
Ethylamine	$CH_3CH_2NH_2$	45.08	17.0
Propyl amine	$CH_3CH_2CH_2NH_2$	59.11	49.0
	2° amines		
Dimethylamine	$(CH_3)_2NH$	45.08	7.0
Diethylamine	$(CH_3CH_2)_2NH$	73.14	56.0
Dipropylamine	$(CH_3CH_2CH_2)_2NH$	101.19	110.0
	3° amines		
Trimethylamine	$(CH_3)_3N$	59.11	2.9
Triethylamine	$(CH_3CH_2)_3N$	101.19	90.0
Tripropylamine	$(CH_3CH_2CH_2)_3N$	143.27	156.0

4.3.8.3 Basicity of Amines

Amines, like ammonia, are strong enough bases and completely protonated in dilute acid solutions to form ammonium salts. For example, amines react with strong mineral acids such as HCl and H_2SO_4 to form amine salts. The amine salt consists of two parts, the cationic ammonium ion and the anionic counter ion.

$$CH_3CH_2CH_2-NH_2 \ + \ HCl \longrightarrow CH_3CH_2CH_2-NH_3^+Cl^-$$

n-Propylamine $\qquad\qquad\qquad$ n-Propylammonium chloride

$$(CH_3CH_2)_3N \ + \ H_2SO_4 \longrightarrow (CH_3CH_2)_3NH^+HSO_4^-$$

Triethylamine $\qquad\qquad\qquad$ Triethylammonium hydrogen sulphate

$$\left(CH_3CH_2\right)_3N+H_2SO_4 \rightarrow \left(CH_3CH_2\right)_3NH^+HSO_4^-$$

Triethylamine $\qquad\qquad$ Triethlammonium hydrogen sulphate

Amines are one of the most common naturally occurring biologically active group of compounds. They can exist as ammonium salts in the aqueous environment of blood and other body fluids; for example, serotonin salt and histamine salt.

Serotonin salt
(A neurotransmitter)

Histamine salt
(Causes allergic reaction)

The nitrogen atom of amines has a lone pair of electrons, and they can react as either bases or nucleophiles. Thus, the basicity and the nucleophilicity of amines (NH_2) are quite similar to those of ammonia (NH_3). Amines are basic compounds. The presence of a nonbonding electron pair on nitrogen makes amines act as Lewis bases (see Section 1.7.1.3). Amines are more basic than analogous alcohols and ethers. Arylamines are much less basic than alkylamines. Since the lone pair of electrons on the nitrogen of aniline are conjugated to the π electrons of the aromatic ring, they are therefore less available for an acid–base reaction (see Section 4.7.11.2).

4.3.8.4 Separation of Amine Salts

Amines are generally volatile, smelly liquids, whereas ammonium salts are crystalline, high melting solids. These ionic solids are soluble in water, but insoluble in organic solvents. The free amines are generally insoluble in water, but soluble in organic solvents. This provides an excellent method for the separation and isolation of amine compounds.

When dilute acid is added, amines react with acid to obtain ammonium salt that is soluble in water. When the solution is made alkaline by adding aqueous NaOH, the now purified free amine is regenerated, which is insoluble in the aqueous solution and therefore precipitates, or can be extracted into an organic solvent such as DCM or $CHCl_3$. This procedure is useful for the purification of all amine containing compounds.

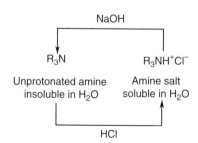

4.3.8.5 Preparations of Amines

Primary amide reacts with Br_2 and base, usually NaOH or KOH, to yield isocyanate. On heating, the isocyanate loses CO_2 (known as decarboxylation) to afford an amine. This reaction is called *Hofmann rearrangement*, not to be confused with *Hofmann degradation or elimination*.

$$R-\overset{\overset{O}{\|}}{C}-NH_2 \xrightarrow[\text{NaOH}]{\text{Br}_2} R-N{=}C{=}O \xrightarrow[\text{Heat}]{\text{H}_2\text{O}} R-NH_2 + CO_2 \uparrow$$

Primary amide   Isocyanate intermediate   Primary amine

Acid chloride reacts with sodium azide (NaN_3) in EtOH to generate acyl azide intermediate, which undergoes *Curtius rearrangement* and loses N_2 gas on heating to give a primary amine.

$$R-\overset{\overset{O}{\|}}{C}-Cl \xrightarrow[\text{EtOH}]{\text{NaN}_3} R-N{=}N{=}N \xrightarrow[\text{Heat}]{\text{H}_2\text{O}} R-NH_2 + N_2 \uparrow$$

Acid chloride   Acyl azide intermediate   Primary amine

Primary amines are synthesized by catalytic hydrogenation or $LiAlH_4$ reduction of primary amines, alkyl azides or alkyl nitriles (see Section 5.9.21). Likewise, secondary and tertiary amides are reduced to corresponding amines by catalytic hydrogenation (H_2/Pd—C or H_2/Pt—C) or $LiAlH_4$ reduction (see Section 5.9.21).

$R-\overset{\overset{O}{\|}}{C}-NH_2$ 1° Amide	H_2/Pd-C or i. LiAlH$_4$ ii. H$_2$O
$RCH_2-\overset{+}{N}{=}\overset{-}{N}{=}NH$ Alkyl azide	H_2/Pd-C or i. LiAlH$_4$ ii. H$_2$O
$R-C{\equiv}N$ Nitrile	$2H_2$/Pd-C or i. LiAlH$_4$ ii. H$_2$O

$\longrightarrow RCH_2 - NH_2$
1° Amine

Catalytic hydrogenation or sodium cyanoborohydride ($NaBH_3CN$) in acetic acid (CH_3CO_2H) reduction of aldehydes and ketones in presence of NH_3, primary (1°) and secondary (2°) amines afford 1°, 2° and 3° amines, respectively. This reaction is known as *reductive amination* (see Section 5.4.3.8). Reduction of the C=N double bond is similar to the reduction of C=O double bond (see Section 5.9.22).

$$R-\overset{\overset{\displaystyle O}{\|}}{C}-Y \quad (Y = H \text{ or } R)$$

$$\xrightarrow[\text{or } H_2/Pd\text{-}C]{\underset{NaBH_3CN}{NH_3}} R-\overset{\overset{\displaystyle NH_2}{|}}{CH}-Y \quad 1°\text{ Amine}$$

$$\xrightarrow[\text{or } H_2/Pd\text{-}C]{\underset{NaBH_3CN}{R'NH2}} R-\overset{\overset{\displaystyle NHR'}{|}}{CH}-Y \quad 2°\text{ Amine}$$

$$\xrightarrow[\text{or } H_2/Pd\text{-}C]{\underset{NaBH_3CN}{R'_2NH}} R-\overset{\overset{\displaystyle NR'_2}{|}}{CH}-Y \quad 3°\text{ Amine}$$

4.3.8.6 Reactivity of Amines

The $-NH$ group is a poor leaving group like the $-OH$ group, and needs to be converted to a better leaving group before nucleophilic substitution can occur. The anion derived from the deprotonation of an amine is the amide ion, NH_2^- and should not be confused with the carboxylic acid derivative amide, $RCONH_2$. Amide ions are important bases in organic reactions.

4.3.8.7 Reactions of Amines

Ammonia (NH_3) primary (RNH_2) and secondary (R_2NH) amines undergo nucleophilic acyl substitution with acid chlorides or anhydrides in pyridine or Et_3N to produce 1°, 2° and 3° amides (see Section 5.6.5.7). Ammonia (NH_3) and primary amines (RNH_2) react with aldehydes or ketones via nucleophilic addition reaction, followed by the loss of H_2O to give imines, also known as *Schiff's bases* (see Section 5.4.3.8). Likewise, secondary amines (R_2NH) react with aldehydes and ketones to yield enamines after loss of a proton (see Section 5.4.3.9).

$$RCH_2-\overset{\overset{\displaystyle NCH_2R'}{\|}}{C}-Y \xleftarrow[Y = H \text{ or } R]{RCH_2CO-Y} R'CH_2-NH_2 \xrightarrow[(RCO)_2O]{RCOCl \text{ or}} RCONHCH_2R'$$

Imine · · · 1° Amine · · · N-subtitued amide (2° amide)

$$RCH=\overset{\overset{\displaystyle NR'_2}{|}}{C}-Y \xleftarrow[Y = H \text{ or } R]{RCH_2CO-Y} R'_2-NH \xrightarrow[(RCO)_2O]{RCOCl \text{ or}} RCONR'_2$$

Enamine · · · 2° Amine · · · N,N-disubstituted amide (3° amide)

The amides derived from sulphonic acids are called *sulphonamides*. They are produced from amines by the reaction with sulphonyl chloride (R′SOCl$_2$) in pyridine.

$$R'CH_2-NH_2 \xrightarrow[\text{Pyridine}]{RSO_2Cl} RSO_2NHCH_2R' \qquad R'_2-NH \xrightarrow[\text{Pyridine}]{RSO_2Cl} RSO_2NR_2$$

1° Amine   *N*-substituted   2° Amine   *N,N*-disubstituted
                            sulphonamide   sulphonamide
                            (2° *amide*)   (3° *amide*)

Amines can readily be converted to quaternary ammonium salt by reacting with excess primary alkyl halides, followed by aqueous silver oxide and heat. The base-catalysed quaternary ammonium salts when heated provide alkenes. This is an E2 elimination reaction, where a proton and a tertiary amine are eliminated to yield alkene. This reaction is known as *Hofmann degradation or elimination*, not to be confused with *Hofmann rearrangement*.

1° Amine   Quarternary ammonium salt   Alkene   3° Amine

4.4 CARBONYL COMPOUNDS

Carbonyl compounds are molecules that contain a carbonyl functional group (C=O), which is a carbon double-bonded to an oxygen atom. An acyl functional group (R—C=O) consists of a carbonyl group attached to an alkyl or an aryl group. A carbonyl group containing compounds can be classified into two broad classes: one group includes compounds that have hydrogen and carbon atoms bonded to the carbonyl carbon, and the other group contains an electronegative atom such as oxygen, chloride and nitrogen bonded to the carbonyl carbon (see Section 4.4.2.9). The carbonyl group is of central importance in organic chemistry because of its abundance. Numerous reactions with aldehydes, ketones, carboxylic acids, esters, acid chlorides and amides have been shown earlier (see Sections 4.3.4.4, 4.3.4.7 and 4.3.8.7).

Carbonyl group   Acyl group (R = alkyl or aryl)

4.4.1 Aldehydes and Ketones

Aldehydes have an acyl group with a hydrogen atom bonded to the carbonyl carbon. The most abundant natural aldehyde is glucose. The simplest aldehyde is formaldehyde (CH_2O), where the carbonyl carbon is bonded to two hydrogen atoms. In all other aldehydes, the carbonyl carbon is bonded to one hydrogen atom and one alkyl or aryl group; for example, acetaldehyde (CH_3CHO) and benzaldehyde (Ph—CHO). Many aldehydes have distinctive strong odours. Benzaldehyde smells like cherries and is found naturally in many fruits such as peaches, grapes and cranberries. It is used as an artificial flavouring in many foods at low levels, but is toxic at higher levels.

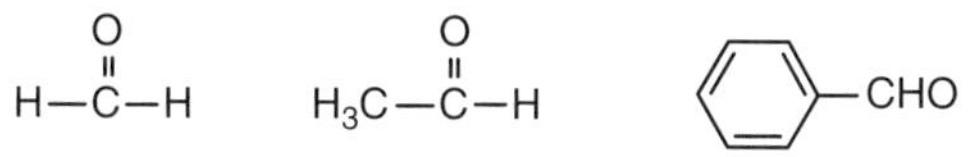

Formaldehyde　　　Ethanal　　Benzaldehyde
Acetadehyde

Ketones have an acyl group with another alkyl or aryl group connected to the carbonyl carbon. The simplest ketone is acetone (CH_3COCH_3), where the carbonyl carbon is bonded to two methyl groups. Acetone and methyl ethyl ketone (MEK) are important solvents and commonly used in organic synthesis. Ketones are used in paints, lacquers, textiles and so on. They are also useful for tanning, preservation and in applications of hydraulic fluids. Many steroid hormones, for example, testosterone and progesterone, contain ketone functionality.

Propanone　　　　　　　　Butanone
Acetone　　　　Methyl ethyl ketone (MEK)

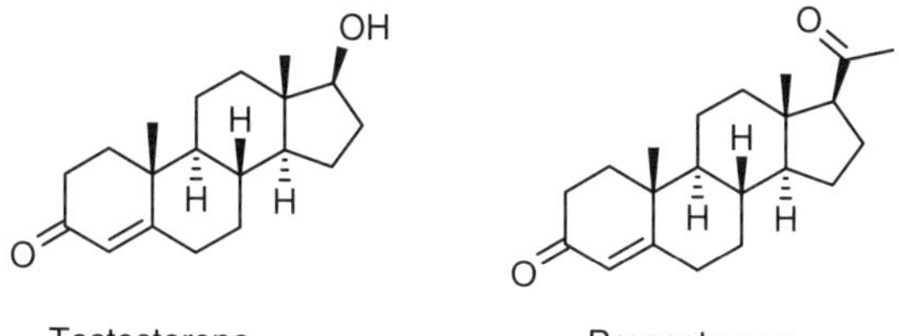

Testosterone　　　　　　　Progesterone

4.4.1.1 Nomenclature of Aldehydes and Ketones

The common names of aldehydes are derived from the corresponding carboxylic acids by replacing *-ic acid* by *-aldehyde*; for example, formic acid and acetic acid give formaldehyde and acetaldehyde, respectively. The simplest ketone has the common name of acetone. In the IUPAC nomenclature of aldehydes, *-e* of alkane is replaced with *-al*, for example, ethan*al* (parent alkane is ethan*e*). Similarly, ketones are named by replacing the *-e* ending of the alkyl name with *-one*, for example, propan*one* (parent alkane is propane). The longest chain carrying the carbonyl group is considered the parent structure. The carbonyl carbon is the first carbon atom of the chain. Other substituents are named using prefixes and their positions are indicated by numbers relative to the carbonyl group. If the aldehyde group is a substituent on a ring, the suffix -*carbaldehyde* is used in the name.

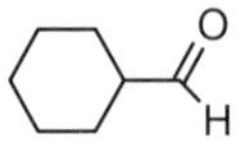

| Cyclohexanecarbaldehyde | 3-Methyl-2-butanone | 4-Chlorocyclohexanone |

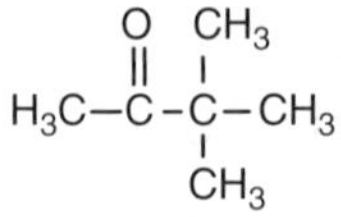

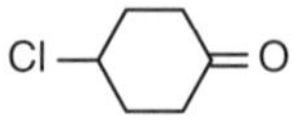

In certain polyfunctional compounds, an aldehyde or ketone group can also be named as a substituent on a molecule with another functional group as its root. The aldehyde carbonyl is given the prefix *formyl-*, and the ketone group is named as a group *oxo-* with a number to show its position in the molecule. Compounds with both an aldehyde and ketone are named as aldehydes, because aldehydes have functional group priority over ketones. A ketone containing a benzene ring is named as a -*phenone*.

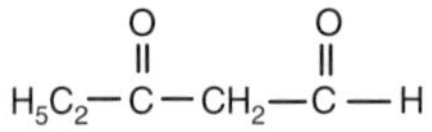

| 3-Oxopentanal | 3-Oxobutanoic acid | 4-Hydroxy-2-butanone |

| 2-Formylbenzoic acid | Acetophenone | Benzophenone |

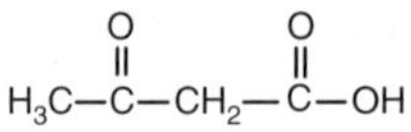

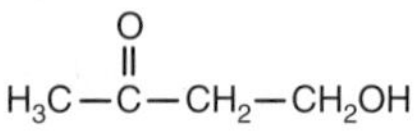

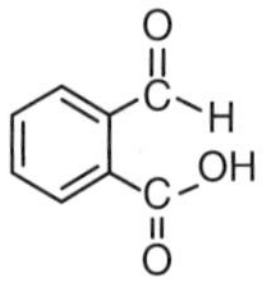

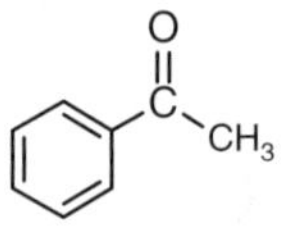

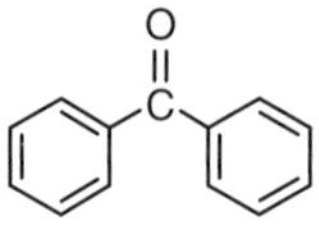

4.4.1.2 Physical Properties of Aldehydes and Ketones

The carbonyl oxygen atom is a Lewis base and can be readily protonated in the presence of an acid. The polar nature of the C=O group is due to the electronegativity difference of the carbon and oxygen atoms (see Section 2.3.3). The C=O group cannot form intermolecular hydrogen bonding, but it can accept hydrogen from hydrogen bond donors; for example, water, alcohols and amines. Therefore, aldehydes and ketones have higher boiling points compared to analogous alkanes and ethers, and much lower boiling points than analogous alcohols. They are much more soluble than alkanes, but less soluble than analogous alcohols in aqueous medium; for example, acetone and acetaldehyde are miscible with water.

Name	Molecular formula	Molecular weight	bp (°C)
n-Butane	$CH_3CH_2CH_2CH_3$	58.12	−6.0
Methyl ethyl ether	$CH_3OCH_2CH_3$	60.10	7.4
Propanal	CH_3CH_2CHO	58.10	48.8
n-Propanol	$CH_3CH_2CH_2OH$	60.10	97.0

Aldehydes and ketones are chemically identified by oxidation reaction. Aldehydes are easily oxidized, whereas ketones are not. In the Tollens' silver mirror test, aldehyde is oxidized to carboxylic acid and produces a silver mirror plate on the side of the test tube as the silver ion is reduced to silver metal (see Section 5.9.9).

4.4.1.3 Preparations of Aldehydes and Ketones

Aldehydes are prepared by selective oxidation of primary alcohols (see Section 5.9.8) or by hydroboration-oxidation of alkynes (see Section 5.4.2.9), by partial and selective reduction of acid chlorides (see Section 5.9.17), esters (see Section 5.9.18) or nitriles (see Section 5.9.21) with lithium tri-*tert*-butoxyaluminium hydride [LiAlH(O–^{t}Bu)$_3$] and diisobutylaluminium hydride (DIBAH), respectively.

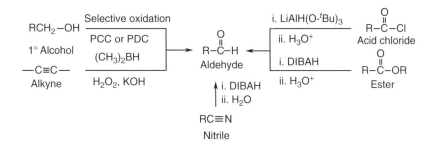

Ketones are prepared by oxidation of secondary alcohols (see Section 5.9.8), and reduction of acid chlorides by the treatment of Gilman reagents (R′$_2$CuLi) followed by hydrolytic work-up (see Section 5.6.5.8). Ketones are also obtained conveniently by the reaction of nitrile and Grignard reagents (R′MgBr) or organolithium (RLi) followed by acidic work-up (see Section 5.4.3.4).

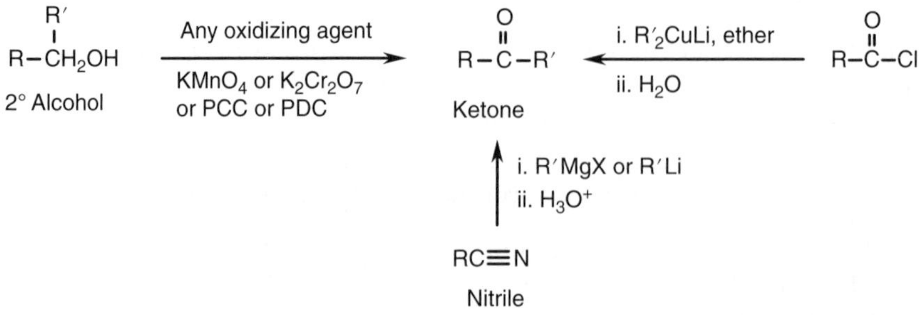

Aldehydes and ketones are produced by ozonolysis of alkenes (see Section 5.9.6), and addition of water or hydration of alkynes (see Section 5.4.2.6).

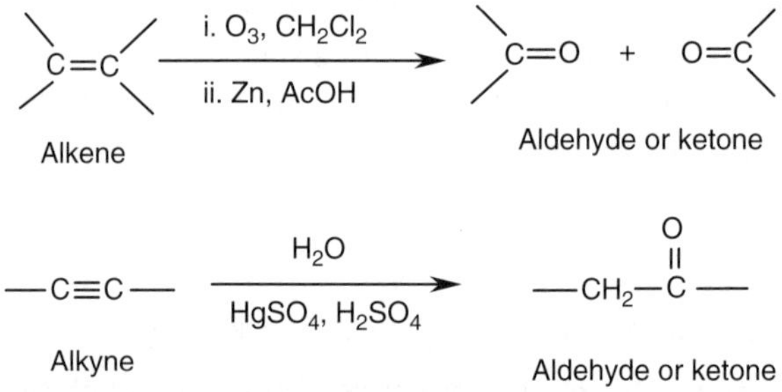

4.4.1.4 Structure and Reactivity of Carbonyl Group

The carbonyl group of aldehydes and ketones is highly polarized, because carbon is less electronegative than oxygen. The carbonyl carbon bears a partial positive charge (δ^+), while the oxygen bears a partial negative charge (δ^-). Therefore, the carbonyl group can function as both a nucleophile and an electrophile.

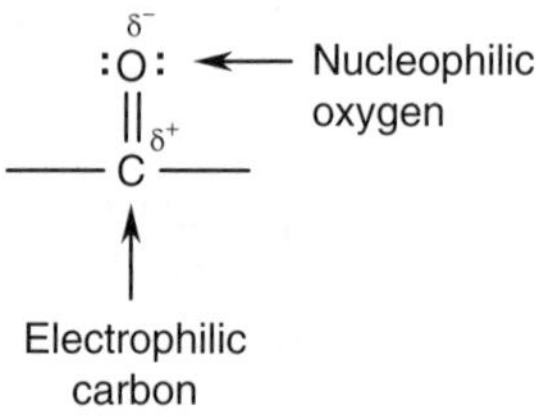

Aldehydes and ketones cannot undergo substitution reaction, because they do not have a leaving group. Therefore, the common carbonyl group reactions are nucleophilic additions. The carbonyl oxygen is weakly basic, so the acid-catalysed reactions occur with the protonation of the nucleophilic oxygen, followed by the attack of the weaker nucleophile on the electrophilic carbonyl carbon. On the other hand, the base-catalysed addition reactions to carbonyl

compounds results from initial attack of a strong nucleophile on the electrophilic carbonyl carbon.

Aldehydes are more reactive than ketones. Two factors that make aldehydes
more reactive than ketones are electronic and steric effects. Ketones have two
alkyl groups, whereas aldehydes have only one. Because alkyl groups are electron
donating, ketones have their effective partial positive charge reduced more than
aldehydes. The electrophilic carbon is the site where the nucleophile approaches
for reaction to occur. In ketones, two alkyl groups create more steric hindrance
than one in aldehydes. As a result, ketones offer more steric resistance towards
the nucleophilic attack than aldehydes.

4.4.1.5 Reactions of Aldehydes and Ketones: Nucleophilic Addition

Carbonyl compounds are an important group of organic compounds owing to their
unique ability to form a range of other derivatives. Grignard reagents, organolithium reagents, lithium aluminium hydride ($LiAlH_4$) and sodium borohydride ($NaBH_4$)
react with carbonyl compounds by nucleophilic addition. As shown earlier, alcohols
can be prepared from aldehydes and ketones by nucleophilic addition of organometallic reagents (see Section 5.4.3.1), catalytic hydrogenation (H_2/Pd—C) and
metal hydride reduction; for example, $NaBH_4$ or $LiAlH_4$ (see Section 5.9.14). Aldehydes and ketones are selectively reduced to alkanes by *Clemmensen reduction*
(see Section 5.9.15) or *Wolff–Kishner reduction* (see Section 5.9.16) and to amines
by reductive amination (see Section 5.9.22). Aldehydes are readily oxidized to
carboxylic acids by a number oxidizing reagents, including CrO_3 in aqueous acid
(see Section 5.9.9).

Aldehydes and ketones are easily hydrated in presence of an acid or a base to
form *gem*-diols and the reaction is reversible (see Section 5.4.3.10). In a similar
fashion, aldehydes and ketones react with alcohols in presence of an acid to form
acetals and ketals, respectively (see Section 5.4.3.11). Nucleophilic addition of HCN
to aldehydes and ketones yield cyanohydrins via the base-catalysed addition mechanism (see Section 5.4.3.7).

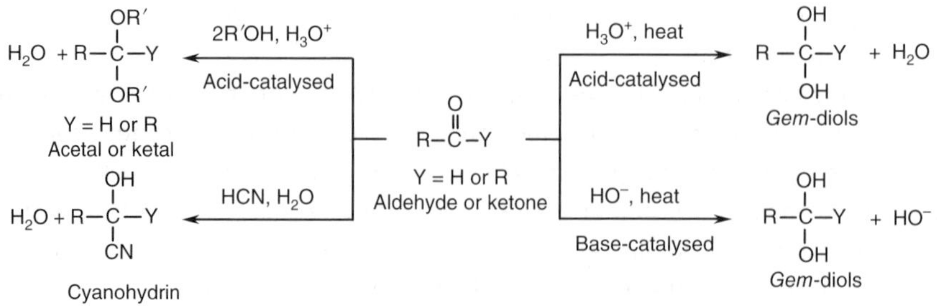

Aldehydes and ketones also condense with other ammonium derivatives, such as ammonia, primary amine, hydroxylamine, hydrazine, pheynylhydrazine and semicarbazide to provide imines, Schiff's bases, oximes, hydrazones and semicarbazones (see Section 5.4.3.8). Generally, these reactions are better than the analogous amine reactions and give superior yields. They are often used in the organic chemistry labs for characterization and identification of the original carbonyl compounds by melting point comparison.

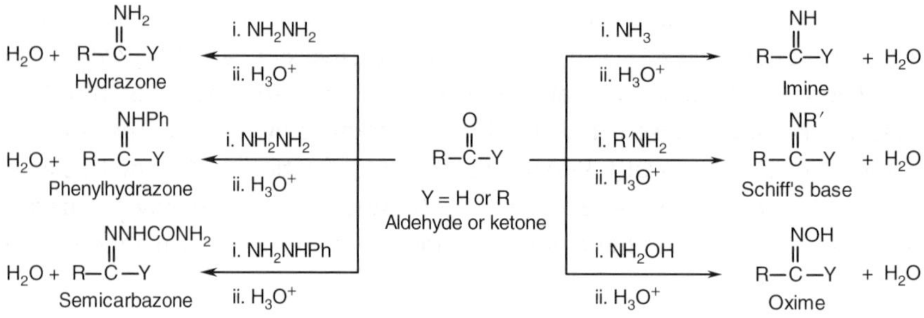

4.4.1.5.1 Aldol Addition: Base-Catalysed

One of the most important reactions of aldehydes and ketones is the *aldol addition*. This is one of the fundamental C—C single bond forming processes of synthetic organic chemistry. In the aldol addition reaction, two molecules of aldehyde or ketone are reacted with a strong base, such as aqueous NaOH and at low temperature (5 °C), to afford β-hydroxyaldehyde or β-hydroxyketone, respectively. An aldol addition reaction is a nucleophilic addition in which an enolate is the nucleophile (see Section 5.4.3.12).

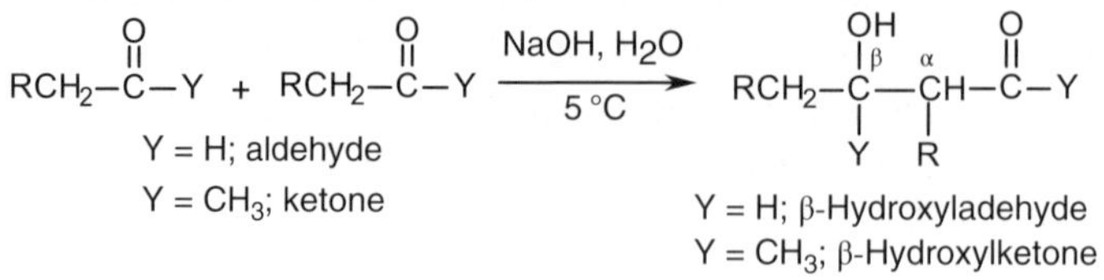

4.4.1.5.2 Aldol Condensation: Base-Catalysed

The aldol condensation is one of the most important synthetic routes for the formation of conjugated aldehydes and ketones. Under a severe basic condition and at higher temperatures, the aldol addition product can be dehydrated to produce α,β-unsaturated aldehyde and ketone, a double bond next to the carbonyl carbon. For example, two molecules of aldehyde or ketone are dehydrated in presence of highly concentrated base, NaOH (1 M) and at higher temperature (80 °C), to yield an α,β-unsaturated aldehydes or ketones. This is an addition–elimination reaction and the overall process is called an *aldol condensation* (see Section 5.4.3.12).

Y = H; aldehyde
Y = CH$_3$; ketone

Y = H; β-Hydroxyaldehyde
Y = CH$_3$; β-Hydroxyketone

α,β-Unsaturated aldehyde or ketone

4.4.1.5.3 Mixed-Aldol Condensation: Base-Catalysed

Generally, a mixed-aldol condensation occurs when two different molecules containing carbonyl groups are combined. This reaction is widely known as the *Claisen–Schmidt condensation*, named after two of its original investigators Rainer Ludwig Claisen and J. G. Schmidt, who independently published the same findings in 1880 and 1881, respectively.

Since ketones are less reactive towards nucleophilic addition, the enolate formed from a ketone can be used to react with an aldehyde. Therefore, a mixed-aldol condensation usually occurs between an aldehyde that has no α-hydrogens, and a ketone that has α-hydrogens, so that the nucleophile is generated solely from the ketone (see Section 5.4.3.12). Since ketones are less reactive towards nucleophilic addition, the enolate formed from a ketone can be used to react with a highly electrophilic aldehyde to give an α,β-unsaturated ketone. The reaction mechanism is similar to the aldol condensation (see Section 5.4.3.12).

Aldehydes with no α-H   Ketone. with α-Hs

α, β-Unsaturated ketone
Y = H, Ph

4.4.2 Carboxylic acids

A *carboxylic acid* is an organic acid that has an acyl group (R—C=O) linked to a hydroxyl group (—OH). In a condensed structural formula, a carboxyl group may be written as —CO_2H, and a carboxylic acid as RCO_2H.

$$\underset{\text{Carboxyl group}}{\underset{\text{—CO}_2\text{H or —COOH}}{-\overset{\overset{\displaystyle O}{\|}}{C}-OH}} \qquad \underset{\text{Carboxylic acid}}{\underset{\text{RCO}_2\text{H or RCOOH}}{R-\overset{\overset{\displaystyle O}{\|}}{C}-OH}}$$

Carboxylic acids are classified as *aliphatic acids*, where an alkyl group is bonded to the carboxyl group, and *aromatic acids*, where an aryl group is bonded to the carboxyl group. The simplest acids are formic acid (HCO_2H), acetic acid (CH_3CO_2H) and benzoic acid (Ph—CO_2H).

$$\underset{\substack{\text{Methanoic acid}\\ \textit{Formic acid}}}{H-\overset{\overset{\displaystyle O}{\|}}{C}-OH} \qquad \underset{\substack{\text{Ethanoic acid}\\ \textit{Acetic acid}}}{H_3C-\overset{\overset{\displaystyle O}{\|}}{C}-OH} \qquad \underset{\substack{\text{Benzenecarboxylic acid}\\ \textit{Benzoic acid}}}{-\overset{\overset{\displaystyle O}{\|}}{C}-OH}$$

4.4.2.1 Nomenclature of Carboxylic Acids

The root name is based on the longest continuous chain of carbon atoms bearing the carboxyl group. The *-e* is replaced by *-oic acid*. The chain is numbered starting with the carboxyl carbon atom. The carboxyl group takes priority over any other functional groups as follows: carboxylic acid > ester > amide > nitrile > aldehyde > ketone > alcohol > amine > alkene > alkyne.

$$\underset{\substack{\text{Propanoic acid}\\ \textit{Propionic acid}}}{CH_3CH_2-\overset{\overset{\displaystyle O}{\|}}{C}-OH} \qquad \underset{\substack{\text{Butanoic acid}\\ \textit{Butyric acid}}}{CH_3CH_2CH_2-\overset{\overset{\displaystyle O}{\|}}{C}-OH} \qquad \underset{\substack{\text{Pentanoic acid}\\ \textit{Valeric acid}}}{CH_3CH_2CH_2CH_2-\overset{\overset{\displaystyle O}{\|}}{C}-OH}$$

$$\underset{\substack{\text{Propenoic acid}\\ \textit{Acrylic acid}}}{H_2C=CH-\overset{\overset{\displaystyle O}{\|}}{C}-OH} \qquad \underset{\text{2-Methoxybutanoic acid}}{CH_3CH_2CH_2-\overset{\overset{\displaystyle O}{\|}}{C}-OH} \qquad \underset{\text{4-Aminobutanoic acid}}{H_2N\diagup\diagdown\diagup\overset{\overset{\displaystyle O}{\|}}{}OH}$$

Cycloalkanes with carboxyl substituents are named as cycloalkane carboxylic acids. Unsaturated acids are named using the name of the alkene with *-e* replaced with -*ioc acid*. The chain is numbered starting with the carboxyl group, a number designates the location of the double bond and then use *Z* or *E*.

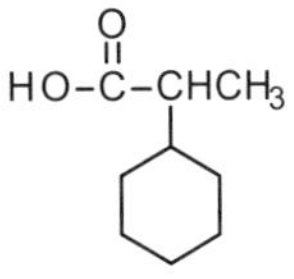

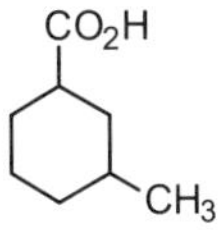

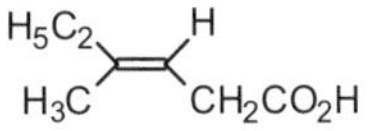

2-Cyclohexylpropanoic acid   3-Methylcyclohexanecarboxylic acid   (*E*)-4-Methyl-3-hexenoic acid

Aromatic acids are named as derivatives of benzoic acids, with *ortho-*, *meta-* and *para-* indicating the location relative to the carboxyl group.

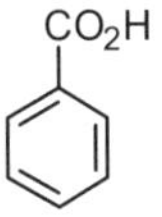

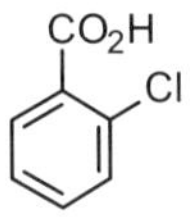

Benzoic acid
Benzenecarboxylic acid

o-Chlorobenzoic acid

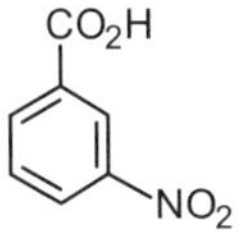

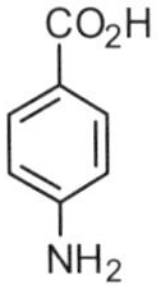

m-Nitrobenzoic acid   *p*-Aminobenzoic acid

Aliphatic dicarboxylic acids are named by simply adding the suffix -*dioic acid* to the root name. The root name comes from the longest carbon chain containing both carboxyl groups. Numbering starts at the end closest to a substituent.

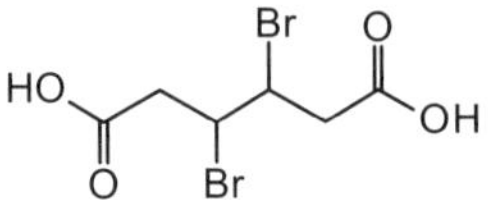

3,4-Dibromohexanedioc acid

4.4.2.2 Structure of the Carboxyl Group

The most stable conformation of a carboxyl group is a planar arrangement of the molecule. The carbon is sp^2-hybridized, and the O—H bond lies in the plane, eclipsing the C=O double bond. This unexpected geometric arrangement can be explained by resonance. The following resonance forms can be written for a carboxyl group.

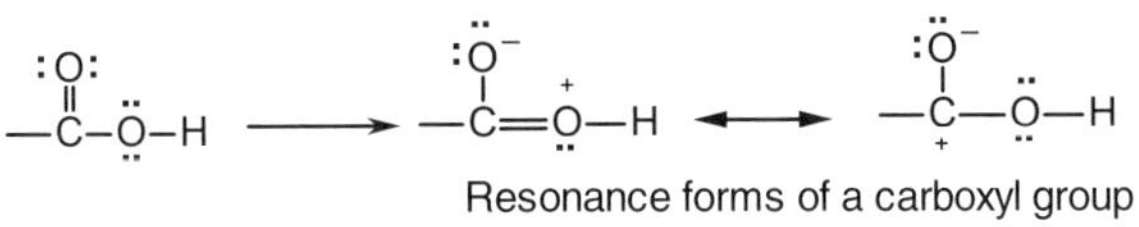

Resonance forms of a carboxyl group

4.4.2.3 Acidity of Carboxylic Acids

Although carboxylic acids are much weaker acids than the strong mineral acids, for example, HCl, H_2SO_4 and HNO_3, they can dissociate in aqueous solution and form carboxylate (RCO_2^-) ion. The equilibrium constant for this process is $K_a = \sim 10^{-5}$ ($pK_a = \sim 5$). Carboxylic acids are more acidic than analogous alcohols. For example, the pK_a values of ethanoic acid and ethanol are 4.74 and 15.9, respectively.

$$R-\overset{\overset{\displaystyle :O:}{\|}}{C}-\ddot{O}-H \;+\; H_2\ddot{O} \;\rightleftharpoons\; R-\overset{\overset{\displaystyle :O:}{\|}}{C}-\ddot{O}^- \;+\; H_3\ddot{O}^+$$

Name	Molecular formula	pK_a	Name	Molecular formula	pK_a
Methanol	CH_3OH	15.5	Formic acid	HCO_2H	3.75
Ethanol	C_2H_5OH	15.9	Acetic acid	CH_3CO_2H	4.76
n-Propanol	C_3H_7OH	16.0	Propionic acid	$C_2H_5CO_2H$	4.87
n-Butanol	C_4H_9OH	16.1	Butyric acid	$C_3H_7CO_2H$	4.82

4.4.2.4 Substituent Effects on the Acidity of Carboxylic Acids

Any substituent that stabilizes a negative charge enhances the dissociation process, that is, increase the acidity. Electronegative elements can enhance the acid strength through inductive effects. The closer the substituent to the anion, the more profound the effects are.

$pK_a = 4.76$ $pK_a = 2.86$ $pK_a = 1.48$ $pK_a = 0.64$

4.4.2.5 Carboxylate Salts

Carboxylic acids are more acidic than alcohols and acetylene. They are readily deprotonated by sodium hydroxide or sodium bicarbonate to form carboxylate salts. Strong aqueous mineral acids readily convert the salt back to the carboxylic acids. Carboxylate salts are soluble in water, but insoluble in non-polar solvents; for example, n-hexane or dichloromethane.

$$R-\overset{\overset{\displaystyle :O:}{\|}}{C}-\ddot{O}-H \;\underset{H^+}{\overset{HO^-\;or\;HCO_3^-}{\rightleftharpoons}}\; R-\overset{\overset{\displaystyle :O:}{\|}}{C}-\ddot{O}^-$$

Carboxylate salt

4.4.2.6 Physical Properties of Carboxylic Acids

Carboxylic acids are polar molecules due to the polar nature of both the O—H and C=O functionalities. They form strong hydrogen bonds with other carboxylic acid molecules (intermolecular hydrogen bonds) or water molecule. Therefore, carboxylic acids have higher boiling points than analogous alcohols as shown.

Name	Molecular formula	Molecular weight	bp (°C)	Solubility in 1 00 g H$_2$O
Formic acid	HCO$_2$H	46.03	100.8	Infinite
Ethanol	CH$_3$CH$_2$OH	46.07	78.3	Infinite
Acetic acid	CH$_3$CO$_2$H	60.05	118.1	Infinite
n-Propanol	CH$_3$CH$_2$CH$_2$OH	60.10	97.0	Infinite
Propionic acid	CH$_3$CH$_2$CO$_2$H	74.08	141.0	Infinite
n-Butanol	CH$_3$CH$_2$CH$_2$CH$_2$OH	74.12	118.0	8.0

Carboxylic acids are highly soluble in aqueous media due to the hydrogen bonding like analogous alcohols.

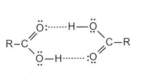

Hydrogen bonds between
two molecule of carboxylic acid

Hydrogen bonds to water in
aqueous solution

4.4.2.7 Preparations of Carboxylic Acids

Acetic acid, the most important of all carboxylic acids, can be prepared by catalytic air oxidation of acetaldehyde.

$$H_3C-\overset{O}{\overset{\|}{C}}-H \quad \xrightarrow[\text{Catalyst}]{O_2} \quad H_3C-\overset{O}{\overset{\|}{C}}-OH$$

Acetaldehyde   Acetic acid

Carboxylic acids are synthesized by hydrolysis of acid chlorides and acid anhydrides (see Section 5.8.1.1), and acidic or basic hydrolysis of esters (see Section 5.8.1.2), primary amides and nitriles (see Section 5.8.1.4).

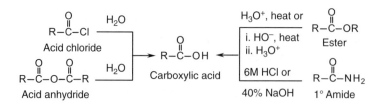

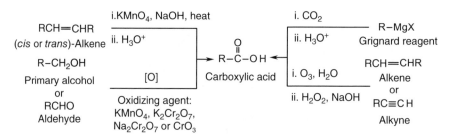

Carboxylic acids can also be prepared from oxidation of alkenes (see Section 5.9.2), alkynes (see Section 5.9.3), primary alcohols (see Section 5.9.8) and aldehydes (see Section 5.9.9), ozonolysis of alkenes and alkynes with oxidative work-up (see Sections 5.9.6 and 5.9.7) and carbonation of Grignard reagents (see Section 5.4.3.3).

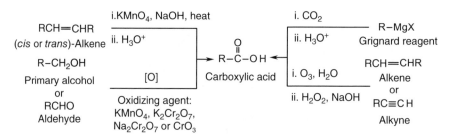

Benzoic acid derivatives are prepared by the oxidation of alkylbenzenes with hot sodium dichromate ($Na_2Cr_2O_7$) or hot potassium chromate ($KMnO_4$). For example, p-chlorotoluene is conveniently oxidized to p-chlorobenzoic acid, using either hot $Na_2Cr_2O_7$ or $KMnO_4$.

CI—⟨benzene⟩—CH₃ $\xrightarrow[\text{or } KMnO_4, H_2O, \text{ heat}]{Na_2Cr_2O_7, H_2SO_4, \text{ heat}}$ CI—⟨benzene⟩—CO₂H

p-Chlorotoluene   p-Chlorobenzoic acid

4.4.2.8 Reactions of Carboxylic Acids

The most important reactions of carboxylic acids are the conversion to various carboxylic acid derivatives; for example, acid chlorides, acid anhydrides and esters. Esters are obtained by the reaction of carboxylic acids and alcohols. The reaction is acid-catalysed and known as *Fischer esterification* (see Section 5.6.5.1).

Acid chlorides are synthesized from carboxylic acids by the treatment of thionyl chloride (SOCl$_2$) or oxalyl chloride [(COCl)$_2$], and acid anhydrides are produced from two carboxylic acids. A summary of the conversion of carboxylic acid is presented next. All these conversions involve *nucleophilic acyl substitutions* (see Section 5.6.5).

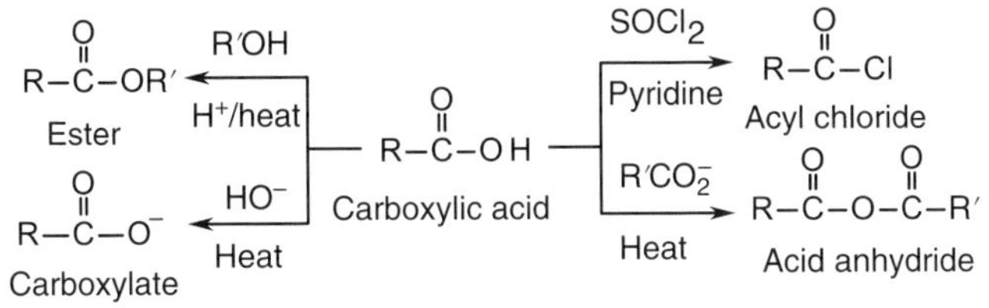

4.4.2.9 Carboxylic Acid Derivatives

Carboxylic acid derivatives are compounds that possess an acyl group (R—C=O) linked to an electronegative atom; for example, —Cl, —CO$_2$R, —OR, —NH$_2$. They can be converted to carboxylic acids via simple acidic or basic hydrolysis. The important acid derivatives are acid chlorides, acid anhydrides, esters and amides. Usually nitriles are also considered as carboxylic acid derivatives. Although nitriles are not directly carboxylic acid derivatives, they are conveniently hydrolysed to carboxylic acids by acid or base catalysts. Moreover, nitriles can be easily prepared through dehydration of amides, which are carboxylic acid derivatives.

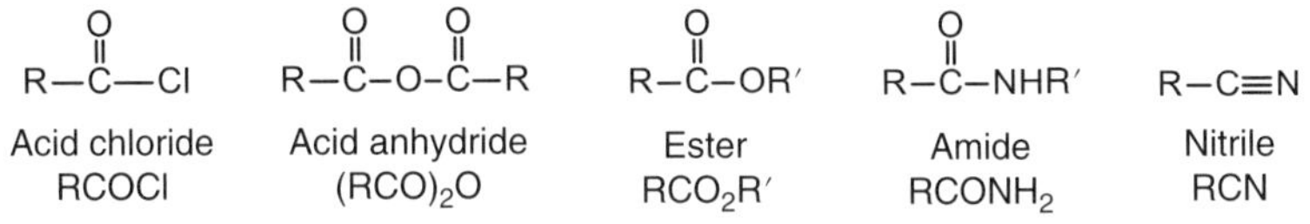

4.4.2.10 Reactivity of Carboxylic Acid Derivatives

The reactivity of carboxylic acid derivatives varies greatly in the nucleophilic acyl substitution reactions, where one nucleophile replaces a leaving group on the acyl carbon. In general, it is easy to convert more reactive derivatives into less reactive derivatives. Therefore, an acid chloride is easily converted to an anhydride, ester or amide, but an amide can only be hydrolysed to a carboxylic acid. Acid chlorides and acid anhydrides are hydrolysed easily, whereas amides are hydrolysed slowly in boiling alkaline water.

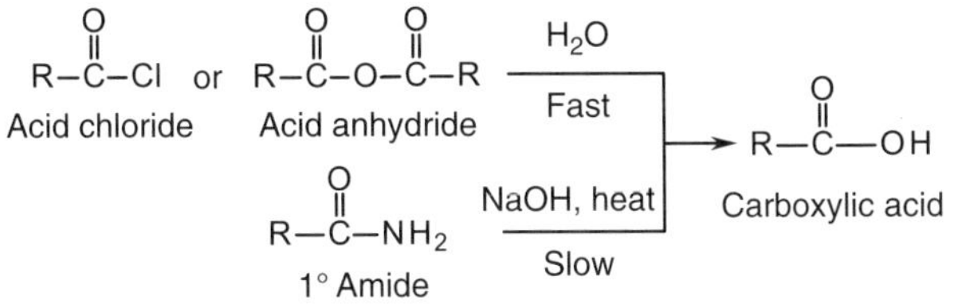

The reactivity of carboxylic acid derivatives depends on basicity of the substituent attached to the acyl group. Therefore, the less basic the substituent, the more reactive is the derivative. In other words, strong bases make poor leaving groups. Carboxylic acid derivatives undergo a variety of reactions under both acidic and basic conditions, and almost all involve nucleophilic acyl substitution mechanism (see Section 5.6.5).

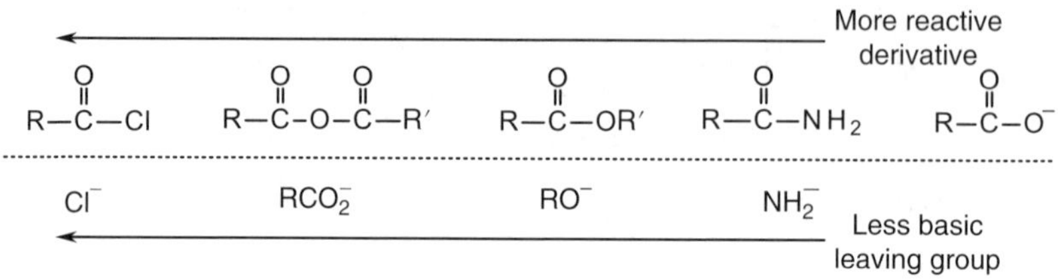

4.4.3 Acid Chlorides

The functional group of an acid chloride, also known as acyl chloride, is an acyl group bonded to a chlorine atom. The simplest member of this family is acetyl chloride (CH_3COCl), where the acyl group is bonded to a chlorine atom.

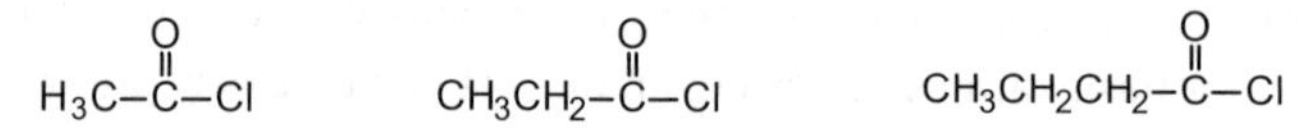

Ethanoyl chloride   Propanoyl chloride   Butanoyl chloride
Acetyl chloride

4.4.3.1 Nomenclature of Acid Chlorides

Acid chlorides are named by replacing the -ic acid ending with -yl chloride or replacing the -carboxylic acid ending with -carbonyl chloride.

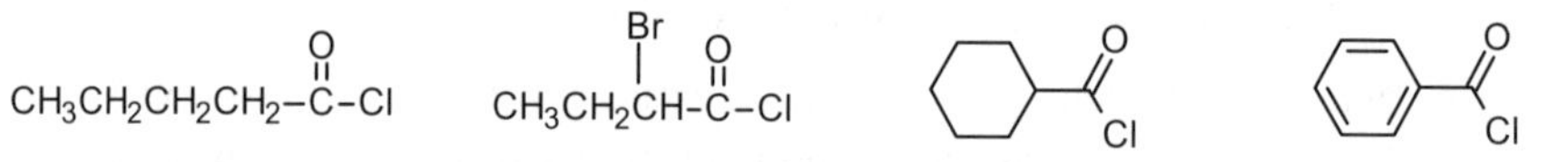

Pentanoyl chloride   3-Bromobutanoyl chloride   Cyclohexanecarbonyl   Benzoyl chloride
chloride

4.4.3.2 Preparations of Acid Chlorides

Acid chlorides are prepared from the corresponding carboxylic acids, most commonly from the reaction with thionyl chloride ($SOCl_2$) or oxalyl chloride, $(COCl)_2$ (see Section 5.6.5). Both these methods are excellent, since they generate gaseous by-products (SO_2, HCl from thionyl chloride and CO_2, CO, HCl from oxalyl chloride) and therefore do not contaminate the acid chloride product.

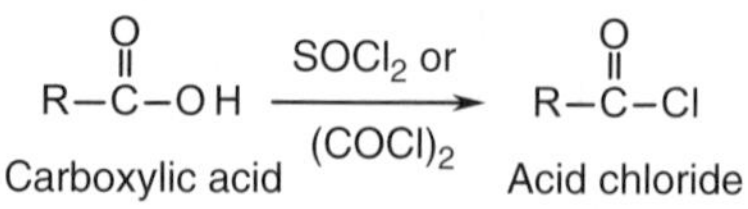

Carboxylic acid   Acid chloride

4.4.3.3 Reactions of Acid Chlorides

Acid chlorides are the most reactive carboxylic acid derivatives, because both the carbonyl oxygen and the chlorine are electron-withdrawing atoms that make the carbonyl carbon very electrophilic. Thus, acid chlorides are highly reactive with nucleophiles and undergo *nucleophilic acyl substitution*. Acid chlorides are conveniently transformed to esters (see Section 5.6.5.5), acid anhydrides (see Section 5.6.5.6) or amides (see Section 5.6.5.7), via nucleophilic acyl substitutions (*see* Section 5.6.5). They are highly reactive with H_2O, and quite readily hydrolysed to carboxylic acid without any catalysts (see Section 5.8.1.1).

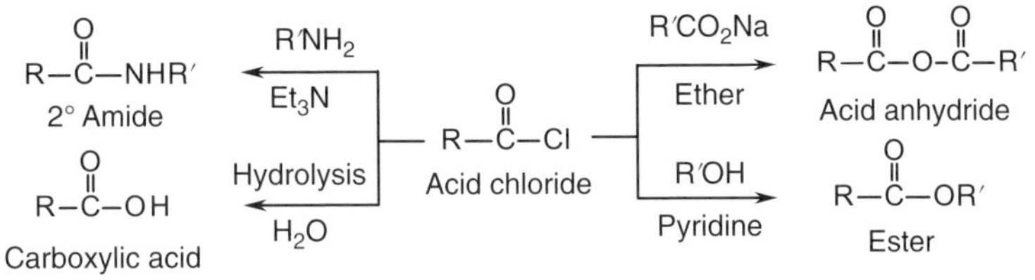

Acid chlorides are easily converted to 1° alcohols and aldehydes (see Section 5.9.17), and 3° alcohols and ketones through the choice of appropriate metal hydride and organometallic reagents (see Sections 5.9.17 and 5.6.5.8).

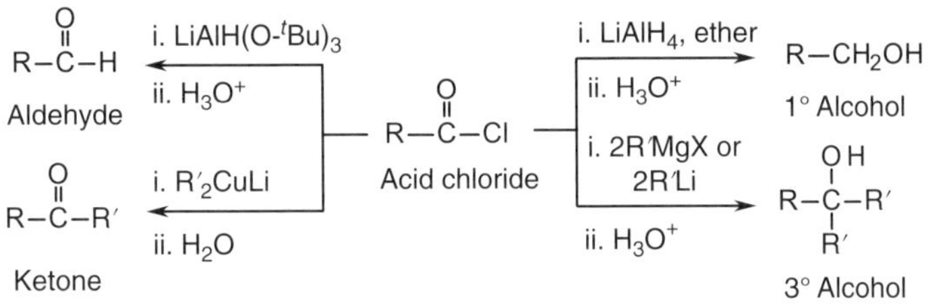

Acetyl chloride reacts with excess benzene in presence of Lewis acid ($AlCl_3$) in Friedel–Crafts (FC) acylation (see Section 5.7.1.5).

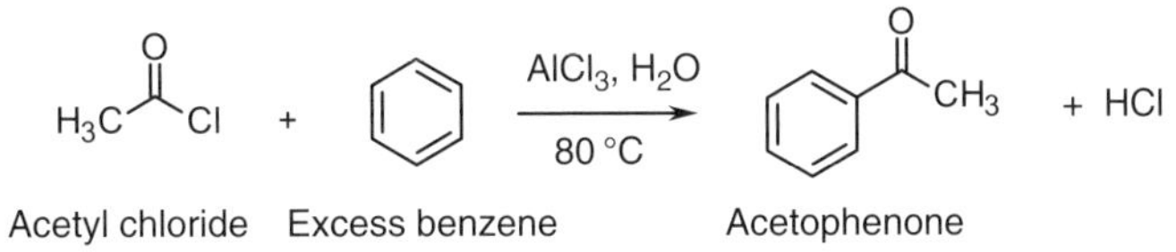

4.4.4 Acid Anhydrides

The functional group of an acid anhydride is two acyl groups bonded to an oxygen atom. These compounds are called *acid anhydrides* or *acyl anhydrides*, because they are condensed from two molecules of carboxylic acid by the loss of a water molecule.

An acid anhydride may be symmetrical, where two acyl groups are identical, or it may be mixed, where two different acyl groups are bonded to an oxygen atom. The simplest member of this family is acetic anhydride, $(CH_3CO)_2O$, where the acyl group (CH_3CO) is bonded to an acetate group (CH_3CO_2).

O   O
‖   ‖
$H_3C-C-O-C-CH_3$

Ethanoic anhydride
Acetic anhydride

O   O
‖   ‖
$H_3C-C-O-C-CH_2CH_3$

Acetic propanoic anhydride

4.4.4.1 Nomenclature of Acid Anhydrides

Symmetrical acid anhydrides are named by replacing the -*acid* suffix of the parent carboxylic acids with the word -*anhydride*. Mixed anhydrides that consist of two different acid-derived parts are named using the names of the two individual acids in alphabetical order.

O   O
‖   ‖
$CH_3CH_2-C-O-C-CH_2CH_3$

Propanoic anhydride

O   O
‖   ‖
$CH_3CH_2-C-O-C-CH_2CH_2CH_3$

Butanoic propanoic anhydride

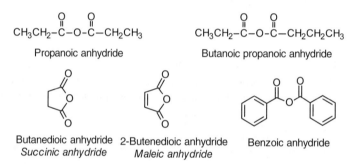

Butanedioic anhydride 2-Butenedioic anhydride   Benzoic anhydride
Succinic anhydride   *Maleic anhydride*

4.4.4.2 Preparations of Acid Anhydrides

Acid anhydrides are prepared most commonly by the reaction of acid chlorides and carboxylic acids or carboxylate salts (see Section 5.6.5.6). Industrially, acetic anhydride is obtained by heating two molecules of acetic acid at 800°C (see Section 5.6.5.3). Five- or six-membered cyclic anhydrides are prepared by heating dicarboxylic acids at high temperatures. For example, succinic anhydride is prepared by heating succinic acid at 200°C.

O
‖   800°C
$2 H_3C-C-OH$ ⟶

Acetic acid

O   O
‖   ‖
$H_3C-C-O-C-CH_3$ + H_2O

Acetic anhydride

Succinic acid   200 °C ⟶ Succinic anhydride + H_2O

4.4.4.3 Reactions of Acid Anhydrides

Acid anhydrides are the second most reactive of the carboxylic acid derivatives. They are fairly readily converted to the other less reactive carboxylic acid derivatives, for example, esters (see Section 5.6.5.5), amides (see Section 5.6.5.7) and carboxylic acids (see Section 5.8.1). Acid anhydrides undergo many other reactions very similar to those of acid chlorides, and often they are used interchangeably.

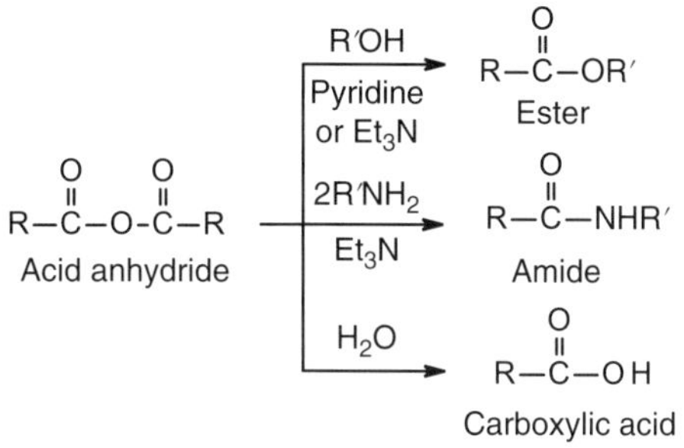

4.4.5 Esters

The functional group of an ester is an acyl group bonded to an alkoxy group (—OR). The simplest members of this family are methyl acetate ($CH_3CO_2CH_3$) and ethyl acetate ($CH_3CO_2C_2H_5$).

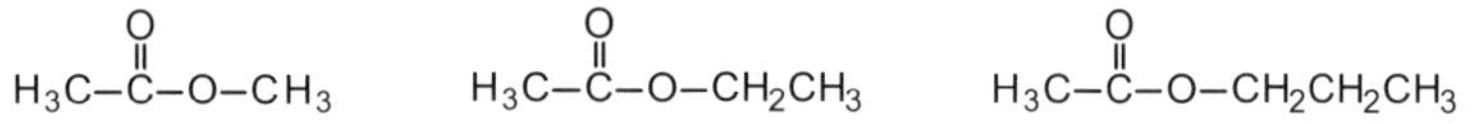

Methyl acetate
Methylethanoate

Ethyl acetate
Ethylethanoate

Propyl acetate
Propylethanoate

4.4.5.1 Nomenclature of Esters

The names of esters originate from the names of the compounds that are used to prepare them. The first word of the name comes from the alkyl group of the alcohol, and the second part comes from the carboxylate group of the carboxylic acid used. A cyclic ester is called a *lactone*, and the IUPAC names of lactones are derived by adding the term *lactone* at the end of the name of the parent carboxylic acid.

Butyl propanoate

Isopentyl acetate

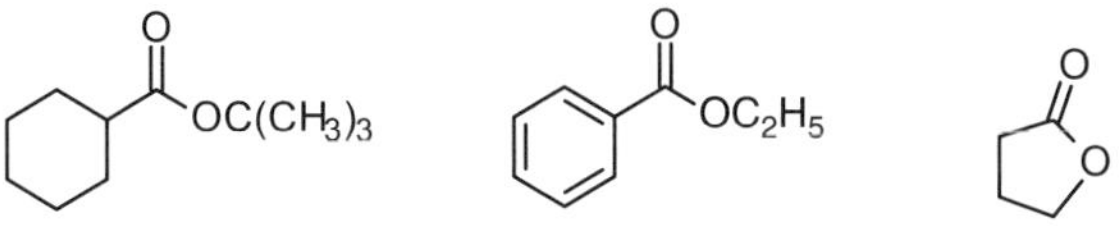

tert-Butylcyclohexanecarboxylate   Ethyl benzoate   4-Hydroxybutanoic acid lactone

4.4.5.2 Preparations of Esters

Esters are synthesized by acid-catalysed reaction of carboxylic acids with alcohols, known as *Fischer esterification* (see Section 5.6.5.1). They also can be produced from acid chlorides, acid anhydrides and other esters. The preparation of esters from other esters in presence of an acid or a base catalyst is called *transesterification* (see Section 5.6.5.1). All these conversions involve nucleophilic acyl substitutions (see Section 5.6.5).

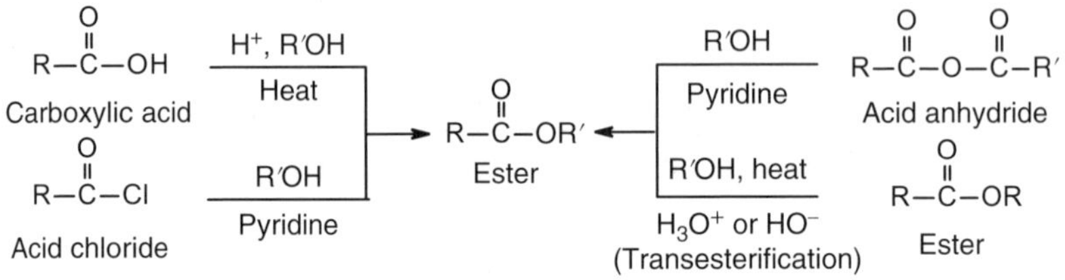

Lactones are made also by the *Fischer esterification*, where the hydroxyl and carboxylic acid groups are present in the same molecule. For example, acid-catalysed dehydration of 4-hydroxybutanoic acid yields 4-hydroxybutanoic acid lactone via cyclization.

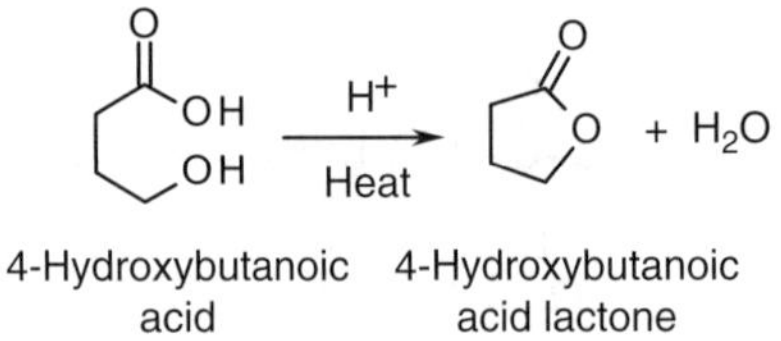

4-Hydroxybutanoic   4-Hydroxybutanoic
       acid   acid lactone

Carboxylic acids are smoothly converted to their methyl esters by the addition of an ethereal solution of diazomethane (CH_2N_2). The reaction is quite clean and straightforward as only the gaseous N_2 by-product is obtained in the reaction.

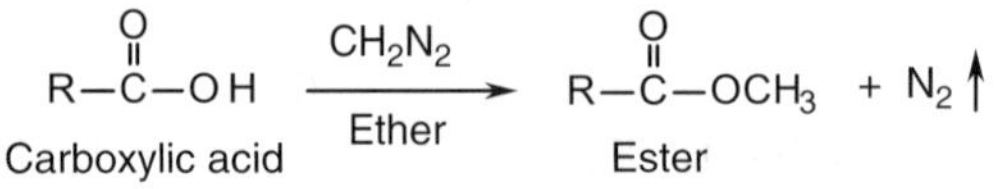

Alcohols react with inorganic acids to form esters; for example, tosylate esters (see Section 5.6.3.5) and phosphate esters. Phosphate esters are important in nature since they link the nucleotide bases together in DNA (see Section 7.1.2.1).

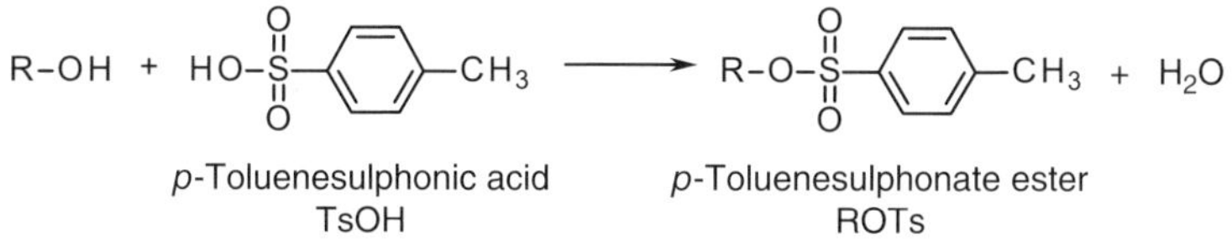

R-OH + HO-S(=O)₂-C₆H₄-CH₃ → R-O-S(=O)₂-C₆H₄-CH₃ + H₂O

p-Toluenesulphonic acid
TsOH

p-Toluenesulphonate ester
ROTs

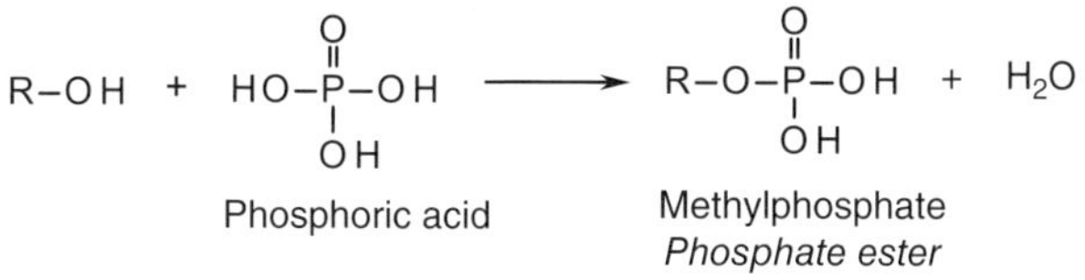

R-OH + HO-P(=O)(OH)-OH → R-O-P(=O)(OH)-OH + H₂O

Phosphoric acid

Methylphosphate
Phosphate ester

4.4.5.3 Reactions of Esters

Esters are less reactive than acid chlorides and acid anhydrides. They are easily converted to carboxylic acids by acid- or base-catalysed hydrolysis (see Section 5.8.1.2) or to another ester by alcoholysis under acidic or basic conditions, known as transesterification (see Section 5.8.1.2).

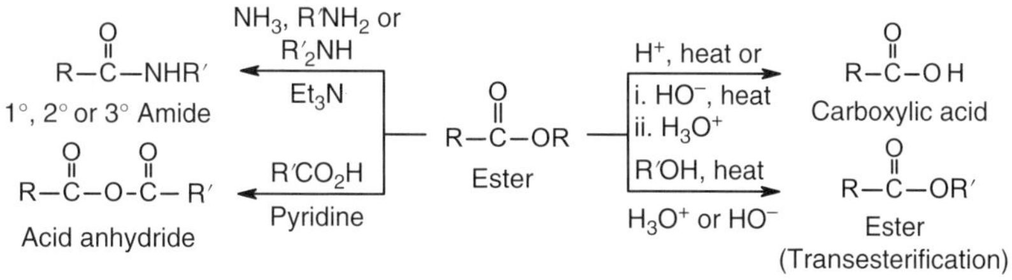

[Reaction scheme with central Ester R-C(=O)-OR]

NH₃, R'NH₂ or R'₂NH / Et₃N → R-C(=O)-NHR' (1°, 2° or 3° Amide)

R'CO₂H / Pyridine → R-C(=O)-O-C(=O)-R' (Acid anhydride)

H⁺, heat or i. HO⁻, heat ii. H₃O⁺ → R-C(=O)-OH (Carboxylic acid)

R'OH, heat / H₃O⁺ or HO⁻ → R-C(=O)-OR' (Ester) (Transesterification)

Primary and tertiary alcohols are conveniently prepared from esters by hydride reductions or two molar equivalent of organometallic reagents (R'MgX or R'Li), respectively (see Sections 5.6.5.8 and 5.9.18). A less powerful reducing agent, diisobutylaluminium hydride (DIBAH) can partially reduce an ester to an aldehyde (see Section 5.9.18).

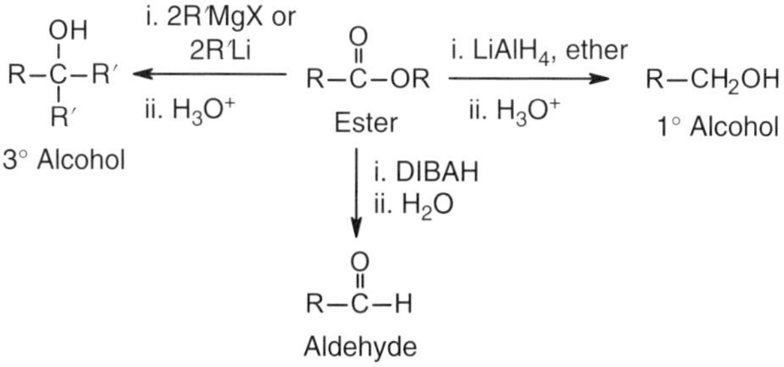

[Reaction scheme with central Ester R-C(=O)-OR]

i. 2R'MgX or 2R'Li / ii. H₃O⁺ → R-C(OH)(R')(R') (3° Alcohol)

i. LiAlH₄, ether / ii. H₃O⁺ → R-CH₂OH (1° Alcohol)

i. DIBAH / ii. H₂O → R-C(=O)-H (Aldehyde)

4.4.5.3.1 Claisen Condensation: Base-Catalysed

Another important reaction of esters is the *Claisen condensation*, not to be confused with the Claisen rearrangement (see Section 5.10.7), which is a C—C bond forming reaction. In fact, Claisen condensation is quite similar to the aldol addition (see Section 5.4.3.12), except it involves making an enolate nucleophile from an ester, not from an aldehyde or a ketone.

In this reaction, an enolate anion is formed between the reaction of an ester and a strong base, such as NaOEt in EtOH. The enotale anion then reacts with another molecule of ester to afford β-ketoester (see Section 5.6.5.9).

$$2RCH_2-\overset{\overset{\displaystyle O}{\|}}{C}-OEt \xrightarrow[\text{EtOH}]{\text{NaOEt}} RCH_2-\overset{\overset{\displaystyle O}{\|}}{\underset{\beta}{C}}-\overset{}{\underset{\underset{R}{|}}{\underset{\alpha}{CH}}}-\overset{\overset{\displaystyle O}{\|}}{C}-OEt$$

Esters

β-Ketoester
1,3 -Dicarbonyl compounds

4.4.5.3.2 Mixed-Claisen Condensation: Base-Catalysed

Like mixed-aldol condensation (see Section 5.4.3.12), it is also possible to achieve *mixed-Claisen condensation;* when one of the esters has no α-hydrogen and the other ester has α-hydrogen so that the nucleophile is generated solely from one molecule. A reactant without an α-hydrogen cannot self-condense because it cannot form an enolate. So, in mixed-Claisen condensation the reaction takes place between an enolate and the ester that has no α-hydrogen to give 1,3-dicabonyl compound (see Section 5.6.5.10).

$$Y-\overset{\overset{\displaystyle O}{\|}}{C}-OR + RCH_2-\overset{\overset{\displaystyle O}{\|}}{C}-OR \xrightarrow[\text{EtOH}]{\text{NaOEt}} Y-\overset{\overset{\displaystyle O}{\|}}{C}-\overset{}{\underset{\underset{R}{|}}{CH}}-\overset{\overset{\displaystyle O}{\|}}{C}-OR$$

Ester with
no α-Hs

Ester with
α-Hs

Y = H or Ph
1,3-Dicarbonyl compound

4.4.6 Amides

The functional group of an amide is an acyl group bonded to a nitrogen atom. The simplest members of this family are formamide (methanamide), acetamide (ethanamide) and propanamide.

$$H-\overset{\overset{\displaystyle O}{\|}}{C}-NH_2 \qquad H_3C-\overset{\overset{\displaystyle O}{\|}}{C}-NH_2 \qquad H_5C_2-\overset{\overset{\displaystyle O}{\|}}{C}-NH_2$$

Methanamide
Formamide

Ethanamide
Acetamide

Propanamide

Amides are usually classified as primary (1°) amide, secondary (2°) or *N*-substituted amide, and tertiary (3°) or *N,N*-disubstituted amide.

$$\underset{\substack{\text{Primary amide}\\(1°)}}{R-\overset{\displaystyle O}{\overset{\|}{C}}-NH_2} \qquad \underset{\substack{\text{Secondary (2°) or}\\ \textit{N}\text{-substituted amide}}}{R-\overset{\displaystyle O}{\overset{\|}{C}}-NHR'} \qquad \underset{\substack{\text{Tertiary (3°) or}\\ \textit{N,N}\text{-disubstituted}\\ \text{amide}}}{R-\overset{\displaystyle O}{\overset{\|}{C}}-NR'_2}$$

4.4.6.1 Nomenclature of Amides

Amides are named by replacing the -*oic* acid or -*ic* acid suffix of the parent carboxylic acids with the word -*amide*, or by replacing the -*carboxylic acid* ending with -*carbox-amide*. Alkyl groups on nitrogen atom are named as substituents, and are prefaced by *N*- or *N,N*- followed by the name(s) of the alkyl group(s).

$$\underset{\textit{N}\text{-Ethylethanamide}}{H_3C-\overset{\displaystyle O}{\overset{\|}{C}}-NHCH_2CH_3} \qquad \underset{\substack{\textit{N,N}\text{-Dimethylmformamide}\\ \textit{DMF}}}{H-\overset{\displaystyle O}{\overset{\|}{C}}-N(CH_3)_2} \qquad \underset{\text{Cyclohexanecarboxamide}}{\text{[cyclohexane]}-\overset{\displaystyle O}{\overset{\|}{C}}-NH_2} \qquad \underset{\text{Benzamide}}{\text{[benzene]}-\overset{\displaystyle O}{\overset{\|}{C}}-NH_2}$$

If the substituent on the nitrogen atom of an amide is a phenyl group, the ends -*amide* is changed to -*anilide*. Cyclic amides are known as *lactams*, and the IUPAC names are derived by adding the term *lactam* at the end of the name of the parent amino acid. They can be of various sizes, for example, α-, β- and γ-lactams, and among them, β-lactam is present the very first group of antibiotics, the penicillins (*see* Section 1.1). Lactam ring derivatives can also display other pharmacological properties.

$$\underset{\text{Acetanilide}}{H_3C-\overset{\displaystyle O}{\overset{\|}{C}}-\underset{H}{N}-\text{[phenyl]}} \qquad \underset{\text{Benzanilide}}{\text{[phenyl]}-\overset{\displaystyle O}{\overset{\|}{C}}-\underset{H}{N}-\text{[phenyl]}} \qquad \underset{\substack{\text{4-Aminobutanoic}\\ \text{acid lactam}}}{\text{[pyrrolidinone ring N-H]}}$$

4.4.6.2 Physical Properties of Amides

Amides are much less basic than their parent amines. The lone pair of electrons on nitrogen atom are delocalized on the carbonyl oxygen, and in presence of a strong acid the oxygen is protonated first. Amides have high boiling points because of their ability to form intermolecular hydrogen bonding. The borderline for solubility in water ranges from five to six carbons for the amides.

4.4.6.3 Preparations of Amides

Amides are the least reactive carboxylic acid derivatives, and are easily afforded from any of the other carboxylic acid derivatives; for example, acid chlorides, acid anhydrides and esters. Carboxylic acids react with ammonia, 1° and 2° amines to offer 1°, 2° and 3° amides, respectively (see Section 5.6.5.4).

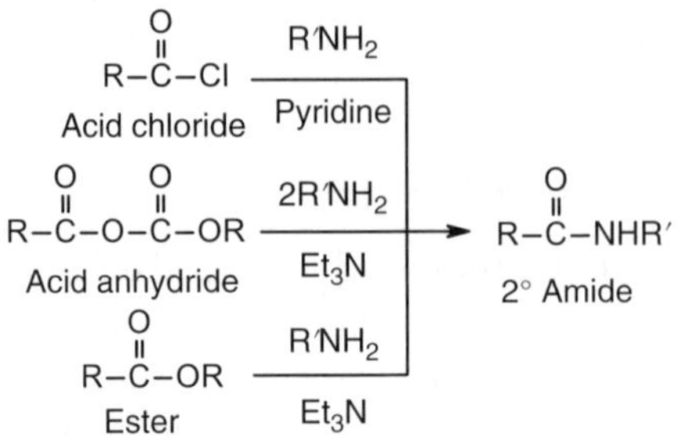

The amides derived from sulphonic acids are called *sulphonamides*. In general, sulphonamides are synthesized by the reaction of amines and sulphonyl chlorides (see Section 4.3.8.7). *Lactams* (cyclic amides) are obtained by heating or dehydrating of amino acids, where the amino and the carboxylic acid groups of the same molecule, react to form an amide linkage.

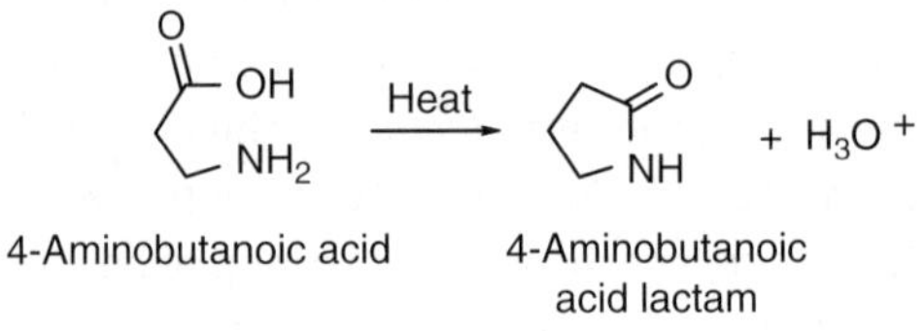

4-Aminobutanoic acid   4-Aminobutanoic acid lactam

4.4.6.4 Reactions of Amides

Amides are the least reactive of the carboxylic acid derivatives, and undergo acidic or basic hydrolysis to produce the parent carboxylic acids, and reduction to appropriate amines (see Section 4.3.8). They can also be dehydrated to nitriles, most commonly with boiling acetic anhydride, $(AcO)_2O$, sulphonyl chloride $(SOCl_2)$ or phosphorous oxychloride $(POCl_3)$ (see Section 4.3.8). Amines (with one less carbon) are prepared from amides by the treatment of halides (Br_2 or Cl_2) in aqueous NaOH or KOH. This reaction is known as *Hofmann rearrangement* (see Section 4.3.8.6).

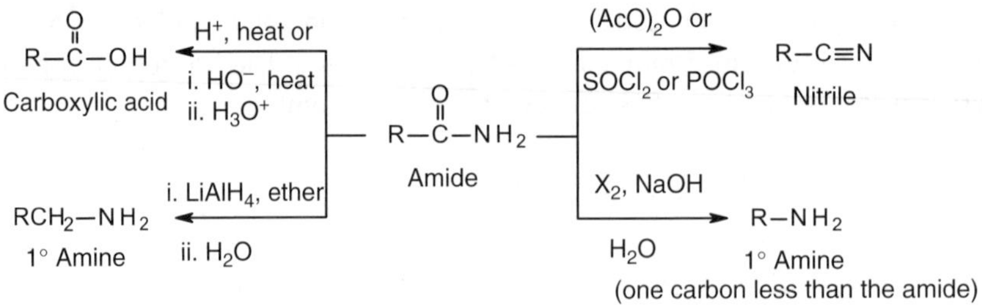

4.4.7 Nitriles

Nitriles are organic compounds that contain a triple bond between a carbon and a nitrogen atom. The functional group in nitriles is the cyano ($-C\equiv N$) group, and they are often named as cyano compounds. Nitriles are not carbonyl compounds, but are often included within them because of the similarities in nitrile and carbonyl chemistry. Nitriles are considered to be acid derivatives, because they can be hydrolysed to form amides and carboxylic acids. The nitriles related to acetic acid and benzoic acid are called acetonitrile (ethanenitrile) and benzonitrile (cyanobenzene), respectively.

$$H_3C-C\equiv N \qquad CH_3CH_2-C\equiv N \qquad \text{(benzene)}-C\equiv N$$

Ethanenitrile   Propanenitrile   Cyanobenzene
Acetonitrile   *Benzonitrile*

4.4.7.1 Nomenclature of Nitriles

The IUPAC requires nitriles to be named on the basis of the name of the alkanes, with the suffix *–nitrile*. For example, 2-bromobutane nitrile and 5-mehoxyhexane nitrile.

$$H_3C-CH_2\underset{\underset{Br}{|}}{CH}-C\equiv N \qquad\qquad H_3C-\underset{\overset{|}{OCH_3}}{CH}-CH_2CH_2CH_2-C\equiv N$$

2-Bromobutane nitrile   5-Methoxyhexane nitrile

4.4.7.2 Preparations of Nitriles

Nitriles are commonly prepared via the conversion of carboxylic acids to primary amides, followed by dehydration with boiling acetic anhydride, or other commonly employed dehydration reagents; for example, $SOCl_2$ or $POCl_3$. This is a useful synthesis for amide, because it is not limited by steric hindrance. Alkyl nitriles can be prepared by the action of metal cyanides on alkyl halides (see Section 5.6.2.6), and aryl nitriles are easily produced by substitution of the aryl diazonium salts with copper cyanide (CuCN).

$$R-X \xrightarrow{\ NaCN\ } R-C\equiv N \underset{SOCl_2 \text{ or } POCl_3}{\overset{(AcO)_2O \text{ or}}{\longleftarrow}} R-\overset{\overset{O}{\|}}{C}-NH_2$$

Alkyl halide   Nitrile   Amide

$$Ar-\overset{+}{N}\equiv N \xrightarrow{\ CuCN\ } Ar-C\equiv N \ + \ N_2\uparrow$$

Aryl diazonium salt   Aryl nitrile

4.4.7.3 Reactions of Nitriles

In nitriles, as nitrogen is more electronegative than carbon, the triple bond is polarized towards the nitrogen, similar to the C=O bond. Therefore, nucleophiles can attack the electrophilic carbon of the nitrile group. Nitriles undergo hydrolysis to 1° amides, and then to carboxylic acids (see Section 5.8.1.4). Reduction of nitriles by $LiAlH_4$ and catalytic hydrogenation provides primary amines (see Section 5.9.21), and reaction of nitriles with Grignard reagent or organolithium yields ketones, after acidic hydrolysis (see Section 5.4.3.4).

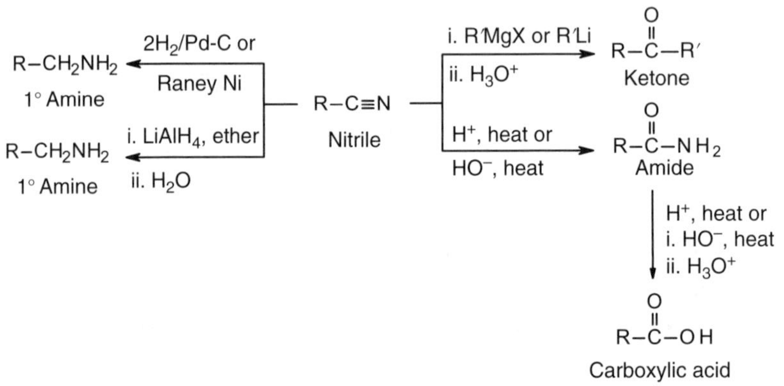

4.5 ALKENES AND THEIR DERIVATIVES

Alkenes are unsaturated hydrocarbons that contain carbon-carbon (C=C) double bonds. They also known as *olefins*. A double bond consists of a σ bond and a π bond. A π bond is weaker than a σ bond, and that makes π bonds more reactive than σ bonds. Thus, π bond is considered to be a functional group. Alkenes form a homologous series with general molecular formula C_nH_{2n}. The simplest members of the series are ethene (C_2H_4), propene (C_3H_6), butene (C_4H_8) and pentene (C_5H_{10}).

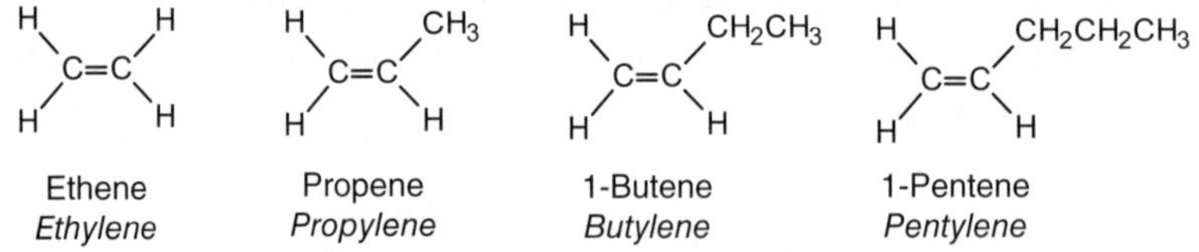

| Ethene | Propene | 1-Butene | 1-Pentene |
| Ethylene | Propylene | Butylene | Pentylene |

Among the cycloalkenes, cyclobutene, cyclopropene and cyclohexene are most common. Cyclobutene is about 4 kcal mol⁻¹ more strained than cyclopentene. The smaller bond angles mean more deviation from 120°, and that makes cyclobutene more reactive than cyclopentene.

Cyclobutene   Cyclopentene   Cyclohexene

4.5.1 Nomenclature of Alkenes

The systematic name of an alkene originates from the name of the alkane corresponding to the longest continuous chain of carbon atoms that contains the double bond. When the chain is longer than three carbons, the atoms are numbered starting from the end nearest to the double bond (C=C). The functional group suffix is -*ene*.

$$\underset{4 \quad \; 3 \quad \; 2 \; 1}{CH_3CH=CHCH_3}$$

2-Butene
(*cis* or *trans*)

$$\underset{5 \quad \; 4 \quad \; 3 \quad \; 2 \; 1}{CH_3CH_2CH=CHCH_3}$$

2-Pentene
(*cis* or *trans*)

$$\underset{1 \; 2 \quad \; 3 \quad \; 4 \quad \; 5 \quad \; 6}{H_3C-CH=CHCH_2CH_2CH_3}$$

2-Hexene
(*cis* or *trans*)

For branches, each alkyl group is given a number, but the double bond still gets preferences when numbering the chain.

$$\overset{\textstyle CH_3}{\underset{1 \; 2 \; 3 \quad 4 \quad 5}{H_2C=CHCHCH_2CH_3}}$$

3-Methyl-1-pentene

A cyclic alkene is named by a prefix *cyclo-* to the name of the acyclic alkene. Double-bonded carbons are considered to occupy positions 1 and 2.

Cyclopentene   1,2-Dichlorocyclopentene   4-Bromo-1,2-cyclohexene

When a geometrical isomer is present, a prefix *cis* (*Z*) or *trans* (*E*) is added. Because of the double bonds, alkenes cannot undergo free rotation. Thus, the rigidity of a π bond generates *geometrical isomers* (see Section 3.3.3). Simple 1,2-alkenes can be described as *cis*- or *trans*- alkenes. When similar groups are present on the same side of the double bond, the alkene is said to be *cis*. When similar groups are present on opposite sides of the double bond it is said to be *trans*. More complex alkenes are best described as *E* or *Z* based on the Cahn–Ingold–Prelog priority rules (see Section 3.2.2.2.8).

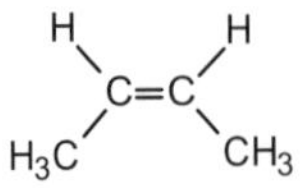

cis-2-Butene

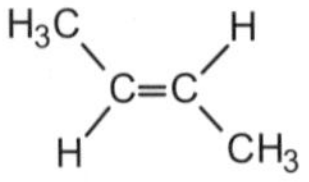

trans-2-Butene

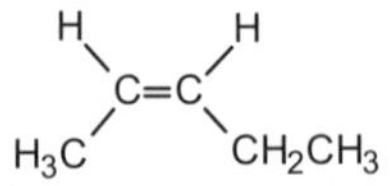

cis-2-Pentene

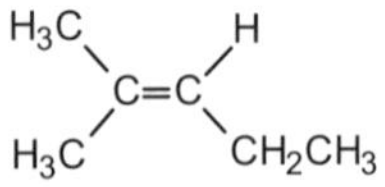

trans-2-Pentene

Compounds with two double bonds are called *dienes*, three double bonds are *trienes* and so on. Where geometrical isomerism exists, each double bond is specified with numbers indicating the positions of all the double bonds.

$H_2C{=}CH{-}CH{=}CH_2$
4 3 2 1

1,3-Butadiene

(2*E*,4*E*)-2,4-Hexadiene

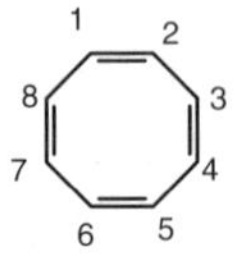

1,3,5,7-Cyclooctatetraene

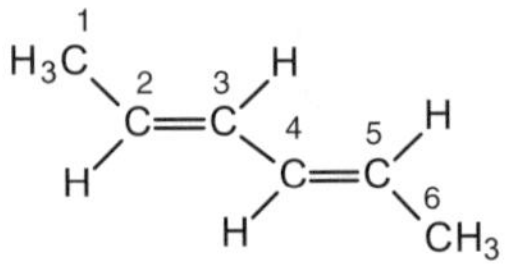

The sp^2 carbon of an alkene is called *vinylic carbon*, and an sp^3 carbon that is adjacent to a vinylic carbon is called an *allylic carbon*. Two unsaturated groups are called *vinyl group* ($CH_2{=}CH{-}$) and *allyl group* ($CH_2CHCH_2{-}$).

$H_2C{=}CH{-}$

Vinyl group

$H_2C{=}CH_2CH_2{-}$

Allyl group

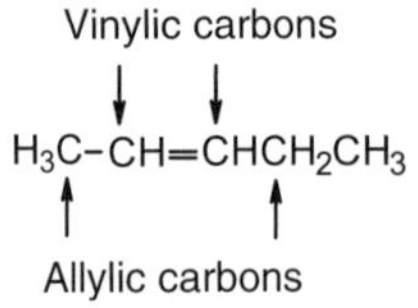

Cycloalkenes must have eight or more carbons before they are large enough to incorporate a *trans* double bond. Thus, cycloalkenes are considered to be *cis* unless otherwise specified. A bridged bicyclic (two rings) compound cannot have a double bond at a bridgehead position unless one of the rings contains at least eight carbon atoms. Bridgehead carbon is part of both rings. A bridged bicyclic compound has at least one carbon in each of the three links between the bridgehead atoms.

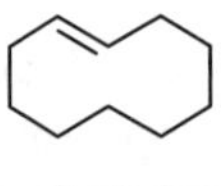

trans-Cyclodecene

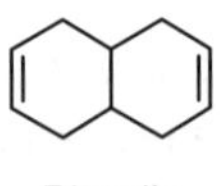

Bicyclic

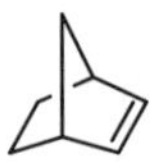

Bridged bicyclic

4.5.2 Physical Properties of Alkenes

As with alkanes, the boiling points and melting points of alkenes inverse with molecular weight but show some variations that depend on the shape of the molecule. Alkenes with same molecular formula are isomers of one another if the position and the stereochemistry of the double bond differ. For example, there are four different acyclic structures that can be drawn for butene (C_4H_8), and they have different boiling points and melting points as follows.

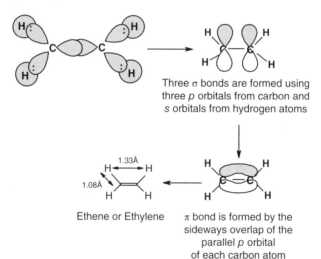

1-Butene	2-Methylpropene	cis-2-Butene	trans-2-Butene
bp: –6 °C	bp: –7 °C	bp: +4 °C	bp: +1 °C
mp: –195 °C	mp: –144 °C	mp: –139 °C	mp: –106 °C

4.5.3 Structure of Alkenes

In ethene (C_2H_4), each carbon atom has three σ bonding electron pairs in its p orbitals to form three σ bonds one with the carbon and the other two with two hydrogen atoms. The π bond in C_2H_4 is formed from the sideways overlap of a parallel p orbital on each carbon atom. The C–C bond in ethene is shorter and stronger than in ethane partly because of the sp^2-sp^2 overlap being stronger than sp^3-sp^3, but especially because of the extra π bond in ethene.

Three σ bonds are formed using three p orbitals from carbon and s orbitals from hydrogen atoms

Ethene or Ethylene

π bond is formed by the sideways overlap of the parallel p orbital of each carbon atom

4.5.4 Industrial uses of Alkenes

Alkenes are useful intermediates in organic synthesis, but their main commercial use is as precursors for polymers. Alkenes can be joined together in the presence of high pressure and suitable catalysts. The reaction involves breaking only the alkenes π bonds, the product formed in this reaction is termed as polymer and the process is known as *addition polymerization*. For example, styrene polymerizes to polystyrene via free radical vinyl polymerization.

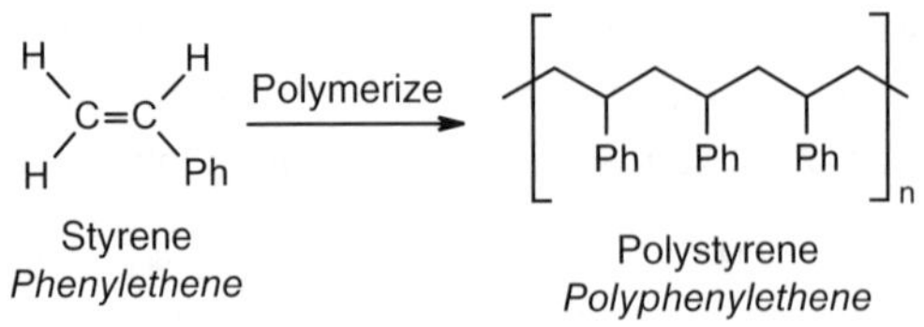

Styrene
Phenylethene

Polystyrene
Polyphenylethene

4.5.5 Preparations of Alkenes

Alkenes are obtained by the transformation of various functional groups; for example, catalytic dehydrogenation of alkanes (see Section 4.3.1.7), dehydration of alcohols (see Section 5.5.3), dehydrohalogenation of alkyl halides (see Section 5.5.5) and dehalogenation (see Section 5.5.5.5).

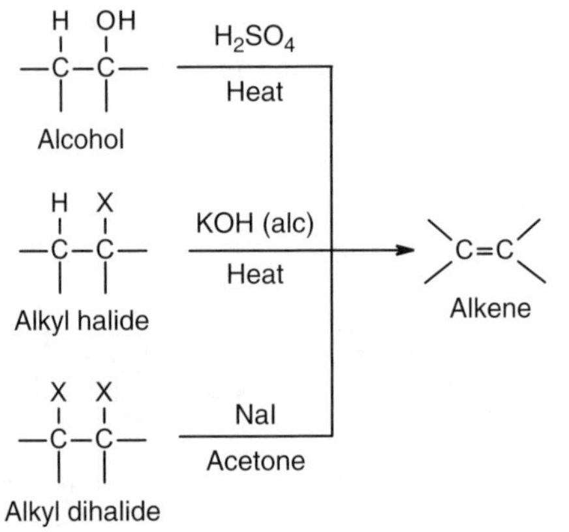

Alcohol

H_2SO_4
Heat

Alkyl halide

KOH (alc)
Heat

C=C

Alkene

Alkyl dihalide

NaI
Acetone

Alkenes can also be prepared from selective hydrogenation of alkynes (see Section 5.4.1.2) and the reaction of a phosphorus ylide (*Wittig reagent*) with an aldehyde or a ketone. This reaction is known as *Wittig reaction* (see Section 5.4.3.6).

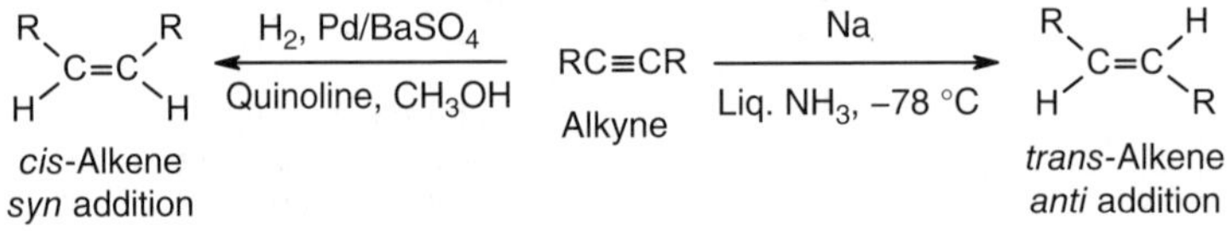

cis-Alkene
syn addition

H_2, Pd/BaSO$_4$
Quinoline, CH$_3$OH

RC≡CR
Alkyne

Na
Liq. NH$_3$, −78 °C

trans-Alkene
anti addition

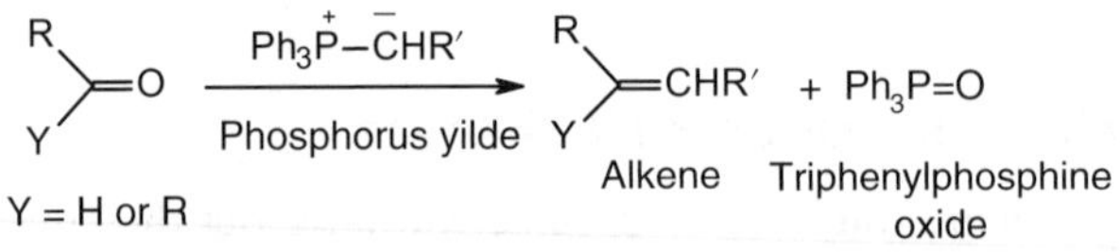

Ph_3P-CHR'
Phosphorus yilde

Y = H or R

=CHR' + Ph$_3$P=O

Alkene   Triphenylphosphine
oxide

4.5.6 Reactivity and Stability of Alkenes

Alkenes are typically nucleophiles, because the double bonds are electron rich and electrons in π bond are loosely held. Electrophiles are attracted to the π electrons.

Thus, alkenes generally undergo addition reactions and addition reactions are typically exothermic. The following three factors influence the stability of alkenes.

i. The degree of substitution: more highly alkylated alkenes are more stable. Thus, the stability follows the order: tetra > tri > di > mono substituted. This is because the alkyl groups stabilize the double bond. The stability arises, because the alkyl groups are electron donating (hyperconjugation) and so donate electron density into the π bond. Moreover, a double bond (sp^2) carbon separates bulky groups better than an sp^3 carbon, thus reducing steric hindrance.

ii. The stereochemistry: *trans* > *cis* due to reduced steric interactions when R groups are on opposite sides of the double bond.

iii. The conjugated alkenes are more stable than isolated alkenes.

4.5.7 Reactions of Alkenes

Alkenes are electron-rich species. The double bond acts as a nucleophile and attacks the electrophile. Since σ bonds are stronger than π bonds, double bonds tend to react to convert the double bond into σ bonds. So, the most important reaction of alkenes is *electrophilic addition* to the double bond (see Sections 5.4.1 and 5.4.2) to produce various other compounds. An outline of the electrophilic addition reactions of alkenes is presented here.

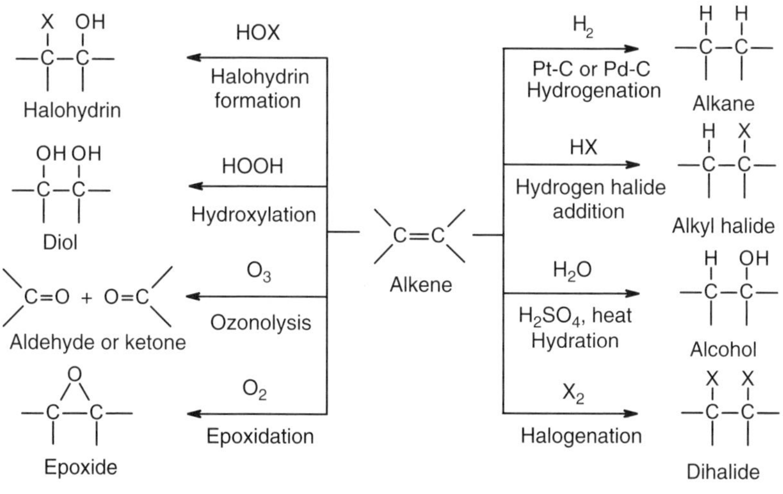

4.6 ALKYNES AND THEIR DERIVATIVES

Alkynes are hydrocarbons that contain a carbon–carbon triple bond. A triple bond consists of a σ bond and two π bonds. The general formula for the alkynes is C_nH_{2n-2}. The triple bond possesses two elements of unsaturation. Alkynes are commonly named as substituted acetylenes. Compounds with triple bonds at the end of a molecule are called *terminal alkynes*. Terminal C–H groups are called acetylenic hydrogens. If the triple bond has two alkyl groups on both sides, it is called an *internal alkyne*. Terminal alkynes are much more acidic than other hydrocarbons. Generally, alkynes are nonpolar and quite soluble in most organic solvents.

HC≡CH	$\underset{4\;\;\;3\quad\;2\;\;\;1}{CH_3CH_2C{\equiv}CH}$	$\underset{5\quad\;4\quad\;\;3\;\;\;2\;1}{CH_3CH_2C{\equiv}CCH_3}$
Ethyne	1-Butyne	2-Pentyne
Acetylene	Ethylacetylene	Ethylmethylacetylene
	(Terminal alkyne)	(Internal alkyne)

The two most important birth-control pills, norethindrone and ethynylestradiol, also contain alkyne functionality (see Section 8.6.6).

Norethindrone
(A synthetic progestin)

Ethynylestradiol
(A synthetic estrogen)

4.6.1 Nomenclature of Alkynes

The IUPAC nomenclature of alkynes is similar to that for alkenes, except the -*ane* ending is replaced with -*yne*. The chain is numbered from the end closest to the triple bond. When additional functional groups are present, the suffixes are combined.

$\underset{5\quad4\quad\;3\;\;2\;\;1}{CH_3CH_2\overset{\underset{\displaystyle|}{OH}}{C}HC{\equiv}CH}$

1-Pentyn-3-ol

$\underset{1\quad\;2\quad\;3\;\;4\;5\;\;\;6\quad\;7}{CH_3\overset{\underset{\displaystyle|}{CH_3}}{C}HC{\equiv}C\overset{\underset{\displaystyle|}{Br}}{C}HCH_2CH_3}$

5-Bromo-2-methyl-3-heptyne

$\underset{1\;\;2\quad\;3\;4\;\;\;5\quad\;\;6}{CH_3C{\equiv}C\overset{\underset{\displaystyle|}{OCH_3}}{C}HCH_2CH_3}$

4-Methoxy-2-hexyne

4.6.2 Structure of Alkynes

The triple bond in acetylene consists of one σ bond and two π bonds. Each carbon is bonded to two other atoms, and there are no nonbonding electrons. Carbon requires two hybrid orbitals to bond to the atoms, thus *sp* hybrids are used. The *sp* orbitals are linear and oriented at 180°. The C–C bond is formed from *sp–sp* overlap. The C–H bond is formed from *sp–s* overlap. The formation of *sp* hybrids leaves two free *p* orbitals, these contribute to the formation of the two other π bonds. The C≡C bond length for ethyne is 1.20 Å, which is shorter than ethane

(1.54 Å) and ethene (1.33 Å). The C—H bond length in ethyne is 1.06 Å, which is also shorter than in ethane (1.09 Å) or ethene (1.08 Å). This is because the C—H bond contains more s character ($sp^3 \rightarrow sp^2 \rightarrow sp$), which gives stronger bonds.

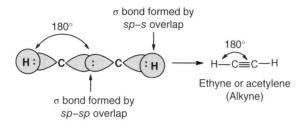

4.6.3 Acidity of Terminal Alkynes

The position of the triple bond can change the acidity and the reactivity of the alkynes. Acetylene has $pK_a = \sim 25$ that makes it a stronger acid than ammonia ($pK_a = 38$), but weaker acid than alcohols ($pK_a = 15.9$). The acidic proton from terminal alkynes can be removed by strong bases, such as $NaNH_2$ or organo-lithiums or Grignard reagents to form metal acetylides (alkynides) in acid-base reactions. Alkynides are strong nucleophiles and bases, and can be protonated in presence of water and acids. Therefore, they must be protected from water and acids.

$$RC\equiv CH \quad \text{(Terminal alkyne)}$$

- $\xrightarrow{NaNH_2}$ $RC\equiv \overset{-}{C}-\overset{+}{Na}$ + NH_3 — Sodium acetylide
- $\xrightarrow{CH_3CH_2Li}$ $RC\equiv \overset{-}{C}-\overset{+}{Li}$ + CH_3CH_3 — Lithium acetylide
- $\xrightarrow{CH_3CH_2MgBr}$ $RC\equiv \overset{-}{C}-\overset{+}{MgBr}$ + CH_3CH_3 — Alkynyl Grignard reagent

4.6.4 Heavy Metal Acetylides: Test for Terminal Alkynes

Acidic alkynes react with certain heavy metal ions, for example, Ag^+ and Cu^+, to form precipitation. Addition of an alkyne to a solution of $AgNO_3$ in alcohol forms a precipitate, which is an indication of hydrogen attached to the triple bonded carbon. Thus, this reaction can be used to differentiate terminal alkynes from internal alkynes.

$$CH_3C\equiv CH \xrightarrow{Ag^+} CH_3C\equiv C-Ag + H^+ \qquad CH_3C\equiv CCH_3 \xrightarrow{Ag^+} \text{No reaction}$$

Terminal alkyne   Precipitate   Internal alkyne

4.6.5 Industrial Uses of Alkynes

Ethynes are industrially used as a starting material for polymers; for example, vinyl flooring, plastic piping, Teflon and acrylics. Polymers are large molecules, which are prepared by linking many small monomers. Polyvinyl chloride, also commonly known as PVC, is a polymer derived from the polymerization of vinyl chloride.

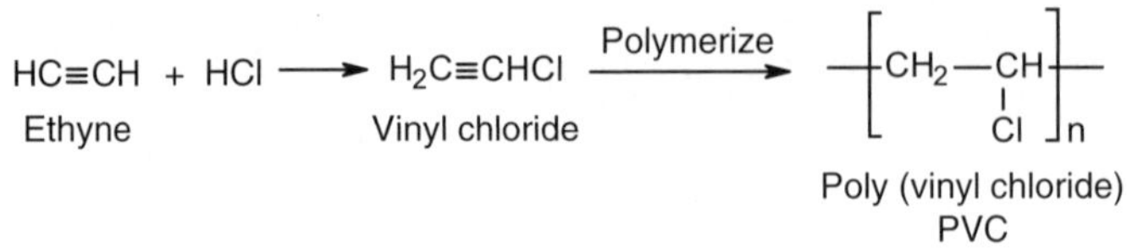

$$HC{\equiv}CH + HCl \longrightarrow H_2C{\equiv}CHCl \xrightarrow{\text{Polymerize}} {-\!\!\left[\!CH_2{-}CH\!\right]\!}_n$$

Ethyne   Vinyl chloride $\overset{|}{Cl}$

Poly (vinyl chloride)
PVC

4.6.6 Preparations of Alkynes

Alkynes are prepared from alkyl dihalides via elimination of atoms or groups from adjacent carbons. Dehydrohalogenation of *vicinal* or *geminal* dihalides is particularly a useful method for the preparation of alkynes (see Section 5.5.5.5).

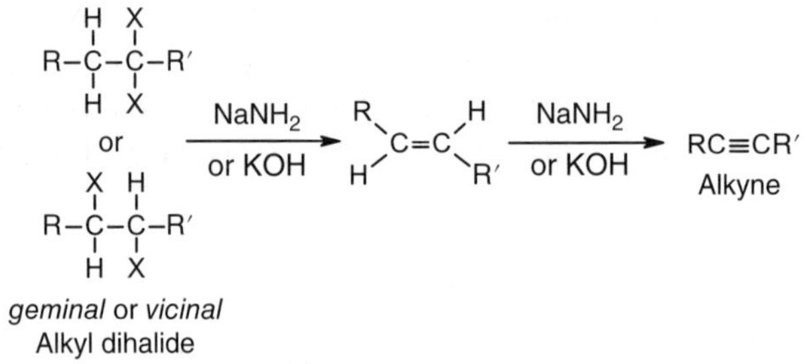

geminal or *vicinal*
Alkyl dihalide

Since alkynides are strong nucleophiles, they react with unhindered primary alkyl halides or tosylates via S_N2 displacements to afford internal alkynes (see Section 5.6.2.8).

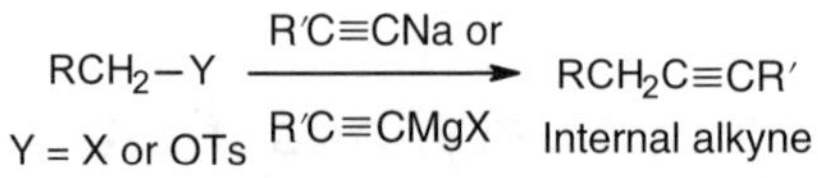

$$RCH_2{-}Y \xrightarrow[R'C{\equiv}CMgX]{R'C{\equiv}CNa \text{ or}} RCH_2C{\equiv}CR'$$

Y = X or OTs   Internal alkyne

4.6.7 Reactions of Alkynes

Alkynes are electron-rich reagents and readily undergo addition reactions. The triple bond acts as a nucleophile and attacks the electrophile. Thus, alkynes undergo *electrophilic addition reactions*, for example, hydrogenation, halogenation and hydrohalogenation, in a similar fashion to alkenes, except that two molecules of reagent are needed for each triple bond for the total addition. It is possible to stop the reaction to the first stage of addition to the formation of alkenes. Therefore, two different halide groups can be introduced in each stage. Terminal alkynes follow Markovnikov's rules in the same way as unsymmetrical alkenes. Internal alkynes show little selectivity since both intermediate cations are equally substituted, thus they do not follow Markovnikov's rule. A summary of electrophilic addition reactions of alkynes (see Sections 5.4.1 and 5.4.2) is presented here.

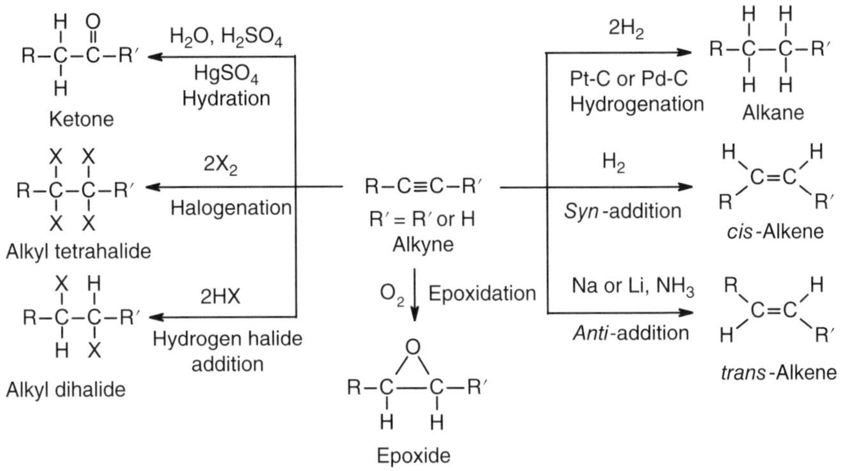

Oxidation of alkynes under mild (cold, neutral) conditions; for example, the reaction of potassium permanganate with an alkyne can generate a diketone (see Section 5.9.3). If the mixture becomes too warm or basic, the oxidation proceeds further to generate two carboxylate anions, which on acidification generates two carboxylic acids (see Section 5.9.3).

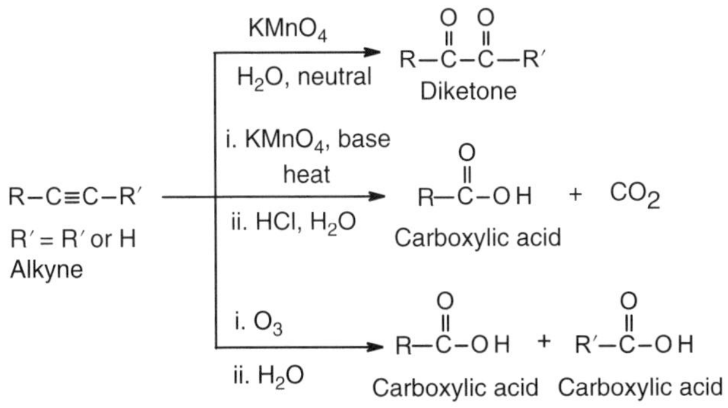

Usually, terminal alkynes are cleaved to provide a carboxylic acid and carbon dioxide (see Section 5.9.3). Ozonolysis of an alkyne followed by hydrolysis yields similar products to those obtained from permanganate oxidative cleavage except terminal alkynes (see Section 5.9.7).

4.6.8 Reactions of Metal Alkynides

Besides electrophilic addition, terminal alkynes can undergo acid–base type reactions owing to the acidic nature of the terminal hydrogen ($pK_a = \sim 25$). The formation of metal alkynides are important reactions of terminal alkynes.

Alkynides anions are strong nucleophiles, which are used for the carbon chain extension by creating new carbon–carbon bonds. The alkylation of alkynides are extremely important in organic synthesis. Metal alkynides ($RC \equiv CM$ or $RC \equiv CMgX$) react with alkyl halides to yield internal alkynes via nucleophilic substitution reaction. This reaction is called *alkylation* (see Section 5.6.2.8). The triple bonds of internal alkynes are readily available for electrophilic additions to a number of other functional groups (see Sections 5.4.2.6, 5.4.2.9 and 5.4.2.14).

Alkynides also undergo acid-catalysed nucleophilic addition with aldehydes and ketones to afford α-alkynyl alcohols (see Section 5.4.3.5). This type of reaction is called *alkynylation*.

4.7 AROMATIC COMPOUNDS AND THEIR DERIVATIVES

All drugs are chemicals and many of them are aromatic compounds. Therefore, in order to understand the chemical nature, physical properties, stability, pharmacological actions and toxicities of a majority of drug molecules, the knowledge of aromatic chemistry is extremely important. Some aromatic compounds are also essential for all living beings; three of the 20 amino acids used to form proteins are aromatic compounds, and all five of the nucleotides that make up DNA and RNA sequences are all aromatic compounds. Before we look into the specific examples of various drugs that belong to this aromatic class, let's try to understand what *aromaticity* really is.

Simply, aromaticity can be defined as a property of conjugated cycloalkenes in which the stabilization of the molecule is enhanced due to the ability of the electrons in the π orbitals to delocalize. Generally, the term 'aromatic compounds' means fragrant substances. Later, benzene and its structural relatives were termed as aromatic. However, there are a number of other non-benzenoid compounds that can be classified as aromatic compounds.

4.7.1 History

In 1825, Michael Faraday, a British scientist, discovered benzene, which is a colourless and highly flammable constituent of crude oil, and is one of the elementary petro-chemicals. He named this as 'bicarburet of hydrogen' because of the equal number of carbon and hydrogen atoms. He isolated benzene from a compressed illuminating gas that was made by pyrolysing whale oil. In 1834, Eilhardt Mitscherlich, a German chemist, synthesized benzene by heating benzoic acid with calcium oxide. In the late nineteenth century, Friedrich August Kekulé, a German organic chemist, first noticed that all early aromatic compounds contain a six-carbon unit that is retained through most chemical transformation and degradation.

4.7.2 Definition: Hückel's Rule

An aromatic compound has a molecular structure containing cyclic clouds of delocal-ized π electrons above and below the plane of the molecule, and the π clouds contain a total of $4n+2$ numbers of π electrons (where n is whole number). This is known as *Hückel's rule*, introduced first by the German physicist, Erich Hückel in 1931. For example, benzene is an aromatic compound that has a sweet smell.

Benzene

If n = 1, we get $4 \times 1 + 2 = 6$, which means that π clouds of any compound contain-ing a total number of six π electrons, is an aromatic compound. In the structure of benzene, there are three double bonds, six π electrons and it is a planar molecule. Thus, benzene follows Hückel's rule and is an aromatic compound.

4.7.3 General Properties of Aromatic Compounds

Aromatic compounds have the following general properties.

i.   They have high degree of unsaturation, but are resistant to addition reactions.

ii. They favour electrophilic substitution reactions.

iii. These compounds are unusually stable.

iv. They have low heat of hydrogenation and low heat of combustion.

v.   They are cyclic compounds.

vi. These compounds are flat and planar.

4.7.4 Classification of Aromatic Compounds

Benzene and its monocyclic derivatives: These aromatic compounds have only one ring, such as benzene and its monosubstituted derivatives; for example, toluene, phenol and aniline.

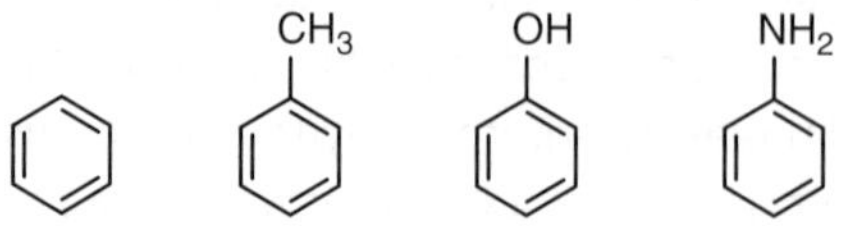

Benzene   Toluene   Phenol   Aniline

Polycyclic benzenoids: These aromatic compounds have two or more benzene rings fused together; for example, naphthalene and anthracene.

Naphthalene   Anthracene

Non-benzenoids: These compounds generally have two or more rings fused together, but none of the rings is a benzene structure and they conform to the Hückel's rule; that is, they have $(4n+2)$ π electrons and are aromatic compounds; for example, azulene.

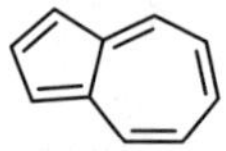

Azulene

In the structure of azulene, there are five conjugated double bonds and 10π electrons, which means that it follows Hückel's rule $(4 \times 2 + 2 - 10)$.

Macrocyclic: These are monocyclic non-benzene structures, and the ring sizes are quite big. There is adequate number of double bonds and π electrons to conform to the Hückel's rule, for example, [14] annulene obeys Hückel's rule and is aromatic.

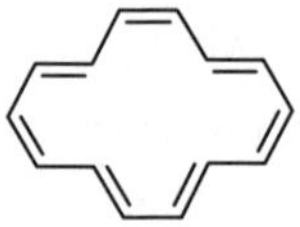

[14] Annulene

Heterocyclic: These are the compounds having at least one hetero atom (any other atom but carbon, e.g. O, N, and S) within the ring, and conform to Hückel's rule. The aromaticity of heterocyclic compounds, for example, pyridine and pyrrole, can be explained as follows.

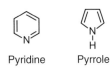

Pyridine   Pyrrole

Pyridine has π electron structure similar to benzene. Each of the five sp^2-hybridized carbons has a p orbital perpendicular to the plane of the ring. Each p orbital has one π electron. The nitrogen atom is also sp^2-hybridized and has one electron in the p orbital.

So, there are six π electrons in the ring.

The nitrogen lone pair electrons are in an sp^2 orbital in the plane of the ring and are not a part of the aromatic π system.

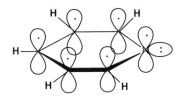

Structure of pydridine with p orbitals

The situation in pyrrole is slightly different. Pyrrole has a π electron system similar to that of cyclopentadienyl anion.

It has four sp^2-hybridized carbons, each of which has a p orbital perpendicular to the ring and contributes one π electron. The nitrogen atom is also sp^2-hybridized and its lone pair electrons occupies a p orbital.

Therefore, there are a total of six π electrons, which make pyrrole an aromatic compound.

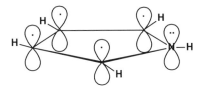

Structure of pyrrole with p orbitals

4.7.5 Pharmaceutical importance of Aromatic Compounds: Some Examples

There are numerous examples of aromatic compounds that are pharmaceutically important as drugs or pharmaceutical additives. Just a few examples of pharmaceutically important aromatic compounds are cited here. Aspirin, a well-known non-narcotic analgesic and antipyretic drug, is a classic example of a pharmaceutically important benzene derivative. Morphine, an aromatic alkaloid, is a narcotic (habit-forming) analgesic drug that is used extensively for the management of post-operative pain. Aromatic compounds, valium is prescribed as a tranquilizer, ibuprofen as an anti-inflammatory and sulpha drugs, for example, sulphamethoxazole, are used as antimicrobial agents. Taxol, one of the best-selling anticancer drugs of modern times, also belongs to the class of aromatic compounds. Saquinavir and crixivan, two anti-HIV drugs (protease inhibitors), also possess aromatic characters.

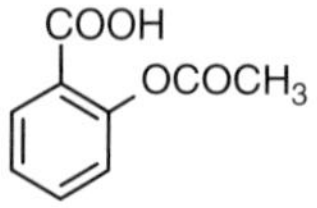

Aspirin
Acetyl salicylic acid

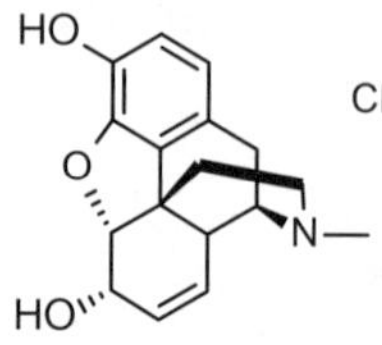

Morphine

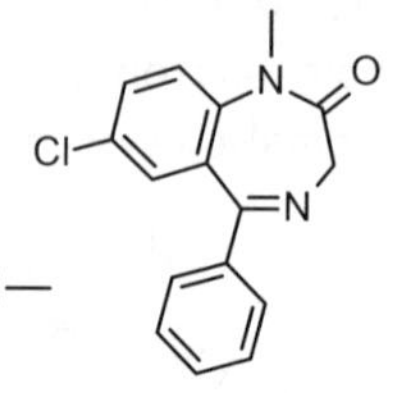

Valium

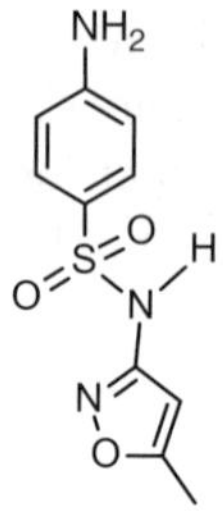

Sulphamethoxazole

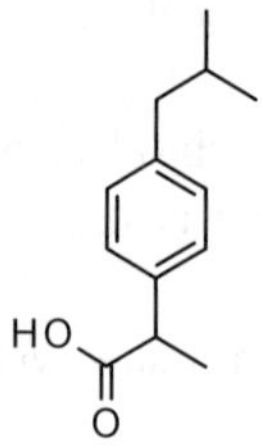

Ibuprofen

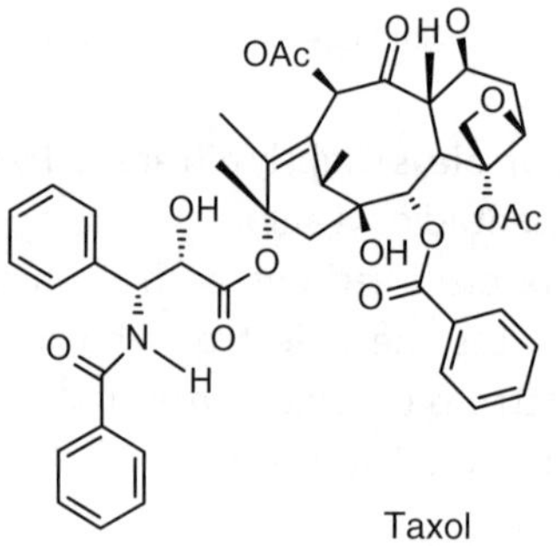

Taxol
(Isolated from *Taxus brevifolia*)

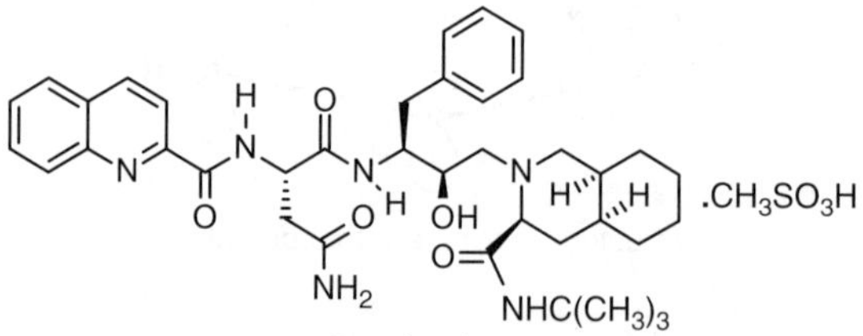

Saquinavir

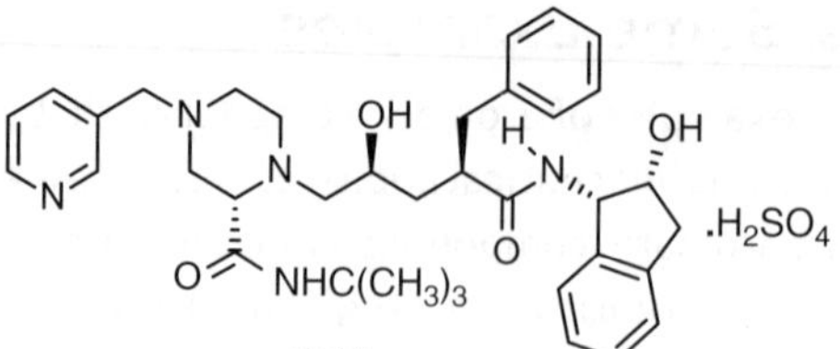

Crixivan

4.7.6 Structure of Benzene: Kekulé Structure of Benzene

In 1865, August Kekulé, a German organic chemist, proposed the structure of benzene (C_6H_6). According to his proposal, in benzene

i. all six carbon atoms are in a ring;

ii. all carbon atoms are bonded to each other by alternating single and double bonds;

iii. one hydrogen atom is attached to each carbon atom;

iv. all hydrogen atoms are equivalent.

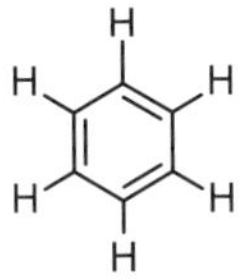

Kekulé structure of benzene

4.7.6.1 Limitations of Kekulé Structure

The Kekulé structure predicts that there should be two different 1,2-dibromobenzenes. In practice, only one 1,2-dibromobenzene has ever been found.

Kekulé proposed that these two forms are in equilibrium, which is established so rapidly that it prevents isolation of the separate compounds. Later, this proposal was proved to be incorrect, because no such equilibrium exists!

Two different 1,2-dibromobenzene as suggested by Kekulé

Benzene cannot be represented accurately by either individual Kekulé structure, and does not oscillate back and forth between two. The Kekulé structure also cannot explain the stability of benzene. Kekulé structure of benzene proposed three alternative double bonds (triene) in the ring and in that case, the expected hydration enthalpy would be $-360\,kJ\ mol^{-1}$. However, actually benzene has the hydration enthalpy of $152\,kJ\ mol^{-1}$, which is much less than the expected value. This means that benzene is actually more stable than the Kekulé structure could predict. Note that when an alkene reacts with hydrogen, the enthalpy change is called enthalpy change of hydration or simply *hydration enthalpy.*

4.7.6.2 The Resonance Explanation of the Structure of Benzene

The resonance theory can be applied successfully to explain the structure of benzene. First of all, let us have a look at the resonance theory. According to this theory

   i. resonance forms are imaginary, not real;

   ii. resonance structures differ only in the positions of their electrons;

   iii. different resonance forms do not have to be equivalent;

   iv. the more resonance structures there are, the more stable the molecule is;

   v. whenever it is possible to draw two or more resonance structures of a molecule, none of the structures will be in complete agreement with the compound's chemical and physical properties;

   vi. the actual molecule or ion is better represented by a hybrid of these structures;

   vii. whenever equivalent resonance structure can be drawn for a molecule, the molecule (or hybrid) is much more stable than any of the resonance structures could be individually if they could exist.

If we consider the Kekulé structure of benzene, it is evident that the two proposed structures differ only in the position of the electrons. Therefore, instead of being two separate molecules in equilibrium, they are indeed two resonance contributors to a picture of the real molecule of benzene.

Two structures of benzene as suggested by Kekulé

If we think of a hybrid of these two structures, then the C—C bonds in benzene are neither single bonds nor double bonds. They should have a bond order between a single (1.47 Å) and a double bond (1.33 Å). It has actually been proven that benzene is a planar molecule, and all of its C—C bonds are of equal length (1.39 Å). The bond order (1.39 Å) is indeed in between a single and a double bond! Thus, instead of drawing the benzene structure using alternative single and double bonds, a hybrid structure can be drawn as follows.

Hybrid structure of benzene

The hybrid structure of benzene is represented by inscribing a circle in the hexagon as depicted previously.

With benzene, the circle represents the six electrons that are delocalized about the six carbon atoms of the benzene ring.

The resonance theory accounts for the much greater stability of benzene (*Resonance energy*) when compared to the hypothetical 1,3,5-cyclohexatriene.

It also explains why there is only one 1,2-dibromobenzene rather than two. Therefore, the structure of benzene is not really a 1,3,5-cyclohexatriene, but a hybrid structure as shown before.

4.7.6.3 The Molecular Orbital Explanation of the Structure of Benzene

The bond angles of the carbon atoms in benzene are 120°. All carbon atoms are sp^2-hybridized, and each carbon atom is with a single unhybridized p orbital perpendicular to the plane of the ring. The carbon sp^2-hybridized orbitals overlap to form the ring of the benzene molecule.

Because the C—C bond lengths are 1.39 Å, the p orbitals are close enough to overlap efficiently and equally all around the ring.

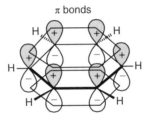

Benzene structure in the light of
molecular orbital theory

The six overlapping p orbitals overlap to form a set of six p molecular orbitals. Six π electrons are completely delocalized around the ring, and form two doughnut-shaped clouds of π electrons, one above and one below the ring.

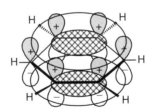

Doughnut-shaped cloud of π electrons

Six p atomic orbitals, one from each carbon of the benzene ring, combine to form six p molecular orbitals. Three of the molecular orbitals have energies lower than that of an isolated p orbital, and are known as *bonding molecular orbitals*.

Another three of the molecular orbitals have energies higher than that of an iso-lated p orbital and are called *antibonding molecular orbitals.* Two of the bonding orbitals have the same energy, as do the antibonding orbitals. Such orbitals are said to be *degenerate.*

4.7.6.4 Stability of Benzene

Benzene has a closed bonding shell of delocalized π electrons. This closed bonding shell partly accounts for the stability of benzene. Benzene is more stable than the Kekulé structure suggests. The stability of benzene can be shown as follows.

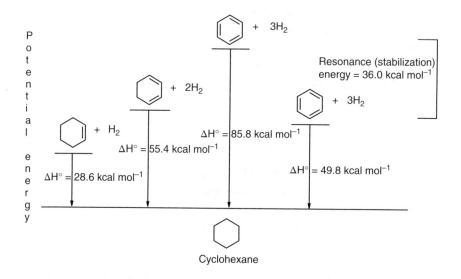

The energy required for the hydrogenation of cyclohexene to cyclohexane is −28.6 kcal mol⁻¹. Therefore, in the case of cyclohexadiene, where there are two double bonds, the energy required for the hydrogenation can be calculated as $2 \times -28.6 = -57.2$ kcal mol⁻¹. In practice, the experimental value is quite close to this calculated value, and is −55.4 kcal mol⁻¹. In this way, if benzene was really a cyclohexatriene as proposed by Kekulé, the calculated required energy would be three times than that of cyclohexene, that is, $3 \times -28.6 = -85.8$ kcal mol⁻¹. In practice, it was found that this required energy for benzene is −49.8 kcal mol⁻¹, which means that there is a clear 36 kcal mol⁻¹ difference between the calculated value and the observed value, and this 36 kcal mol⁻¹ is known as the *stabilization energy* or *resonance energy.* This explains the stability of benzene. Due to this *stabilization energy,* benzene does not undergo similar reactions as a cycloalkene. This can be depicted in an example as follows.

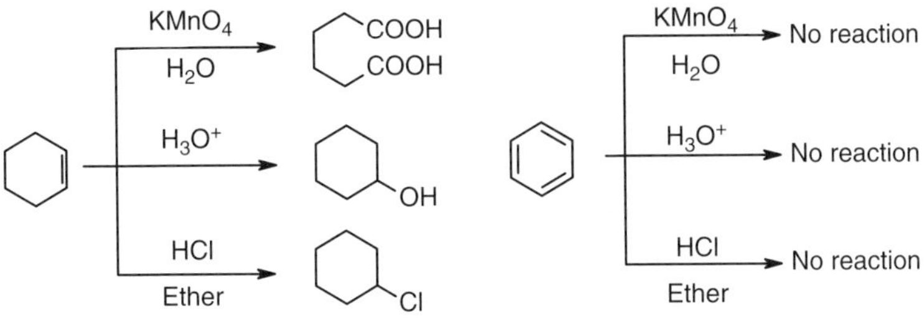

Cyclohexene versus benzene

4.7.7 Nomenclature of Benzene Derivatives

Benzene derivatives are named by prefixing the name of the substituent group to the word *benzene*; for example, chlorobenzene and nitrobenzene. Many benzene derivatives have their trivial names, which may show no resemblance to the name of the attached substituent group; for example, phenol, toluene and aniline.

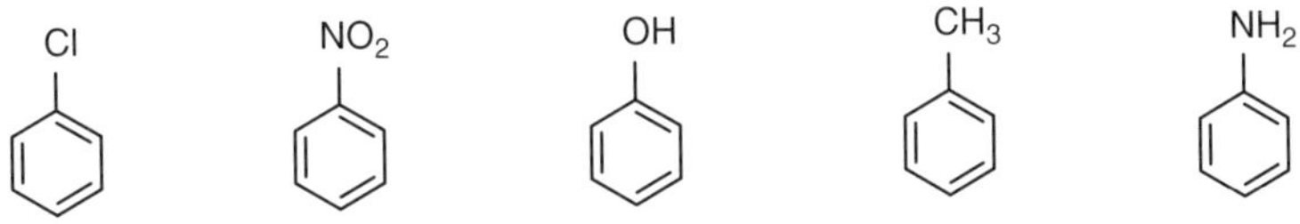

When two groups are attached to the benzene ring, their relative positions have to be identified. The three possible isomers of a disubstituted benzene are differentiated by the use of the names *ortho*, *meta* and *para*, abbreviated as *o-*, *m-* and *p-*, respectively.

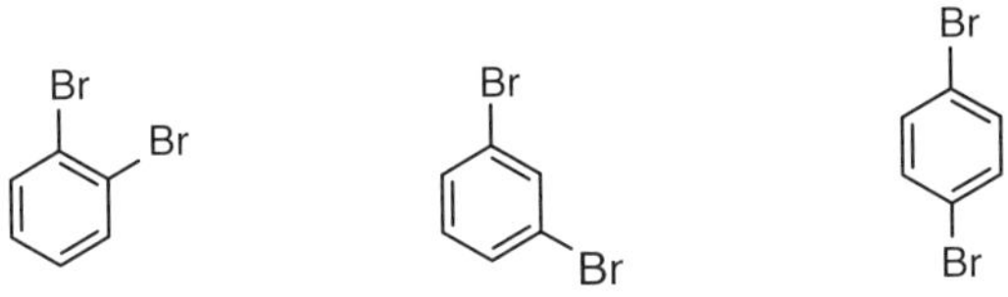

If the two groups are different and neither is a group that offers a trivial name to the molecule, the two groups are named successively and the word *benzene* is added. If one of the two groups is the kind that gives a trivial name to the molecule, then the compound is named as a derivative of that compound. In both cases, relative position should also be designated.

Chapter 4: Organic Functional Groups   **183**

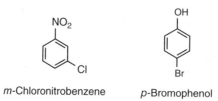

m-Chloronitrobenzene   *p*-Bromophenol

When more than two groups are attached to the benzene ring, numbers are used to indicate their relative positions. If the groups are the same, each is given a number, the sequence being the one that gives the lowest combination of numbers; if the groups are different, then the last-named group is understood to be in position 1 and the other numbers conform to that. If one of the groups that attribute a trivial name is present, then the compound is named as having the special group in position 1.

3-Bromo-5-chloronitrobenzene   1,2,4-Tribromobenzene

4.7.8 Electrophilic Substitution of Benzene

Benzene is susceptible to electrophilic attack and, unlike any alkene, it undergoes substitution reactions rather than addition reactions. Before we go into any details of such reactions, let us try to understand the following terms.

Arenes. Aromatic hydrocarbons, as a class, are called arenes.

Aryl group. Aromatic hydrocarbon with a hydrogen atom removed is called aryl group, designated by Ar—.

Phenyl group. The benzene ring with one hydrogen atom removed (C_6H_5—) is called the phenyl group, designated by Ph—.

Electrophile. Electron-loving. Cations, E⁺ or electron-deficient species. For example, Cl^+ or Br^+ (halonium ion) and $^+NO_2$ (nitronium ion).

An electrophile (E⁺) reacts with benzene ring and substitutes for one of its six hydrogen atoms. Cloud of π electrons exists above and below the plane of benzene ring. These π electrons are available to electrophiles. Benzene's closed shell of six π electrons renders it a special stability.

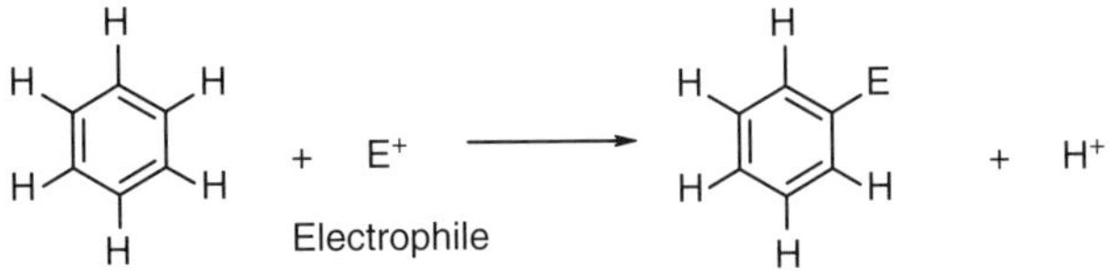

Substitution reactions allow the aromatic sextet of π electrons to be regenerated after attack by the electrophile has occurred. Electrophiles attack the π system of benzene to form a delocalized non-aromatic carbocation (*arenium ion* or σ complex). Some specific examples of electrophilic substitution reactions of benzene are summarized here.

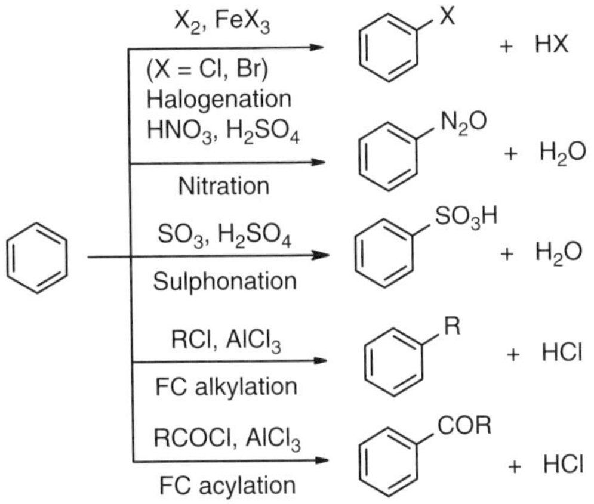

4.7.8.1 Reactivity and Orientation in Electrophilic Substitution of Substituted Benzene

When substituted benzene undergoes electrophilic attack, groups already on the ring affect reactivity of the benzene ring as well as the orientation of the reaction. A summary of these effects of substituents on reactivity and orientation of electrophilic substitution of substituted benzene is presented here.

Substituent	Reactivity	Orientation	Inductive effect	Resonance effect
$-CH_3$	Activating	*ortho, para*	Weak electron donating	None
$-OH, -NH_2$	Activating	*ortho, para*	Weak electron withdrawing	Strong electron donating
$-F, -Cl, -Br, -I$	Deactivating	*ortho, para*	Strong electron withdrawing	Weak electron donating

Substituent	Reactivity	Orientation	Inductive effect	Resonance effect
$-N^+(CH_3)_3$	Deactivating	*meta*	Strong electron withdrawing	None
$-NO_2, -CN,$ $-CHO, -COOCH_3,$ $-COCH_3, -COOH$	Deactivating	*meta*	Strong electron withdrawing	Strong electron withdrawing

4.7.8.1.1 *Reactivity of Substituted Benzene*

Groups already present on the benzene ring may activate the ring (*activating groups*), making it more reactive towards electrophilic substitution than benzene, for example, a −OH substituent makes the ring 1000 times more reactive than benzene, or may deactivate the ring (*deactivating groups*) making it less reactive than benzene, for example, a −NO$_2$ substituent makes the ring more than 10 million times less reactive. Relative rate of reaction depends on whether the substituent group (−S) withdraws or releases electrons relative to hydrogen. When −S is an electron-releasing group, the reaction is faster, whereas when this group is an electron-withdrawing group, a slower rate of reaction is observed.

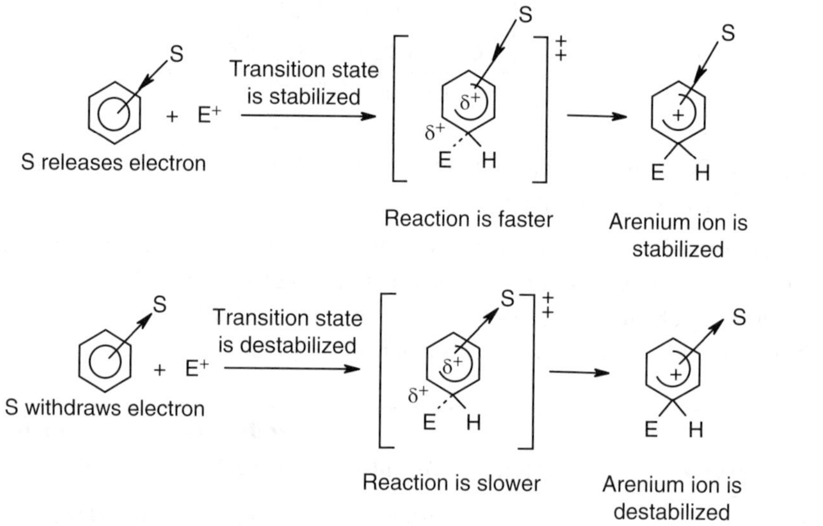

4.7.8.1.2 *Orientation of Substituted Benzene*

Similarly, groups already present on the benzene ring direct the orientation of the new substituent to *ortho, para* or *meta* positions. For example, nitration of chlorobenzene yields *ortho*-nitrochlorobenzene (30%) and *para*-nitrochloroben-zene (70%).

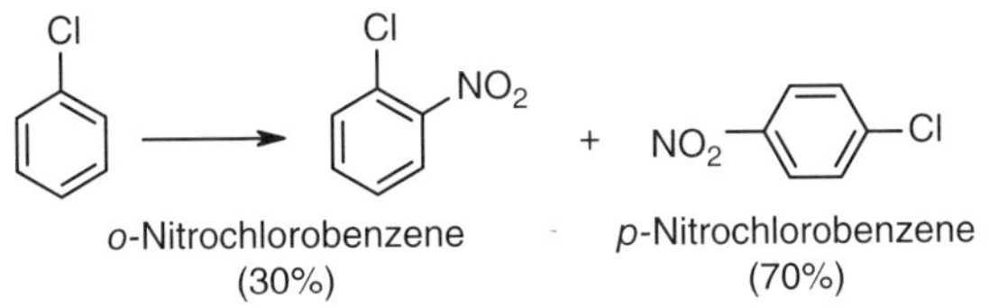

o-Nitrochlorobenzene
(30%)

p-Nitrochlorobenzene
(70%)

All activating groups are *ortho* and *para* directing, and all deactivating groups other than halogens are *meta* directing. The halogens are unique in being deactivating but *ortho* and *para* directing. A summary of various groups and their effects on the benzene ring in relation to reactivity and orientation is presented here.

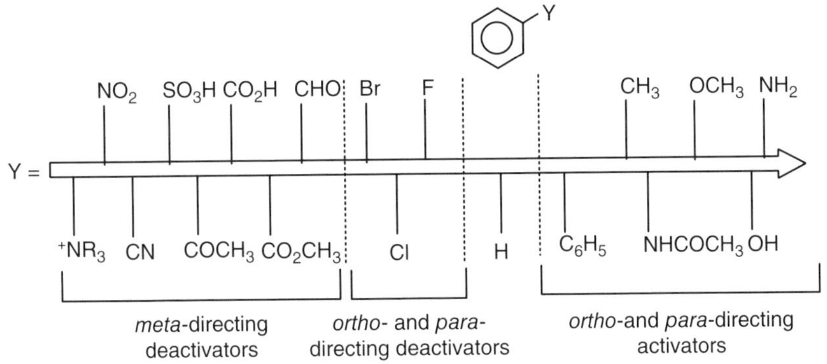

4.7.8.2 Inductive Effect of Substituted Benzene

An *inductive effect* is the withdrawal or donation of electrons through a σ bond due to electronegativity and the polarity of bonds in functional groups (*electrostatic interaction*). When the substituent (—S) bonded to a benzene ring is more electronegative atom or group than carbon, for example, F, Cl or Br, the benzene ring will be at the positive end of the dipole. These substituents will withdraw electron from the ring. As a consequence, an electrophilic attack will be less favoured because of an additional full positive charge on the ring.

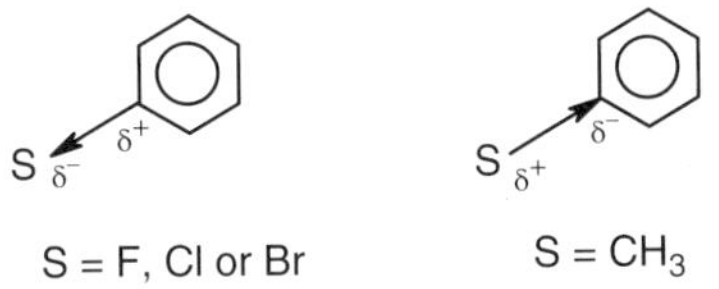

If a substituent (—S) bonded to a benzene ring is less electron-withdrawing than a hydrogen, the electrons in the σ bond that attaches the substituent to the benzene ring will move towards the ring more readily than will those in the S bond that attaches a hydrogen to the ring. Such a substituent (e.g. CH_3), compared to a hydrogen atom, donates electrons inductively into the ring. Inductive electron donation makes the ring more reactive towards electrophilic substitution because of the increased availability of electrons.

4.7.8.3 Resonance Effect of Substituted Benzene

A resonance effect is the withdrawal (e.g. by $-CO$, $-CN$ and $-NO_2$) or donation (e.g. by $-X$, $-OH$ and $-OR$) of electrons through a π bond due to the overlap of a p orbital on the substituent with a p orbital on the aromatic ring. The presence of a substituent may increase or decrease the resonance stabilization of the intermediate arenium ion complex.

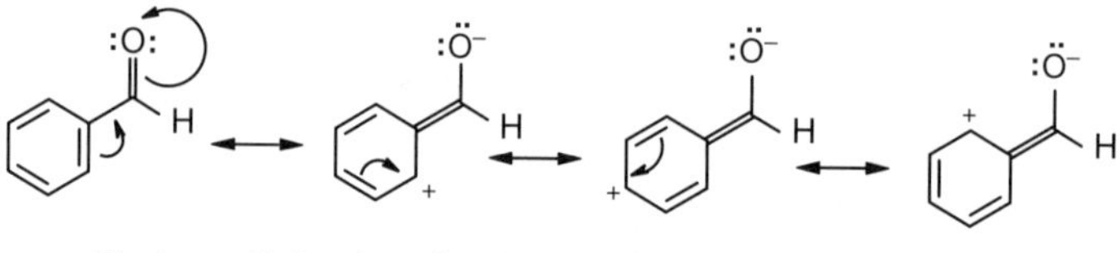

Electron withdrawing effect by an aldehyde (CHO) group

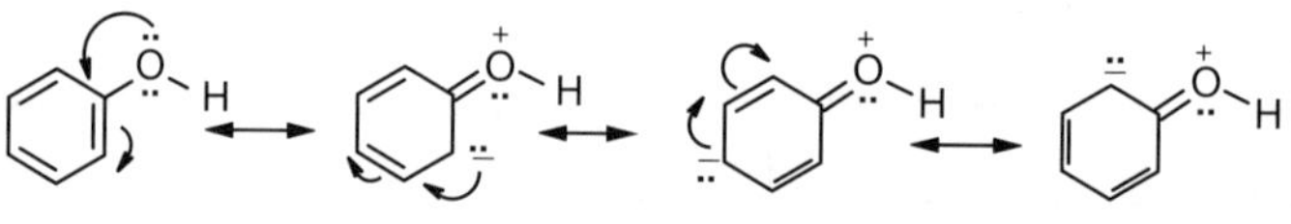

Electron donating phenolic hydroxyl group (OH)

The electron donating resonance effect applies with decreasing strength in the following order:

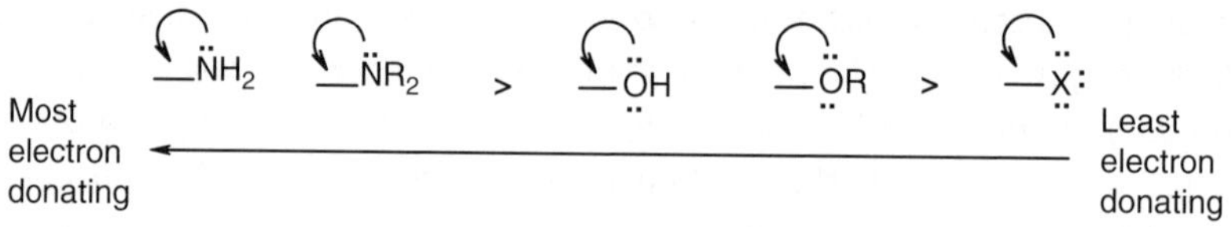

Most electron donating → Least electron donating

4.7.8.3.1 Why the $-CF_3$ Group is Meta-Directing

All *meta* directing groups have either a partial positive charge or a full positive charge on the atom directly linked to the benzene ring. In trifluoromethyl group ($-CF_3$), there are three electronegative fluorine atoms, which make this group strongly electron withdrawing. As a result $-CF_3$ deactivates benzene ring, and directs further substitutions to *meta* positions. The *ortho* and *para* electrophilic attacks in trifluoromethyl benzene result in one highly unstable contributing resonance structure of the arenium ion, but no such highly unstable resonance structure is formed from *meta* attack. In the case of *ortho* and *para* attacks, the positive charge in one of the resulting contributing resonance structures, is located on the ring carbon that bears the electron-withdrawing group. The arenium ion formed from *meta* attack is the most stable among the three, and thus the substitution in *meta* position is favoured. Therefore, trifluoromethyl group is a *meta*-directing group.

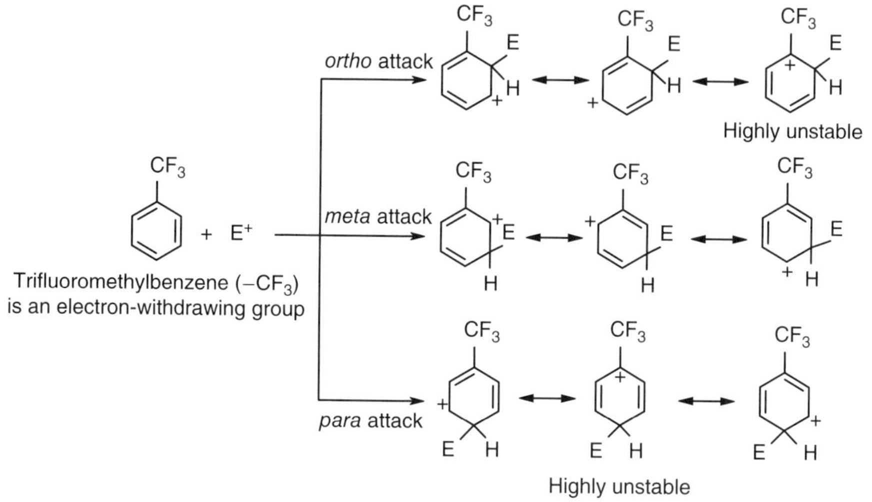

Trifluoromethylbenzene (−CF₃) is an electron-withdrawing group

4.7.8.3.2 Why the −CH₃ Group is Ortho-Para Directing

The stability of the carbocation intermediate formed in the rate-determining step actually is the underlying factor for a substituent to direct an incoming electrophile to a particular position, *ortho, meta* or *para*. The methyl group (−CH₃) donates electron inductively and, in the presence of this electron-donating group, the resonance contributors formed from *ortho, meta* and *para* attacks are shown next. In the most stable contributors arising from *ortho* and *para* attacks, the methyl group is attached directly to the positively charged carbon, which can be stabilized by donation of electron through inductive effect. From a *meta* attack no such stable contributor is formed. Thus, the substitutions in *ortho* and *para* positions are favoured. Therefore, methyl group is an *ortho* and *para* directing group.

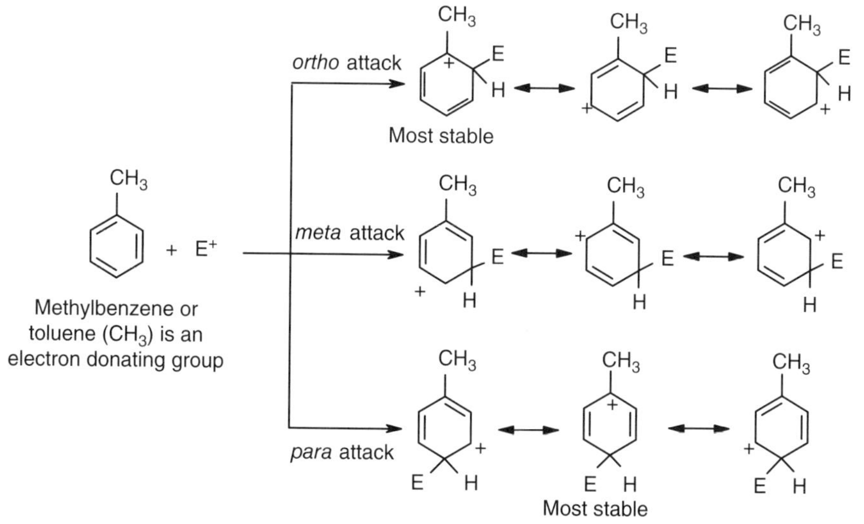

Methylbenzene or toluene (CH₃) is an electron donating group

4.7.8.3.3 Why Halogens are Ortho-Para Directing

Halogens are the only deactivating substituents that are *ortho-para* directors. However, they are the weakest of the deactivators. Halogens withdraw electrons from the ring through inductive effect more strongly than they donate electrons by resonance. It is the resonance-aided electron donating effect that causes halogens to be *ortho-para* directing groups. Halogens can stabilize the transition states leading to reaction at the *ortho* and *para* positions. On the other hand, the electron withdrawing inductive effect of halogens influences the reactivity of halobenzenes. A halogen atom, for example, Cl, donates an unshared pair of electrons, which give rise to relatively stable resonance structures contributing to the hybrids for the *ortho* and *para*-substituted arenium ions. Thus, despite being deactivators, halogens are *ortho* and *para*-director. The resonance contributors formed from *ortho*, *meta* and *para* attacks on the chlorobenzene are shown next.

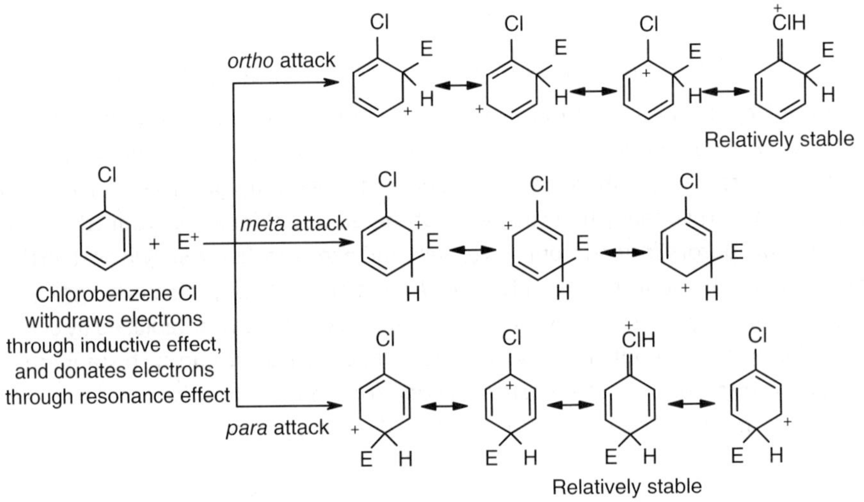

4.7.9 Alkylbenzene: Toluene

Toluene, also known as methylbenzene, is the simplest member of the series known as *alkylbenzenes* where an alkyl group, for example, CH_3, is directly attached to the benzene ring. It is a colourless, water-insoluble liquid with the smell associated with paint thinners. As the use of benzene as a nonpolar solvent has long been prohibited because of its adverse effect on the central nervous system (CNS) and on bone marrow, as well as its carcinogenic property, toluene has replaced benzene as a nonpolar solvent. Although it has CNS depressant property like benzene, it does not cause leukaemia or aplastic anaemia. The effects of toluene on the CNS in most cases are temporary, and include headaches, dizziness and unconsciousness, but effects, such as incoordination, cognitive impairment and vision and hearing loss, may become permanent with prolonged and repeated exposure, especially at concentrations associated with solvent abuse. High levels of toluene exposure during pregnancy may affect the normal mental and physical growth in children.

4.7.9.1 Preparations of Toluene

Generally, toluene is prepared from benzene by Friedel–Crafts alkylation, where benzene reacts with CH_3Cl in presence of a strong Lewis acid ($AlCl_3$) to produce toluene. The reaction mechanism is similar to that of the isopropylbenzene (see Section 5.7.1.4).

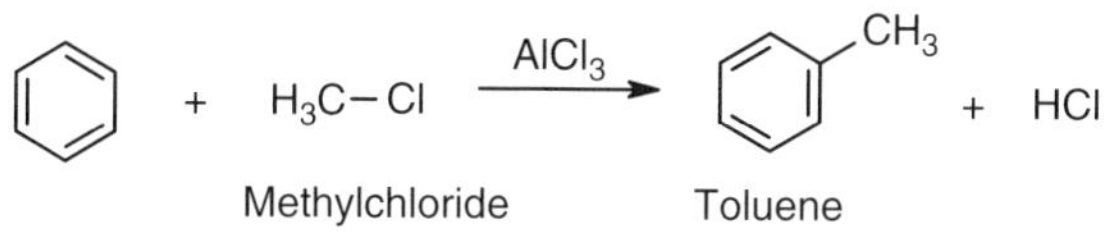

Methylchloride   Toluene

4.7.9.2 Reactions of Toluene

Like benzene, toluene undergoes electrophilic substitutions, where the substitutions take place in *ortho* and *para* positions. As $-CH_3$ group is an activating group, the reaction rate is much faster than usually observed with benzene. For example, the nitration of toluene provides *o*-nitrotoluene (61%) and *p*-nitrotoluene (39%).

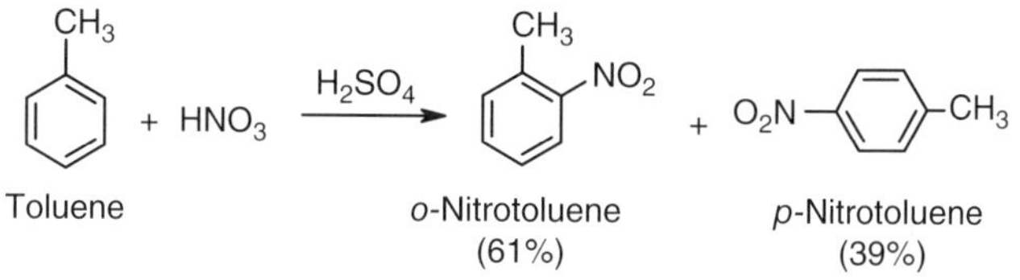

Toluene   o-Nitrotoluene   p-Nitrotoluene
                    (61%)   (39%)

    Apart from the usual electrophilic aromatic substitution reactions, other reactions can be carried out involving the methyl group in toluene; for example, oxidation and halogenation of the alkyl group.

4.7.9.2.1 Oxidation of Toluene

Regardless of the length of the alkyl substituent in any alkylbenzene, it can be oxidized to a carboxylic acid provided that it has a hydrogen atom bonded to the benzylic carbon. So, reaction can occur with 1° and 2°, but not 3° alkyl side chains.
    Toluene is oxidized to benzoic acid.

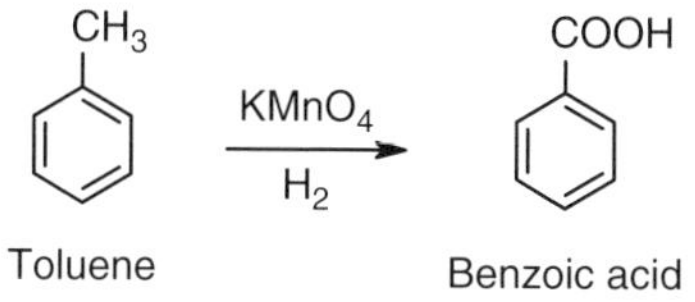

Toluene   Benzoic acid

4.7.9.2.2 Benzylic Bromination of Toluene

Bromine selectively substitutes for a benzylic hydrogen in toluene in a radical substitution reaction to afford bromomethylbenzene or benzylbromide. *N*-bromosuccinimide is used to carry out benzylic bromination of toluene.

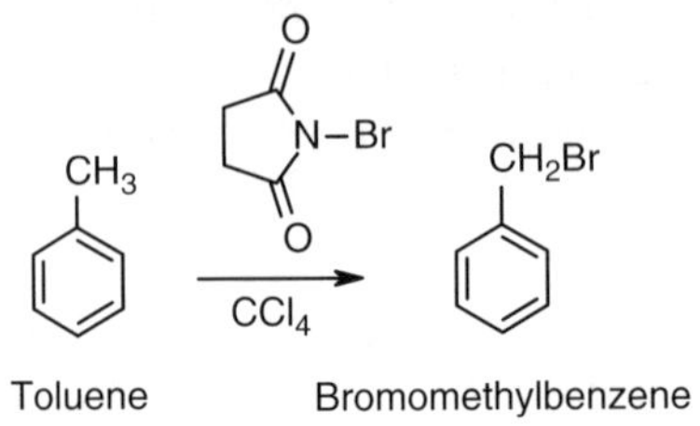

| Toluene | Bromomethylbenzene |

Bromomethylbenzene or benzylbromide can be subjected to further nucleophilic reactions. Bromine can be replaced by several nucleophiles by means of a nucleophilic substitution reaction resulting in various monosubstituted benzenes.

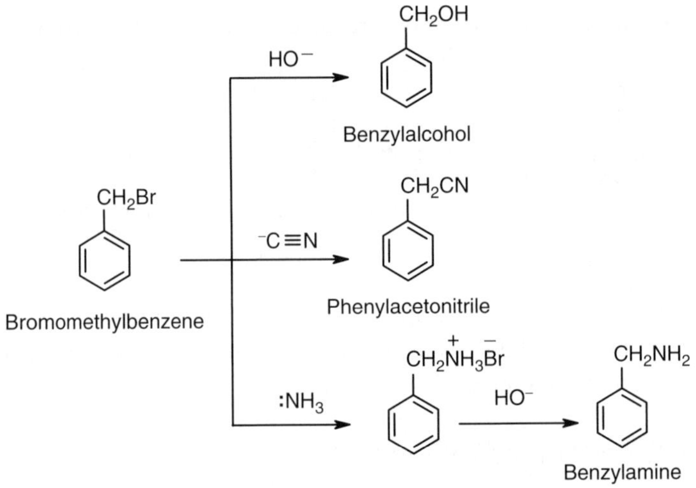

4.7.10 Phenols

Phenols are compounds of the general formula ArOH, where Ar is a phenyl, a substituted phenyl, or one of the other aryl groups, for example, naphthyl. Phenols differ from alcohols in having the —OH group attached directly to an aromatic ring. Hydroxybenzene, the simplest member of the phenols, is generally referred to as *phenol*.

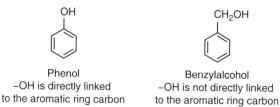

Phenol
−OH is directly linked
to the aromatic ring carbon

Benzylalcohol
−OH is not directly linked
to the aromatic ring carbon

Many pharmaceutically and pharmacologically important compounds, either of natural or synthetic origin, belong to this class of compounds; for example, salicylic acid and quercetin.

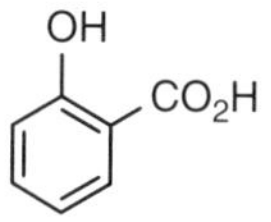

Salicylic acid
(An analgesic and a
precusor for aspirin)

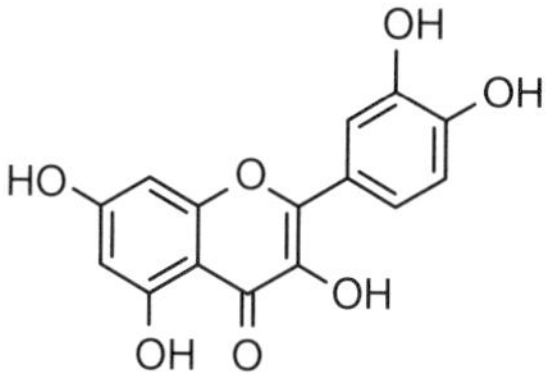

Quercetin
(A natural antioxidant)

4.7.10.1 Nomenclature of Phenols

Phenols are generally named as derivatives of the simplest member of the family, phenol, for example, o-chlorophenol. Sometimes trivial or special names are also used; for example, m-cresol. Occasionally, phenols are named as hydroxyl-compounds; for example, para-hydroxybenzoic acid. Numbering is often used to denote the position(s) of the substituent(s) on a phenol skeleton; for example, 2,4-dinitrophenol.

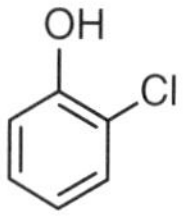

o-Chlorophenol 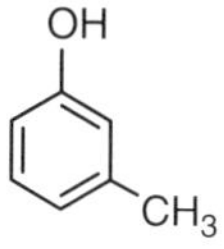   m-Cresol 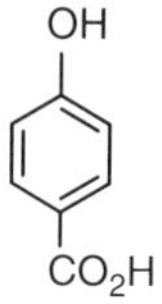   p-Hydroxybenzoic acid 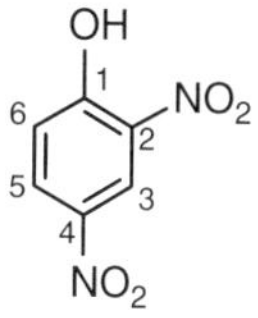   2,4-Dinitrophenol

4.7.10.2 Physical Properties of Phenols

The simplest phenols are liquids or low-melting solids. Because of hydrogen bonding, phenols have quite high boiling point (e.g. boiling point of m-cresol is 201 °C). Phenol itself is somewhat soluble in water (9 g/100 g of water), because of hydrogen bonding with water. Most other phenols are insoluble in water. Generally, phenols themselves are colourless. However, they are easily oxidized to form coloured substances. Phenols are acidic in nature and most phenols have K_a values of ~10^{-10}.

4.7.10.3 Physical Properties of Nitrophenols

Physical properties of o-, m- and p-nitrophenols differ considerably.

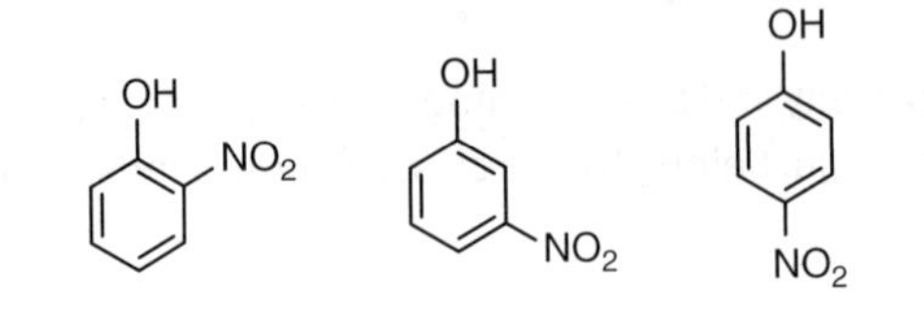

o-Nitrophenol   m-Nitrophenol   p-Nitrophenol

Nitrophenols	Boiling point at 70 mm in °C	Solubility in g/100 g H_2O
o-Nitrophenol	100	0.2
m-Nitrophenol	194	1.35
p-Nitrophenol	Decompose	1.69

Among the nitrophenols, *meta-* and *para-*nitrophenols have high boiling points, because of *intermolecular hydrogen bonding* as shown.

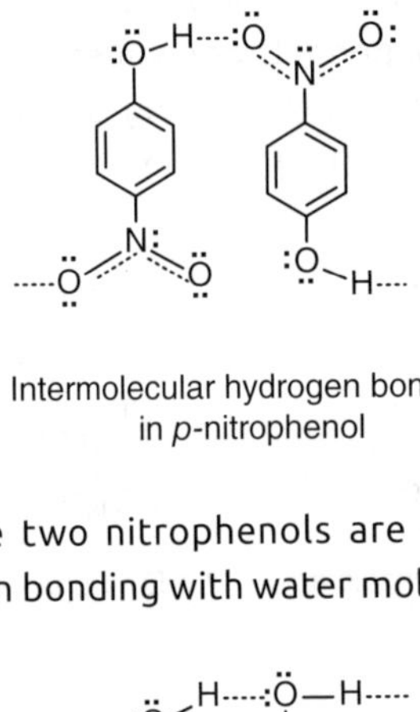

Intermolecular hydrogen bonding
in p-nitrophenol

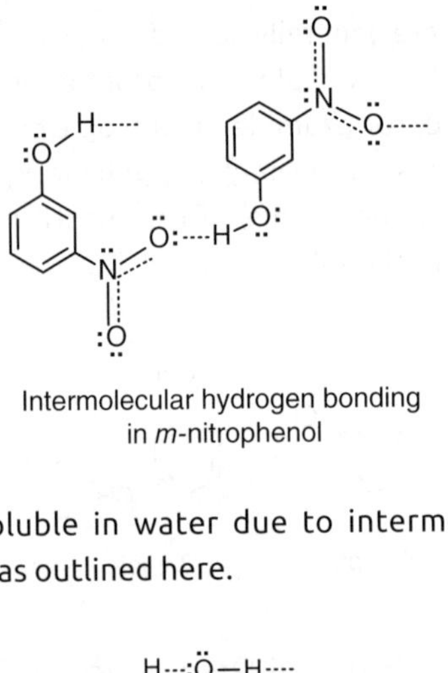

Intermolecular hydrogen bonding
in m-nitrophenol

These two nitrophenols are also soluble in water due to intermolecular hydrogen bonding with water molecules as outlined here.

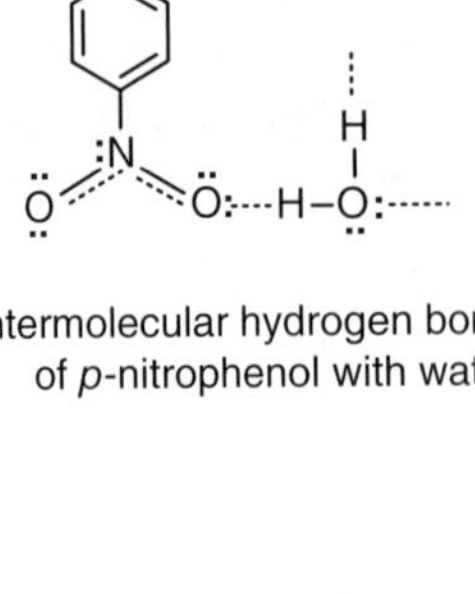

Intermolecular hydrogen bonding
of p-nitrophenol with water

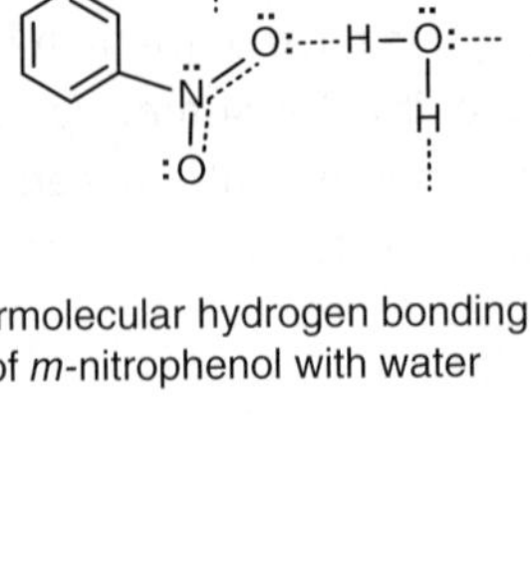

Intermolecular hydrogen bonding
of m-nitrophenol with water

However, in the case of *ortho*-nitrophenol, the —NO$_2$ and —OH groups are located exactly right for the formation of a hydrogen bond within a single molecule, that is, *intramolecular hydrogen bonding* as shown next. This intramolecular hydrogen bonding takes the place of intermolecular hydrogen bonding with other phenol molecules or water molecules.

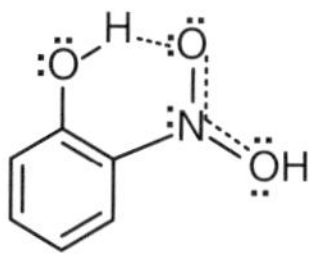

Intramolecular hydrogen
bonding in *o*-nitrophenol

As a consequence, *ortho*-nitrophenol has lower boiling point than *meta*- and *para*-nitrophenols, and extremely poor solubility in water compared to that of *meta*- and *para*-nitrophenols.

4.7.10.4 Acidity of Phenols

Phenols are fairly acidic compounds. They are more acidic than related alcohols.

Aqueous hydroxides, for example, NaOH, convert phenols into sodium phenoxide (not by aqueous bicarbonates).

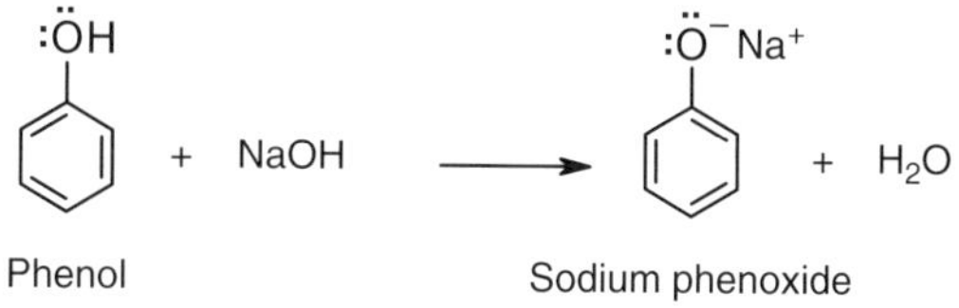

Aqueous mineral acids, carboxylic acids or carbonic acid, convert the salts back to free phenols.

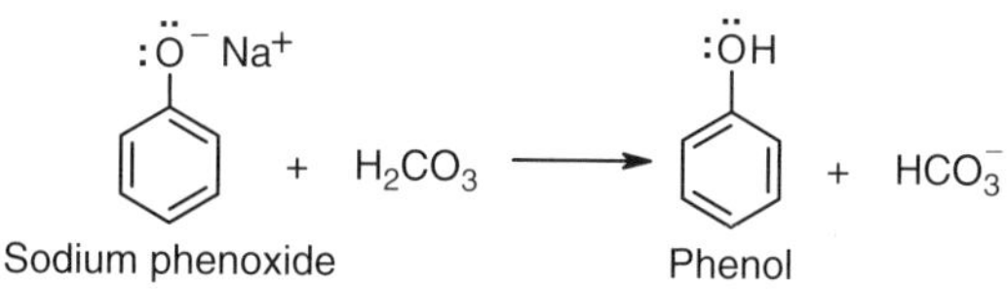

Most phenols (K_a values of ~10^{-10}) are considerably weaker acids than carboxylic acids ($K_a = $~$10^{-5}$).

Although weaker than carboxylic acids, phenols are more acidic than alcohols ($K_a = $~$10^{-16}$ to 10^{-18}).

The benzene ring of a phenol acts as if it were an electron-withdrawing group. It withdraws electrons from the —OH group and makes the oxygen positive. The acidity of phenols is mainly due to an electrical charge distribution in phenols that causes the —OH oxygen to be more positive. As a result, the proton is held less strongly, and phenols can easily give this loosely held proton away to form a *phenoxide ion*. Thus, phenoxide ion has *resonance stabilization* since the negative charge can be *delocalized* over four atoms (three carbons and one oxygen), making it more stable as outlined here.

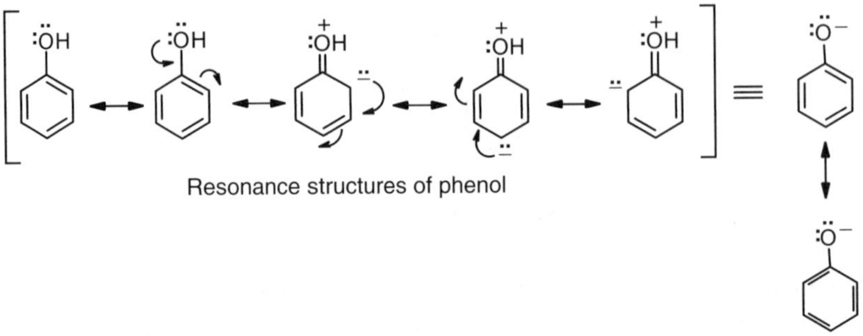

Resonance structures of phenol

Phenoxide ion

4.7.10.5 Preparations of Phenols

In the laboratory, phenols are predominantly prepared by either hydrolysis of aryl diazonium salts or alkali fusion of onates.

4.7.10.5.1 *Hydrolysis of Aryl Diazonium Salts*

Aryl diazonium salts react with water in presence of mineral acids to yield phenols.

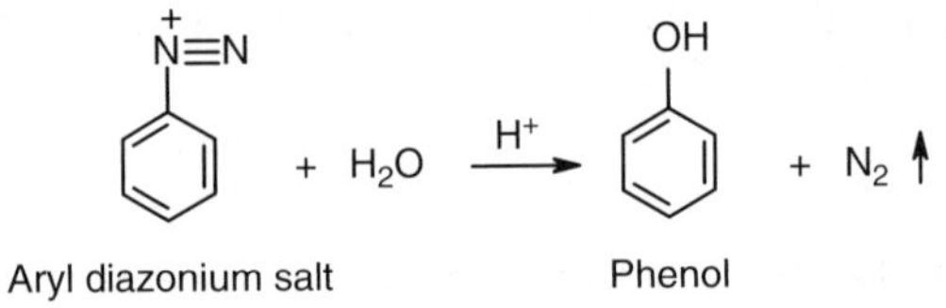

Aryl diazonium salt   Phenol

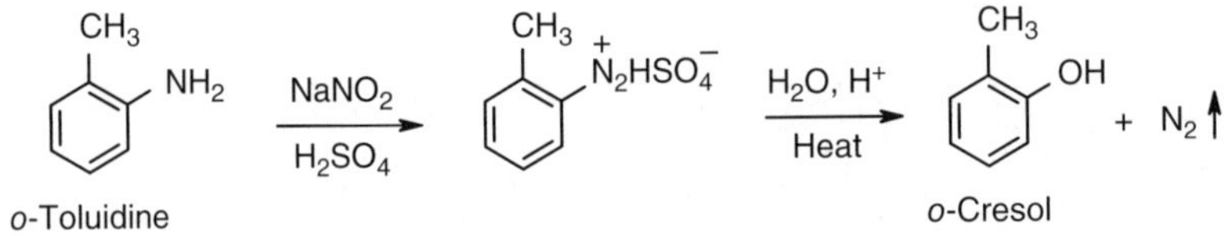

o-Toluidine   o-Cresol

4.7.10.5.2 *Alkali Fusion of Sulphonates*

Phenols can be prepared from the corresponding sulphonic acids by fusion with alkali.

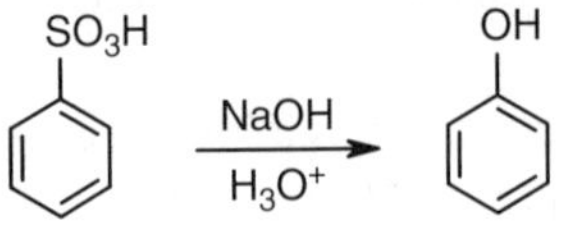

4.7.10.6 Reactions of Phenols

Phenols undergo electrophilic substitutions. In phenol, the substitutions take place in *ortho* and *para* positions. As —OH group is an activating group, the reaction rate is much faster than usually observed with benzene. For example, the bromination of phenol produces *ortho*-bromophenol (12%) and *para*-bromophenol (88%).

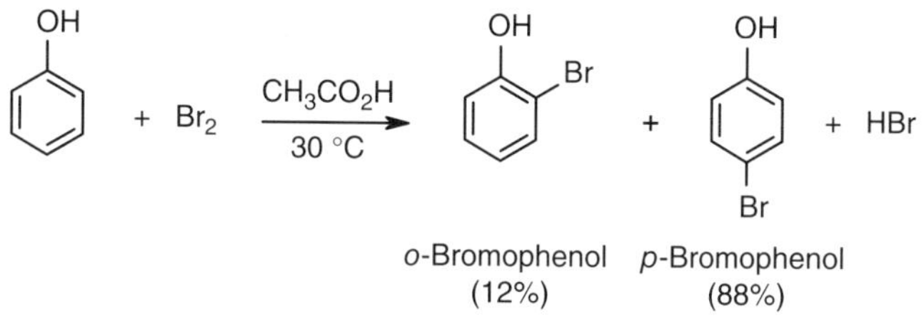

o-Bromophenol p-Bromophenol
(12%)   (88%)

A number of other reactions can also be carried out with phenols as follows.

4.7.10.6.1 Salt Formation

Phenol is acidic in nature, and can form a salt with alkali, for example, NaOH.

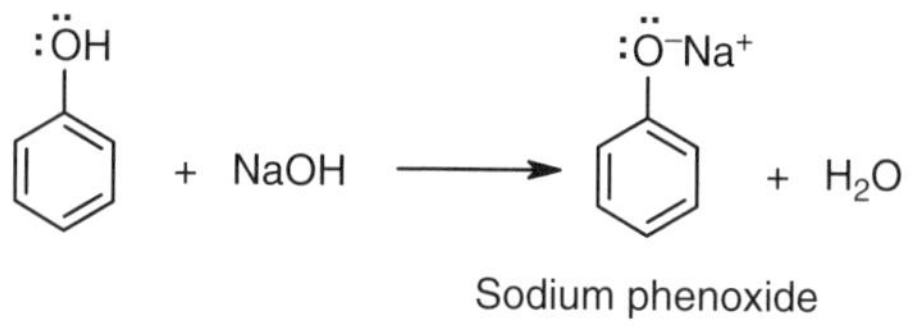

Sodium phenoxide

4.7.10.6.2 Ether Formation

Phenol reacts with ethyliodide (C_2H_5I), in the presence of aqueous NaOH, to give ethylphenylether, also known as *phenetole*.

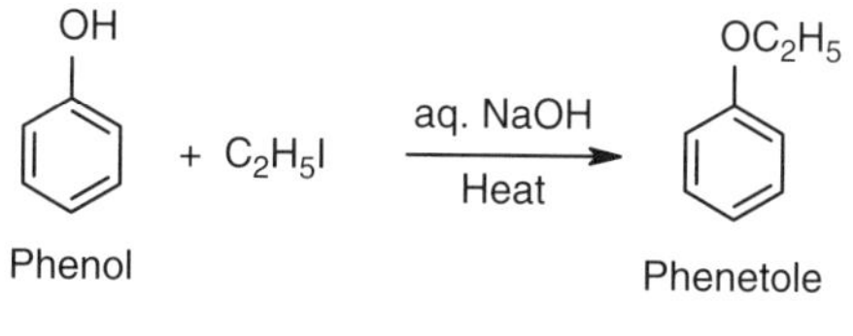

Phenol   Phenetole

4.7.10.6.3 Ester Formation

Phenols can undergo esterification, and yield corresponding esters. For example, phenol reacts with benzoylchloride to afford phenylbenzoate, and bromophenol reacts with toluenethionyl chloride to produce bromophenyltoluene sulphonate.

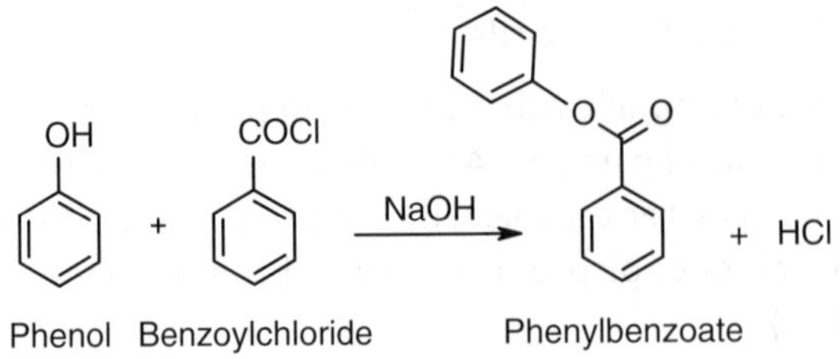

Phenol   Benzoylchloride   Phenylbenzoate

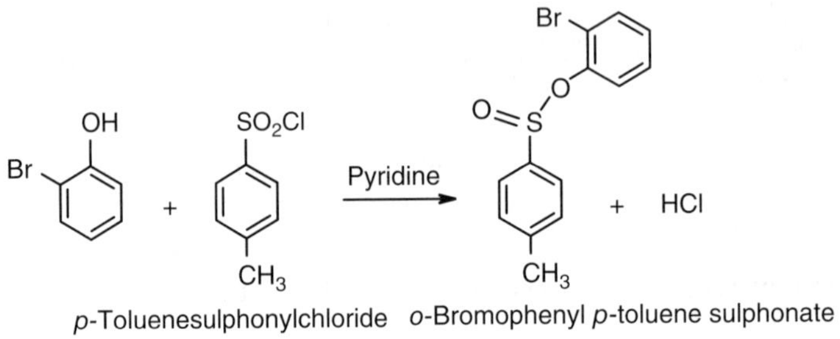

p-Toluenesulphonylchloride   o-Bromophenyl p-toluene sulphonate

4.7.10.6.4 Carbonation: Kolbe Reaction

Treatment of salt of a phenol with CO_2 replaces a ring hydrogen with a carboxyl group. This reaction is applied in the conversion of phenol itself into *ortho*-hydroxybenzoic acid, known as salicylic acid. Acetylation of salicylic acid provides acetylsalicylic acid (aspirin), which is the most popular painkiller in use today.

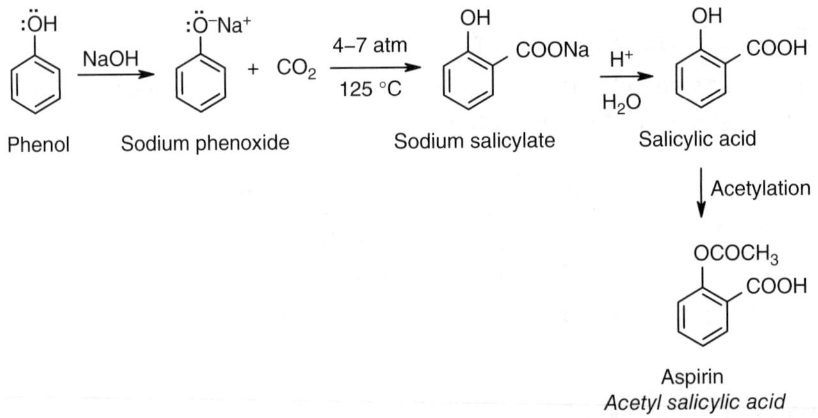

Phenol   Sodium phenoxide   Sodium salicylate   Salicylic acid

Aspirin
Acetyl salicylic acid

4.7.10.6.5 Aldehyde Formation: Reimer–Tiemann Reaction

Treatment of a phenol with chloroform ($CHCl_3$) and aqueous hydroxide, introduces an aldehyde group (—CHO), onto the aromatic ring, generally *ortho* to the —OH group. A substituted benzalchloride is initially formed, but is hydrolysed by the alkaline medium. Salicylaldehyde can be obtained from phenol by this reaction. Again, salicylaldehyde could be oxidized to salicylic acid, which could be acetylated to aspirin.

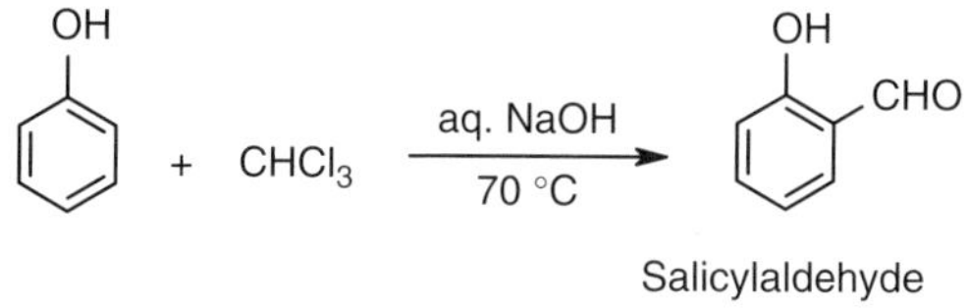

Salicylaldehyde

4.7.10.6.6 Polymer Formation

Phenol reacts with formaldehyde (HCHO) to produce *o*-hydroxymethylphenol, which reacts with phenol to yield *o*-(*p*-hydroxybenzyl)-phenol. This reaction continues to form a polymer.

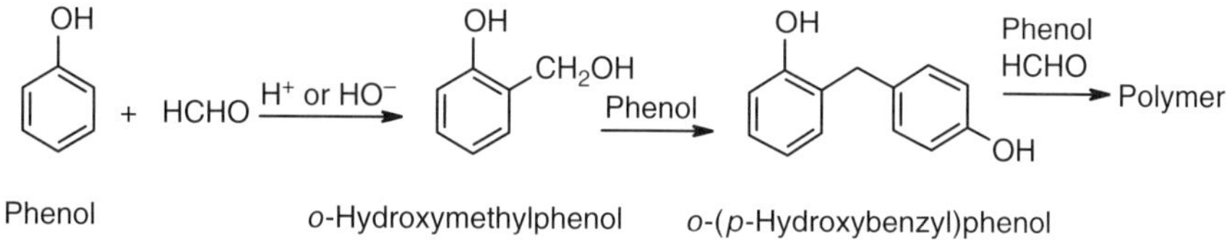

Phenol   *o*-Hydroxymethylphenol   *o*-(*p*-Hydroxybenzyl)phenol

4.7.10.6.7 Amide Formation: Paracetamol from 4-Aminophenol

Paracetamol is prepared by reacting 4-aminophenol with acetic anhydride at room temperature. This reaction forms an amide bond with 4-aminophenol and an acetic acid is produced as a by-product.

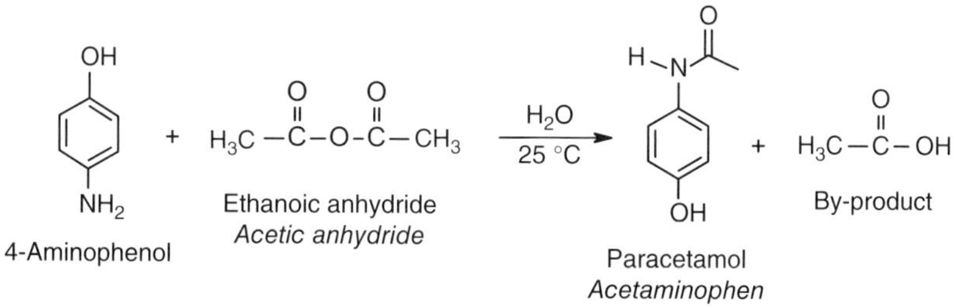

4-Aminophenol   Ethanoic anhydride   Paracetamol   By-product
                       Acetic anhydride   Acetaminophen

4.7.11 Aromatic Amines: Aniline

An amine has the general formula: RNH_2 (1° amine), R_2NH (2° amine) or R_3N (3° amine), where R = alkyl or aryl group, for example, methylamine CH_3NH_2, dimethylamine $(CH_3)_2NH$ and trimethylamine $(CH_3)_3N$.

When an amino group (—NH$_2$) is directly attached to the benzene ring, the compound is known as *aniline*.

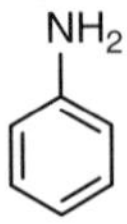

Aniline
—NH$_2$ group is attached
directly to the benzene ring

Aniline, first introduced in 1826 by the destructive distillation of indigo, is an aromatic organic base generally used to make dyes, drugs, explosives, plastics and photographic and rubber chemicals. Its name is taken from the scientific name of the indigo-yielding plant *Indigofera anil* (also known as *Indigofera suffruticosa*).

4.7.11.1 Physical Properties of Aniline

Aniline is a highly poisonous, oily and colourless substance with a pleasant odour. It is a polar compound, and can form *intermolecular hydrogen bonding* between two aniline molecules. Aniline has higher boiling point (bp: 184 °C) than non-polar compounds of the same molecular weight. It also forms hydrogen bonds with water. This hydrogen bonding accounts for the solubility of aniline in water (3.7 g/100 g water).

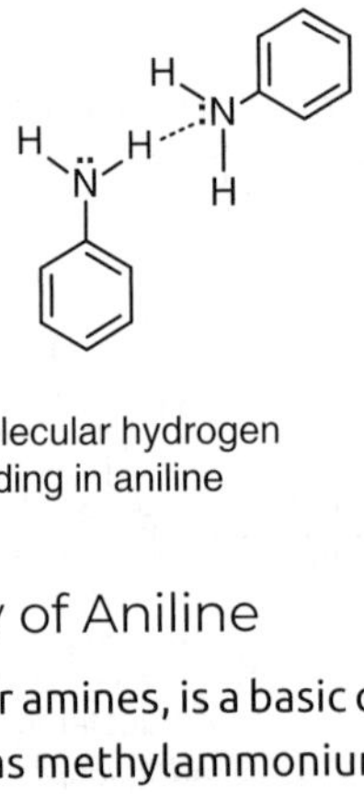

Intermolecular hydrogen
bonding in aniline

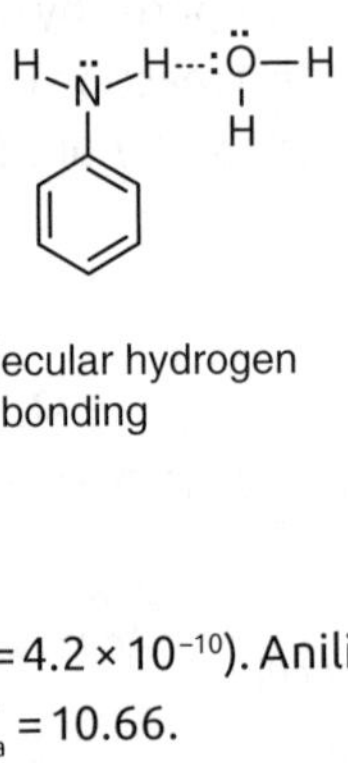

Intermolecular hydrogen
bonding

4.7.11.2 Basicity of Aniline

Aniline, like all other amines, is a basic compound ($K_b = 4.2 \times 10^{-10}$). Anilinium ion has a p$K_a = 4.63$, whereas methylammonium ion has a p$K_a = 10.66$.

Arylamines, for example, aniline, are less basic than alkylamines, because the nitrogen lone pair electrons are *delocalized* by interaction with the aromatic ring π electron system and are less available for bonding to H$^+$. Arylamines are stabilized relative to alkylamines because of the five resonance structures as shown next. Resonance stabilization is lost on protonation, because only two resonance structures are possible for the arylammonium ion.

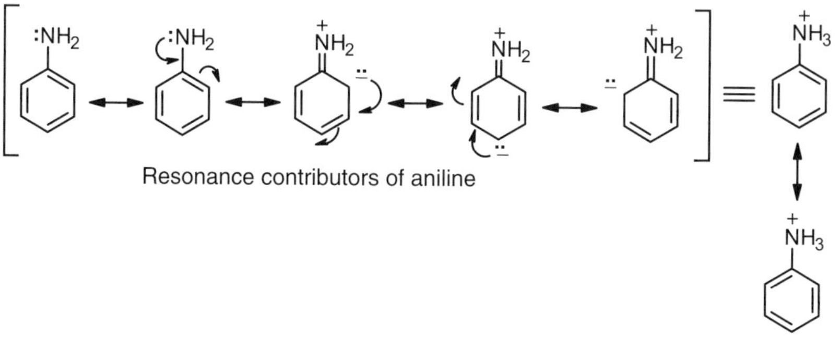

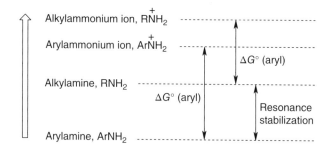

Resonance contributors of aniline

Arylammonium ion

The energy difference −G° between protonated and non-protonated forms, as shown in the following diagram, is higher for arylamines than it is for alkylamines. This is why arylamines are less basic.

Alkylammonium ion, $\overset{+}{R}NH_2$

Arylammonium ion, $Ar\overset{+}{N}H_2$

$\Delta G°$ (aryl)

Alkylamine, RNH_2

$\Delta G°$ (aryl)

Resonance stabilization

Arylamine, $ArNH_2$

4.7.11.2.1 *Effect of Substituents on the Basicity of Aniline*

The effect of substituent(s) on the basicity of aniline is summarized next. Electron donating substituents (Y = −CH₃, −NH₂, −OCH₃) activate the ring, and increase the basicity of aniline, whereas electron-withdrawing substituents (Y = −Cl, −NO₂, −CN) deactivate the ring and decrease the basicity.

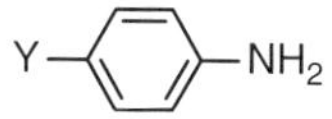

$$Y\text{—}\langle\bigcirc\rangle\text{—}NH_2$$

	Substituent Y	pK_a	Effect on reactivity
Stronger base	−NH₂	6.15	Activating
	−OCH₃	5.34	Activating
	−CH₃	5.08	Activating
	−H	4.63	No effect
	−Cl	3.98	Deactivating
	−CN	1.74	Deactivating
Weaker base	−NO₂	1.00	Deactivating

4.7.11.3 Preparations of Aniline

4.7.11.3.1 Reduction of Nitrobenzene

Aniline can be prepared from nitrobenzene by either chemical reduction using acid and metal or catalytic hydrogenation using molecular hydrogen.

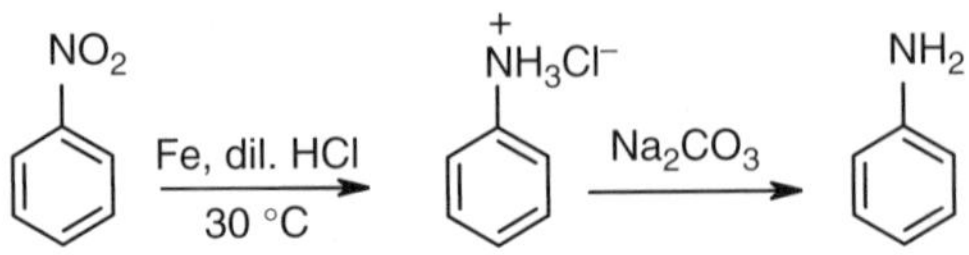

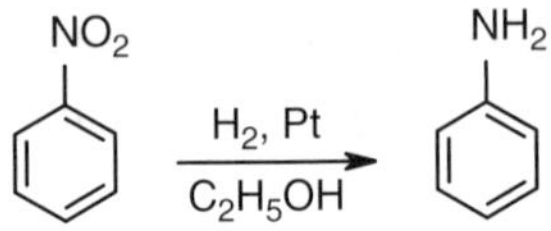

4.7.11.3.2 From Chlorobenzene

Treatment of chlorobenzene with ammonia (NH_3) at high temperature and high pressure in the presence of a catalyst affords aniline.

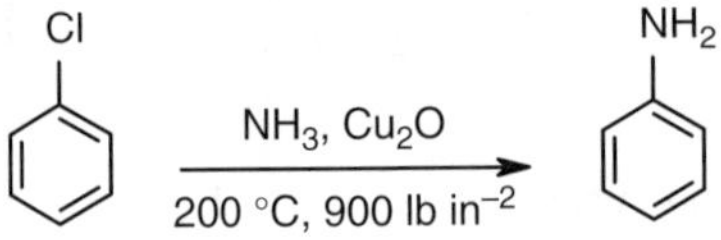

4.7.11.3.3 Hofmann Degradation of Benzamide

This reaction produces aniline, which contains one less carbon than the starting material benzamide, when reacts with sodium hypochlorite. The group (phenyl) attached to the carbonyl carbon in the amide (benzamide) is found joined to nitrogen in the product (aniline). This is an example of *molecular rearrangement*, also known as *Hofmann rearrangement*.

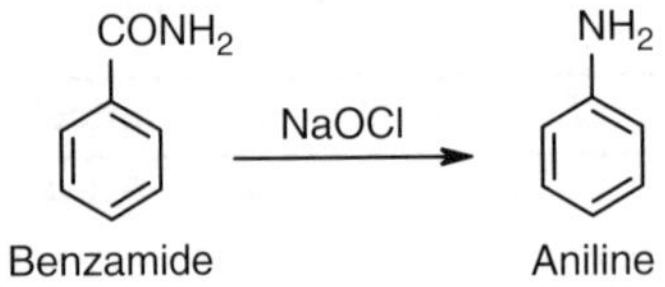

Benzamide   Aniline

Substituted benzamides give substituted anilines, and these show the following order of reactivity:

$$Y = -OCH_3 > -CH_3 > -H > -Cl > -NO_2$$

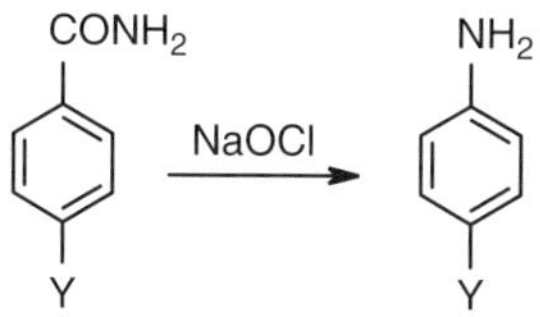

4.7.11.4 Reactions of Aniline

Aniline undergoes electrophilic substitutions. In aniline, the substitutions take place in *ortho* and *para* positions. As —NH$_2$ group is a strong activating group, the reaction rate is much faster than usually observed with benzene. A number of other types of reactions can also be carried out with aniline. Some of these reactions are discussed here.

4.7.11.4.1 Salt Formation

As aniline is a base, it forms salt with mineral acids.

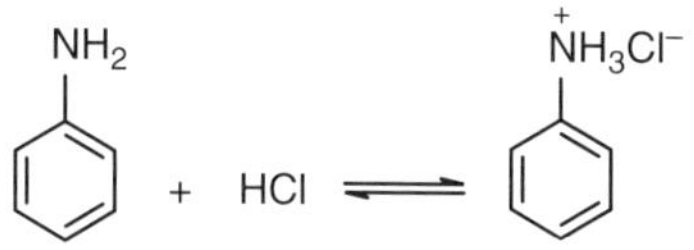

Anilinium chloride

4.7.11.4.2 N-Alkylation

The hydrogen atoms of the amino group in aniline can be replaced by alkyl substituent to obtain *N*-alkylated aniline.

Aniline reacts with CH$_3$Cl to yield *N*-methylaniline, *N,N*-dimethylaniline and *N,N,N*-trimethylaniline salt.

N-Methylaniline   *N,N*-Dimethylaniline   *N,N,N*-Trimethylaniline salt

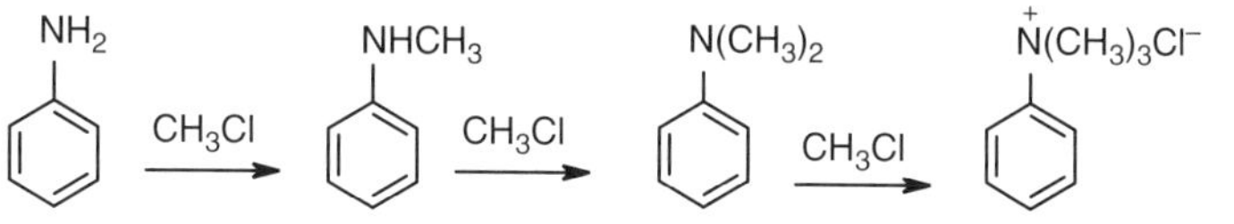

The alkyl halide, such as CH$_3$Cl, undergoes nucleophilic substitution with the basic aniline serving as the nucleophilic reagent. One of the hydrogen atoms attached to the nitrogen is replaced by an alkyl group. The final stage of this reaction involves the formation of a quaternary ammonium salt, where four organic groups are covalently bonded to nitrogen, and the positive charge of this ion is balanced by the negative chloride (Cl$^-$) ion.

4.7.11.4.3 Formation of Amide

Aniline reacts with acid chloride to form corresponding amide. For example, when aniline is treated with benzoylchloride in presence of pyridine, it produces benzanilide.

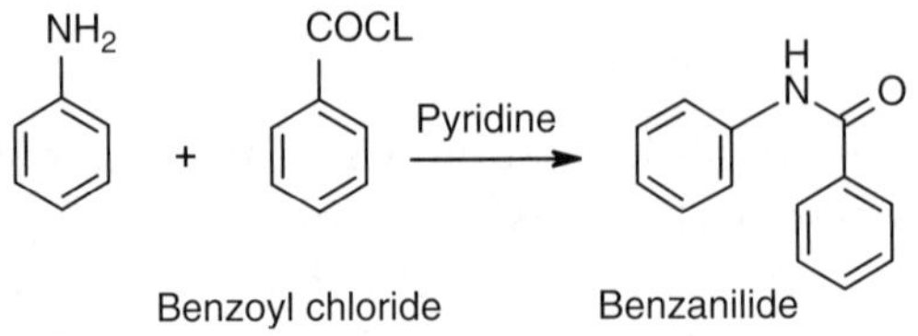

Benzoyl chloride   Benzanilide

4.7.11.4.4 Formation of Sulphonamide

Aniline reacts with sulphonylchloride to form corresponding *sulphonamide*. For example, when aniline is treated with benzenesulphonylchloride in the presence of a base, it gives the sulphonamide, *N*-phenylbenzenesulphonamide.

Benzenesulphonylchloride   *N*-Phenylbenzenesulphonamide

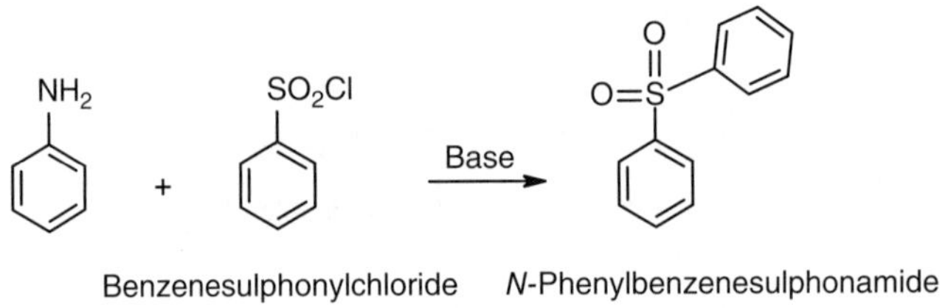

4.7.11.4.5 Application in Reductive Amination

Aniline can be used in reductive amination reactions. For example, aniline can be converted to *N*-isopropylaniline by the reaction with acetone (CH_3COCH_3) in presence of the reducing agent, sodium borohydride ($NaBH_4$).

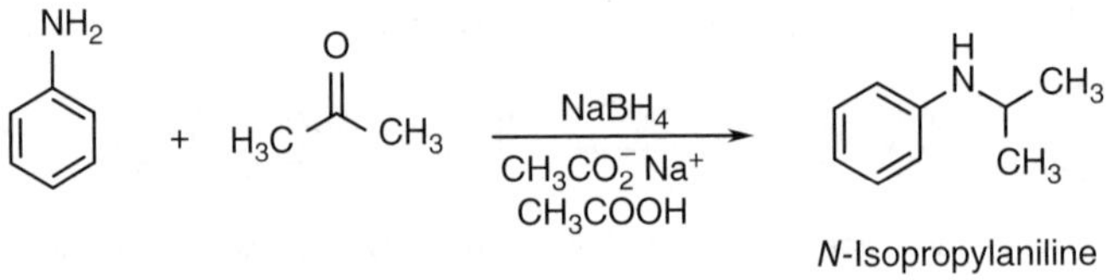

N-Isopropylaniline

4.7.11.4.6 Benzene Diazonium Salt Formation

Primary arylamines react with nitrous acid (HNO_2) to yield stable aryl diazonium salts, $ArN^+\equiv NX^-$. Alkylamines also react with nitrous acid, but the alkane diazonium salts are so reactive that they cannot be isolated.

Aniline is a primary arylamine, and it reacts with nitrous HNO_2 to generate stable benzene diazonium salt, ($Ph-N^+\equiv NX^-$).

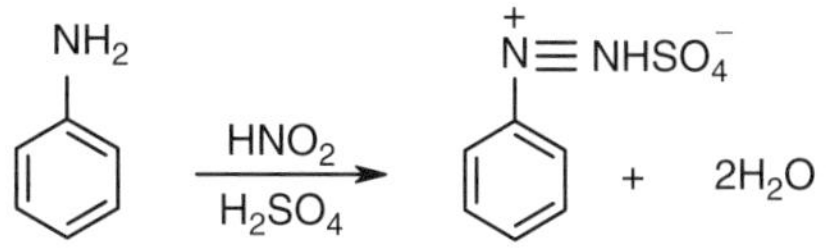

$\overset{+}{N}\equiv NHSO_4^-$ (benzene diazonium salt)

Benzene diazonium salt (with NH_2 → $\overset{+}{N}\equiv NHSO_4^-$) via HNO_2 / H_2SO_4, $+\ 2H_2O$

The drive to form a molecule of stable nitrogen gas causes the leaving group of a diazonium ion to be easily displaced by a wide variety of nucleophiles (Nu:⁻).

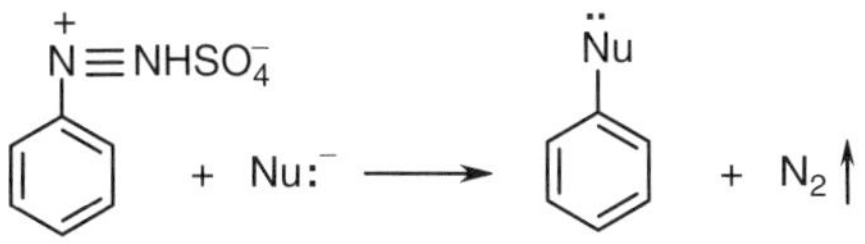

Benzene diazonium salt $\overset{+}{N}\equiv NHSO_4^-$ + Nu:⁻ ⟶ Nu (on ring) + $N_2\uparrow$

The mechanism by which a nucleophile displaces the diazonium group depends on the nucleophile. While some displacements involve phenyl cations, the other involves radicals. Nucleophiles, for example, CN⁻, Cl⁻ and Br⁻, replace the diazonium group if the appropriate cuprous salt is added to the solution containing the arene diazonium salt. The reaction of an arene diazonium salt with cuprous salt is known as a *Sandmeyer reaction*.

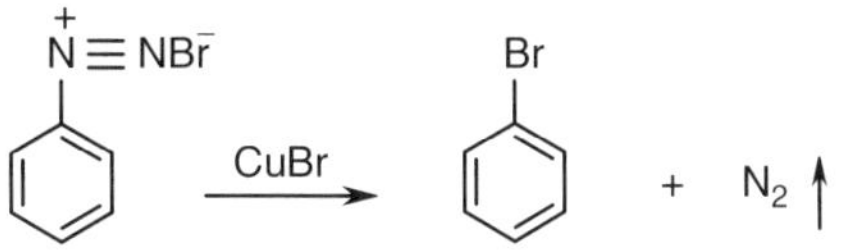

$\overset{+}{N}\equiv NBr^-$ → (CuBr) → Br (Bromobenzene) + $N_2\uparrow$

This diazotization reaction is compatible with the presence of a wide variety of substituents on the benzene ring.

Arenediazonium salts are extremely important in synthetic chemistry, because the diazonio group (N≡N) can be replaced by a nucleophile in a radical substitution reaction, for example, preparation of phenol, chlorobenzene and bromobenzene. Under proper conditions, arenediazonium salts react with certain aromatic compounds to afford products of the general formula Ar—N≡N—Ar′, called *azo compounds*. In this coupling reaction, the nitrogen of the diazonium group is retained in the product.

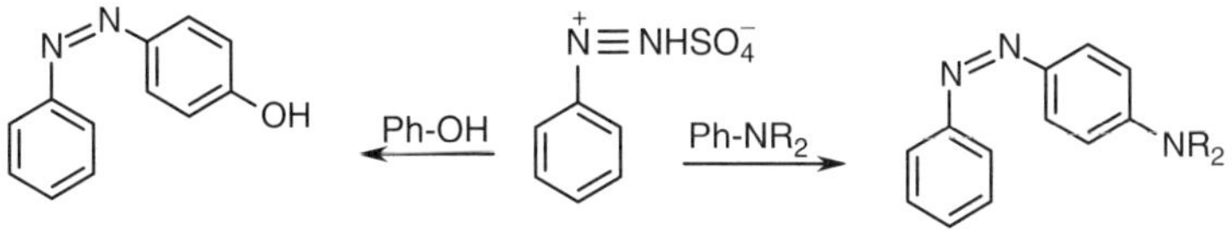

N=N–C₆H₄–OH ← (Ph-OH) — $\overset{+}{N}\equiv NHSO_4^-$ — (Ph-NR₂) → N=N–C₆H₄–NR₂

4.7.11.5 Synthesis of Sulpha Drugs from Aniline

Antimicrobial sulpha drugs, for example, sulphanilamide, are the amide of sulpha-nilic acid and certain related substituted amides. Sulphanilamide, the first of the sulpha drugs prepared by the Austrian chemist Paul Josef Jakob Gelmo in 1908 and patented in 1909, acts by inhibiting the bacterial enzyme than incorporate *para*-aminobenzoic acid (PABA) into folic acid. It is a bacteriostatic drug, that is, inhibits the further growth of the bacteria. Multistep synthesis, starting from aniline, as depicted in the following scheme, can achieve the product, a sulpha drug.

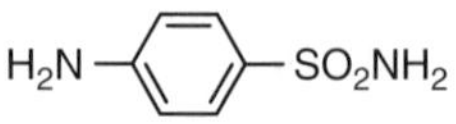

H_2N — ⟨ ⟩ — SO_2NH_2

p-Aminobenzenesulphonamide
Sulphanilamide

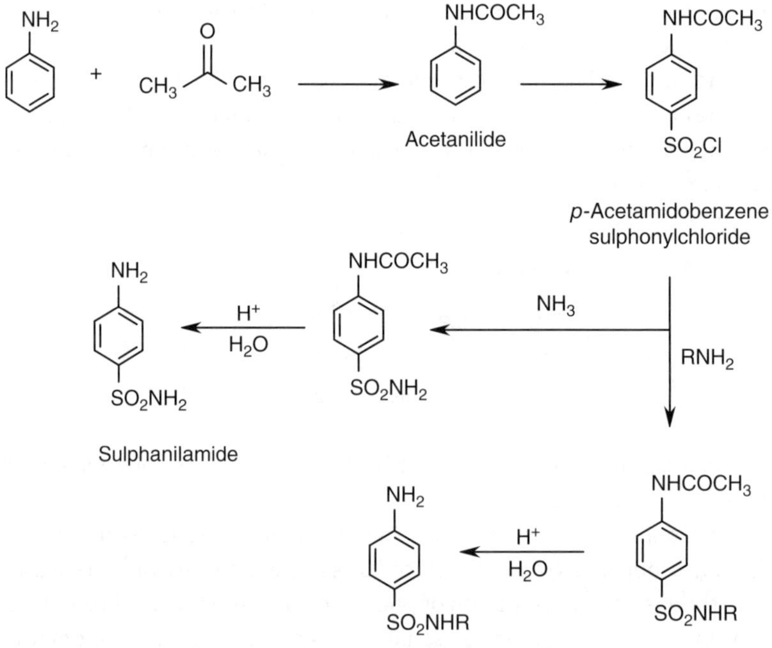

Acetanilide

p-Acetamidobenzene sulphonylchloride

Sulphanilamide

Substituted sulphanilamide

4.7.11.6 Separation of Aniline from a Mixture of Compounds by Solvent Extraction

If a mixture contains an aniline and a neutral compound, both the constituents can easily be separated and purified by solvent extraction method. To purify these compounds, the mixture is dissolved in DCM, HCl and H_2O are added, and the solution is shaken in a separating funnel. Once two layers, aqueous and ether layers, are formed, they are separated. The lower layer (aqueous) contains the salt of aniline, and the ether layer has the neutral compound. Ether is evaporated from the ether layer using a rotary evaporator to get purified neutral compound. Into the aqueous layer, sodium hydroxide and ether are added and the resulting solution is shaken in a separating funnel. Two layers are separated. The ether layer (top layer) contains free aniline, and the aqueous layer (bottom layer) has the salt, sodium chloride. Ether is evaporated from the ether layer using a rotary evaporator to get purified aniline.

Aniline + Neutral compound

      Dissolve in ether, add HCl and water
      Shake in a separting funnel

Ether layer   Aqueous layer
(Contains neutral compound)   (Contains salt of aniline)

   Evaporation of ether   Add NaOH adn ether
             Shake in a separting funnel

Purified neutral compound

Ether layer   Aqueous layer
(Contains aniline)   (Contains NaCl)

   Evaporation of ether

Purified aniline

4.7.12 Polycyclic Benzenoids

Two or more benzene rings fused together form a number of polycyclic benzenoid aromatic compounds, naphthalene, anthracene and phenanthrene, and their derivatives. All these hydrocarbons are obtained from coal tar. Naphthalene is the most abundant (5%) of all constituents of coal tar.

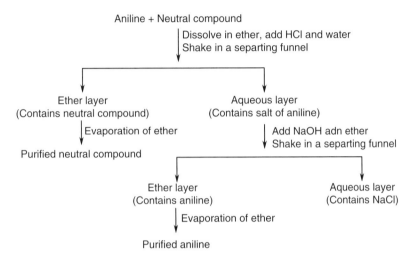

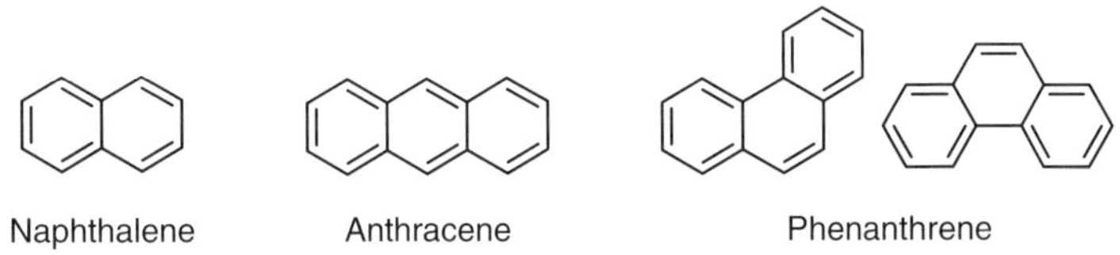

Naphthalene   Anthracene   Phenanthrene

4.7.12.1 Synthesis of Naphthalene from Benzene: Haworth Synthesis

Naphthalene can be synthesized from benzene through multi-step synthesis involving, notably, Friedel–Crafts (FC) acylation, the Clemmensen reduction and aromatization reactions as outlined in the following scheme.

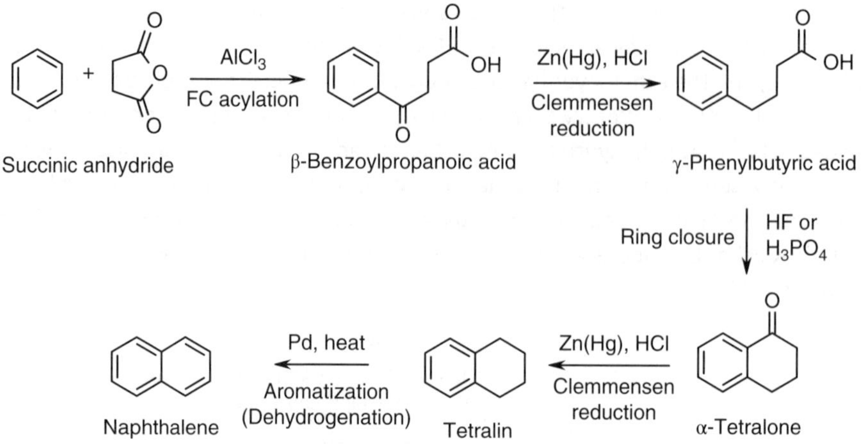

Succinic anhydride   β-Benzoylpropanoic acid   γ-Phenylbutyric acid

AlCl₃ / FC acylation

Zn(Hg), HCl / Clemmensen reduction

Ring closure | HF or H₃PO₄

Naphthalene   (Dehydrogenation) Aromatization ← Pd, heat   Tetralin ← Clemmensen reduction Zn(Hg), HCl   α-Tetralone

4.7.12.2 Reactions of Naphthalene

Naphthalene undergoes electrophilic substitutions on the ring resulting in several of its derivatives. In addition to the usual electrophilic substitutions, naphthalene can also undergo oxidation and reduction reactions under specific conditions as outlined next.

4.7.12.2.1 Oxidation

Oxidation of naphthalene by oxygen in the presence of vanadium pentoxide (V_2O_5) destroys one ring and produces phthalic anhydride (an important industrial process). However, oxidation of in presence of CrO_3 and acetic acid (AcOH) destroys the aromatic character of one ring and yields naphthoquinone (a di-keto compound).

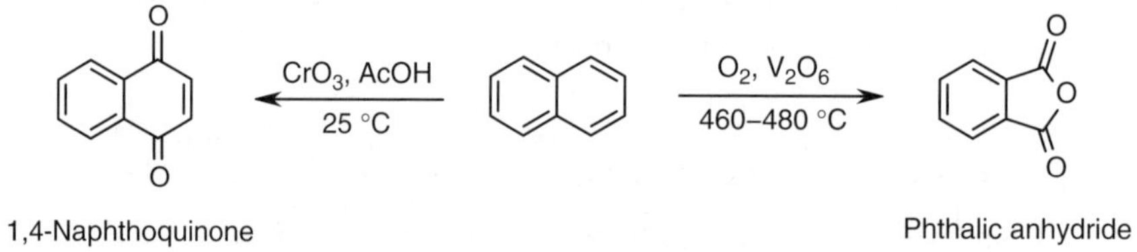

1,4-Naphthoquinone CrO_3, AcOH / 25 °C O_2, V_2O_6 / 460–480 °C   Phthalic anhydride

4.7.12.2.2 Reduction

One or both rings of naphthalene can be reduced partially or completely, depending upon the reagents and reaction conditions.

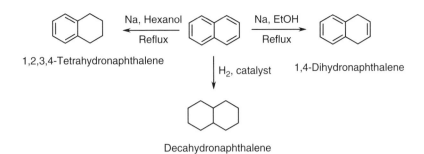

1,2,3,4-Tetrahydronaphthalene   Na, Hexanol / Reflux   Na, EtOH / Reflux   1,4-Dihydronaphthalene

H$_2$, catalyst

Decahydronaphthalene

4.8 IMPORTANCE OF FUNCTIONAL GROUPS IN DETERMINING DRUG ACTIONS AND TOXICITY

In Chapter 2, you have already learned that most drugs bind to the appropriate receptor molecules to exhibit their pharmacological actions, and also toxicity, which in fact is the adverse pharmacological action. A drug's pharmacological activity is inherently related to its chemical structure. Various functional groups present in the drug molecules are involved in the drug-receptor binding or interaction. For example, drugs containing hydroxyl or amino groups tend to be involved in hydrogen bonding with the receptor.

Any changes in the functional groups in a drug molecule can render significant changes in the activity and toxicity, and this is the basis of any structure-activity-relationships (SARs) study of drug molecules. The SAR study is the study for understanding the relationship of chemical structure to activity. The activity can be a pharmacological response, binding, toxicity or any other quantifiable event. In SAR studies, essential functional groups or structural features of a drug molecule, which are responsible for the optimum pharmacological actions with minimum toxicity index, are identified or optimized. These essential functional groups for the pharmacological activities are called *pharmacophores.*

By changing the functional groups of any drug, several analogues are usually synthesized in an effort to improve its binding to the receptor, facilitate absorption by the human body, increase specificity for different organs/tissue types, broaden the spectrum of activity or to reduce the toxicity/side effects.

Toxicity prevents many compounds from being developed and approved. A number of approved drugs have also been forced to be withdrawn from the market because of toxicities. For example, in 2004, Merck's arthritis drug Vioxx was withdrawn owing to severe cardiovascular side effects, and the Parke–Davis and Warner–Lambert antidiabetic drug troglitazone (Rezulin) was withdrawn from the market in 2000 after it was found to cause severe liver toxicity. The drug industries expend considerable time and effort trying to avoid or minimize toxic effects by identifying and altering the functional groups responsible for toxic effects. A change in functional groups leading to toxicity can be demonstrated by paracetamol toxicity.

The sulpha drugs and the penicillin group of antibacterial agents can be the ideal examples for demonstrating the importance of functional groups in drug actions and effectiveness. In Chapter 6, you will also see how a small change in the functional group(s) of steroidal molecules can render remarkable changes in their pharmacological and hormonal functions.

4.8.1 Structure-Activity Relationships of Sulpha Drugs

To date, over 10000 structural analogues of sulphanilamide, the parent of all sulpha drugs, have been synthesized and used in the SAR studies. However, about 40 of them have ever been used as prescribed drugs. Sulpha drugs are *bacteriostatic*, that is, they inhibit bacterial growth but do not actively kill bacteria. These drugs act on the biosynthetic pathway of tetrahydrofolic acid, inhibit dihydropteroate synthetase and mimic the shape of PABA (*para*-aminobenzoic acid).

Sulphanilamide
(the first sulpha drug)

General structure of sulphonamides
R = SO$_2$NHR' or SO$_3$H

Prontosil

From numerous studies, it has now been established that the amino functional groups (—NH$_2$) is essential for the activity. In addition, the following structural features have to be present in sulpha drugs for the optimum antibacterial activity.

i. The amino and the sulphonyl groups have to be *para* to each other, that is, *para*-di substituted benzene ring is essential.

ii. The anilino (Ph—NH$_2$) amino group may be substituted, but optimum activity is observed with unsubstituted from.

iii. Replacement of the central benzene ring (aromatic) or additional functional groups on the benzene ring diminishes activity.

iv. *N'*-monosubstitution on SO$_2$NH$_2$ increase potency, especially with hetero-aromatic groups.

v. *N'*-disubstitution on SO$_2$NH$_2$ leads to inactive compounds.

The structure of prontosil, an azo dye, is quite similar to the structure of sulphanilamide with the modification that the —NH$_2$ is substituted. As result, it does not have any *in vitro* antibacterial activity, but, *in vivo*, prontosil is converted via reduction of the —N=N— linkage its active metabolite sulphanilamide.

N-Heterocyclic derivatives of sulphanilamide, for example, sulphadiazine, sulphathiazole, and sulphoxazole, have broad-spectrum antimicrobial activity. They are generally more water-soluble and thus are better absorbed and retained; that is, excreted more slowly.

Sulphadiazine R =

SO$_2$NHR

NH$_2$

Sulphoxazole R =

Me   Me

Sulphathiazole R =

4.8.2 Structure-Activity Relationships of Penicillins

Penicillin group of antibiotics, also known as β-lactam antibiotics, have revolutionized the history of modern medicine by their effectiveness against several pathogenic bacterial species that cause various forms of infections. Penicillin G, the parent of all these antibiotics, was first isolated from a fungal species, *Penicillium notatum*. Since the discovery of this antibiotic, several modifications have been introduced to the parent structure in order to enhance the activity, increase the acid-resistance, facilitate bioavailability and reduce toxicity. Penicillin G is rather a complex molecule, and possesses various types of functional groups; for example, phenyl, alkyl, amide, carboxylic acid and β-lactam.

All penicillins are susceptible to attack in acidic solution via intramolecular attack of the amide carbonyl oxygen on the β-lactam carbonyl leading to the complete destruction of the β-lactam ring, and thus the antibacterial activity. Similarly, penicillins are unstable in basic solution because of β-lactam ring opening by free basic nucleophiles. Thus, for the antibacterial activity, the stability of the β-lactam functional group in penicillins is of paramount importance.

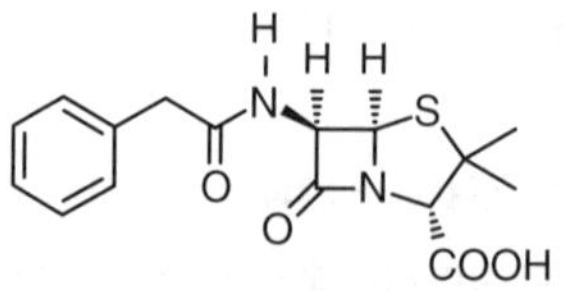

Penicillin G
(The first penicillin of the penicillin group of antibiotics)

The degree of instability of the β-lactam ring depends on the availability of the electrons for attack. So, modification of penicillins with the addition of electron withdrawing groups near the amide carbonyl decreases the availability of these electrons and significantly improves acid stability. For example, the amino group of amoxicillin and ampicillin makes these molecules acid stable.

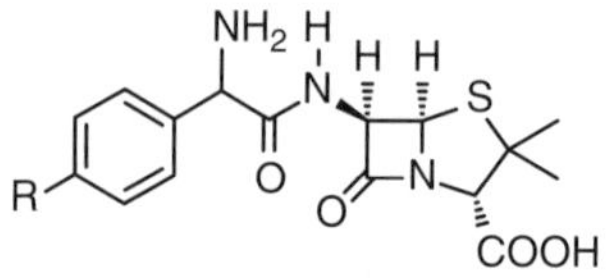

Amoxicillin R = OH
Ampicillin R = H
(Stable in acidic condition)

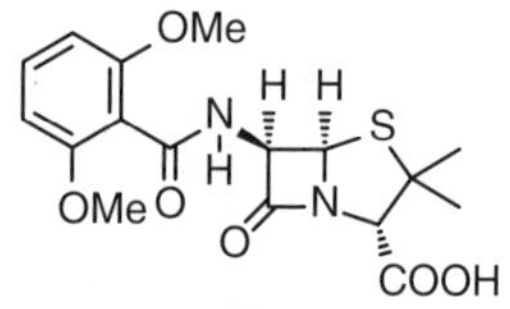

Methicillin

From numerous studies with semisynthetic penicillins, it has been established that the penicillins, which contain more polar groups are able to cross easily the Gram-negative cell wall and will have a greater spectrum of antibacterial activity. For example, the amino group in amoxicillin offers the molecule polarity, and makes it effective against both Gram-positive and Gram-negative bacteria.

The SAR of penicillin can be summarized as follows.

i. Oxidation of the sulphur to a sulphone or sulphoxide decreases the activity of penicillins but provides better acid stability.

ii. The β-lactam carbonyl and nitrogen are absolutely necessary for activity.

iii. The amide carbonyl is essential for activity.

iv. The group attached to the amide carbonyl (the R group) is the basis for the changes in activity, acid stability and susceptibility to resistance.

v. Any other changes generally decrease activity.

vi. A bulky group directly adjacent to the amide carbonyl usually offers β-lactamases resistant property.

A bulky group directly adjacent to the amide carbonyl will prevent the penicillin from entering the active site of penicillin-destroying enzymes, for example, β-lactamases but still allow them to enter the active site of penicillin binding proteins. For example, methicillin has a bulky group directly adjacent to the amide carbonyl, and is β-lactamase resistance.

The addition of polar groups to the R group, that is, the group directly linked to the amide carbonyl, generally allows the penicillin molecule, for example, amoxicillin, to more easily pass through the Gram-negative cell wall, and thus increases antibacterial activity.

4.8.3 Paracetamol Toxicity

Bioactivation is a classic toxicity mechanism, where the functional group or the chemical structure of the drug molecule is altered by enzymatic reactions. For example, the enzymatic breakdown of the analgesic acetaminophen (paracetamol), where the aromatic nature and the hydroxyl functionality in paracetamol are lost, yields *N*-acetyl-*p*-benzoquinone imine, a hepatotoxic agent. Paracetamol can cause liver damage and even liver failure, especially when combined with alcohol.

Paracetamol
Acetaminophen

Bioactivation

N-Acetyl-*p*-benzoquinone imine
(The hepatotoxic metabolite)

4.9 IMPORTANCE OF FUNCTIONAL GROUPS IN DETERMINING STABILITY OF DRUGS

You have already seen that just introducing new functional group on penicillin molecule, the acid stability of penicillins can be improved remarkably (see Section 7.3), and similarly the introduction of bulky functional groups in penicillin offers stability against β-lactamases. Thus, functional groups play a vital role in the stability of drugs.

Certain functional groups in drug molecules are prone to chemical degradation. Drugs with ester functional groups are easily hydrolysed, for example, aspirin is easily hydrolysed to salicylic acid and acetic acid. Similarly, many drug molecules are susceptible to oxidation because of certain oxidizable functional groups; for example, alcohol.

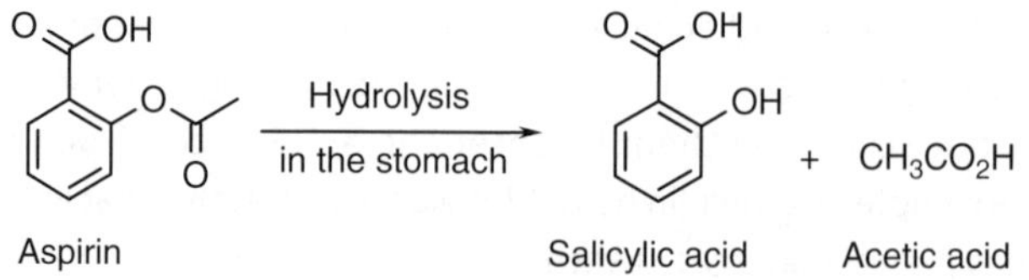

Aspirin — Hydrolysis in the stomach → Salicylic acid + Acetic acid (CH_3CO_2H)

Insulin is a protein and contains amide linkages, which makes this compound unstable in an acidic medium and unsuitable for oral administration. Like any other proteins in the gastrointestinal tract, insulin is reduced to its amino acid components and the activity is totally lost. Many drugs possessing olefinic double bonds exhibit *trans–cis* isomerism in the presence of light. Similarly, because of the presence of certain functional groups or the chemical structure, a drug can be sensitive to heat. Many monoterpenyl or sesquiterpenyl drugs are unstable at high temperatures.

Chapter 5
Organic Reactions

Learning objectives

After completing this chapter, students should be able to

- understand various types of organic reactions occur with functional groups;
- recognize various types of arrows used in chemical reactions and mechanisms;
- describe the mechanisms and examples of radical, addition, elimination, substitution, oxidation-reduction, hydrolysis and pericyclic reactions.

5.1 TYPES OF ORGANIC REACTIONS OCCUR WITH FUNCTIONAL GROUPS

Simply, organic chemical reactions are reactions that occur between organic compounds. They are fundamental to various essential biochemical processes in all living beings, and also to drug discovery, development and delivery as well as the metabolic processes that the drugs undergo inside the body. Most of the major classes of organic reactions can be categorized as follows.

Chemistry for Pharmacy Students: General, Organic and Natural Product Chemistry,
Second Edition. Lutfun Nahar and Satyajit Sarker.
© 2019 John Wiley & Sons Ltd. Published 2019 by John Wiley & Sons Ltd.

Reaction types	Brief definition	Where the reaction occurs
Radical reactions	New bond is formed using radical from each reactant.	Alkanes and alkenes.
Addition reactions	Addition means two systems combine to a single entity.	Alkenes, alkynes, aldehydes and ketones.
Elimination reactions	Elimination refers to the loss of water, hydrogen halide or halogens from a molecule.	Alcohols, alkyl halides and alkyl dihalides.
Substitution reactions	Substitution implies that one group replaces the other.	Alkyl halides, alcohols, epoxides, carboxylic acid and its derivatives, and benzene and its derivatives.
Oxidation-reduction reactions	Oxidation = loss of electrons.	Alkenes, alkynes, 1° and 2° alcohols, aldehydes and ketones.
	Reduction = gain of electrons.	Alkene, alkyne, aldehydes, ketones, alkyl halides, nitriles, carboxylic acid and its derivatives and benzene and its derivatives.
Hydrolysis reactions	Cleavage of chemical bonds with water	Carboxylic acids and its derivatives.
Pericyclic reactions	Concerted reaction that takes place as a result of a cyclic rearrangement of electrons.	Conjugated dienes and α,β-unsaturated carbonyl compounds.

5.2 REACTION MECHANISMS AND TYPES OF ARROW IN CHEMICAL REACTIONS

A *reaction mechanism* shows a complete step by step description in which old bonds are broken and new bonds are formed in a chemical reaction. Different types of arrow can be seen in organic reactions and mechanisms, which are depicted with their uses in the following table.

Name	Use	Arrow
Single-headed curved arrow	Drawn for the movement of a single electron	
Double-headed curved arrow	Drawn for the movement of an electron pair	
Reaction arrow	Drawn for the reaction to product formation	
Double-headed reaction arrow	Drawn for the reaction equilibrium	
Double reaction arrow	Drawn between resonance structures	

5.3 FREE RADICAL REACTIONS: CHAIN REACTIONS

A *radical,* often called a *free radical,* is a highly reactive and short-lived atomic or molecular species with an unpaired electron on an otherwise open shell configuration. Free radicals are electron-deficient species, but usually uncharged. So, their chemistry is different from the chemistry of even-electron and electron-deficient species; for example, carbocations and carbenes (see Section 5.3.2). A free radical behaves like an electrophile, as it requires only a single electron to complete its octet. They are linked to ageing, cancer and a host of other diseases.

Radical reactions are often called *chain reactions.* All chain reactions have three steps: chain initiation, chain propagation and chain termination. For example, the halogenation of alkane is a chain reaction. A chain reaction is also known as a *radical substitution* reaction, because radicals are involved as intermediates, and the end result is the substitution of a halogen atom for one of the hydrogen atoms of alkane.

5.3.1 Free Radical Chain Reaction of Alkanes

5.3.1.1 Preparation of Alkyl Halides

Molecular chlorine (Cl_2) or bromine (Br_2) reacts with alkanes in the presence of light ($h\nu$) or high temperatures to give alkyl halides. However, Cl_2 is more reactive towards the alkane than Br_2. Usually this method produces mixtures of halogenated compounds containing mono-, di-, tri- and tetra-halides. This is an important reaction of alkanes as it is the only way to convert inert alkanes to reactive alkyl halides. The simplest example is the reaction of methane with Cl_2 to yield a mixture of chlorinated methane derivatives.

$$CH_4 \; + \; Cl_2 \; \xrightarrow{h\nu} \; CH_3Cl \; + \; CH_2Cl_2 \; + \; CHCl_3 \; + \; CCl_4$$

| Methane | | Methyl chloride | Dichloro methane | Chloroform | Carbon tetrachloride |

Mechanism
The high temperature or light supplies the energy to break the chlorine–chlorine bond homolytically. In the *homolytic bond cleavage,* one electron of covalent bond goes to each atom. A single headed arrow indicates the movement of one electron. The chlorine molecule (Cl_2) dissociates into two chlorine radicals in the first step, known as the *initiation step,* leading to the substitution reaction of chlorine atoms for hydrogen atoms in methane.

Initiation:
This first step in a free radical reaction generates a reactive intermediate. A chlorine atom is highly reactive because of the presence of an unpaired electron in its valence shell. It is electrophilic, seeking a single electron to complete the octet. It acquires this electron by abstracting a hydrogen atom from methane.

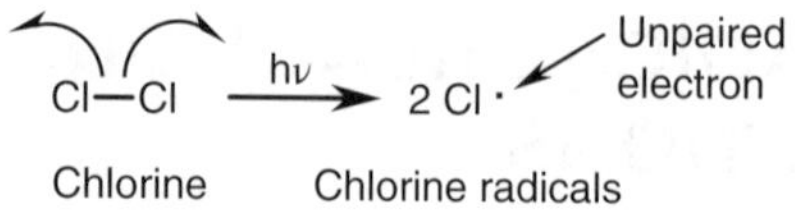

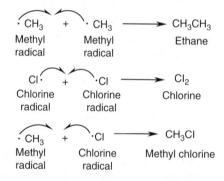

$$Cl-Cl \xrightarrow{h\nu} 2\,Cl\cdot$$

Unpaired electron

Chlorine   Chlorine radicals

Propagation:

In this step, the intermediate reacts with a stable molecule to furnish another reactive intermediate and a product molecule. The propagation step affords a new electrophilic species, the methyl radical, which has an unpaired electron. In a second propagation step, the methyl radical abstracts a chlorine atom from a chloromethane molecule, and generates a chlorine radical.

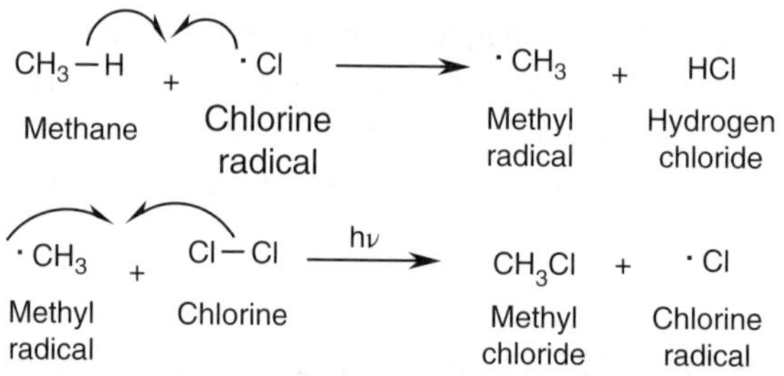

$$CH_3-H + \cdot Cl \longrightarrow \cdot CH_3 + HCl$$

Methane   Chlorine radical   Methyl radical   Hydrogen chloride

$$\cdot CH_3 + Cl-Cl \xrightarrow{h\nu} CH_3Cl + \cdot Cl$$

Methyl radical   Chlorine   Methyl chloride   Chlorine radical

Termination:

Various reactions between the possible pairs of radicals allow for the formation of ethane, Cl_2 or the methyl chloride. In this step, the reactive particles are consumed, but not generated.

$$\cdot CH_3 + \cdot CH_3 \longrightarrow CH_3CH_3$$

Methyl radical   Methyl radical   Ethane

$$Cl\cdot + \cdot Cl \longrightarrow Cl_2$$

Chlorine radical   Chlorine radical   Chlorine

$$\cdot CH_3 + \cdot Cl \longrightarrow CH_3Cl$$

Methyl radical   Chlorine radical   Methyl chlorine

To maximize the formation of a mono-halogenated product, a radical substitution reaction must be carried out in the presence of excess alkane. For example, when large excess of methane is used, the product is almost completely methyl chloride (chloromethane).

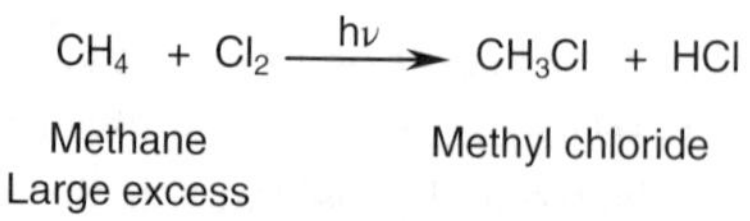

$$CH_4 + Cl_2 \xrightarrow{h\nu} CH_3Cl + HCl$$

Methane   Methyl chloride
Large excess

Similarly, when a large excess of cyclopentane is heated with chlorine at 250°C, the major product is chlorocyclopentane (95%), along with small amounts of 1,2- and 1,3-dichlorocyclopentanes.

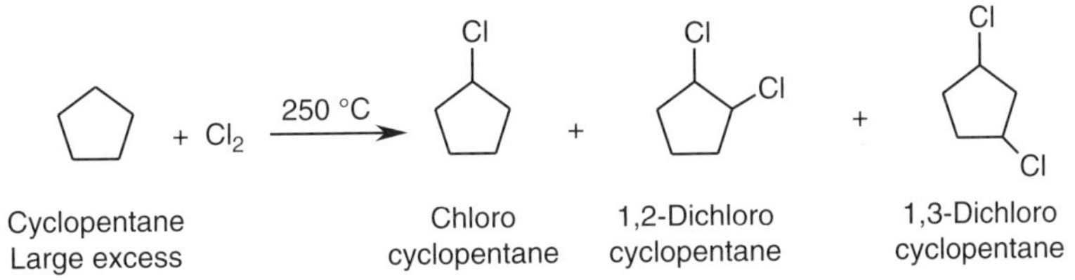

| Cyclopentane Large excess | | Chloro cyclopentane | 1,2-Dichloro cyclopentane | 1,3-Dichloro cyclopentane |

Bromination of alkanes follows the same mechanism as chlorination. The only difference is the reactivity of the radical; that is, the chlorine radical is much more reactive than the bromine radical. Thus, the chlorine radical is much less selective than bromine radical, and it is a useful reaction when there is only one kind of hydrogen in the molecule. If a radical substitution reaction yields a product with a chiral centre, the major product is a racemic mixture. For example, radical chlorination of n-butane produces 71% racemic mixture of 2-chlorobutane, and bromination of n-butane gives 98% racemic mixture of 2-bromobutane.

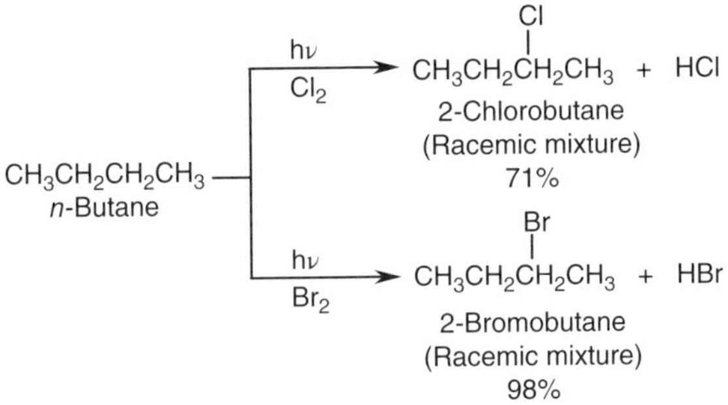

Among the alkanes, the reactivity decreases going from tertiary to secondary to primary alkanes. Again, selectivity is observed in the case of bromination of t-butane, which provides t-butyl bromide almost quantitatively.

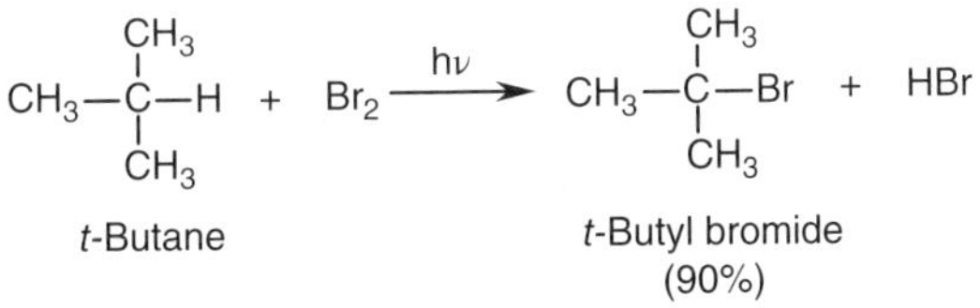

5.3.2 Relative Stabilities of Carbocations, Carbanions, Radicals and Carbenes

A species containing a positively charged carbon atom is called a *carbocation*. Carbocations are electron-deficient species, known as *electrophiles*. The term *electrophile* literally means 'electron-loving', and is an electron-deficient species that can accept an electron pair. They are strong Lewis acids and react rapidly with Lewis bases or species that are capable of donating electrons.

Carbocations are classified according to the number of alkyl groups that are bonded to the positively charged carbon. A primary (1°) carbocation has one alkyl group, a secondary (2°) has two and a tertiary (3°) has three alkyl groups. Alkyl groups are able to decrease the concentration of positive charge on the carbocation by donating electrons inductively, which increases the stability of the carbocation. The greater the number of alkyl groups bonded to the positively charged carbon, the more stable the carbocation. Therefore, a 3° carbocation is more stable than a 2° carbocation, and a 2° carbocation is more stable than a 1° carbocation, which in turn is more stable than a methyl cation.

In molecular orbital terms, alkyl groups can stabilize a carbocation by *hyperconjugation*. It is the overlap of the filled σ orbitals of the C—H or C—C bonds adjacent to the carbocation with an empty *p* orbital on the positively charged carbon atom. As a result, the positive charge is delocalized onto more than one atom, and thus increases the stability of the system. The more alkyl groups attached to the carbocation, the more σ bonds are available for hyperconjugation, and the more stable the carbocation.

$$
\underset{\text{3° carbocation}}{R-\overset{\displaystyle R}{\underset{\displaystyle R}{C^+}}} \quad > \quad \underset{\text{2° carbocation}}{R-\overset{\displaystyle H}{\underset{\displaystyle R}{C^+}}} \quad > \quad \underset{\text{1° carbocation}}{R-\overset{\displaystyle H}{\underset{\displaystyle H}{C^+}}} \quad > \quad \underset{\text{Methyl cation}}{H-\overset{\displaystyle H}{\underset{\displaystyle H}{C^+}}}
$$

A species containing a negatively charged carbon atom is called a *carbanion*. Carbanions are electron-rich species, known as nucleophiles. A nucleophile is an electron rich species that reacts with an electrophile. They are strong Lewis bases and donate an electron pair to Lewis acids such as H^+ and other electropositive atoms and groups.

Carbanions have a trivalent carbon that bears a negative charge. Like amines, carbanions are nucleophilic and basic, but the negative charge on carbon makes carbanions more nucleophilic and more basic than amines. Thus, carbanions are sufficiently basic to remove a proton from amines. The stability order of carbanions (electron rich species) reflects their high electron density and the stability order is reversed from radicals and carbocations (both electron deficient species). Typically, carbanions are stabilized by either resonance or inductive effects.

$$
\underset{\text{3° carbanion}}{R-\overset{\displaystyle R}{\underset{\displaystyle R}{\ddot{C}{}^-}}} \quad < \quad \underset{\text{2° carbanion}}{R-\overset{\displaystyle H}{\underset{\displaystyle R}{\ddot{C}{}^-}}} \quad < \quad \underset{\text{1° carbanion}}{R-\overset{\displaystyle H}{\underset{\displaystyle H}{\ddot{C}{}^-}}} \quad < \quad \underset{\text{Methyl anion}}{H-\overset{\displaystyle H}{\underset{\displaystyle H}{\ddot{C}{}^-}}}
$$

The formation of different radicals from the same starting compound offers a way to estimate relative radical stabilities, which correspond directly to the stabilities of the corresponding carbocations. Thus, the relative stabilities of radicals follow the same trend as for carbocations. Like carbocations, radicals are also electron deficient species and are stabilized by hyperconjugation. Therefore, the most substituted radical is most stable. For example, a 3° alkyl radical is more stable than a 2° alkyl radical, which in turn is more stable than a 1° alkyl radical. Allyl and benzyl radicals are more stable than alkyl radicals, because their unpaired electrons are delocalized. Electron delocalization increases the stability of a molecule. The more stable a radical, the faster it can be formed. Therefore, a hydrogen atom, bonded to either an allylic carbon or a benzylic carbon, is substituted more selectively in the halogenation reaction. The percentage of substitution at an allylic and benzylic carbon is greater in the case of bromination than in the case of chlorination, because a bromine radical is more selective.

| $CH_2=CH\dot{C}H_2$ = | | | | | |
| Allyl radical | Benzyl radical | 3° radical | 2° radical | 1° radical | Methyl radical |

A *carbene* is a species containing a divalent carbon atom linked to two adjacent groups by a covalent bond. Carbenes have a lone pair and a vacant orbital means they display both nucleophilic and electrophilic properties. They are uncharged species and highly reactive like radicals (see Section 5.4.2.16). Carbenes can dimerize to give alkenes and the reaction proceeds quite fast.

5.3.3 Allylic Bromination

5.3.3.1 Preparation of Alkene Halides

Under high temperature or UV light and in the gas phase, cyclohexene can undergo substitution by halogens. A common reagent for allylic bromination is N-bromo-succinimide (NBS), because it continually generates small amounts of Br_2 through the reaction with HBr. The bromination of cyclohexene yields 3-bromocyclohexene. An allylic hydrogen atom is substituted for a bromine atom. Allylic means the substituent is adjacent to a carbon–carbon double bond.

| Cyclohexene | NBS | 3-Bromocyclohexene (80%) | Succinimide |

Mechanism

Homolytic cleavage of the *N*–Br bond of NBS generates radicals.

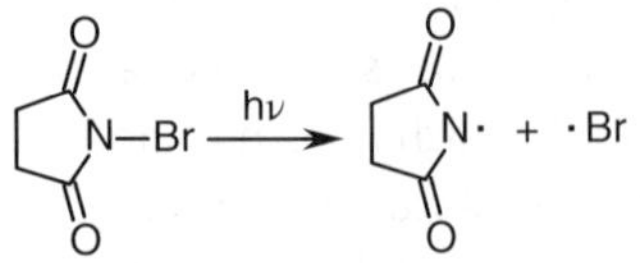

The bromine radical abstracts an allylic hydrogen atom of the cyclohexene, and forms a resonance stabilized allylic radical and hydrogen bromide.

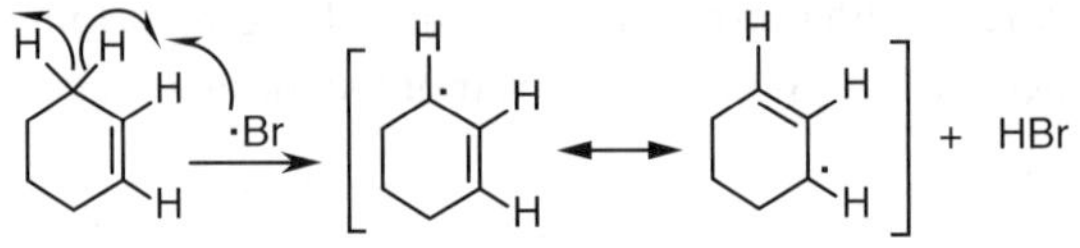

Hydrogen bromide reacts with NBS to afford a Br_2 molecule, which reacts with the allylic radical to form 3-bromocyclohexene, and a bromine radical is produced to continue the chain.

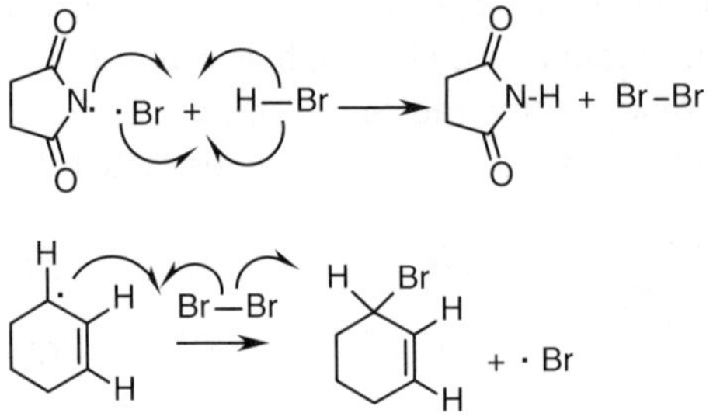

5.3.4 Radical Inhibitors

Radical inhibitors are compounds that can slow down or stop any radical reaction. Radical inhibitors are used as antioxidants or preservatives. They preserve food by preventing unwanted radical reactions. Butylated hydroxyanisol (BHA) and butylated hydroxytoluene (BHT) are synthetic preservatives that are added to many packaged foods. However, it is now known that BHT can also be obtained from some natural sources; certain phytoplanktons, including the green algae *Botryococcus braunii*, and three different cyanobacteria, *Cylindrospermopsis raciborskii*, *Microcystis aeruginosa* and *Oscillatoria* sp., produce BHT. Note that phytoplanktons are the self-feeding components of the plankton community (the diverse collection of organisms that live in large bodies of water and are unable to swim against a current) and a key part of oceans, seas and freshwater basin ecosystems.

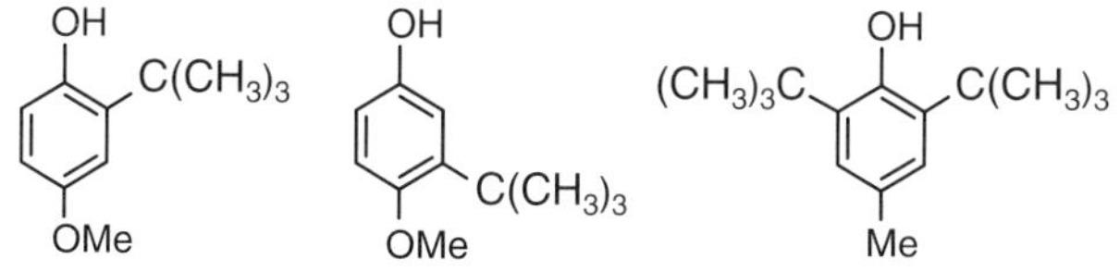

Butylated hydroxyanisole (BHA)   Butylated hydroxytoluene (BHT)

The natural antioxidants vitamin C (ascorbic acid) and vitamin E (α-tocopherol) are examples of two most common radical inhibitors that are present in biological systems (see Section 2.5). While vitamin C, a water-soluble vitamin, is found in many foods, particularly in fruits (blackcurrants, guava, lemon, lime, oranges and tomatoes) and vegetables (cauliflower, broccoli, chilli, kale, parsley, spinach and thyme), vitamin E, a fat-soluble vitamin, can be sourced from foods like avocados, nuts (almonds, peanuts and hazelnuts/filberts), seeds (sunflower seeds), vegetable oils (sunflower, safflower, corn and soybean oils) and some green vegetables (broccoli and spinach).

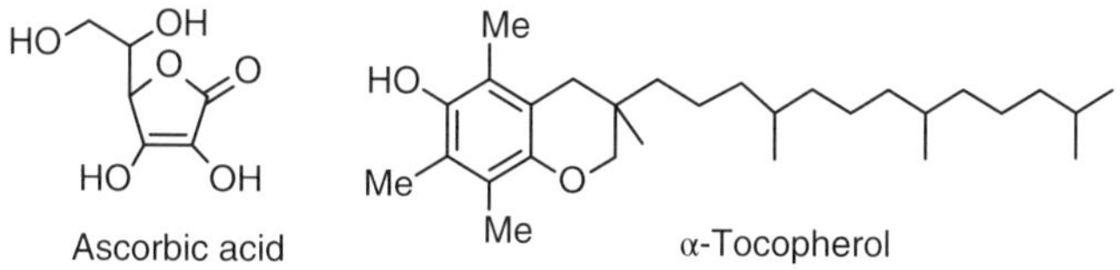

Ascorbic acid
Vitamin C
(A natural antioxidant)

α-Tocopherol
Vitamin E
(A natural antioxidant)

5.4 ADDITION REACTIONS

Addition reactions occur in compounds having π electrons in C—C double (alkenes), triple bonds (alkynes) or carbon-oxygen double bonds (aldehydes and ketones). Addition reactions are of two types: *electrophilic addition* to alkenes and alkynes, and *nucleophilic addition* to aldehydes and ketones. The mechanisms of these two addition reactions are quite different (see Sections 5.4.1 and 5.4.3). In an addition reaction, the product contains all of the elements of the two reacting species.

5.4.1 Electrophilic Additions to Alkenes and Alkynes

Alkenes and alkynes readily undergo electrophilic addition reactions. They are nucleophilic and commonly react with electrophiles. The π bonds of alkenes and alkynes are involved in the reaction and reagents are added to the double or triple bonds. In the case of alkynes, two molecules of reagent are needed for each triple bond for the total addition.

An alkyne is less reactive than an alkene. A vinyl cation is less able to accommodate a positive charge, as the hyperconjugation is less effective in stabilizing the positive charge on a vinyl cation than on an alkyl cation. The vinyl cation is more stable with a positive charge on the more substituted carbon. Electrophilic addition reactions allow the conversion of alkenes and alkynes into a variety of other functional groups.

General reaction and mechanism

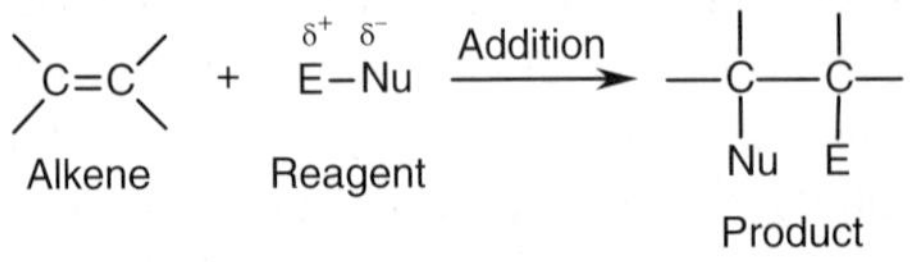

The π electrons attack the electrophile, the positive part of the reagent, usually the H^+, and form a carbocation intermediate.

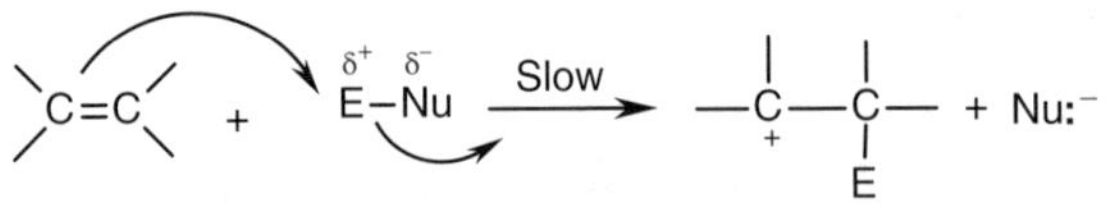

The nucleophile ($Nu:^-$), the negative part of the reagent, usually X^-, HO^- and so on, attacks the carbocation to form the product.

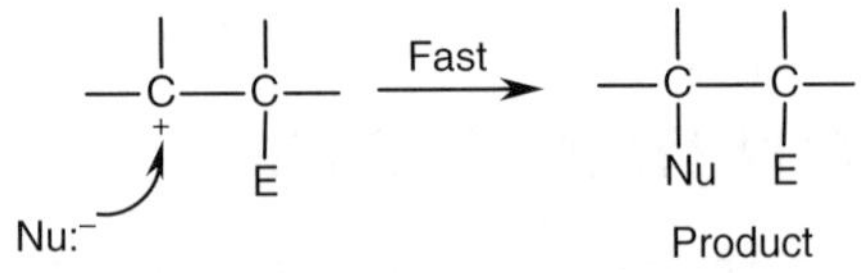

5.4.1.1 Addition of Hydrogen to Alkenes: Catalytic Hydrogenation

Preparation of Alkanes

Addition of hydrogen atoms in the presence of a metal catalyst to double or triple bonds is known as *hydrogenation* or *catalytic hydrogenation*. Alkenes and alkynes are reduced to alkanes by the treatment with H_2 over a finely divided metal catalyst such as platinum (Pt–C), palladium (Pd–C) or Raney nickel (Ni). The platinum catalyst is also frequently used in the form of PtO_2, which is known as *Adams' catalyst*. The catalytic hydrogenation reaction is a *reduction reaction*.

In the catalytic hydrogenation, two new C–H σ bonds are formed simultaneously from H atoms absorbed into the metal surface. Thus, catalytic hydrogenation is stereospecific, giving only the *syn*-addition product. If the atoms are added on the same side of the molecule, the addition is known as *syn*-addition. If the atoms are added on opposite sides of the molecule, the addition is called an *anti*-addition. For example, 2-butene reacts with H_2 in the presence of a metal catalyst to provide *n*-butane.

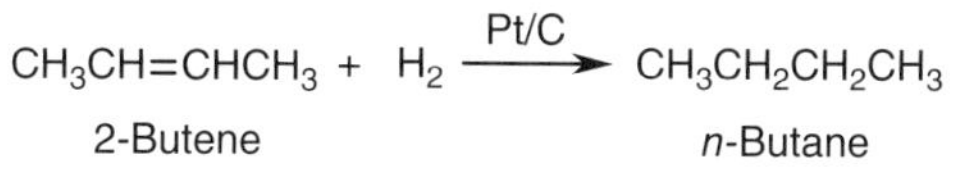

$$CH_3CH=CHCH_3 + H_2 \xrightarrow{Pt/C} CH_3CH_2CH_2CH_3$$

2-Butene   *n*-Butane

Similarly, 2-methyl-1-butene and 3-methylcyclohexene react with H_2 in the presence of a metal catalyst to yield 2-methylbutane and methylcyclohexane, respectively.

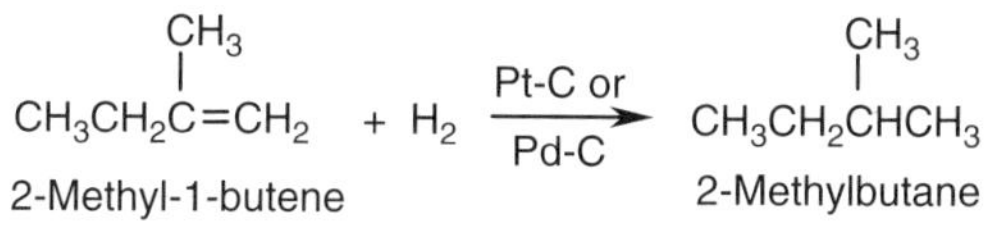

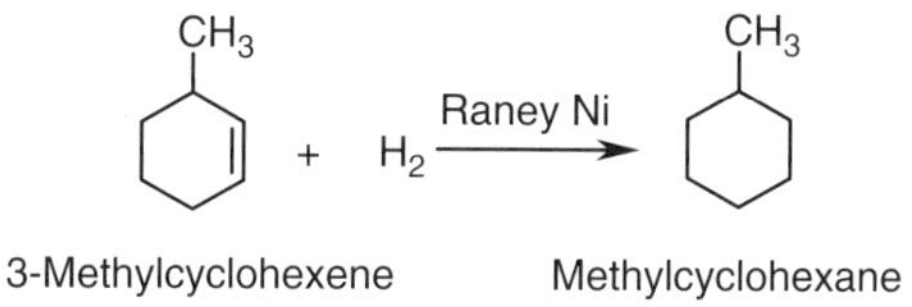

$$CH_3CH_2\overset{\overset{\displaystyle CH_3}{|}}{C}=CH_2 + H_2 \xrightarrow[Pd-C]{Pt-C\ or} CH_3CH_2\overset{\overset{\displaystyle CH_3}{|}}{C}HCH_3$$

2-Methyl-1-butene   2-Methylbutane

3-Methylcyclohexene   Methylcyclohexane

5.4.1.2 Addition of Hydrogen to Alkynes: Catalytic Hydrogenation

Preparation of Alkanes

Hydrogen adds twice to alkynes in the presence of a catalyst to generate alkanes. For example, acetylene reacts with hydrogen in the presence of a metal catalyst to afford ethane.

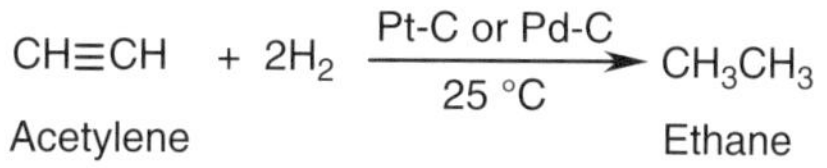

$$CH{\equiv}CH + 2H_2 \xrightarrow[25\ °C]{Pt-C\ or\ Pd-C} CH_3CH_3$$

Acetylene   Ethane

The reduction of alkynes occurs in two steps: addition of 1 mol of hydrogen atom to form alkenes, and then addition of the second mole of hydrogen to alkenes to form alkanes. This reaction proceeds through a *cis*-alkene intermediate, but cannot be stopped at this stage except with the use of a special catalyst.

5.4.1.3 Selective Hydrogenation to Alkynes

5.4.1.3.1 *Preparation of cis-Alkenes: syn-Addition*

Lindlar's catalyst, which is also known as a *poisoned catalyst*, consists of barium sulphate, palladium and quinoline, and is used in selective and partial hydrogenation of alkynes to produce *cis*-alkenes. Hydrogen atoms are delivered simultaneously to the same side of the alkyne, resulting in *syn*-addition (*cis*-alkenes). Thus, the *syn*-addition of alkyne follows the same procedure as the catalytic hydrogenation of alkyne.

$$R-C{\equiv}C-R \xrightarrow[\text{Quinoline, CH}_3\text{OH}]{\text{H}_2,\ \text{Pd/BaSO}_4} $$

R—C=C—R (cis-Alkene, R and R on same side with H)

cis-Alkene
syn-addition

5.4.1.3.2 Preparation of trans-Alkenes: Anti-addition

The *anti*-addition (*trans*-alkenes) is achieved in the presence of an alkali metal, for example sodium or lithium, in ammonia at −78 °C.

$$R-C{\equiv}C-R \xrightarrow[\text{Liq. NH}_3]{\text{Na}} $$

trans-Alkene
anti-addition

5.4.2 Symmetrical and Unsymmetrical Addition to Alkenes and Alkynes

When the same substituents are at each end of the double or triple bond, this is called *symmetrical*. Unsymmetrical means different substituents are at each end of the double or triple bond. Electrophilic addition of unsymmetrical reagents to an unsymmetrical double or triple bond follows *Markovnikov's rule*. According to the Markovnikov's rule, addition of unsymmetrical reagents, for example HX, H_2O or ROH, to an unsymmetrical alkene proceeds in such a way that the hydrogen atom adds to the carbon that already has the most hydrogen atoms. The reaction is not stereoselective, since it proceeds via a planar carbocation intermediate. However, when reaction proceeds via a cyclic carbocation intermediate, it provides stereospecific (e.g. bromination of alkenes) and regiospecific (e.g. Markonikov or anti-Markonikov) products. A reaction in which one stereoisomer is formed predominantly is called a *stereoselective reaction*. A *regioselective reaction* is a reaction that can potentially afford two or more constitutional isomers (see Section 3.2.1), but actually yields only one isomer.

$$R_2C{=}CH \quad + \quad HX \longrightarrow R-\underset{X}{\overset{R}{C}}-\underset{H}{\overset{R}{C}}-H$$

Unsymmetrical alkene   Markovnikov addition

The modern Markovnikov's rule states that, in the ionic addition of an unsymmetrical reagent to a double bond, the positive portion of the adding reagent adds to a carbon atom of the double bond to give the more stable carbocation as an intermediate. Thus, Markovnikov addition to unsymmetrical π bonds produces regioselective products.

5.4.2.1 Addition of Hydrogen Halides to Alkenes

Preparation of Alkyl Halides: Addition to Alkenes

Alkenes are converted to alkyl halides by the addition of HX (HCl, HBr or HI). Addition of HX to unsymmetrical alkenes follows Markovnikov's rule. The reaction is regioselective and occurs via the most stable carbocation intermediate. For example, addition of hydrogen bromide (HBr) to propene yields 2-bromopropane as the major product.

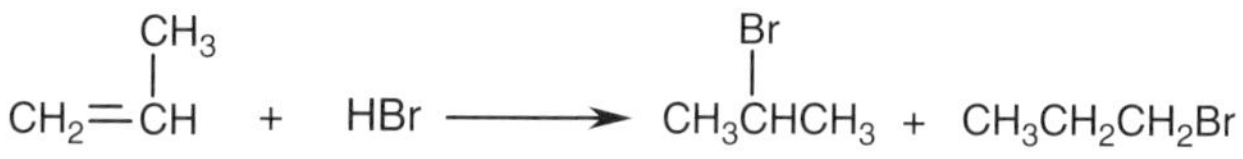

$$CH_2{=}CH{-}CH_3 \quad + \quad HBr \longrightarrow CH_3CHCH_3 \;(Br) \;+\; CH_3CH_2CH_2Br$$

Propene   2-Bromopropane   1-Bromopropane
       (Major product)   (Minor product)

Mechanism

The double bond π electrons attack the electrophile. Protonation of the double bond affords a secondary carbocation intermediate. The bromine nucleophile attacks the carbocation to form 2-bromopropane.

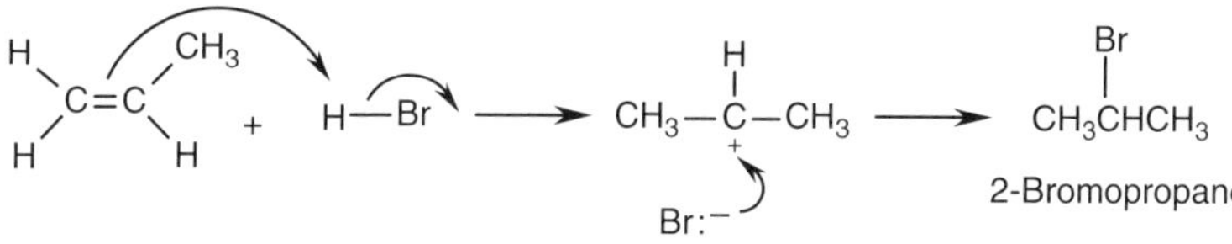

Addition of HBr to 2-methylpropene gives mainly *tert*-butyl bromide, because the product with the more stable carbocation intermediate always predominates in this type of reaction.

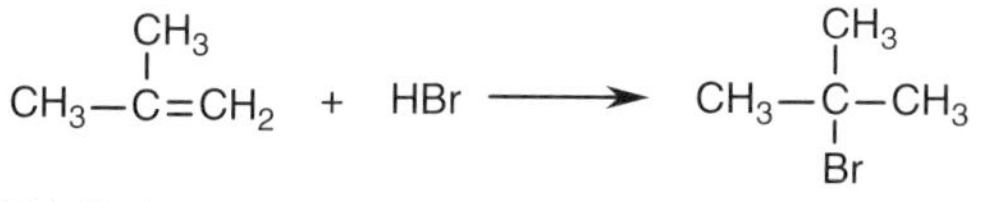

$$CH_3{-}C({=}CH_2){-}CH_3 \;+\; HBr \longrightarrow CH_3{-}C(CH_3)(Br){-}CH_3$$

2-Methylpropene   *tert*-Butyl bromide
Isobutylene   (Major product)

Mechanism

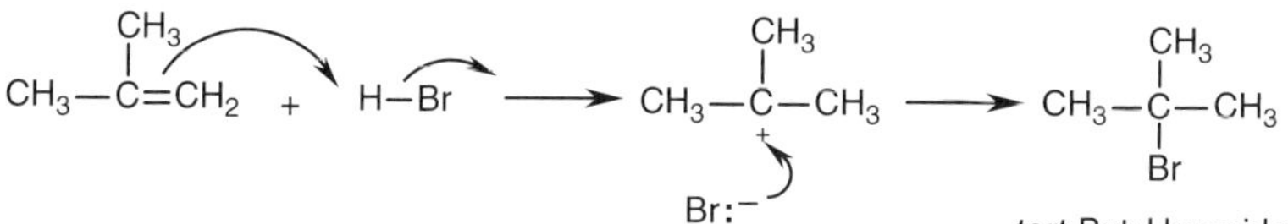

Addition of HBr to 1-butene gives a chiral molecule. The reaction is *regioselective* and a racemic mixture is formed.

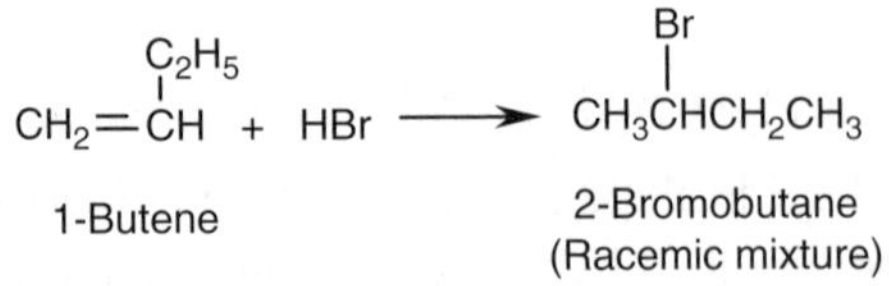

CH₂=CH (with C₂H₅) + HBr ⟶ CH₃CHCH₂CH₃ (with Br)

1-Butene   2-Bromobutane
(Racemic mixture)

Mechanism

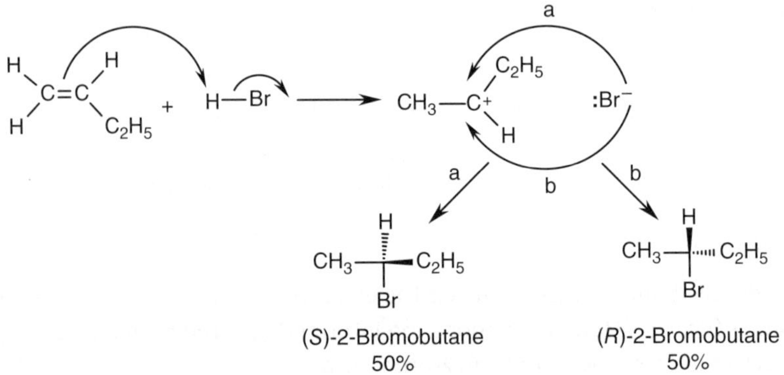

(S)-2-Bromobutane
50%

(R)-2-Bromobutane
50%

5.4.2.2 Addition of Hydrogen Halides to Alkynes

Preparation of Alkyl Dihalides

Electrophilic addition to terminal alkynes (unsymmetrical) is regioselective and follows Markovnikov's rule. Hydrogen halides can be added to alkynes just like alkenes, first to form the vinyl halide, and then the *geminal* alkyl dihalide. The addition of HX to an alkyne can be stopped after the first addition of HX. A second addition takes place when excess HX is present. For example, 1-propyne reacts with one equivalent of HCl to provide 2-chloropropene, and a second addition of HCl gives 2,2-dichloropropane, a *geminal* dihalide.

$$CH_3-C\equiv CH + 2HCl \longrightarrow CH_3-\underset{\underset{Cl}{|}}{\overset{\overset{Cl}{|}}{C}}-CH_3$$

1-Propyne

2,2-Dichloropropane
(A *geminal* dihalide)

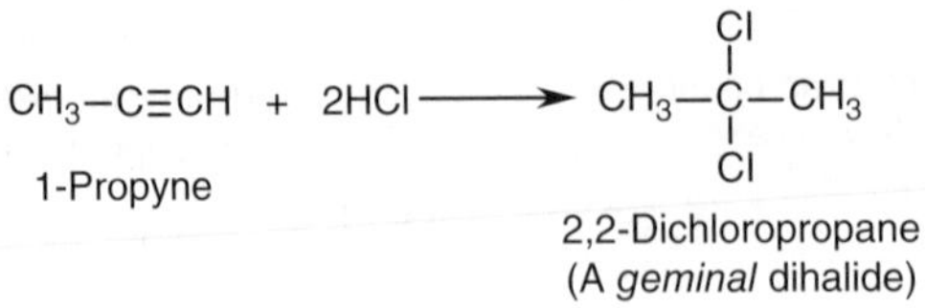

Mechanism

The vinyl cation is more stable with positive charge on the more substituted carbon, because a secondary vinylic cation is more stable than a primary vinylic cation.

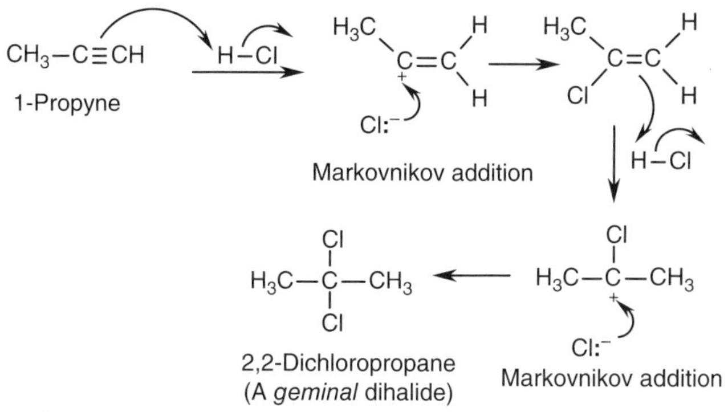

Markovnikov addition

2,2-Dichloropropane
(A *geminal* dihalide)   Markovnikov addition

Addition of hydrogen halides to an internal alkyne is not regioselective. When the internal alkyne has identical groups attached to the *sp* carbons, only one *geminal* dihalide is produced.

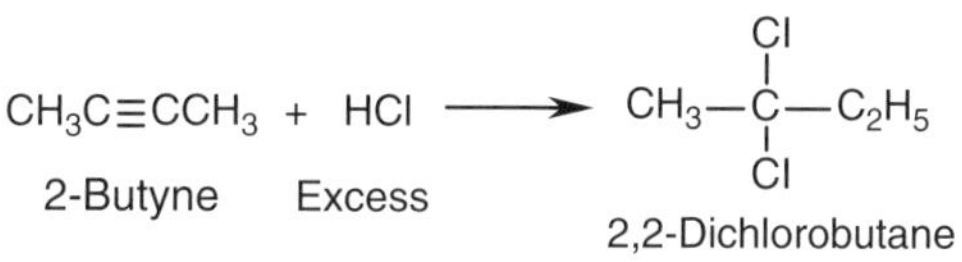

When the internal alkyne has different groups attached to the *sp* carbons, two *geminal* dihalides are formed, since both intermediate cations are substituted. For example, 2-pentyne reacts with excess HBr to yield 3,3-dibromopentane and 2,2-dibromopentane.

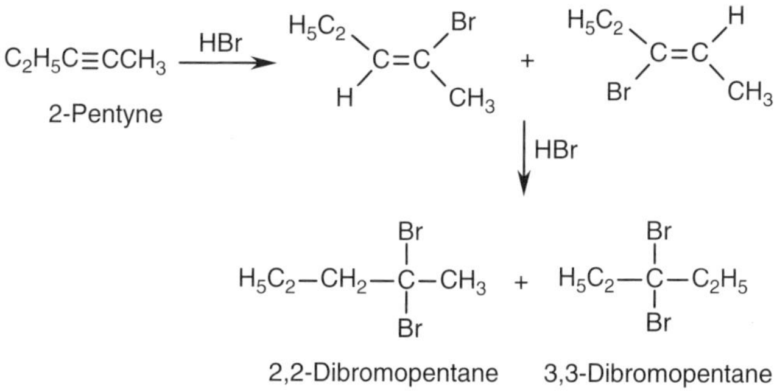

5.4.2.3 Free Radical Addition of HBr to Alkenes: The Peroxide Effect

Preparation of Alkyl Halides

It is possible to obtain anti-Markovnikov products when HBr is added to alkenes in the presence of *free radical initiators*; for example, hydrogen peroxide (HOOH) or alkyl peroxide (ROOR). The free radical initiators change the mechanism of addition from an electrophilic addition to a free radical addition. This change of mechanism gives rise to the anti-Markovnikov regiochemistry. For example, 2-methyl propene reacts with HBr in the presence of peroxide (ROOR) to form 1-bromo-2-methyl propane, which is an anti-Markovnikov product. Radical additions do not proceed with HCl or HI.

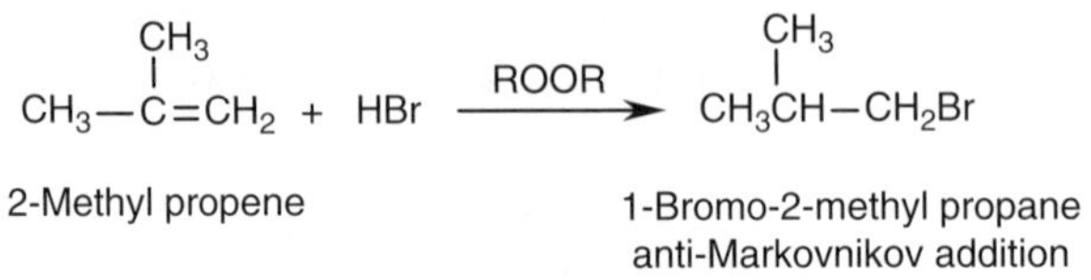

$$CH_3-\underset{\underset{CH_3}{|}}{C}=CH_2 \ + \ HBr \ \xrightarrow{ROOR} \ CH_3\underset{\underset{CH_3}{|}}{CH}-CH_2Br$$

2-Methyl propene   1-Bromo-2-methyl propane
                                          anti-Markovnikov addition

Initiation:

The oxygen-oxygen bond is weak and is easily homolytically cleaved to generate two alkoxy radicals, which in turn abstract hydrogen to generate bromine radicals.

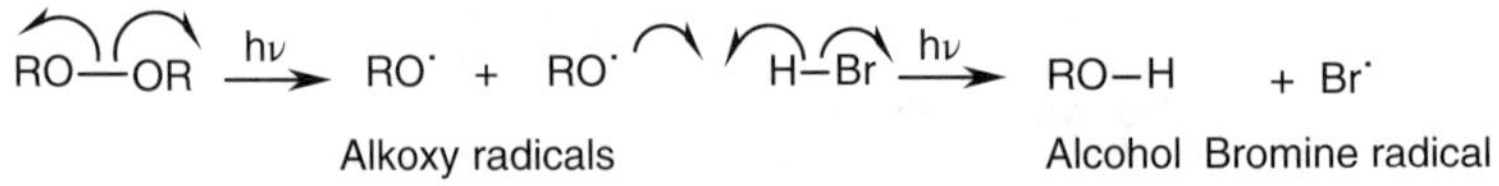

$$RO-OR \ \xrightarrow{h\nu} \ RO^{\cdot} \ + \ RO^{\cdot} \quad H-Br \ \xrightarrow{h\nu} \ RO-H \ + \ Br^{\cdot}$$

Alkoxy radicals   Alcohol Bromine radical

Propagation:

The bromine radical is electron deficient and electrophilic. The radical adds to the double bond, generating a carbon-centred radical. This radical abstracts hydrogen from HBr, giving the product and another bromine radical. The orientation of this reaction is anti-Markovnikov. The reversal of regiochemistry through the use of peroxides is called the *peroxide effect.*

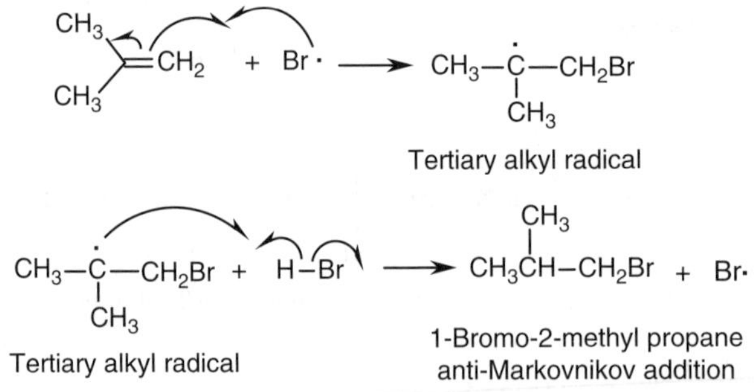

$$\underset{CH_3}{\overset{CH_3}{>}}{=}CH_2 \ + \ Br^{\cdot} \ \longrightarrow \ CH_3-\underset{\underset{CH_3}{|}}{\overset{\cdot}{C}}-CH_2Br$$

                                       Tertiary alkyl radical

$$CH_3-\underset{\underset{CH_3}{|}}{\overset{\cdot}{C}}-CH_2Br \ + \ H-Br \ \longrightarrow \ CH_3\underset{\underset{CH_3}{|}}{CH}-CH_2Br \ + \ Br^{\cdot}$$

Tertiary alkyl radical   1-Bromo-2-methyl propane
                                       anti-Markovnikov addition

Termination:

Any two radicals present in the reaction mixture can combine in a termination step and end the radical chain reaction. Thus, radical reactions yield a mixture of products.

Br· + Br· $\longrightarrow$ Br$_2$

Bromine radicals   Bromine molecule

CH$_3$—Ċ—CH$_2$Br + Br· $\longrightarrow$ CH$_3$—C—CH$_2$Br
$\quad\quad\quad$|$\quad\quad\quad\quad\quad\quad\quad\quad\quad\quad\quad\quad$|
$\quad\quad\quad$CH$_3\quad$ Bromine radical$\quad\quad\quad\quad\quad$Br

CH$_3$
|
CH$_3$—C—CH$_2$Br

Tertiary alkyl radical$\quad\quad\quad\quad\quad\quad\quad$1,2-Dibromo-2-methyl propane

CH$_3$—Ċ—CH$_2$Br + CH$_3$—Ċ—CH$_2$Br $\longrightarrow$ BrCH$_2$—C—C—CH$_2$Br
$\quad\quad\quad$|$\quad\quad\quad\quad\quad\quad\quad\quad$|$\quad\quad\quad\quad\quad\quad\quad\quad\quad$|$\quad$|
$\quad\quad\quad$CH$_3\quad\quad\quad\quad\quad\quad\quad$CH$_3\quad\quad\quad\quad\quad\quad\quad\quad$CH$_3$ CH$_3$

Tertiary alkyl radicals$\quad\quad\quad\quad$1,5-Dibromo-3,3,4,4-tetramethyl butane

5.4.2.4 Free Radical Addition of HBr to Alkynes: Peroxide Effect Preparation of Bromoalkenes

The peroxide effect is also observed with the addition of HBr to alkynes. Peroxides (ROOR) generate anti-Markovnikov products, for example, 1-butyne reacts with HBr in the presence of peroxide to form 1-bromobutene.

C$_2$H$_5$C≡CH + HBr $\xrightarrow{\text{ROOR}}$

H$_5$C$_2$ Br
$\quad$\C=C/
H$\quad\quad$H

1-Bromobutene
anti-Markovnikov addition

1-Butyne

5.4.2.5 Addition of Water to Alkenes: Acid-Catalysed Hydration Preparation of Alcohols

Addition of water is known as a *hydration reaction*. The hydration reaction occurs when alkenes are treated with aqueous acids, most commonly H_2SO_4, to produce alcohols. This is called *acid-catalysed hydration* of alkenes, which is the reverse of the *acid-catalysed dehydration* of an alcohol (see Section 5.5.3).

Addition of water to an unsymmetrical alkene follows Markovnikov's rule. The reaction is highly regiospecific. According to Markovnikov's rule, in the addition of water (H—OH) to alkene, the hydrogen atom adds to the least sub-0stituted carbon of the double bond. For example, 2-methylpropene reacts with H_2O in the presence of dilute H_2SO_4 to form *tert*-butyl alcohol. The reaction proceeds via protonation to produce the more stable tertiary carbocation intermediate. The mechanism is the reverse of that for dehydration of an alcohol (see Section 5.5.3).

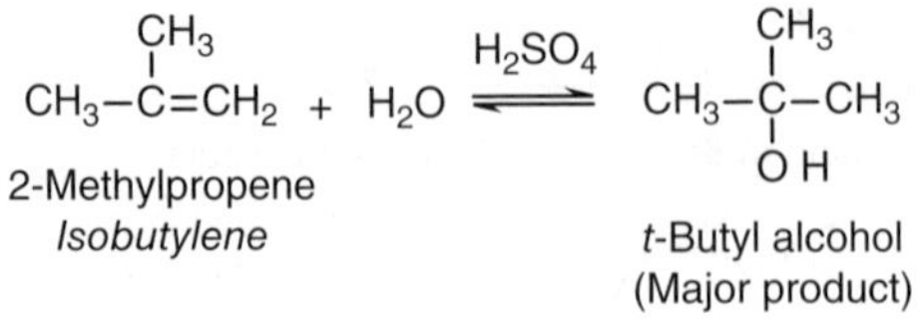

$$CH_3-\underset{\underset{CH_3}{|}}{C}=CH_2 \;+\; H_2O \;\underset{}{\overset{H_2SO_4}{\rightleftharpoons}}\; CH_3-\underset{\underset{OH}{|}}{\overset{\overset{CH_3}{|}}{C}}-CH_3$$

2-Methylpropene
Isobutylene

t-Butyl alcohol
(Major product)

Mechanism

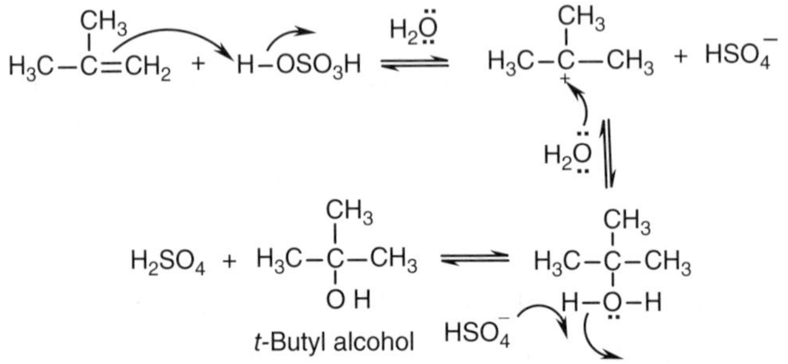

$$H_3C-\underset{\underset{CH_3}{|}}{C}=CH_2 \;+\; H-OSO_3H \;\rightleftharpoons\; H_3C-\overset{+}{\underset{\underset{CH_3}{|}}{C}}-CH_3 \;+\; HSO_4^-$$

$H_2\ddot{O}$

$$H_2SO_4 \;+\; H_3C-\underset{\underset{OH}{|}}{\overset{\overset{CH_3}{|}}{C}}-CH_3 \;\rightleftharpoons\; H_3C-\underset{\underset{H-O-H}{|}}{\overset{\overset{CH_3}{|}}{C}}-CH_3$$

t-Butyl alcohol HSO_4^-

5.4.2.6 Addition of Water to Alkynes: Acid-Catalysed Hydration

Preparation of Aldehydes and Ketones

Internal alkynes undergo acid-catalysed addition of water in the same way as alkenes, except the product is an enol. Enols are unstable and tautomerize readily to more stable keto forms. Thus, enols are always in equilibrium with their keto form. This is an example of *keto-enol tautomerism*. Keto-enol tautomerism exists in various naturally occurring bioactive compounds; for example, in curcumin, as shown next.

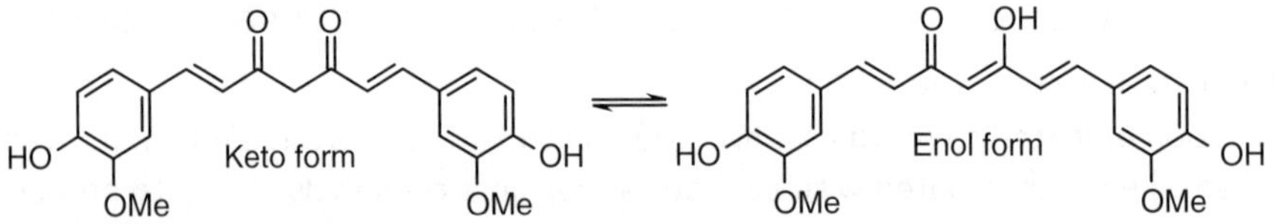

Keto form   OH   HO   Enol form   OH
HO   OH
OMe   OMe   OMe   OMe

Keto-enol tautomerism in curcumin, a natural antioxidant isolated from *Curcuma longa*

Addition of water to an internal alkyne is not regioselective. When the internal alkyne has identical groups attached to the *sp* carbons, only one ketone is obtained. For example, 2-butyne reacts with water in the presence of acid catalyst to afford 2-butanone.

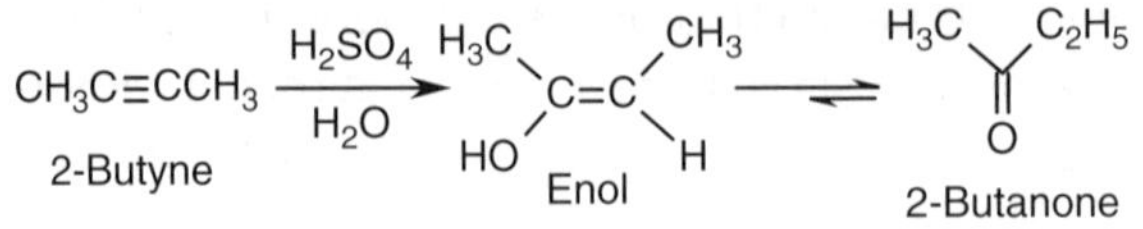

$$CH_3C{\equiv}CCH_3 \;\xrightarrow[H_2O]{H_2SO_4}\; \underset{HO}{\overset{H_3C}{}}C=C\underset{H}{\overset{CH_3}{}} \;\rightleftharpoons\; \underset{O}{\overset{H_3C\quad C_2H_5}{}}$$

2-Butyne   Enol   2-Butanone

When the internal alkyne has different groups attached to the *sp* carbons, two ketones are formed since both intermediate cations are substituted. For example, 2-pentyne reacts with water in the presence of acid catalyst to form 3-pentanone and 2-pentanone.

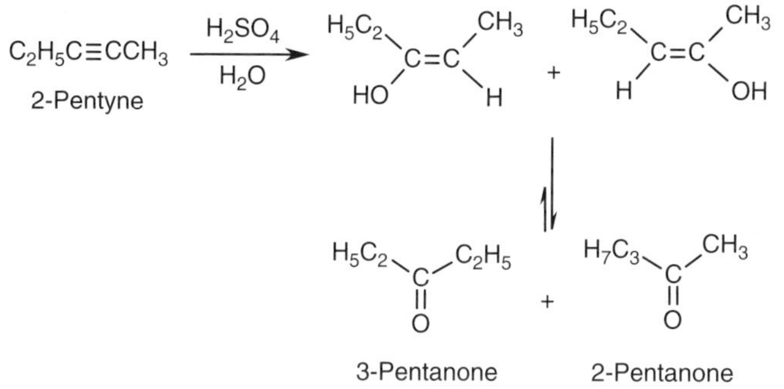

3-Pentanone   2-Pentanone

Terminal alkynes are less reactive than internal alkynes towards the acid-catalysed addition of water. Therefore, terminal alkynes require a Hg salt ($HgSO_4$) catalyst for the addition of water to yield aldehydes and ketones. Addition of water to acetylene affords acetaldehyde, and all other terminal alkynes give ketones. The reaction is regioselective and follows Markovnikov addition. For example, 1-butyne reacts with water in the presence of H_2SO_4 and $HgSO_4$ to produce 2-butanone.

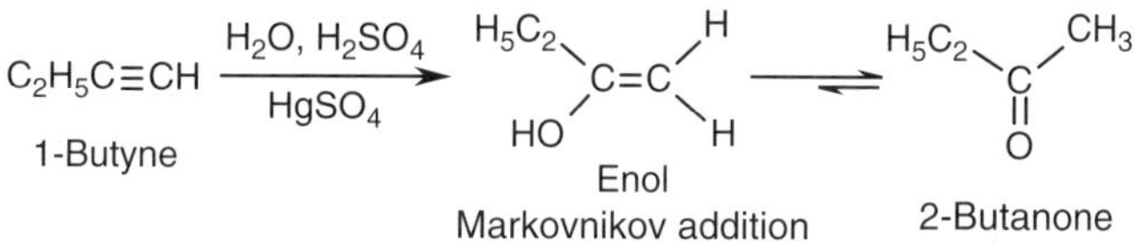

Enol
Markovnikov addition   2-Butanone

Mechanism
Addition of $HgSO_4$ generates a cyclic mercurinium ion, which is attacked by a nucleophilic water molecule on the more substituted carbon. Oxygen loses a proton to form a mercuric enol, which under work-up provides enol (vinyl alcohol). The enol is rapidly converted to 2-butanone.

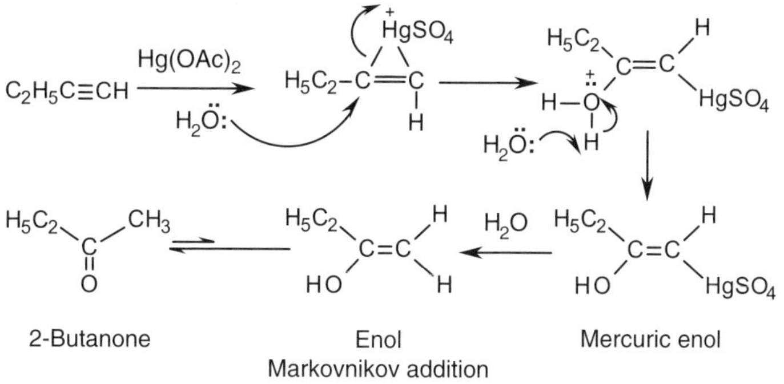

2-Butanone   Enol   Mercuric enol
Markovnikov addition

5.4.2.7 Addition of Water to Alkenes: Oxymercuration-Reduction

Preparation of Alcohols

Addition of water to alkenes by an oxymercuration-reduction results in the formation of alcohols via Markovnikov addition. This addition is similar to the acid-catalysed addition of water. Oxymercuration is regiospecific and *anti*-stereospecific. In the addition reaction, Hg(OAc) bonds to the less substituted carbon, and the OH to the more substituted carbon of the double bond. For example, propene reacts with mercuric acetate in the presence of an aqueous tetrahydrofuran (THF) to give a hydroxy-mercurial compound, followed by reduction with sodium borohydride (NaBH$_4$) to produce 2-propanol.

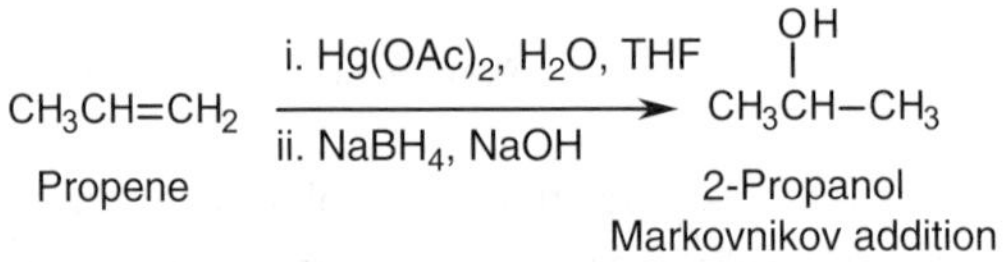

$$CH_3CH=CH_2 \xrightarrow[\text{ii. NaBH}_4,\ \text{NaOH}]{\text{i. Hg(OAc)}_2,\ H_2O,\ THF} \overset{\displaystyle OH}{CH_3CH-CH_3}$$

Propene   2-Propanol

Markovnikov addition

Mechanism

The reaction is analogous to the addition of bromine molecules to an alkene. The electrophilic mercury of mercuric acetate adds to the double bond and forms a cyclic mercurinium ion intermediate rather than a planer carbocation. In the next step, water attacks the most substituted carbon of the mercurinium ion to yield the addition product. The hydroxy-mercurial compound is reduced in situ using NaBH$_4$ to give alcohol. The removal of Hg(OAc) in the second step is called *demercuratrion*. Therefore, the reaction is also known as *oxymercuration-demercuratrion*.

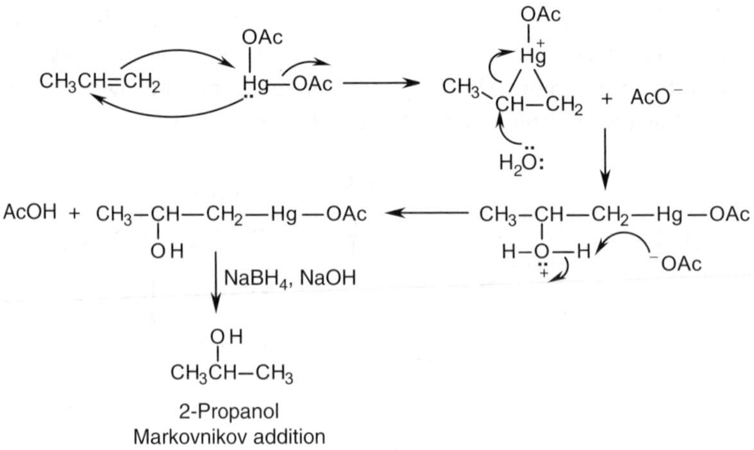

2-Propanol
Markovnikov addition

5.4.2.8 Addition of Water to Alkenes: Hydroboration-Oxidation

Preparation of Alcohols

Addition of water to alkenes by hydroboration-oxidation affords alcohols via anti-Markovnikov addition. This addition is opposite to the acid-catalysed addition of water. Hydroboration is regioselective and *syn*-stereospecific. In the addition reaction, borone bonds to the less substituted carbon and hydrogen to the more substituted carbon of the double bond. For example, propene reacts with borane and THF complex, followed by oxidation with basic hydrogen peroxide (H_2O_2), to yield propanol.

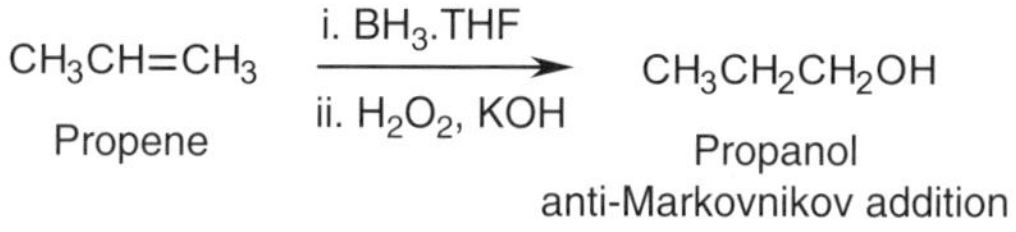

$$CH_3CH=CH_3 \xrightarrow[\text{ii. } H_2O_2, \text{ KOH}]{\text{i. } BH_3.THF} CH_3CH_2CH_2OH$$

Propene   Propanol
anti-Markovnikov addition

Mechanism

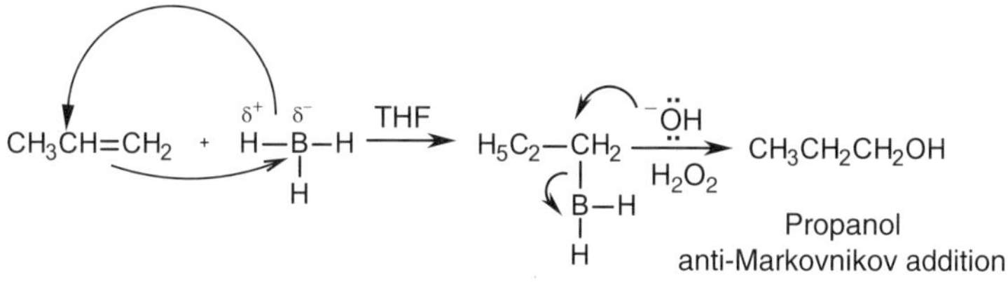

5.4.2.9 Addition of Water to Alkynes: Hydroboration-Oxidation

Preparation of Aldehydes and Ketones

Hydroboration-oxidation of terminal alkynes provides *syn*-addition of water across the triple bond. The reaction is regioselective and follows anti-Markovnikov addition. Terminal alkynes are converted to aldehydes and all other alkynes are converted to ketones. A sterically hindered dialkylborane must be used to prevent the addition of two borane molecules. A vinyl borane is produced with anti-Markovnikov orientation, which is oxidized by basic hydrogen peroxide to an enol. This enol tautomerizes readily to a more stable keto form.

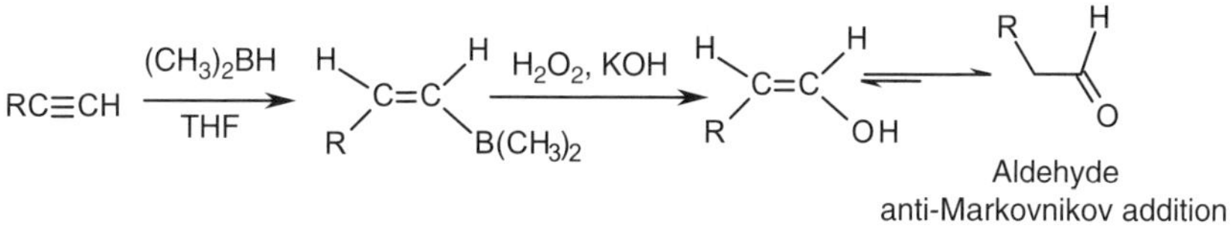

Aldehyde
anti-Markovnikov addition

Mechanism

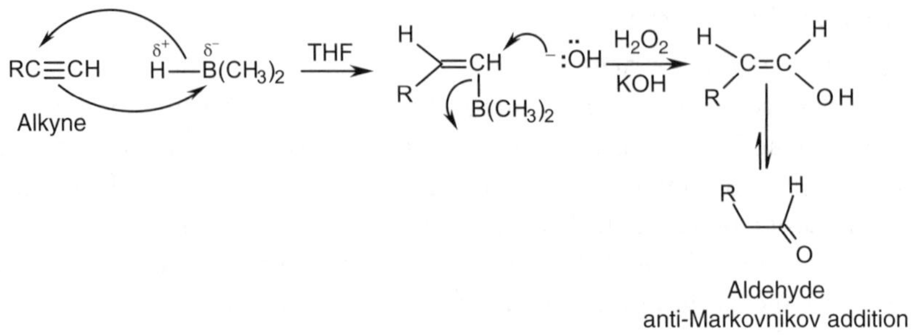

Aldehyde
anti-Markovnikov addition

5.4.2.10 Addition of Alcohols to Alkenes: Acid-Catalysed Preparation of Ethers

Alcohols react with alkenes in the same way as water does. The addition of alcohols in the presence of an acid catalyst, most commonly using aqueous H_2SO_4, yields ethers. Addition of alcohol to an unsymmetrical alkene follows Markovnikov's rule. The reaction proceeds via protonation to form the more stable carbocation intermediate. The mechanism is the reverse of that for dehydration of alcohol (see Section 5.5.3).

For example, 2-methylpropane reacts with methanol (CH_3OH) in the presence of aqueous H_2SO_4 to form methyl *tert*-butyl ether.

2-Methylpropene

Methyl *tert*-butyl ether
Markovnikov addition

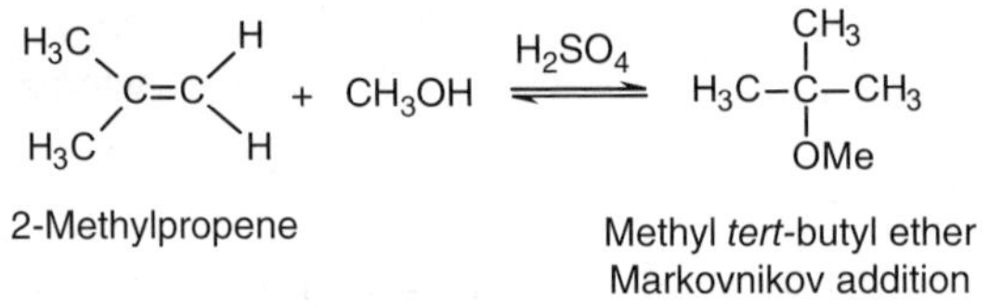

5.4.2.11 Addition of Alcohols to Alkenes: Alkoxymercuration-Reduction

Preparation of Ethers

Addition of alcohol to alkenes by alkoxymercuration-reduction produces ethers via Markovnikov addition. This addition is similar to the acid-catalysed addition of an alcohol. For example, propene reacts with mercuric acetate in aqueous THF, followed by the reduction with $NaBH_4$, to afford methyl propyl ether. The second step is known as *demercuration*, where Hg(OAc) is removed by $NaBH_4$. Therefore, this reaction is also called *alkoxymercuration-demercuration*. The reaction mechanism is exactly the same as the oxymercuration-reduction of alkenes.

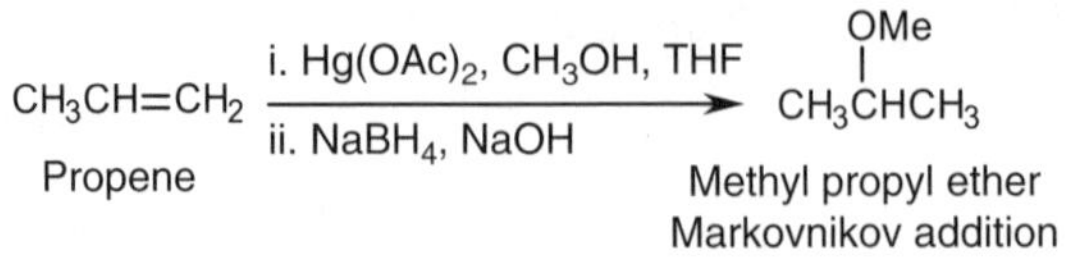

$$CH_3CH{=}CH_2 \xrightarrow[\text{ii. NaBH}_4\text{, NaOH}]{\text{i. Hg(OAc)}_2\text{, CH}_3\text{OH, THF}} CH_3CHCH_3$$

Propene

Methyl propyl ether
Markovnikov addition

5.4.2.12 Addition of Sulphuric Acid to Alkenes

Preparation of Alcohols

Addition of concentrated H_2SO_4 to alkenes yields acid-soluble alkyl hydrogen sulphates. The addition follows Markovnikov's rule. The sulphate is hydrolysed to obtain the alcohol. The net result is Markovnikov addition of acid-catalysed hydration to an alkene. The reaction mechanism of H_2SO_4 addition is similar to acid-catalysed hydration.

$$CH_2{=}CHCH_3 \xrightarrow[\text{No heat}]{\text{Conc } H_2SO_4} \underset{\substack{\text{Isopropayl hydrogen} \\ \text{sulphate} \\ \text{Markovnikov addition}}}{CH_3\overset{OSO_3H}{\underset{|}{C}}HCH_3} \xrightarrow[\text{Heat}]{H_2O} \underset{\text{Propanol}}{CH_3\overset{OH}{\underset{|}{C}}HCH_3} + H_2SO_4$$

Propene

5.4.2.13 Addition of Halides to Alkenes: Stereospecific Addition

Preparation of Alkyl Dihalides

Addition of X_2 (Br_2 and Cl_2) to alkenes gives *vicinal*-dihalides. This reaction is used as a test for unsaturation (π bonds), because the red colour of the bromine reagent disappears when an alkene or alkyne is present. For example, when ethylene is treated with Br_2 in CCl_4 in the dark at room temperature, the red colour of Br_2 disappears rapidly forming 1,2-dibromoethane, a colourless product.

$$\underset{\text{(Colourless)}}{H_2C{=}CH_2} + \underset{\text{(Red colour)}}{Br_2} \xrightarrow[\text{Dark, r.t.}]{CCl_4} \underset{\substack{\text{1,2-Dibromoethane} \\ \text{(colourless)}}}{H_2C\underset{\underset{Br}{|}}{-}CH_2\underset{\underset{Br}{|}}{\,}}$$

Mechanism

When Br_2 approaches to the double bond it becomes polarized. The positive part of the bromine molecule is attacked by the electron rich π bond, and forms a cyclic bromonium ion. The negative part of bromine is the nucleophile, which attacks the less substituted carbon to open up the cyclic bromonium ion and produces 1,2-dibromoethane (*vicinal*-dihalide).

$$H_2C{=}CH_2 \xrightarrow[\text{Dark, r.t.}]{CCl_4} \overset{+}{Br}\ \ CH_2{-}CH_2 \quad {:}Br^- \longrightarrow \underset{\substack{\text{1,2-Dibromoethane} \\ \text{(colourless)}}}{H_2C\underset{\underset{Br}{|}}{-}CH_2\underset{\underset{Br}{|}}{\,}}$$

Halogenation of double bonds is *stereospecific*. A reaction is stereospecific when a particular stereoisomeric form of the starting material generates a specific stereoisomeric form of the product. For example, the halogenation of *cis-* and *trans*-2-butene yields a racemic mixture of 2,3-dibromobutane and *meso*-2,3-dibromobutane, respectively.

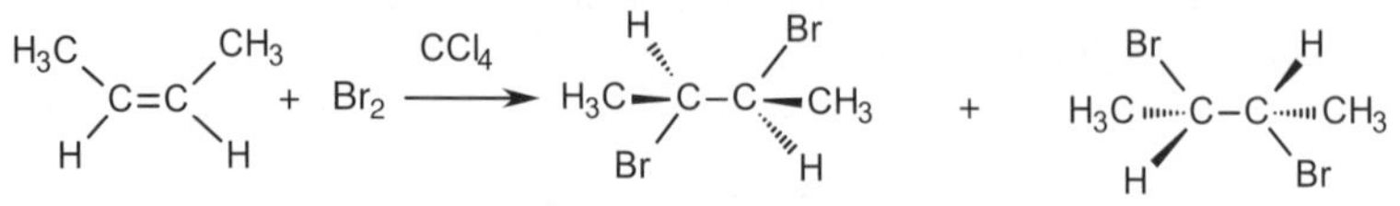

cis-2-Butene   (2R,3R)-2,3-Dibromobutane (2S,3S)-2,3-Dibromobutane

Racemic mixture

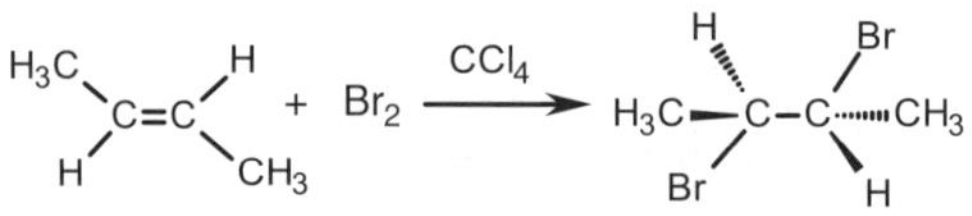

trans-2-Butene   (2R,3S)-2,3-Dibromobutane
                        (Meso compound)

When cyclopentene reacts with Br_2, the product is a racemic mixture of *trans*-1,2-dibromocyclopentane. Addition of Br_2 to cycloalkenes produces a cyclic bromonium ion intermediate instead of the planar carbocation. The reaction is stereospecific and only gives *anti*-addition of dihalides.

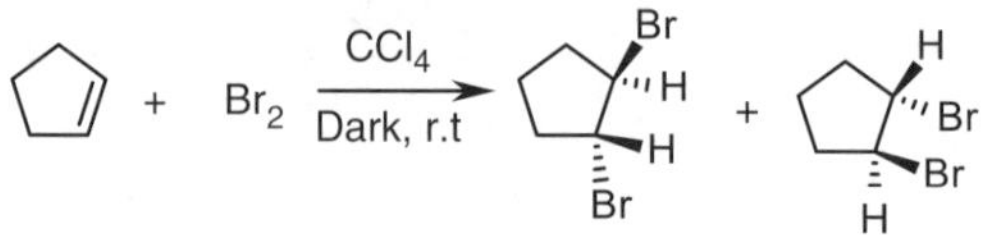

trans-1,2-Dibromocyclopentane

Mechanism

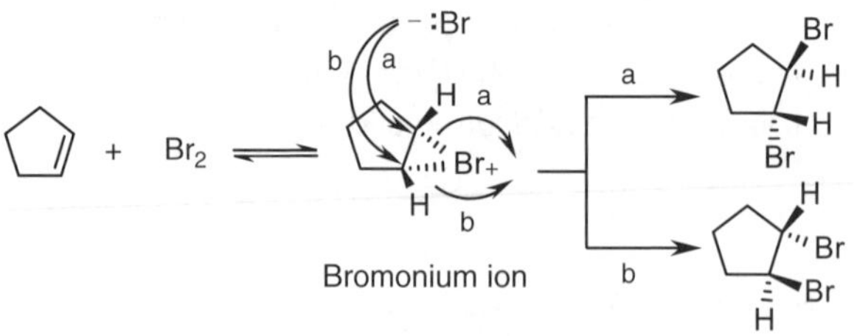

Bromonium ion

trans-1,2-Dibromocyclopentane

5.4.2.14 Addition of Halides to Alkynes

Preparation of Alkyl Dihalides and Tetrahalides

Halides (Cl_2 or Br_2) add to alkynes in an analogous fashion to alkenes. When 1 mol of halogen is added, a dihaloalkene is obtained and a mixture of *syn-addition* and *anti*-addition is observed.

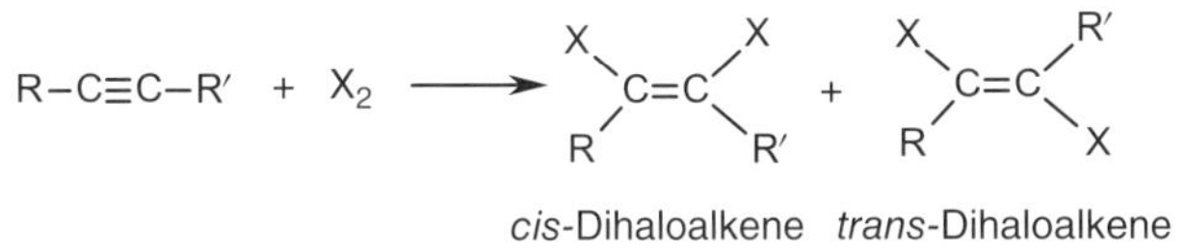

cis-Dihaloalkene   trans-Dihaloalkene

It is usually hard to control the addition of just one equivalent of halogen and it is more common to add two equivalents to generate tetrahaloalkane.

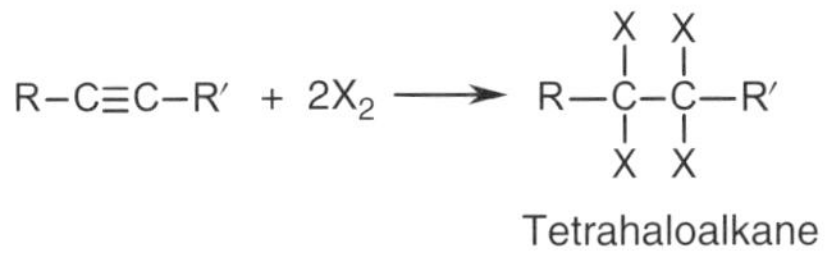

Tetrahaloalkane

Acetylene undergoes an electrophilic addition reaction with bromine in the dark. Bromine adds successively to each of the two π bonds of the alkyne. In the first stage of the reaction, acetylene is converted to an alkene, 1,2-dibromoeth-ene. In the final stage, another molecule of bromine is added to the π bond of this alkene to afford 1,1,2,2-tetrabromoethane.

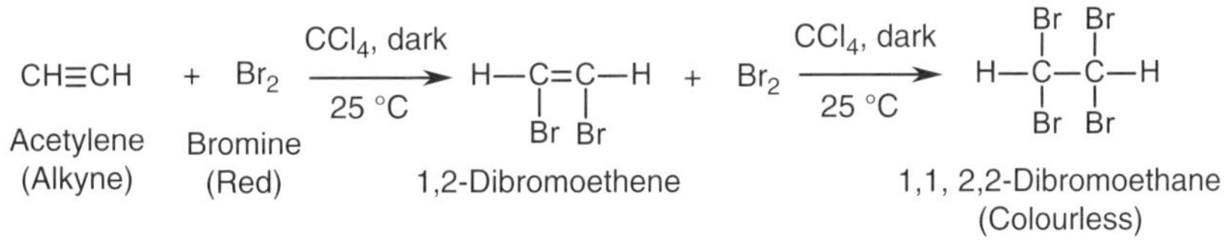

5.4.2.15 Addition of Halides and Water to Alkenes: Regiospecific Addition

Preparation of Halohydrins

When halogenation of alkenes is carried out in aqueous solvent, a *vicinal* halohy-drin is obtained. The reaction is *regioselective* and follows the Markovnikov rule. The halide adds to the less substituted carbon atom via a *bridged halonium ion* intermediate and the hydroxyl adds to the more substituted carbon atom. The reaction mechanism is similar to the halogenation of alkenes, except that instead of the halide nucleophile, the water attacks as a nucleophile.

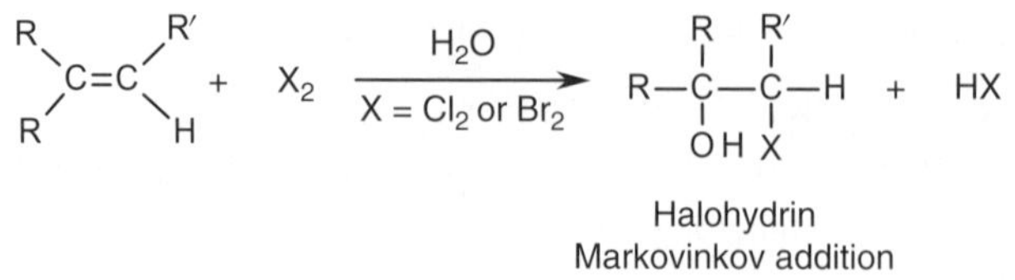

Halohydrin
Markovinkov addition

Mechanism

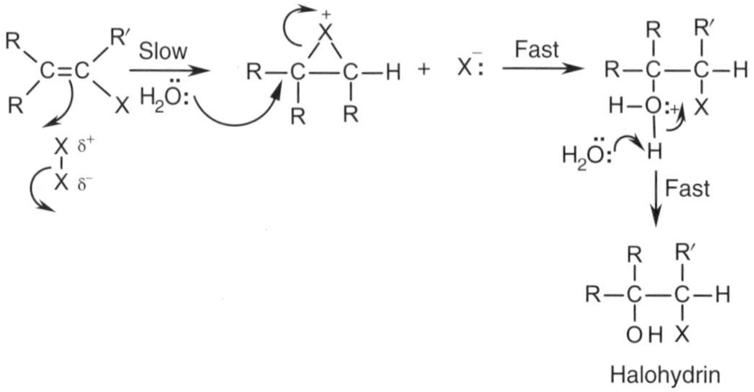

Halohydrin

5.4.2.16 Addition of Carbenes to Alkenes

Preparation of Cyclopropanes

Carbenes are divalent carbon compounds, also known as *methylene*. They have neutral carbons with a lone pair of electrons and are highly reactive. Methylene can be prepared by heat or light initiated decomposition of diazomethane (explosive and toxic gas).

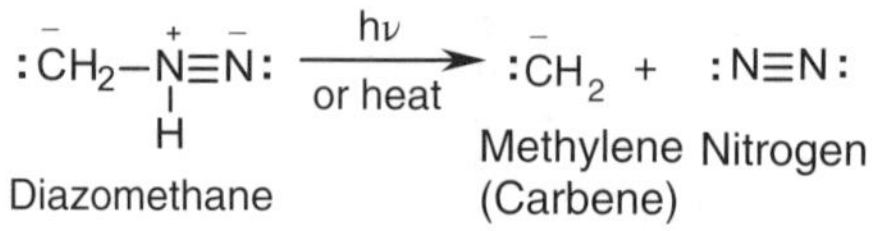

Diazomethane

Methylene Nitrogen
(Carbene)

Addition of methylene (CH_2) to alkenes gives substituted cyclopropanes. For example, methylene reacts with ethylene to form cyclopropane.

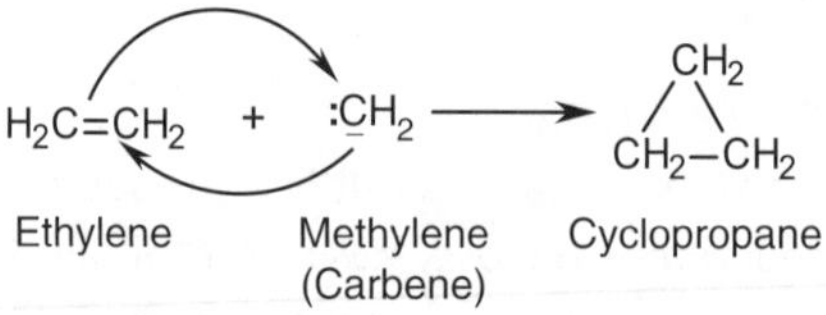

Ethylene   Methylene   Cyclopropane
             (Carbene)

5.4.3 Nucleophilic Addition to Aldehydes and Ketones

The most common reaction of aldehyde and ketone is *nucleophilic addition*. Aldehyde generally undergoes nucleophilic addition more readily than ketone. In the nucleophilic addition reaction, carbonyl compound can behave as both a Lewis acid and Lewis base, depending on the reagents. The carbonyl group is strongly polarized with the oxygen bearing a partial negative charge (δ^-) and the carbon bearing a partial positive charge (δ^+). So the carbon is electrophilic and therefore readily attacked by the nucleophile. The attacking nucleophile can be either negatively charged (Nu:$^-$) or a neutral (Nu:) molecule. Aldehydes and ketones react with nucleophiles to form addition products followed by protonation.

If the nucleophile is a negatively charged anion (good nucleophile, e.g. HO$^-$, RO$^-$ and H$^-$), it readily attacks the carbonyl carbon and forms an alkoxide tetrahedral intermediate, which is usually protonated in a subsequent step either by the solvent or by added aqueous acid.

General reaction mechanism

$$R\overset{\delta^-}{\underset{\delta^+}{C}}{=}O,\ Y \quad \xrightarrow{\ Nu:^-\ } \quad R\underset{Nu:}{\overset{:\ddot{O}:^-}{C}}Y \quad \underset{\text{Alkoxide}\atop\text{Tetrahedral intermediate}}{\overset{H_3O^+}{\rightleftharpoons}} \quad R\underset{Nu:}{\overset{:\ddot{O}H}{C}}Y$$

Alkoxide Tetrahedral intermediate — Alcohol — Y = H or R

If the nucleophile is a neutral molecule with a lone pair of electrons (weaker nucleophile, e.g. water and alcohol), it requires an acid catalyst. The carbonyl oxygen is protonated by acid that increases the susceptibility of the carbonyl carbon to nucleophilic attack.

$$R\overset{\delta^-}{\underset{\delta^+}{C}}{=}O,\ Y \quad \underset{Y = H\ or\ R}{\overset{H^+}{\rightleftharpoons}} \quad R\overset{^+:OH}{\underset{Nu:^-}{C}}{=}Y \quad \longrightarrow \quad R\underset{Nu:}{\overset{:\ddot{O}H}{C}}Y$$

Y = H or R — Alcohol

If the attacking nucleophile has a pair of nonbonding electrons available in the addition product, water is eliminated in the presence of anhydrous acid from the addition product. This is known as a *nucleophilic addition-elimination* reaction.

$$R\underset{Nu:}{\overset{:\ddot{O}H}{C}}Y \quad \underset{}{\overset{Dry\ H+}{\rightleftharpoons}} \quad R\underset{Nu:}{\overset{^+\ddot{O}H_2}{C}}Y \quad \rightleftharpoons \quad R\underset{\underset{+}{Nu}}{\overset{}{C}}{=}Y\ +\ H_2O$$

Nu: = ROH, RNH$_2$ — Y = H or R

5.4.3.1 Addition of Organometallics to Aldehydes and Ketones

Nucleophilic addition of organometallics to the carbonyl (C=O) group is a versatile and useful synthetic reaction. Aldehydes and ketones react with organometallic reagents to form different classes of alcohols depending on the starting carbonyl compound. Nucleophilic addition of Grignard (RMgX) or organolithium (RLi) reagents to the carbonyl (C=O) group is a versatile and useful synthetic reaction. These reagents add to the carbonyl and are protonated in a separate step by the solvent or by added acid.

General reaction mechanism

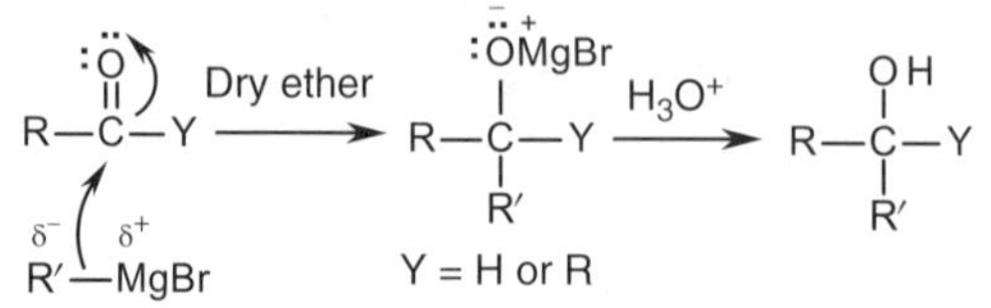

5.4.3.1.1 *Preparation of primary alcohols*

Formaldehyde is the simplest aldehyde. It reacts with a Grignard (RMgX) or organo-lithium (RLi) reagents to generate a primary alcohol, which contains one more carbon atom than the original Grignard (RMgX) or organolithium (RLi) reagents. For example, formaldehyde reacts with methyl magnesium bromide to afford ethanol.

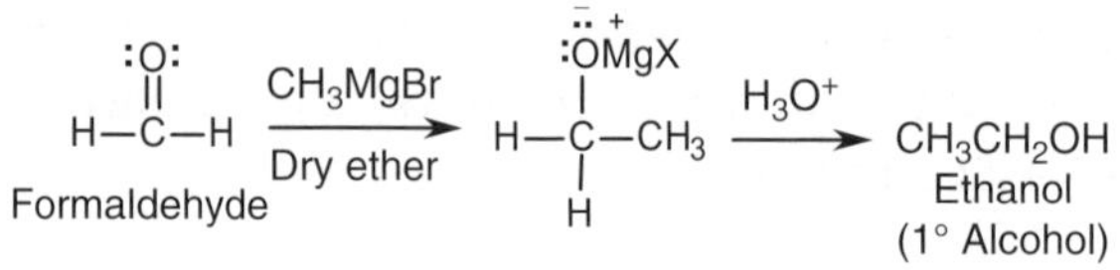

5.4.3.1.2 *Preparation of secondary alcohols*

Reaction of an aldehyde with a Grignard reagent (RMgX) or organolithium (RLi) reagents generates a secondary alcohol. For example, acetaldehyde reacts with methyl magnesium bromide to give 2-propanol.

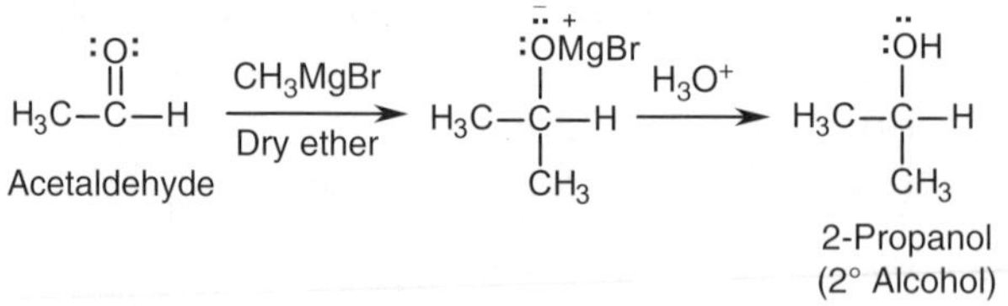

5.4.3.1.3 *Preparation of tertiary alcohols*

Addition of Grignard (RMgX) or organolithium (RLi) reagents to a ketone produces tertiary alcohol. For example, acetone reacts with methyl magnesium bromide to yield *tert*-butanol.

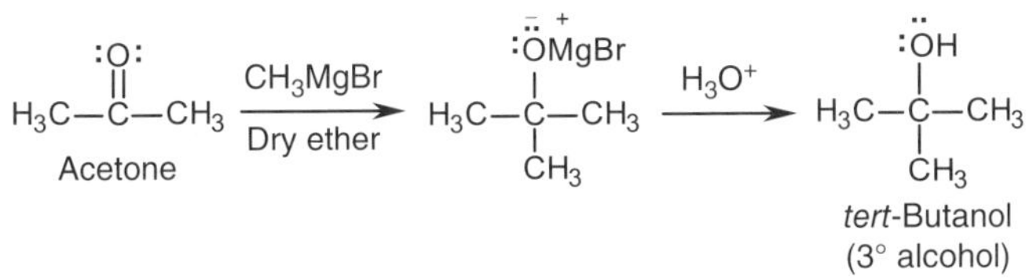

5.4.3.2 Addition of Organometallics to Acid Chlorides and Esters

Preparation of Tertiary Alcohols

Acid chlorides and esters are derivatives of carboxylic acids, where the −OH group is replaced either by a chlorine atom or an alkoxy group. Acid chlorides and esters react with two equivalents of Grignard (RMgX) or organolithium (RLi) reagents to provide tertiary alcohols.

The addition of the first Grignard reagent forms a tetrahedral intermediate, which is unstable. The leaving group is expelled (either chloride or alkoxide) as the carbonyl group is reformed and a ketone is obtained. The ketone then reacts with the second equivalent of Grignard reagent to give a tertiary alcohol. This is a good route to produce tertiary alcohols with two identical alkyl substituents; for example, 1,1-diphenylethanol.

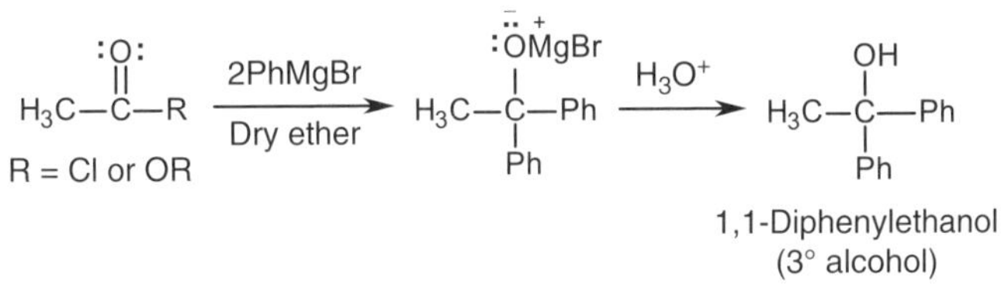

5.4.3.3 Carbonation of Grignard Reagent

Preparation of Carboxylic Acids

Grignard reagent reacts with CO_2 to afford magnesium salts of carboxylic acid. Addition of aqueous acid produces carboxylic acid.

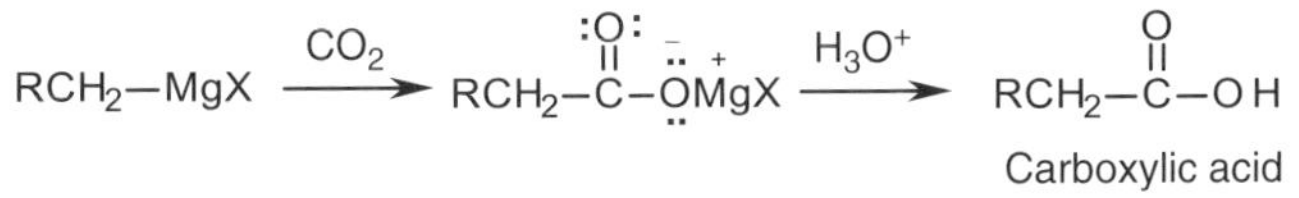

5.4.3.4 Addition of Organometallics to Nitriles

Preparation of Ketones

Grignard (RMgX) or organolithium (RLi) reagent attacks a nitrile to generate the magnesium or lithium salt of imine. Acid hydrolysis of this salt forms a ketone. Since the ketone is not formed until the work-up, the organometallic reagent does not get the opportunity to react with the ketone.

$$RC{\equiv}N \ + \ R'MgBr \longrightarrow \left[\underset{R}{\overset{R'}{>}}C{=}N{-}MgBr \right] \xrightarrow{H_3O^+} \underset{R}{\overset{R'}{>}}C{=}O \ + \ NH_4{+}$$

Nitrile   Ketone

5.4.3.5 Addition of Alkynides to Aldehydes and Ketones
Preparation of Alcohols: Alkynylation

Alkynides are strong nucleophiles. They react with carbonyl group to produce alkoxides, which under acidic work-up afford alcohols. The reaction mechanism is similar to the addition of organometallics to aldehydes and ketones (see Section 5.4.3.1).

$$R{-}\overset{\overset{\displaystyle :O:}{\|}}{C}{-}Y \ \xrightarrow[\substack{R'C{\equiv}CMgX \\ (M = Na, Li)}]{R'C{\equiv}CM \ or} \ R{-}\overset{\overset{\displaystyle :\ddot{O}^-}{|}}{\underset{Y}{C}}{-}C{\equiv}C{-}R' \ \xrightarrow{H_3O^+} \ R{-}\overset{\overset{\displaystyle :\ddot{O}H}{|}}{\underset{Y}{C}}{-}C{\equiv}C{-}R'$$

Y = H or R

5.4.3.6 Addition of Phosphorus Ylide to Aldehydes and Ketones

5.4.3.6.1 Wittig Reaction

Georg Wittig (1954) discovered that the addition of a phosphorus ylide (stabilized anion) to an aldehyde or a ketone generates an alkene, not an alcohol. This reaction is known as the *Wittig reaction*.

$$\underset{Y}{\overset{R}{>}}C{=}O \ + \ \underset{R'}{\overset{R'}{>}}\overset{-}{C}{-}\overset{+}{P}(Ph)_3 \longrightarrow \underset{Y}{\overset{R}{>}}C{=}C\underset{R'}{\overset{R'}{<}} \ + \ Ph_3P{=}O$$

Y = H or R   Phosphorous ylide   Alkene   Triphenylphosphine oxide

5.4.3.6.2 Preparation of Phosphorus Ylide

Phosphorus ylides are produced from the reaction of triphenylphosphine and alkyl halides. *Phosphorus ylide* is a molecule that is neutral overall, but exists as a carbanion bonded to a positively charged phosphorous. The ylide can also be written in the double-bonded form, because phosphorous can have more than eight valence electrons.

$$RCH_2{-}X \ \xrightarrow[\substack{\text{i. (Ph)}_3P: \\ \text{ii. BuLi, THF}}]{} \ (Ph)_3\overset{+}{P}{-}\overset{-}{C}HR \ \longleftrightarrow \ (Ph)_3P{=}CHR$$

Alkyl halide   Phosphorous ylide

Mechanism

In the first step of the reaction, the nucleophilic attack of the phosphorus on the primary alkyl halide generates an alkyl triphenylphosphonium salt. Treatment of this salt with a strong base, for example butyllithium, removes a proton to generate the ylide. The carbanionic character of the ylide makes it a powerful nucleophile.

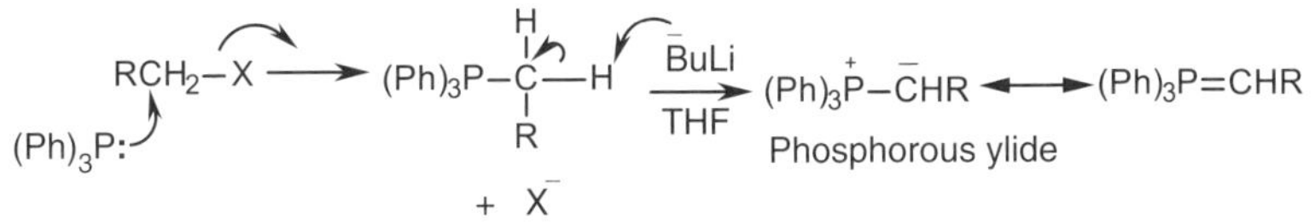

5.4.3.6.3 Preparation of Alkenes

Aldehydes and ketones react with phosphorous yilide to afford alkenes. By dividing a target molecule at the double bond, one can decide which of the two components should best come from the carbonyl and which from the ylide. In general, the ylide should come from an unhindered alkyl halide, since triphenyl phosphine is bulky.

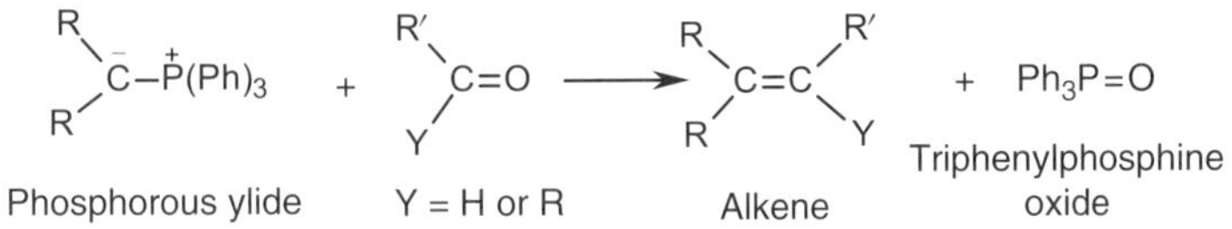

Mechanism

Phosphorous yilde reacts rapidly with aldehydes and ketones to give an intermediate called a *betaine*. Betaines are unusual since they contain negatively charged oxygen and positively charged phosphorus.

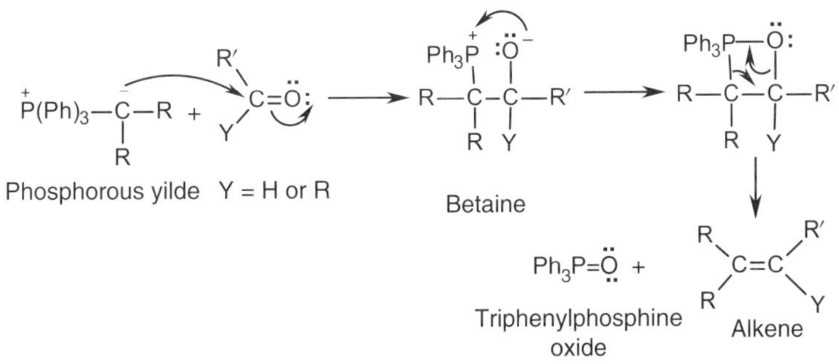

Phosphorus and oxygen always form strong bonds and these groups therefore combine to generate a four-membered ring, an oxaphosphetane ring. This four-membered ring quickly decomposes to generate an alkene and a stable triphenyl phosphine oxide (Ph$_3$P=O). The net result is replacement of the carbonyl oxygen atom by the $R_2C=$ group, which originally bonded to the phosphorous atom. This is a good synthetic route to make alkenes from aldehydes and ketones.

5.4.3.7 Addition of Hydrogen Cyanide to Aldehydes and Ketones

Preparation of Cyanohydrins

Addition of hydrogen cyanide to an aldehyde and ketone forms cyanohydrin. The reaction is usually carried out using sodium or potassium cyanide with HCl. Hydrogen cyanide is a toxic volatile liquid and a weak acid. Therefore, the best way to carry out this reaction is to generate it in situ by adding HCl to a mixture of aldehydes or ketones and excess sodium or potassium cyanide. Cyanohydrins are useful in organic reaction, because the cyano group can be converted easily to an amine, amide or a carboxylic acid.

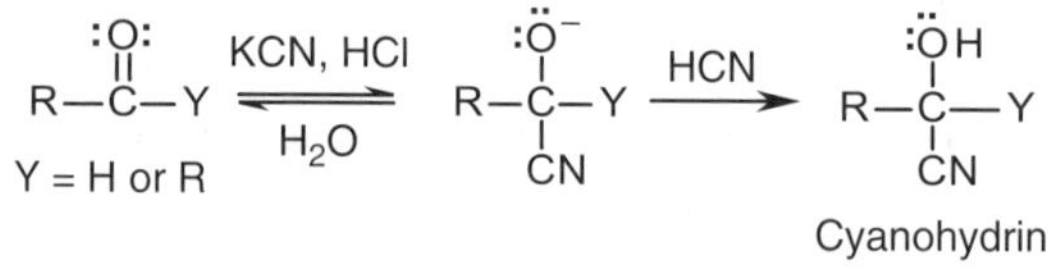

5.4.3.8 Addition of Ammonia and Its Derivatives to Aldehydes and Ketones

5.4.3.8.1 *Preparation of Oximes and Imine Derivatives: Schiff's Bases*

Ammonia and its derivatives, such as primary amine (RNH_2), hydroxylamine (NH_2OH), hydrazine (NH_2NH_2) and semicarbazide ($NH_2NHCONH_2$), react with aldehydes and ketones in the presence of an acid catalyst to generate imines or substituted imines. An imine is a nitrogen analogue of an aldehyde or a ketone with a C=N nitrogen double bond instead of a C=O. Imines are nucleophilic and basic. Imines derived from ammonia do not have a substituent other than a hydrogen atom bonded to the nitrogen. They are relatively unstable to be isolated, but can be reduced in situ to primary amines.

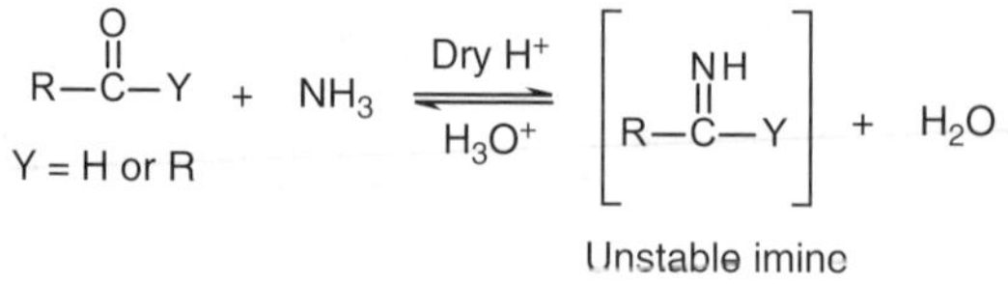

Unstable imine

Imines obtained from hydoxylamines are known as *oximes*, and imines obtained from primary amines are called *Schiff's bases*. They are nitrogen analogues of aldehydes or ketones in which the C=O groups are replaced by C=N—R groups.

Reaction scheme (top):

$$RCH_2-\underset{\overset{\|}{:O:}}{C}-Y \quad (Y = H \text{ or } R)$$

$$\xrightarrow[\text{Dry } H^+]{NH_2\text{-}OH} \quad RCH_2-\underset{\overset{\|}{N\text{-}OH}}{C}-Y \quad \text{Oxime}$$

$$\xrightarrow[\text{Dry } H^+]{R'NH_2} \quad RCH_2-\underset{\overset{\|}{N\text{-}R'}}{C}-Y \quad \text{Imine (Schiff's base)}$$

$$\xrightarrow[\text{Dry } H^+]{R'_2NH} \quad RCH_2-\underset{\overset{\|}{\overset{+}{N}R'_2}}{C}-Y \xrightarrow{-H^+} R'CH=\underset{\overset{|}{NR_2}}{C}-Y$$

Iminium salt → Enamine

The formation of Schiff's bases from aldehydes or ketones generally takes place under anhydrous acid. The reaction is reversible and the formation of all imines (Schiff's base, oxime, hydrazone and semicarbazide) follows the same mechanism. In aqueous acidic solution, imines are easily hydrolysed back to the parent aldehydes or ketones and amines.

Mechanism
The neutral amine nucleophile attacks the carbonyl carbon to form a dipolar tetrahedral intermediate. The intramolecular proton transfer from nitrogen and oxygen yields a neutral carbinolamine tetrahedral intermediate. The hydroxyl group is protonated and the dehydration of the protonated carbinolamine produces an iminium ion and water. Loss of proton to water affords the imine and regenerates the acid catalyst.

Mechanism scheme:

$$R-\underset{\overset{\|}{:O:}}{C}-Y \underset{H_3O^+}{\overset{R'NH_2}{\rightleftharpoons}} R-\underset{\underset{\overset{|}{H}}{\overset{+}{N}HR'}}{\overset{:\overset{..}{O}:^-}{C}}-Y \underset{}{\overset{\pm H^+}{\rightleftharpoons}} R-\underset{:NHR'}{\overset{:\overset{..}{O}H}{C}}-Y \xrightarrow[\text{Dry } H^+]{} R-\underset{:NHR'}{\overset{\overset{+}{:}\overset{..}{O}H_2}{C}}-Y$$

Neutral carbinolamine, A tetrahedral intermediate

$$\downarrow H_3O^+$$

$$H_3\overset{..}{O}^+ + R-\underset{\overset{\|}{NR'}}{C}-Y \rightleftharpoons H_2\overset{..}{O} + R-\underset{\underset{\overset{|}{H-N-R'}}{\overset{+}{}}}{\overset{\|}{C}}-Y$$

5.4.3.8.2 Preparation of Hydrazone and Semicarbazones

Imines produced from hydrazines are known as *hydrazones*, and imines obtained from semicarbazides are called *semicarbazones*.

$$R-\underset{\overset{\|}{N\text{-}NH_2}}{C}-Y \xleftarrow[\text{Dry } H^+]{NH_2NH_2} R-\underset{\overset{\|}{:O:}}{C}-Y \xrightarrow[\text{Dry } H^+]{NH_2NHCONH_2} R-\underset{\overset{\|}{N\text{-}NHCONH_2}}{C}-Y$$

Hydrazone $Y = H$ or R Semicarbazone

Mechanism
The hydrazine nucleophile attacks the carbonyl carbon and forms a dipolar tetrahedral intermediate. Intramolecular protons transfer forms a neutral

tetrahedral intermediate. The hydroxyl group is protonated and the dehydration affords an ionic hydrazone and water. Loss of proton to water produces the hydrazone and regenerates the acid catalyst.

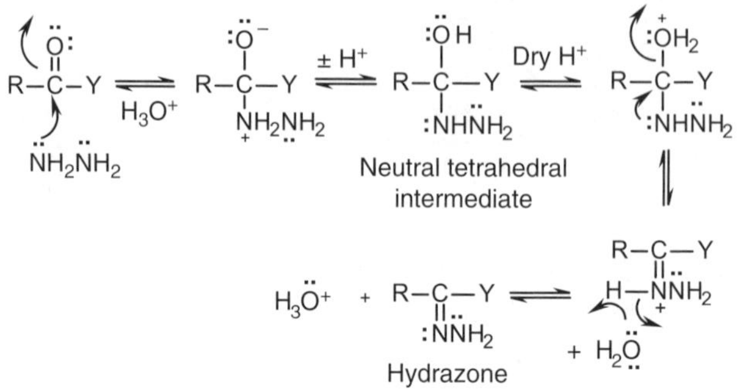

Neutral tetrahedral intermediate

Hydrazone

5.4.3.9 Addition of Secondary Amine to Aldehydes and Ketones

Preparation of Enamines

A secondary amine reacts with aldehydes and ketones to give an enamine. An *enamine* is an α,β-unsaturated tertiary amine. Enamine formation is a reversible reaction and the mechanism is exactly the same as the mechanism for imine formation, except for the last step of the reaction.

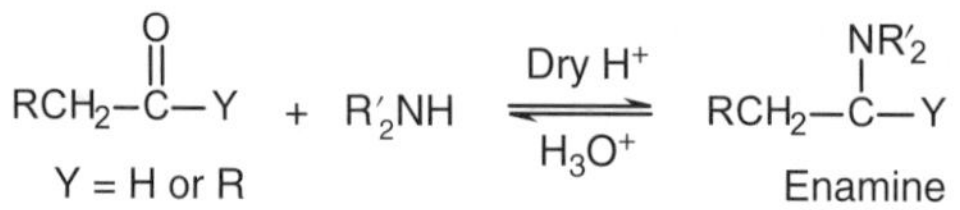

$Y = H$ or R

Enamine

Mechanism

Neutral tetrahedral intermediate

Enamine

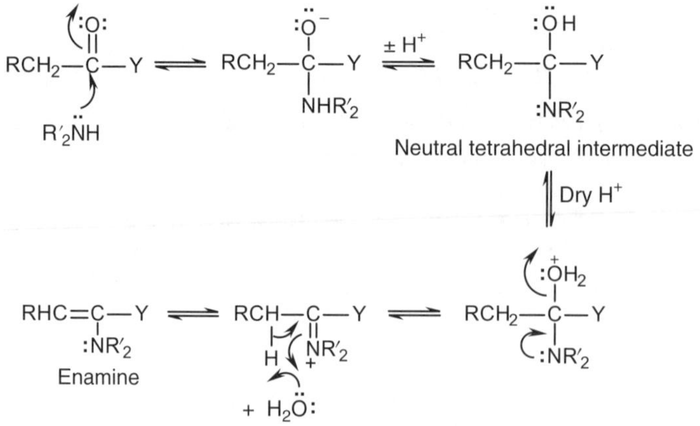

5.4.3.10 Addition of Water to Aldehydes and Ketones: Acid- or Base-Catalysed

Preparation of Diols

Aldehydes and ketones react with water in the presence of aqueous acid or base to form hydrates. The addition reaction is reversible. A *hydrate* is a molecule with two hydroxyl groups on the same carbon. It is also called a *gem*-diol. The addition reaction is highly regioselective. Addition always occurs with oxygen adding to the carbonyl *carbon* atom. Hydration proceeds through the two classic nucleophilic addition mechanisms with water in the acid condition or hydroxide in the basic condition. Hydrates of aldehydes or ketones are generally unstable to isolate. Addition of water to carbonyl compounds under acidic conditions is analogous to addition of water to alkenes (see Section 5.4.2.5).

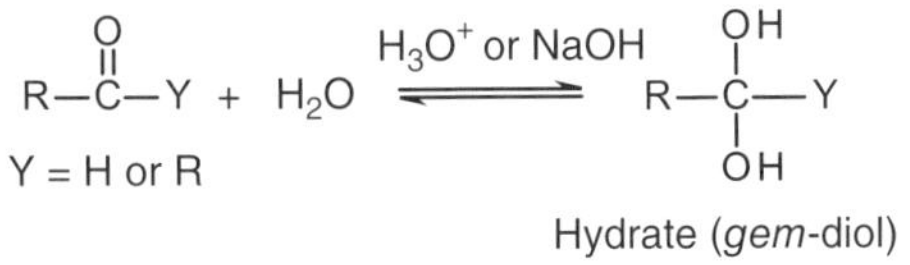

Acid-catalysed mechanism

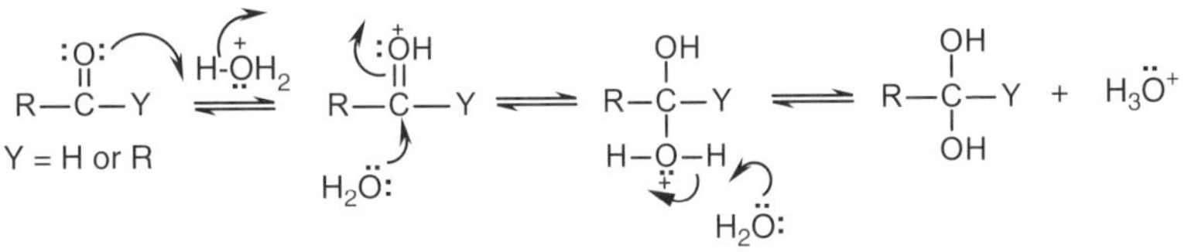

Base-catalysed mechanism

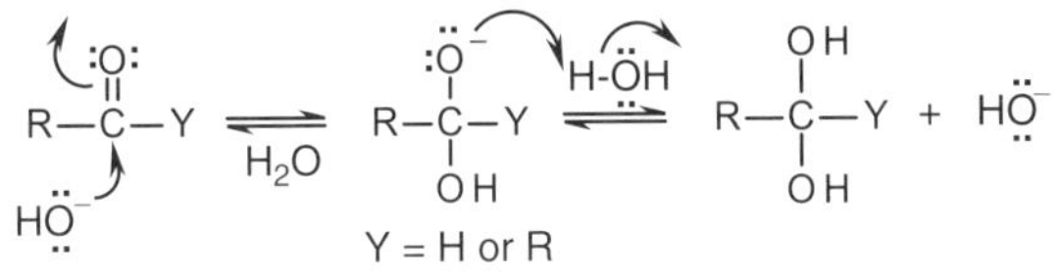

5.4.3.11 Addition of Alcohol to Aldehydes and Ketones

5.4.3.11.1 *Preparation of Acetals and Ketals*

In a similar fashion to the formation of hydrate with water, aldehydes and ketones react with alcohol to form acetal and ketal, respectively. In the formation of an acetal, two molecules of alcohol add to the aldehyde and 1 mol of water is eliminated. An alcohol, like water, is a poor nucleophile. Therefore, the acetal formation only takes place in the presence of anhydrous acid catalyst. Acetal or ketal formation is a reversible reaction and the formation follows the same mechanism. The equilibrium lies towards the formation of acetal when an excess of alcohol is used. In hot aqueous acidic solution, acetals or ketals are hydrolysed back to the carbonyl compounds and alcohols.

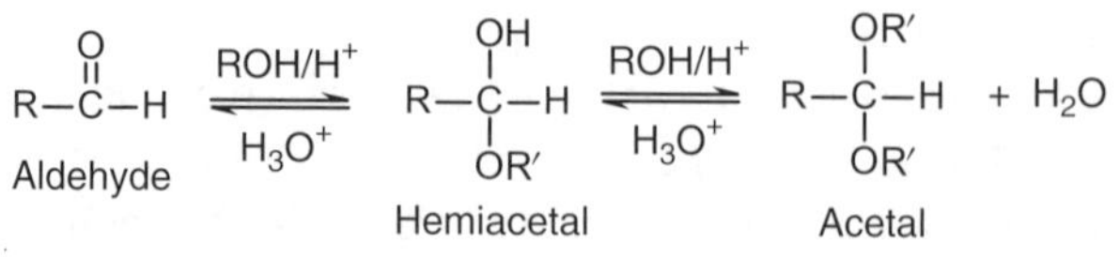

When an alcohol adds to an aldehyde or ketone, hemiacetal is formed. Hemiacetals are unstable and can't be isolated in most cases, and they react further with alcohols under *acidic* conditions to afford acetals.

Acid-catalysed mechanism
The first step is the typical acid-catalysed addition to the carbonyl group. Then the alcohol nucleophile attacks the carbonyl carbon and forms a tetrahedral intermediate. Intramolecular proton transfer from nitrogen and oxygen produces a hemiacetal tetrahedral intermediate. The hydroxyl group is protonated, followed by its leaving as water to form hemiacetal, which reacts further to yield a more stable acetal.

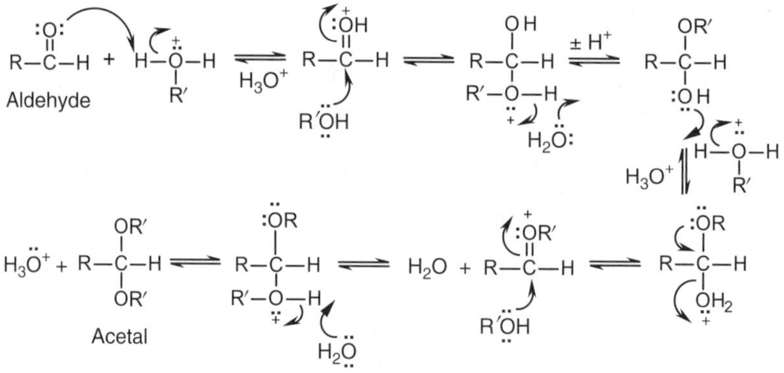

Instead of two molecules of alcohols, a diol is often used to synthesize cyclic acetals. 1,2-Ethanediol (ethylene glycol) is usually the diol of choice and the products are called ethylene acetals.

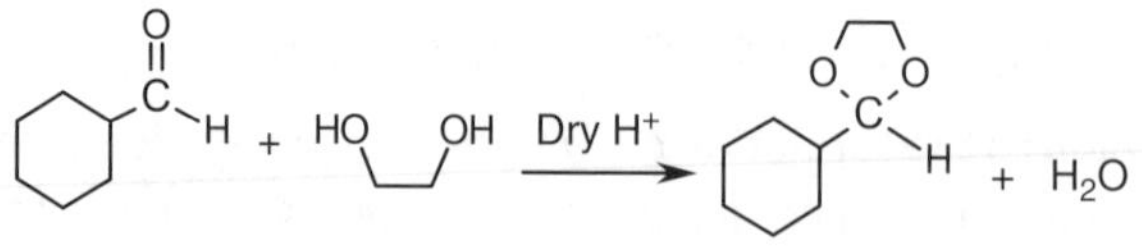

5.4.3.11.2 *Acetal as a Protecting Group*

A protecting group converts a reactive functional group into a different group that is inert to the reaction conditions in which the reaction is carried out. Later, the protecting group is removed. Acetals are hydrolysable under acidic conditions, but are stable to strong bases and nucleophiles. These characteristics make acetals ideal protecting groups for aldehydes and ketones. They are also easily formed from aldehydes and ketones and also easily converted back to the parent carbonyl compounds. They can be used to protect aldehydes and ketones from reacting with strong bases and nucleophiles; for example, Grignard reagents and metal hydrides.

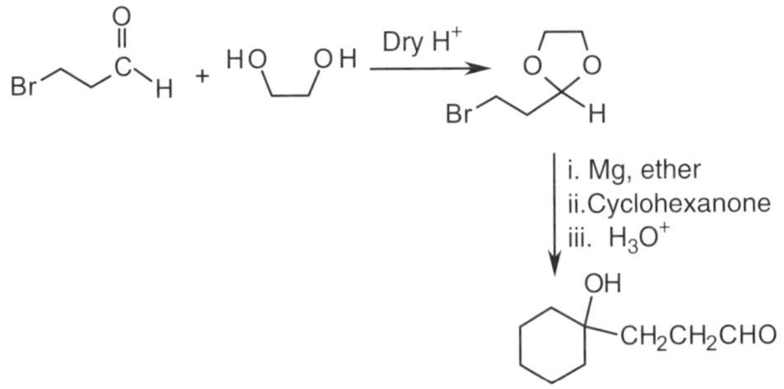

Aldehydes are more reactive than ketones. Therefore, aldehydes react with ethylene glycol to form acetals preferentially over ketones. Thus, aldehydes can be protected selectively. This is a useful way to perform reactions on ketone functionalities in molecules that contain both aldehyde and ketone groups.

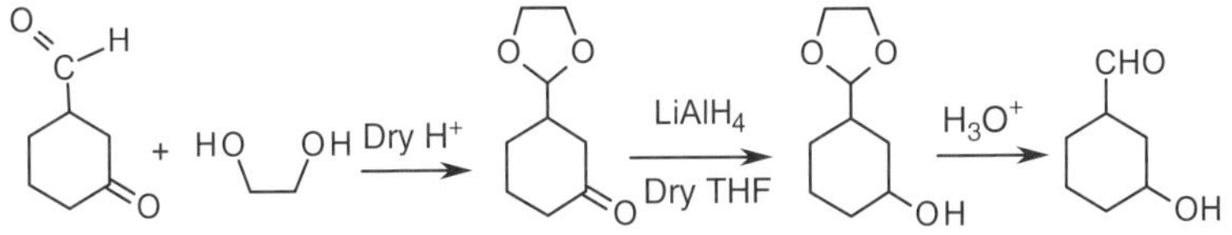

5.4.3.12 Addition of Enolates to Aldehydes and Ketones

The chemistry of enolates is widely exploited for the generation of carbon–carbon bonds in various reaction to produce a large molecule via nucleophilic addition reaction. There are several types of reactions involving enolates, such as aldol addition, aldol condensation and mixed-aldol condensation (Claisen–Schmidt condensation).

5.4.3.12.1 Aldol Addition: Base-Catalysed
Preparation of β-hydroxyaldehydes or β-hydroxyketones
Generally, aldol addition occurs between two same carbonyl compounds that have an acidic proton (α-H is on the carbon adjacent to the C=O group) to form a large molecule. Thus, the aldol reaction requires an aldehyde or ketone that contains at least one α-hydrogen, since the α-hydrogen is required in order to form enolate anions.

$$\alpha \quad RCH_2-\overset{\overset{\displaystyle O}{\|}}{C}-H$$

Aldehyde with two α-Hs

$$\alpha \quad RCH_2-\overset{\overset{\displaystyle O}{\|}}{C}-R'$$

Ketone with two α-Hs

In an aldol addition reaction, the enolate anion of one carbonyl compound reacts as a nucleophile, and attacks the electrophilic carbonyl group of another carbonyl compound to form a larger molecule. The aldol addition reaction may be either acid-catalysed or base-catalysed. However, base catalysis is more common and the aldol audition is reversible and more favourable for aldehydes than ketones.

The α-hydrogen (bonded to a carbon adjacent to a carbonyl carbon, known as α-carbon) is acidic enough to be removed by a strong base, such as aqueous NaOH, to form an enolate anion, which is highly nucleophilic. The aldol addition is reversible and more favourable for aldehydes than ketones. For example, two acetaldehyde molecules condense together in the presence of aqueous NaOH and at low temperature (5 °C), to yield 3-hydroxybutanal. In a similar fashion, two acetone molecules are condensed together to afford 4-hydroxy-4-methyl-2-pentanone.

Y = H; acetaldehyde
Y = CH₃; acetone

Y = H; 3-hydroxybutanal
Y = CH₃; 4-hydroxy-4-methyl-2-pentanone

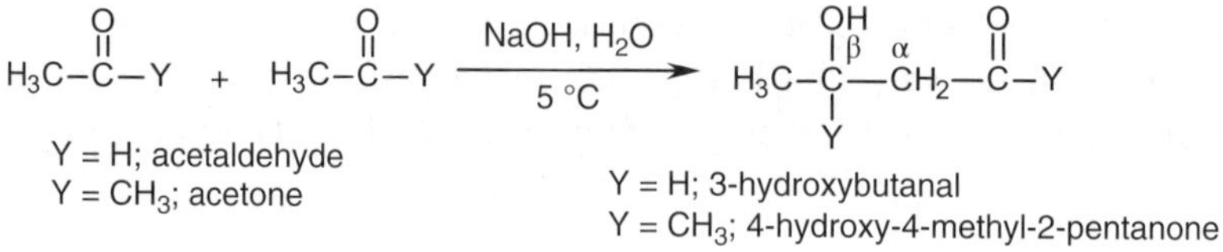

Base-catalysed mechanism

Removal of an acidic α-hydrogen from one acetaldehyde or acetone by a strong base (HO⁻) and a low temperature gives an enolate anion, which adds to another molecule of acetaldehyde or acetone to form an alkoxide intermediate. The resulting alkoxide is protonated by H_2O and yields either a β-hydroxyaldehyde or a β-hydroxyketone and regenerates the base (HO⁻). The overall process demonstrates that how carbonyl compound can react as both as an electrophile and a nucleophile.

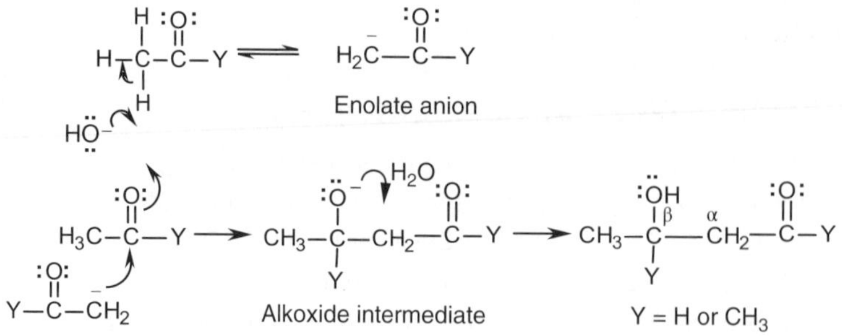

Enolate anion

Alkoxide intermediate

Y = H or CH₃

5.4.3.12.2 Aldol Condensation: Base-Catalysed

Preparation of α,β-Unsaturated Aldehydes or Ketones

An aldol addition product can undergo dehydration (loss of water) to afford conjugate aldehyde or ketone in the presence of a highly concentrated base and at a higher temperature. An aldol condensation is a self-condensation reaction of a β-hydroxyaldehyde or β-hydroxyketone under a severe basic condition and at a higher temperature. For example, two acetaldehyde or acetone molecules are dehydrated in the presence of highly concentrated NaOH (1 M) and at a higher temperature (80 °C) to provide 2-butenal or 4-methyl-3-penten-2-one, respectively.

Y = H; acetaldehyde   Y = H; 3-hydroxybutanal   Y = H; 2-butenal
Y = CH$_3$; acetone   Y = CH$_3$; 4-hydroxy-4-methyl-   Y = CH$_3$; 4-methyl-3-penten-2-one
                   2-pentanone

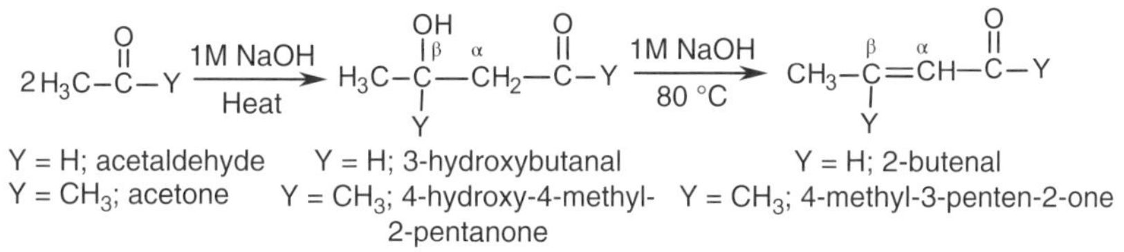

Base-catalysed mechanism

Removal of an acidic α-hydrogen from one acetaldehyde or acetone by a highly concentrated strong base (HO⁻) and at high temperature forms an enolate anion that then adds to another molecule of acetaldehyde or acetone to form an alkoxide intermediate. The resulting alkoxide is protonated by H$_2$O to yield aldol products either β-hydroxyaldehyde or β-hydroxyketone and regenerates the base (HO⁻).

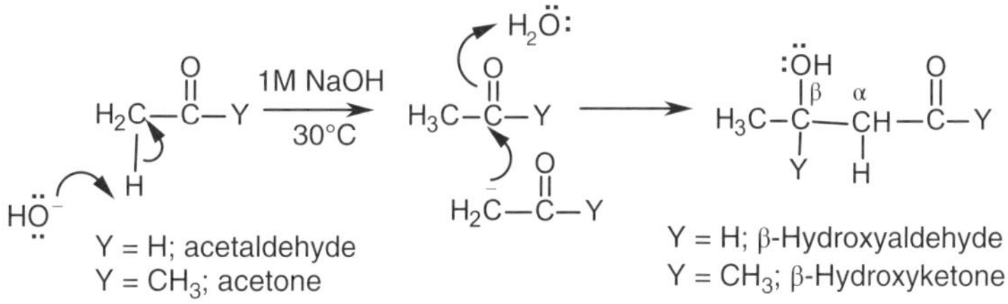

Y = H; acetaldehyde   Y = H; β-Hydroxyaldehyde
Y = CH$_3$; acetone   Y = CH$_3$; β-Hydroxyketone

Due to the highly concentrated base and higher temperature, the base (HO⁻) deprotonates another acidic α-hydrogen from the aldol products and produces a new enolate anion, which then activates the elimination of the water molecule and affords 2-butenal or 4-methyl-3-penten-2-one, respectively.

Y = H; 2-butenal
Y = CH$_3$; 4-methyl-3-penten-2-one

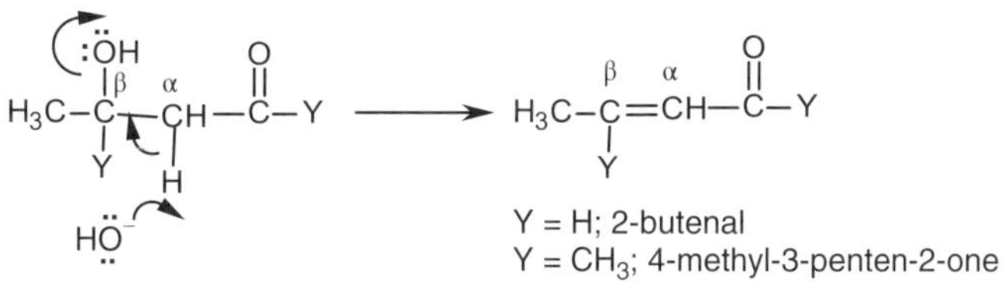

5.4.3.12.3 Mixed-Aldol Condensation: Base-Catalysed Preparation of α,β-Unsaturated Ketones

Mixed-aldol condensation is a variation of aldol condensation. It is also possible to achieve mixed-aldol condensation reactions with two different molecules. This reaction is known as the *Claisen–Schmidt condensation* (see Section 4.4.1.5). Generally, mixed-aldol condensation or Claisen–Schmidt condensation occurs between two different carbonyl compounds in the presence of a highly concentrated strong base and at a higher temperature. Since ketones are less reactive towards nucleophilic addition, the enolate formed from a ketone can be used to react with a highly electrophilic aldehyde. For example, formaldehyde reacts with acetone in the presence of a highly concentrated base (1 M NaOH) and at higher temperature to give methyl vinyl ketone (but-3-en-2-one).

Formaldehyde with no α-H   Acetone with three α-Hs

Methyl vinyl ketone
But-3-en-2-one

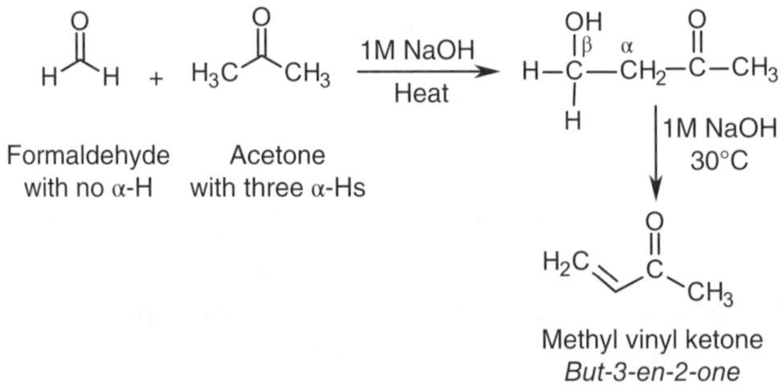

Similarly, benzaldehyde reacts with acetone in the presence of a highly concentrated base (1 M NaOH) and at a higher temperature to afford benzalacetone (4-pheny-3-buten-2-one). The reaction mechanism is similar to the aldol condensation.

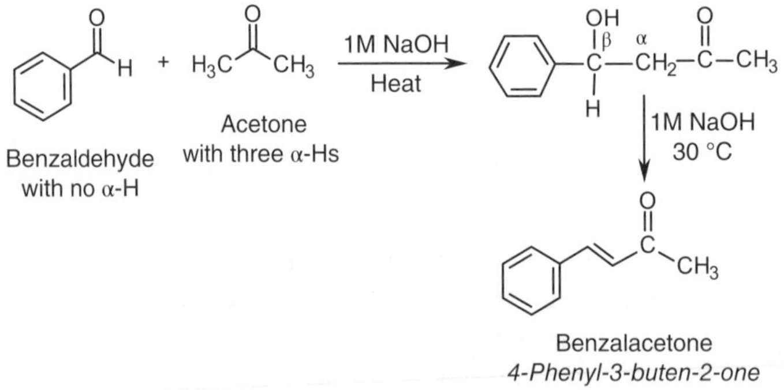

Benzaldehyde with no α-H   Acetone with three α-Hs

Benzalacetone
4-Phenyl-3-buten-2-one

5.5 ELIMINATION REACTIONS: 1,2-ELIMINATION OR β-ELIMINATION

The term *elimination* can be defined as the electronegative atom or leaving group being removed along with a hydrogen atom from adjacent carbons in the presence of strong acids or strong bases and high temperatures forming a multiple bond. Simply, an *elimination reaction* is the removal of a pair of atoms or groups of atoms from a molecule. These reactions are just the reverse of addition reactions (see Section 5.4). Elimination reactions transform organic compounds containing only single carbon–carbon bonds (saturated compounds) to compounds containing double or triple carbon–carbon bonds (unsaturated compounds).

Elimination reactions are generally named by the kind of atoms or groups leaving the molecule; for example, *dehydrohalogenation* where a hydrogen atom and a halogen atom are removed, *dehydration* involves the elimination of a water molecule (usually from an alcohol) and *dehydrogenation* where both removed atoms are hydrogen.

Alkenes can be prepared from alcohols or alkyl halides by elimination reactions. The two most important methods for the preparation of alkenes are dehydration ($-H_2O$) of alcohols, and dehydrohalogenation ($-HX$) of alkyl halides. These reactions are reverse of the *electrophilic addition* of water and hydrogen halides to alkenes (see Section 5.4.2).

$$\beta \text{ Carbon} \quad \alpha \text{ Carbon}$$

$$CH_3CH_2OH \xrightarrow[\text{Heat}]{\text{Conc. } H_2SO_4} CH_2\!=\!CH_2 + H_2O$$

Ethyl alcohol   Ethylene

$$CH_3CH_2Cl \xrightarrow[\text{Heat}]{\text{Alcoholic KOH}} CH_2\!=\!CH_2 + HCl$$

Ethyl chloride   Ethylene

In 1,2-elimination, for example dehydrohalogenation of alkyl halide, the atoms are removed from adjacent carbons. This is also called *β-elimination* because a proton is removed from a β-carbon. The carbon to which the functional group is attached is called the α-carbon. A carbon adjacent to the α-carbon is called a β-carbon.

Depending on the relative timing of the bond breaking and bond formation, different pathways are possible: *E1 reaction* or unimolecular elimination and *E2 reaction* or bimolecular elimination.

$$H_3C\!-\!\underset{\underset{CH_3}{|}}{\overset{\overset{H}{|}}{C}}\!-\!\underset{\underset{CH_3}{|}}{\overset{\overset{X}{|}}{C}}\!-\!CH_3 + B\!:\ \xrightarrow{\text{Heat}}\ \underset{H_3C}{\overset{H_3C}{>}}C\!=\!C\overset{CH_3}{\underset{CH_3}{<}} + B\text{-}H + X\!:\!-$$

Alkyl halide   Base   Alkene

5.5.1 E1 Reaction or First Order Elimination

E1 reaction or *first order elimination* results from the loss of a leaving group to form a carbocation intermediate, followed by the removal of a proton to generate the C=C bond. This reaction is most common with good leaving groups, stable carbocations and weak bases (strong acids). For example, 3-bromo-3-methyl pentane reacts with methanol to give 3-methyl-2-pentene. This reaction is unimolecular, that is, the rate-determining step involves one molecule, and it has the slow ionization to produce a carbocation. The second step is the fast removal of a proton by the base (solvent) to form the C=C bond. In fact, any base in the reaction mixture (ROH, H_2O, HSO_4^-) can remove the proton in an elimination reaction. The E1 is not particularly useful from a synthetic point of view, and occurs in competition with S_N1 reaction of tertiary alkyl halides. Primary and secondary alkyl halides do not usually react with this mechanism.

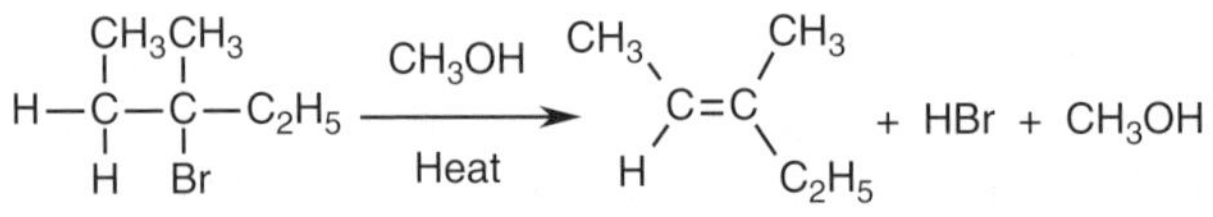

3-Bromo-3-methyl pentane 3-Methyl-2-pentene

Mechanism

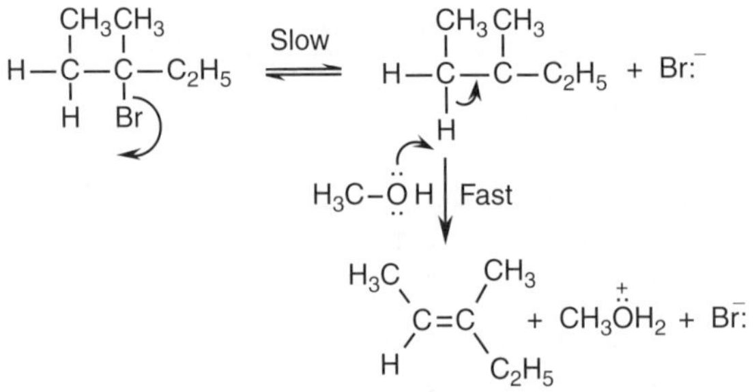

3-Methyl-2-pentene

5.5.2 E2 Reaction or Second Order Elimination

E2 elimination or *second order elimination* takes place through the removal of a proton and simultaneous loss of a leaving group to form the C=C bond. This reaction is most common with a high concentration of strong bases (weak acids), poor leaving groups and less stable carbocations. For example, 3-chloro-3-methyl pentane reacts with sodium methoxide to afford 3-methyl-2-pentene. The bromide and the proton are lost simultaneously to form the alkene. The E2 reaction is the most effective for the synthesis of alkenes from primary alkyl halides.

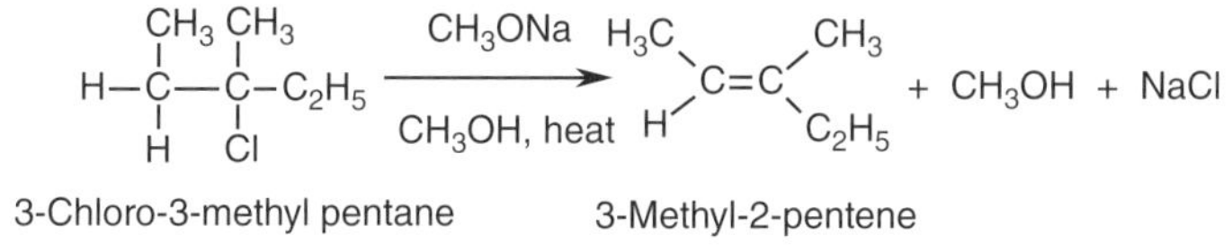

3-Chloro-3-methyl pentane → 3-Methyl-2-pentene

Mechanism

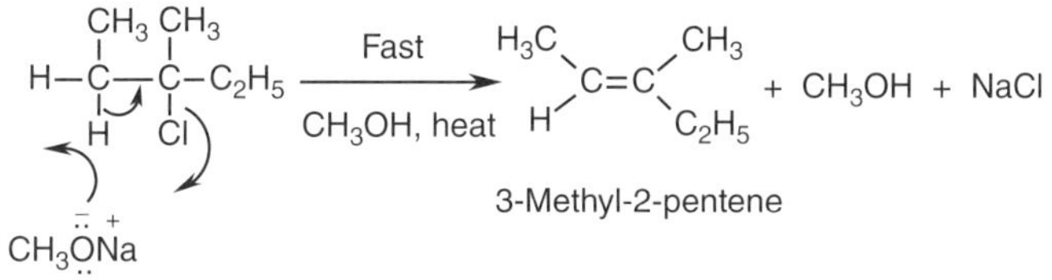

3-Methyl-2-pentene

5.5.3 Dehydration of Alcohols

Preparation of Alkenes

The dehydration of alcohols is a useful synthetic route to alkenes. As stated earlier, a dehydration reaction is the process of removing a proton from one carbon and a hydroxyl group from an adjacent atom to form a water molecule and a C—C double bond. Alcohols typically undergo elimination reactions when heated with strong acid catalysts; for example, H_2SO_4 or phosphoric acid (H_3PO_4), to generate an alkene and water. The hydroxyl group is not a good leaving group, but under acidic conditions it can be protonated. The ionization generates a molecule of water and a cation, which then easily deprotonates to provide an alkene. For example, the dehydration of 2-butanol predominantly provides (*E*)-2-butene. The reaction is reversible and the following equilibrium does exist.

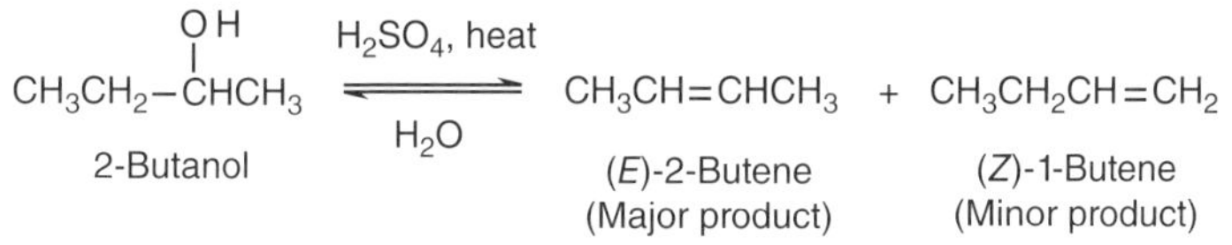

2-Butanol

(*E*)-2-Butene (Major product)

(*Z*)-1-Butene (Minor product)

Mechanism

2-Butanol

(*E*)-2-Butene (Major product)

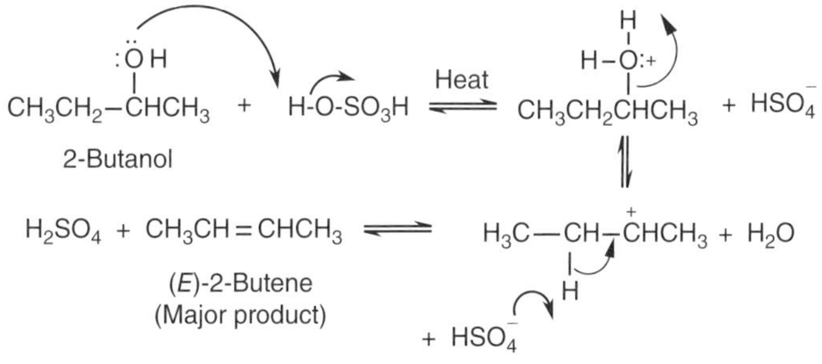

Similarly, the dehydration of 2,3 dimethylbut-2-ol yields predominantly 2,3-dimethylbutene via an E1 reaction.

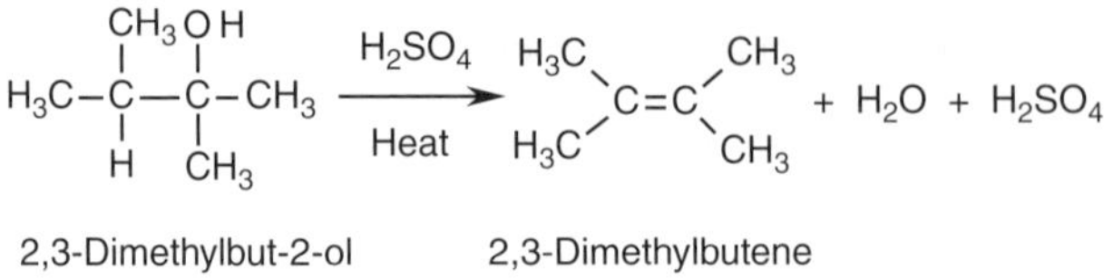

2,3-Dimethylbut-2-ol   2,3-Dimethylbutene

Mechanism

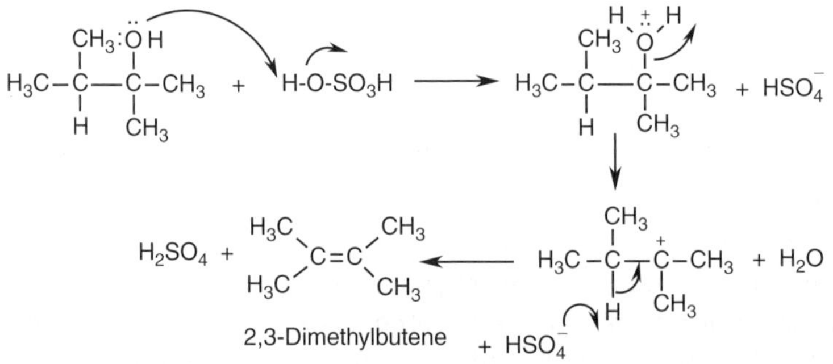

2,3-Dimethylbutene   + HSO$_4^-$

While dehydration of 2° and 3° alcohols is an E1 reaction, dehydration of 1° alcohols is an E2 reaction. Dehydration of 2° and 3° alcohols involves the formation of a carbocation intermediate, but formation of a primary carbocation is rather difficult and unstable. For example, dehydration of propanol produces propene via E2.

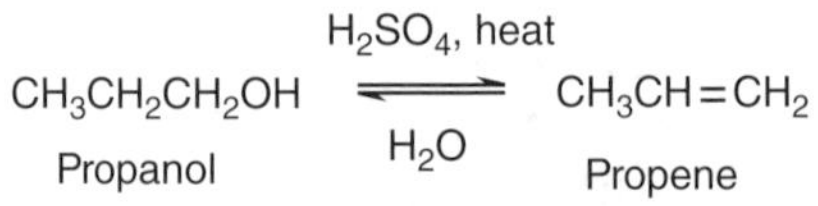

$$CH_3CH_2CH_2OH \underset{H_2O}{\overset{H_2SO_4,\ heat}{\rightleftharpoons}} CH_3CH=CH_2$$

Propanol   Propene

Mechanism

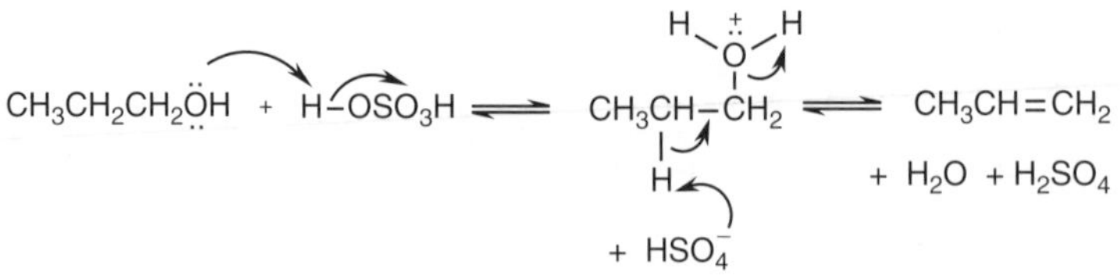

An E2 reaction occurs in one step: first, the acid protonates the oxygen of the alcohol, a proton is removed by a base (HSO$_4^-$) and simultaneously C—C double bond is formed via the departure of the water molecule.

258   Chemistry for Pharmacy Students

Use of concentrated acid and high temperature favours alkene formation, but use of dilute aqueous acid favours alcohol formation. To prevent the alcohol formation, alkene can be removed by distillation as it is formed because it has much lower boiling point than the alcohol. When two elimination products are produced, the major product is generally the more substituted alkene.

5.5.4 Dehydration of Diols: Pinacol Rearrangement

Preparation of Pinacolone

Pinacol rearrangement is a dehydration of 1,2-diol to form a ketone. 2,3-Dimethyl-2,3-butanediol has the common name *pinacol* (a symmetrical diol). When it is treated with strong acid, for example H_2SO_4, it provides 3,3-dimethyl-2-butanone (methyl *tert*-butyl ketone), also commonly known as *pinacolone*. The product results from the loss of water and molecular rearrangement. In the rearrangement of pinacol, equivalent carbocations are formed no matter which hydroxyl group is protonated and leaves.

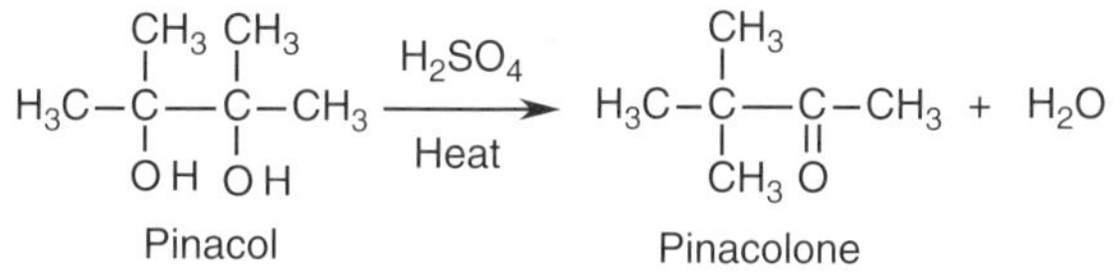

Mechanism
The protonation of OH, followed by the loss of H_2O from the protonated diol yields a tertiary carbocation, which rearranges with a 1,2-methyl shift to form a protonated pinacolone. The rearranged product is deprotonated by the base to give pinacolone.

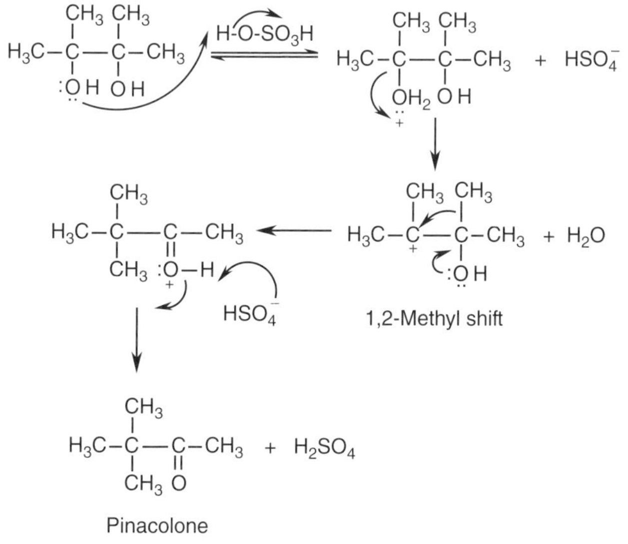

5.5.5 Base-Catalysed Dehydrohalogenation of Alkyl Halides

Alkyl halides typically undergo elimination reactions when heated with strong bases, typically hydroxides and alkoxides, to generate alkenes. Removal of a proton and a halide ion is called *dehydrohalogenation*. Any base in the reaction mixture (H_2O, HSO_4^-) can remove the proton in the elimination reaction.

5.5.5.1 E1 elimination of Hydrogen Halides

Preparation of Alkenes

The E1 reaction involves the formation of a planar carbocation intermediate. Therefore, both *syn*-elimination and *anti*-elimination can occur. If an elimination reaction removes two substituents from the same side of the C—C bond, the reaction is called a *syn*-elimination. When the substituents are removed from opposite sides of the C—C bond, the reaction is called an *anti*-elimination. Thus, depending on the substrates E1 reaction forms a mixture of *cis* (*Z*) and *trans* (*E*) products. For example, *tert*-butyl bromide (3° alkyl halide) reacts with water to form 2-methylpropene following an E1 mechanism. The reaction requires a good ionizing solvent and a weak base. When the carbocation is formed, S_N1 and E1 processes compete with each other, and often mixtures of elimination and substitution products occur. The reaction of *tert*-butyl bromide and ethanol affords major product via E1 and minor product via S_N1.

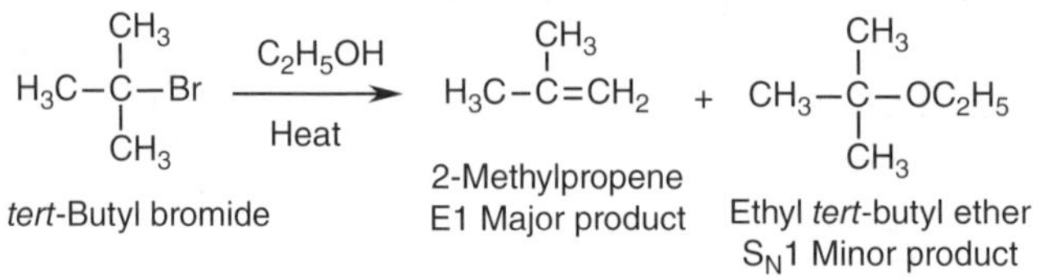

tert-Butyl bromide

2-Methylpropene
E1 Major product

Ethyl *tert*-butyl ether
S_N1 Minor product

Mechanism

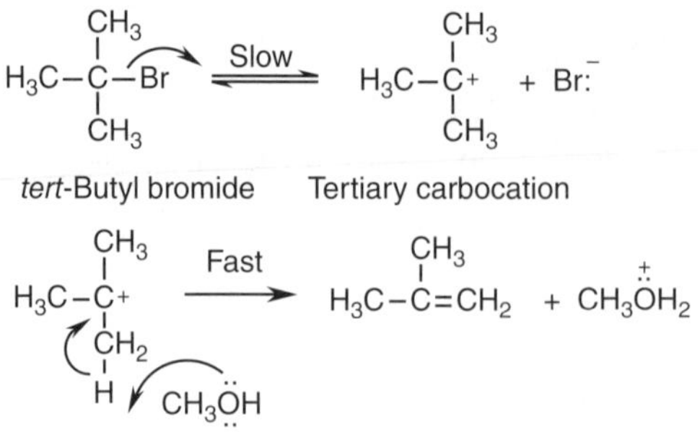

tert-Butyl bromide   Tertiary carbocation

5.5.5.2 E2 Elimination of Hydrogen Halides

Preparation of Alkenes

Dehydrohalogenation of 2° and 3° alkyl halides undergo both E1 and E2 reactions. However, 1° halides undergo only E2 reactions. They cannot undergo E1 reaction because of the difficulty of forming primary cabocations. E2 elimination is stereo-specific and it requires an *anti-periplanar* (180°) arrangement of the groups being eliminated. Since only *anti*-elimination can take place, E2 reaction predominately forms one product.

The elimination reaction may proceed to alkenes that are constitutional iso-mers with one formed in excess of the other, described as *regioselectivity*. Simi-larly, eliminations often favour the more stable *trans*-product over the *cis*-product, described as *stereoselectivity*. For example, bromopropane reacts with sodium eth-oxide (EtONa) to produce only propene.

$$CH_3CH_2CH_2-Br \xrightarrow[\text{EtOH, heat}]{C_2H_5ONa} CH_3CH=CH_2 + CH_3CH_2OH + NaBr$$

Mechanism

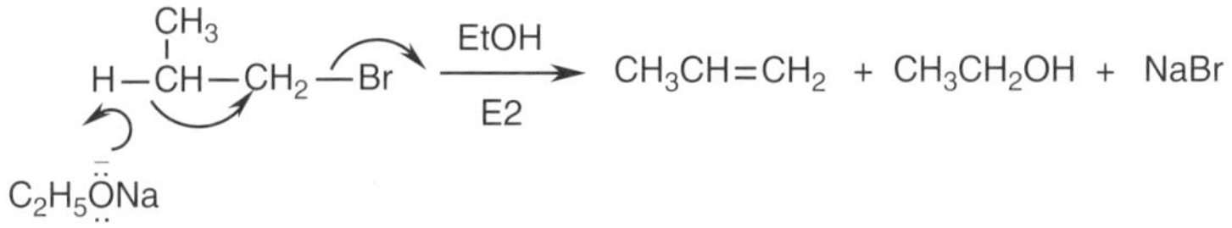

The E2 elimination can be an excellent synthetic method for the preparation of an alkene when a 3° alkyl halide and a strong base, for example alcoholic KOH, is used. This method is not suitable for a S_N2 reaction.

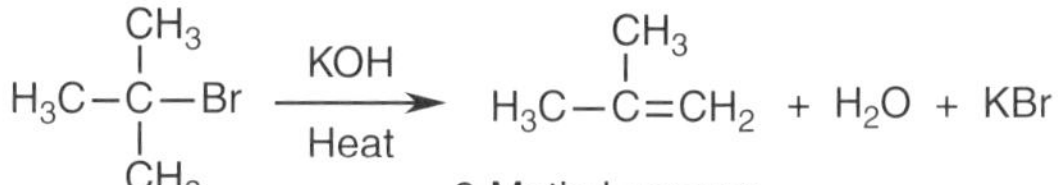

tert-Butyl bromide

2-Methylpropene
(>90%)

A bulky base (a good base, but poor nucleophile) can further discourage unde-sired substitution reactions. Most common bulky bases are potassium-*tert*-butox-ide (*t*-BuOK), diisopropylamine and 2,6-dimethylpyridine.

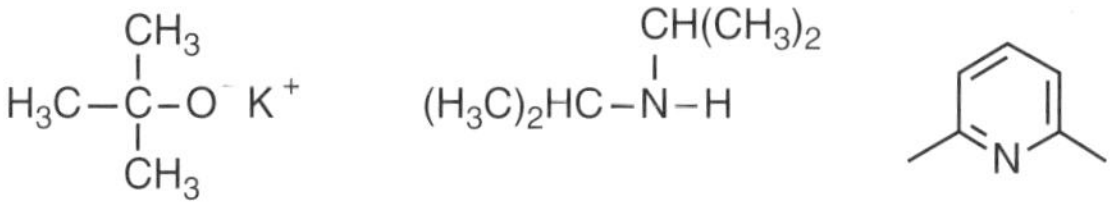

Potassium *tert*-butoxide   Diisopropylamine   2,6-Dimethylpyridine

Cyclohexene can be synthesized from bromocyclohexane in a high yield using diisopropylamine.

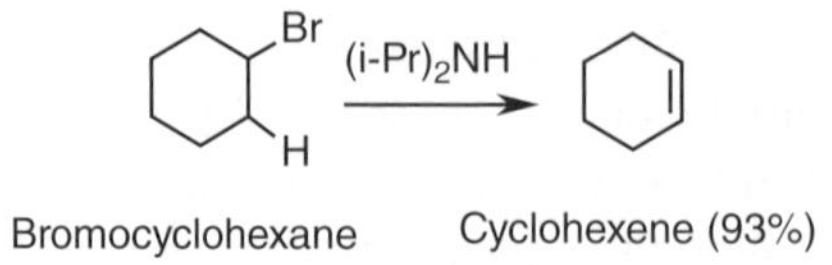

Bromocyclohexane   Cyclohexene (93%)

Generally, E2 reactions occur with a strong base, which eliminates a proton quicker than the substrate can ionize. Normally, the S_N2 reaction does not compete with E2 since there is steric hindrance around the C—X bond, which retards the S_N2 process.

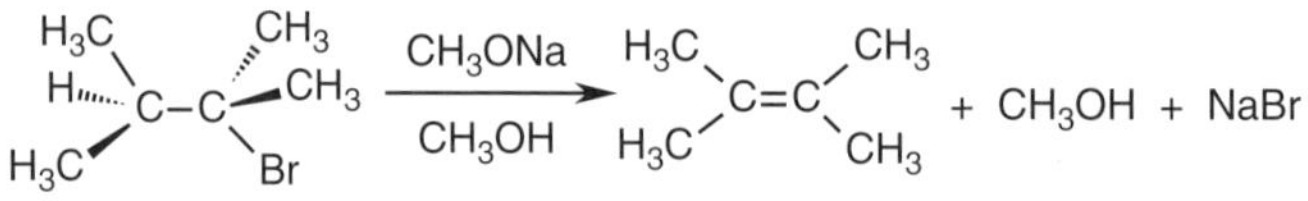

Mechanism

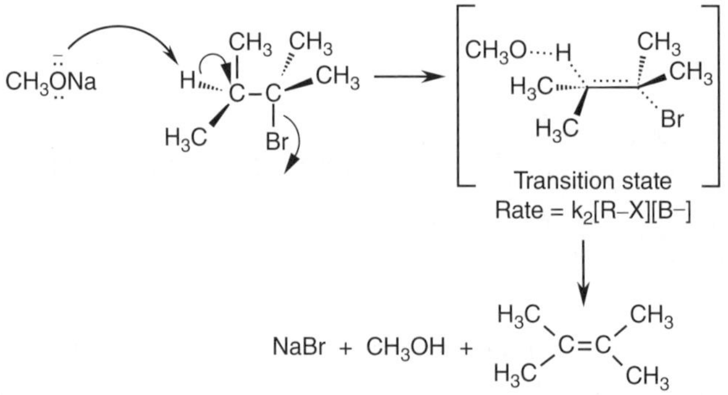

Transition state
Rate = k_2[R–X][B–]

NaBr + CH$_3$OH +

The methoxide (CH$_3$O⁻) acts as a base rather than a nucleophile. The reaction occurs in one concerted step, with the C—H and C—Br bonds breaking as the CH$_3$O—H and C=C bonds are forming. The rate is related to the concentrations of the substrate and the base, giving a second order rate equation. The elimination requires a hydrogen atom adjacent to the leaving group. If there are two or more possibilities of adjacent hydrogen atoms, mixtures of products are formed as shown in the following example.

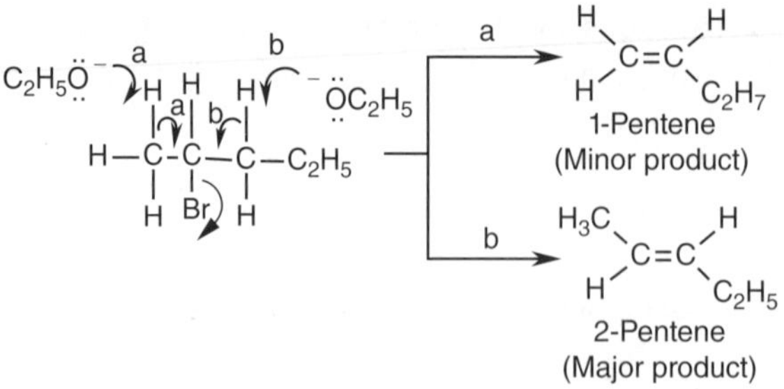

1-Pentene
(Minor product)

2-Pentene
(Major product)

The major product of elimination is the one with the most highly substituted double bond and follows the following order.

$$R_2C = CR_2 > R_2C = CRH > RHC = CHR \text{ and } R_2C = CH_2 > RCH = CH_2$$

5.5.5.3 Stereochemical Considerations in the E2 Reactions

The E2 follows a concerted mechanism where removal of a proton and formation of a double bond occur at the same time. The partial π bond in the transition state requires the parallel alignment or coplanar arrangement of the p orbitals. When the hydrogen and leaving group eclipse each other (0°), this is known as the *syn-coplanar* conformation.

syn-Coplanar (0°)   syn-Elimination   anti-Coplanar (180°)   anti-Elimination

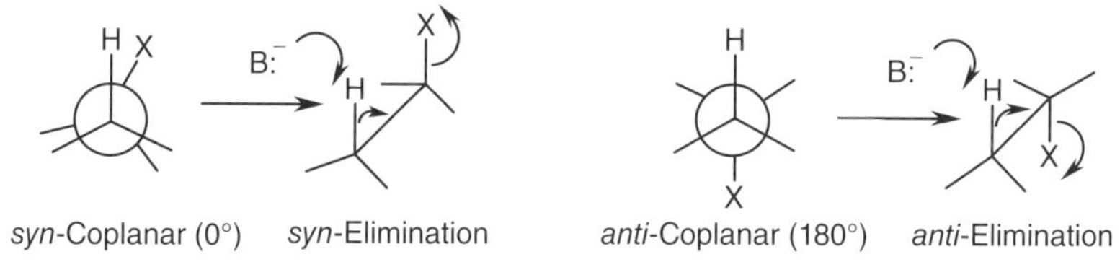

When the leaving group and hydrogen atom are *anti* to each other (180°), this is called the *anti*-coplanar conformation. The *anti*-coplanar conformation is of lower energy and is by far the most common. In the *anti*-coplanar conformation, the base and leaving group are well separated, thus removing electron repulsions. The *syn*-coplanar conformation requires the base to approach much closer to the leaving group, which is energetically unfavourable.

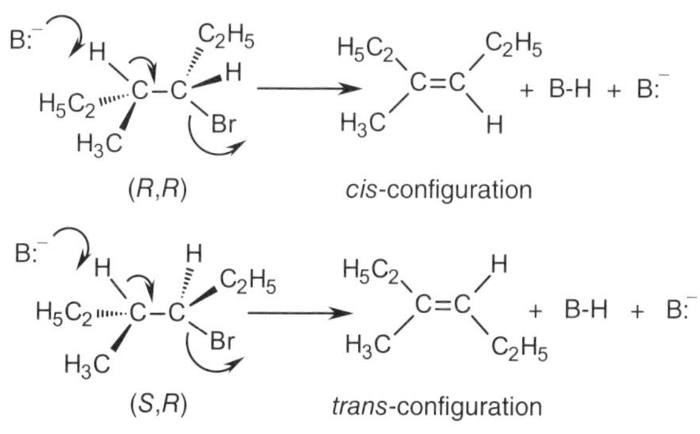

(R,R)   cis-configuration

(S,R)   trans-configuration

The E2 reaction is a stereospecific reaction; that is, a particular stereoisomer reacts to form one specific stereoisomer. It is stereospecific, since it prefers the *anti*-coplanar transition state for elimination. Therefore, the (R,R) diastereomer offers a *cis*-alkene and the (S,R) diastereomer gives a *trans*-alkene.

5.5.5.4 E2 Elimination of Hydrogen Halides in Cyclohexane System

Almost all cyclohexane systems are most stable in the chair conformations. In a chair, adjacent axial positions are in an *anti*-coplanar arrangement, ideal for E2 eliminations. Adjacent axial positions are said to be in a *trans*-diaxial arrangement. E2 reactions only proceed in chair conformations from *trans*-diaxial positions and chair–chair interconversions allow the hydrogen and the leaving group to attain the *trans*-diaxial arrangement. The elimination of HBr from bromocyclohexane produces cyclohexene. The bromine must be in an axial position before it can leave.

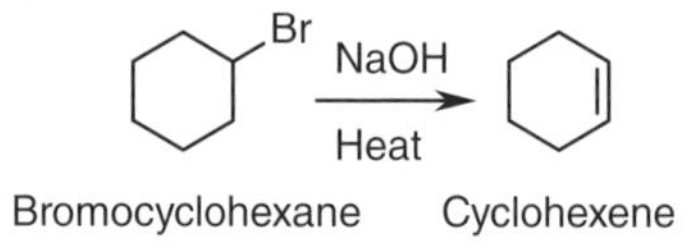

Bromocyclohexane   Cyclohexene

Mechanism

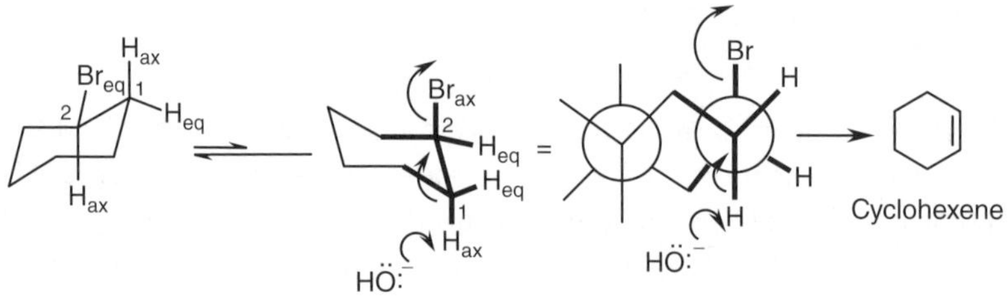

Cyclohexene

5.5.5.5 E2 Elimination of Halogens
5.5.5.5.1 Preparation of Alkenes

Dehalogenation of alkyl dihalides (*vicinal* dihalides) with NaI in acetone results in the formation of an alkene via E2 reactions.

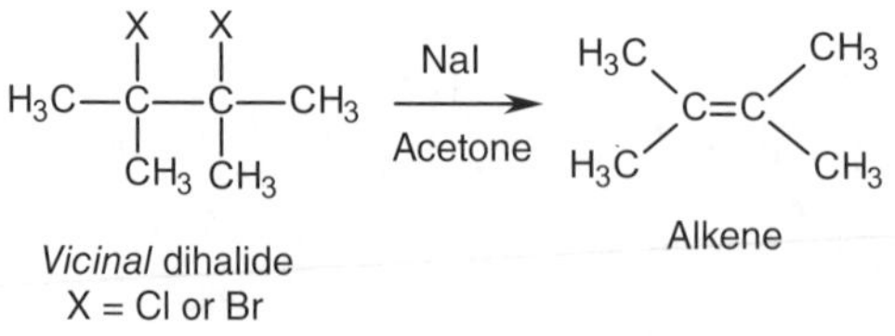

Vicinal dihalide
X = Cl or Br

Alkene

5.5.5.5.2 Preparation of Alkynes

Alkynes can be synthesized by elimination of 2 mol of HX from a *geminal* (halides on the same carbon) or *vicinal* (halides on the adjacent carbons) dihalide at high temperatures. Stronger bases (KOH or NaNH$_2$) are used for the formation of alkyne via two consecutive E2 dehydrohalogenation steps. Under mild conditions, dehydrohalogenation stops at the vinylic halide stage. For example, 2-butyne is obtained from *geminal-*or *vicinal*-dibromobutane.

geminal- or vicinal-Dibromobutane

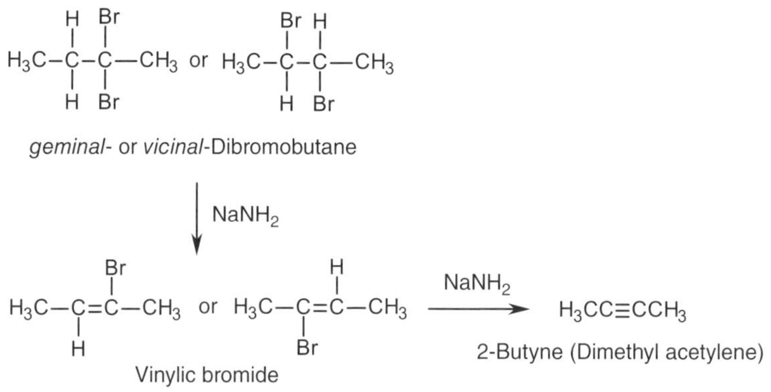

5.5.5.6 E1 Versus E2 Mechanism

Some key features of the E1 and E2 mechanisms are compared and contrasted in the following table.

Criteria	E1	E2
Substrate	Tertiary > secondary > primary	Primary > secondary > tertiary
Rate of reaction	Depends only on the substrate	Depends on both sub- strate and base
Carbocation	More stable carbocation	Less stable carbocation
Rearrangement	Rearrangements are common	No rearrangements
Geometry	No special geometry required	*Anti*-coplanarity required
Leaving group	Good leaving group	Poor leaving group
Base strength	Weak base	Strong and more concen- trated base

5.6 SUBSTITUTION REACTIONS

The word *substitution* implies the replacement of one atom or group by another. For example, the reaction, in which the chlorine atom in the chloromethane molecule is displaced by the hydroxide ion forming methanol, is a substitution reaction. There are two types of substitution reactions: *nucleophilic substitution* and *electrophilic substitution*. The mechanisms of these two substitution reactions are completely different.

Nucleophilic substitution reactions mainly occur with alkyl halides, alcohols, ethers and epoxides (see Sections 5.6.2–5.6.4). However, it can also take place with carboxylic acid derivatives, and is known as *nucleophilic acyl substitution* (see Section 5.6.5).

Electrophilic substitution reactions are those where an electrophile displaces another group, usually a hydrogen atom. Electrophilic substitution reactions only take place in aromatic compounds (see Section 5.7.1).

5.6.1 Nucleophilic Substitutions

Alkyl halides (RX) are good substrates for substitution reactions. The nucleophile (Nu:⁻) displaces the leaving group (X:⁻) from the carbon atom by using its electron pair or lone pair to form a new σ bond to the carbon atom. Two different mechanisms for nucleophilic substitution are S_N1 and S_N2 mechanisms. In fact, the preference between S_N1 and S_N2 mechanisms depends on the structure of the alkyl halide, the reactivity and structure of the nucleophile, the concentration of the nucleophile and the solvent in which the reaction is carried out.

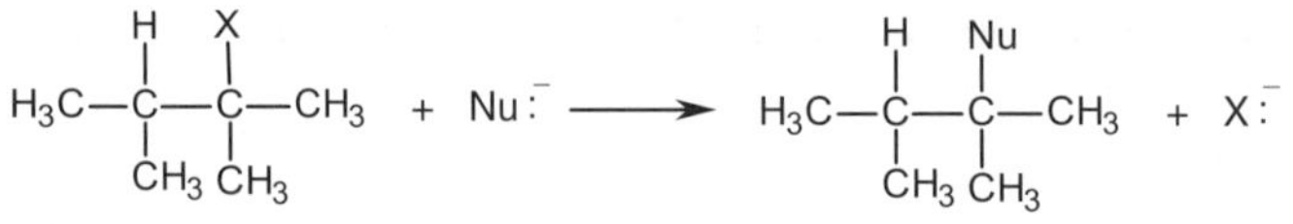

5.6.1.1 First Order Nucleophilic Substitution: S_N1 Reactions

S_N1 reaction means a unimolecular nucleophilic substitution. The S_N1 reaction occurs in two steps, with the first being a slow ionization reaction generating a carbocation. Thus, the rate of an S_N1 reaction depends only on the concentration of the alkyl halide. First, the C—X bond breaks without any help from nucleophile, and then the quick nucleophilic attack by the nucleophile on the carbocation. When water or alcohol is the nucleophile, a quick loss of a proton by the solvent forms the final product. For example, the reaction of *tert*-butylbromide and methanol gives *tert*-butyl methyl ether.

tert-Butyl bromide   tert-Butyl methylether

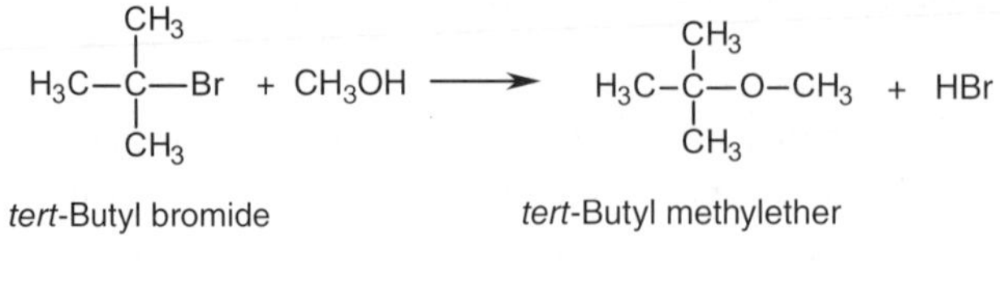

Mechanism

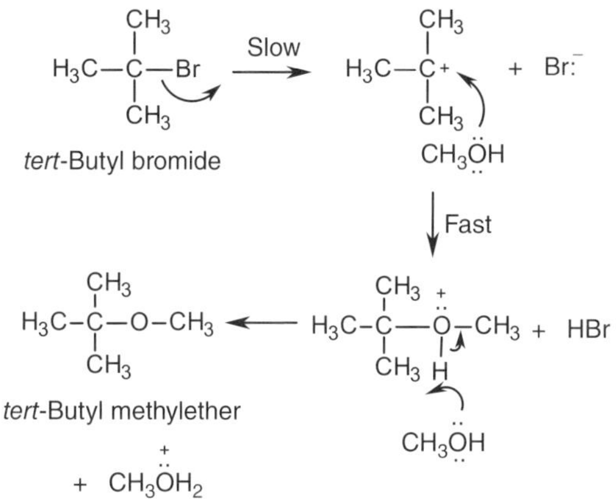

tert-Butyl bromide

CH₃ÖH

Fast

tert-Butyl methylether

+ CH₃ÖH₂

The rate of reaction depends only on the concentration of *tert*-butylbromide. Therefore, the rate is first order or unimolecular overall.

$$\text{Rate} = k_1[(CH_3)_3C\text{-Br}]$$

5.6.1.1.1 Substituent effects: S_N1 reactions

Carbocations are formed in S_N1 reactions. The more stable the carbocation, the faster it is formed. Thus, the rate depends on carbocation stability, since alkyl groups are known to stabilize carbocations through inductive effects and hyper-conjugation (see Section 5.3.1). The reactivity of S_N1 reactions decreases in the order of 3° carbocation > 2° carbocation > 1° carbocation > methyl cation. The primary carbocation and methyl cation are so unstable that a primary alkyl halide and methyl halide do not undergo S_N1 reactions. This is the opposite of S_N2 reactivity.

5.6.1.1.2 Strength of nucleophiles: S_N1 reactions

The rate of the S_N1 reaction does not depend on the nature of the nucleophiles, since the nucleophiles come into play after the rate-determining steps. Therefore, the reactivity of the nucleophiles has no effect on the rate of the S_N1 reaction. Sometimes in an S_N1 reaction the solvent is the nucleophile; for example, water and alcohol. When the solvent is the nucleophile, the reaction is called *solvolysis*.

5.6.1.1.3 Leaving group effects: S_N1 reactions

Good leaving groups are essential for S_N1 reactions. In the S_N1 reaction, a highly polarizable leaving group helps to stabilize the negative charge through partial bonding as it leaves. Negative charges are most stable on more electronegative elements. The leaving group should be stable after it has left with the bonding electrons and can also be a weak base. The leaving group starts to take on a partial negative charge as the cation starts to form. Most common leaving groups are:

$$\text{Anions}: Cl^-, Br^-, I^-, RSO_3^- \text{ (sulphonate)}, RSO_4^- \text{ (sulphate)},$$
$$RPO_4^- \text{ (phosphate)}.$$

$$\text{Neutral species}: H_2O, ROH, R_3N, R_3P.$$

5.6.1.1.4 Solvent effects: S$_N$1 reactions

Protic solvents are especially useful since the hydrogen bonding stabilizes the anionic leaving group after ionization. Note that a protic solvent has a hydrogen atom bound to an oxygen (as in a hydroxyl group) or a nitrogen (as in an amine group) and, simply, any solvent that contains a labile H^+ is called a *protic solvent*. Ionization requires the stabilization of both positive and negative charges. Solvents with a higher dielectric constant (ε), which is a measure of a solvent's polarity, have faster rates for S$_N$1 reactions.

5.6.1.2 Stereochemistry of the S$_N$1 Reactions

The S$_N$1 reaction is not stereospecific. The carbocation produced is planar and sp^2-hybridized. For example, the reaction of (S)-2-bromobutane and ethanol gives a racemic mixture, (S)-2-butanol and (R)-2-butanol.

The nucleophile may attack either from the top or the bottom face. If the nucleophile attacks from the top face, from which the leaving group departed, the product displays retention of configuration. If the nucleophile attacks from the bottom face, the backside of the leaving group, the product displays an inversion of configuration. A combination of inversion and retention is called *racemization*. Often, complete racemization is not achieved since the leaving group will partially block one face of the molecule as it ionizes, thus giving a major product of inversion.

5.6.1.2.1 Carbocation Rearrangements in S_N1 Reactions Through a 1,2-Hydride Shift

Carbocations often undergo rearrangements, producing more stable ions. This rearrangement produces a more stable tertiary cation instead of a less stable secondary cation. Rearrangements occur when a more stable cation can be produced by a *1,2-hydride shift* where hydrogen moves with its bonding electrons to the next carbon. For example, the S_N1 reaction of 2-bromo-3-methylbutane and ethanol affords a mixture of structural isomers, the expected product and a rearranged product.

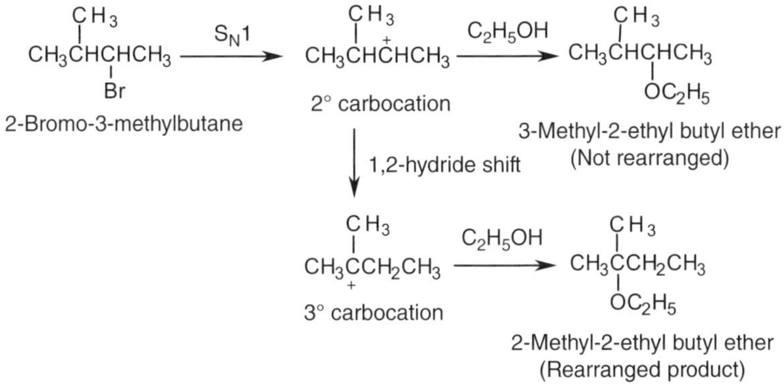

2-Bromo-3-methylbutane

2° carbocation

1,2-hydride shift

3-Methyl-2-ethyl butyl ether
(Not rearranged)

3° carbocation

2-Methyl-2-ethyl butyl ether
(Rearranged product)

5.6.1.2.2 Carbocation Rearrangements in S_N1 Reactions Through a 1,2-Methyl Shift

Carbocation rearrangements often occur when a more stable cation can be produced by an alkyl group or methyl shift. For example, 2,2-dimethyl propyl bromide exclusively provides a rearranged product, which results from a *1,2-methyl shift*, where methyl moves with its bonding electrons to the next carbon. This rearrangement produces a more stable tertiary cation instead of an unstable primary cation. Rearrangements *do not occur* in S_N2 reactions since carbocations are not formed.

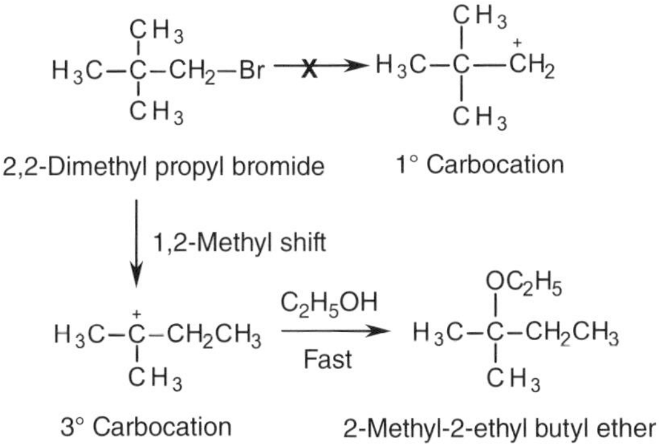

2,2-Dimethyl propyl bromide

1° Carbocation

1,2-Methyl shift

3° Carbocation

Fast

2-Methyl-2-ethyl butyl ether

5.6.1.3 Second Order Nucleophilic Substitution: S_N2 Reactions

S_N2 means bimolecular nucleophilic substitution. For example, the reaction of hydroxide ion with methyl iodide yields methanol. The hydroxide ion is a good nucleophile, since the oxygen atom has a negative charge and a pair of unshared electrons. The carbon atom is electrophilic, since it is bonded to a more electronegative halogen. The halogen pulls electron density away from the carbon, thus polarizing the bond with a carbon bearing partial positive charge and a halogen bearing partial negative charge. The nucleophile attacks the electrophilic carbon through donation of two electrons.

Typically, S_N2 reaction requires a backside attack. The C—X bond weakens as the nucleophile approaches. All these occur in one step. This is a *concerted reaction*, as it takes place in a single step with the new bond forming as the old bond is breaking. The S_N2 reaction is *stereospecific*, always proceeding with an inversion of stereochemistry. The inversion of configuration resembles the way an umbrella turns inside out in the wind. For example, the reaction between ethyl iodide and hydroxide ion produces ethanol, which is an S_N2 reaction.

$$C_2H_5-I \ + \ H\ddot{O}^- \longrightarrow C_2H_5-OH \ + \ I^{..}$$

Ethyl iodide   Ethanol

Mechanism

Ethyl iodide   Transition state   Ethanol

Rate = k_2[R–X][HO⁻]

The reaction rate is doubled when the concentration of ethyl iodide [C_2H_5I] is doubled, and also doubled when the concentration of hydroxide ion [HO⁻] is doubled. The rate is first order with respect to both reactants and is second order overall.

$$\text{Rate} = k_2[C_2H_5I][HO^-]$$

5.6.1.3.1 *Strength of Nucleophiles: S_N2 Reactions*

The rate of the S_N2 reaction strongly depends on the nature of the nucleophile; that is, a good nucleophile (nucleophile with a negative charge) offers faster rates than a poor nucleophile (neutral molecules with lone pair of electrons). Generally, negatively charged species are better nucleophiles than analogous neutral species. For example, methanol (CH_3OH) and sodium methoxide (CH_3ONa) react with CH_3I to produce dimethyl ether in both cases. CH_3ONa reacts about a million times faster than CH_3OH in S_N2 reactions.

5.6.1.3.2 Basicity and Nucleophilicity: S_N2 Reactions

Basicity is defined by the equilibrium constant for abstracting a proton. Nucleophilicity is defined by the rate of attack on an electrophilic carbon atom. A base forms a new bond with a proton. On the other hand, a nucleophile forms a new bond with an atom other than a proton. Species with a negative charge are stronger nucleophiles than analogous species without a negative charge. The stronger base is also a stronger nucleophile than its conjugate acids.

$$HO^- > H_2O \quad HS^- > H_2S \quad {}^-NH_2 > NH_3 \quad CH_3O^- > CH_3OH$$

Nucleophilicity decreases from left to right across the periodic table. The more electronegative elements hold on more tightly to their nonbonding electrons.

$$HO^- > F^- \quad NH_3 > H_2O \quad (CH_3CH_2)_3P > (CH_3CH_2)_2S$$

Nucleophilicity increases down the periodic table with the increase in polarizability and size of the elements.

$$I^- > Br^- > Cl^- > F^- \quad HSe^- > HS^- > HO^- \quad (C_2H_5)_3P > (C_2H_5)_3N$$

As the size of an atom increases, its outer electrons get further from the attractive force of the nucleus. The electrons are held less tightly and are said to be more polarizable. Fluoride is a nucleophile having hard or low polarizability with its electrons held close to the nucleus and it must approach the carbon nucleus closely before orbital overlap can occur. The outer shell of the soft iodide has loosely held electrons, and these can easily shift and overlap with the carbon atom from a relatively far distance.

5.6.1.3.3 Solvent Effects: S_N2 Reactions

Different solvents have different effects on the nucleophilicity of a species. Solvents with acidic protons are called *protic solvents*, usually O—H or N—H groups. Polar protic solvents, for example, dimethyl sulphoxide (DMSO), dimethyl formamide (DMF), acetonitrile (CH_3CN) and acetone (CH_3COCH_3) are often used in S_N2 reactions, since the polar reactants (nucleophile and alkyl halide) generally dissolve well in them.

Small anions are more strongly solvated than larger anions and sometimes this can have an adverse effect. Certain anions, for example F⁻, can be solvated so well in polar protic solvents that their nucleophilicity is reduced by the solvation. For efficient S_N2 reactions with small anions, it is usual to use polar aprotic solvents that do not have any O—H or N—H bonds to form hydrogen bonds to the small anions.

5.6.1.3.4 Steric Effects: S_N2 Reactions

A base strength is relatively unaffected by steric effect, because a base removes a relatively unhindered proton. Thus, the strength of a base depends only on how well the base shares its electrons with a proton. On the other hand, nucleophilicity is affected by steric effects, also known as *steric hindrance*. A bulky nucleophile has difficulty in getting near the backside of the sp^3 carbon. Therefore, large groups tend to hinder this process.

$$CH_3$$
$$H_3C-\overset{\overset{\displaystyle CH_3}{|}}{\underset{\underset{\displaystyle CH_3}{|}}{C}}-\ddot{\overset{..}{O}}^- \qquad\qquad H_5C_2-\ddot{\overset{..}{O}}^-$$

tert-Butoxide

A weak nucleophile and strong base

Ethoxide

A strong nucleophile and weak base

5.6.1.3.5 Steric Effects of the Substrate: S_n2 Reactions

Large groups on the electrophile hinder the approach of the nucleophile. Generally, one alkyl group slows the reaction, two alkyl groups make it difficult and three alkyl groups make it close to impossible. Thus, steric hindrance occurs when a large group in a molecule restricts the reaction site. Generally, an S_N2 reaction does not happen on carbon atoms that have three substituents.

Relative rates for S_N2: Methyl halides $> 1° > 2° > 3°$ alkyl halides

5.6.1.3.6 Leaving Group Effects: S_n2 Reactions

A good leaving group must be a weak base, and it should be stable after it has left with the bonding electrons. Thus, the weaker the base, the better it is as a leaving group. Good leaving groups are essential for both S_N1 and S_N2 reactions.

5.6.1.4 Stereochemistry of the S_N2 Reactions

A nucleophile donates its electron pairs to the C—X bond on the backside of the leaving group, since the leaving group itself blocks attack from any other direction. Inversion of stereochemistry is observed in the product of an S_N2 reaction. The reaction is stereospecific, since a certain stereoisomer reacts to form one specific stereoisomer as product. For example, (S)-2-bromobutane provides (R)-2-Butanol via an S_N2 reaction.

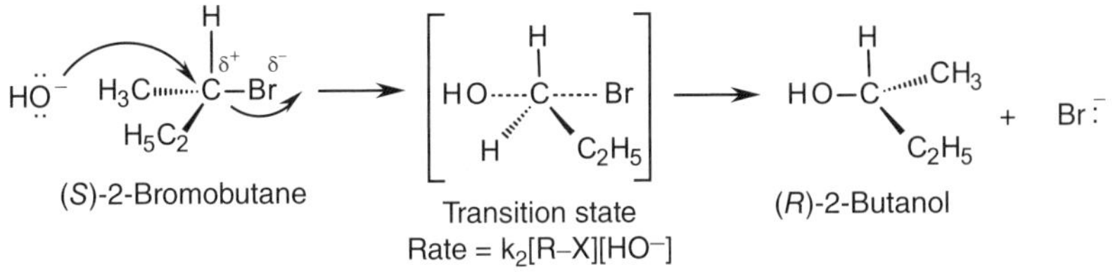

(S)-2-Bromobutane

Transition state
Rate = $k_2[R\text{–}X][HO^-]$

(R)-2-Butanol

5.6.2 Nucleophilic Substitutions of Alkyl Halides

The most important reaction for alkyl halides is the nucleophilic substitution reaction. Alkyl halides undergo a variety of transformation through S_N2 reactions with a wide range of strong nucleophiles and bases, such as organometallics (RMgX, RLi or R_2CuLi), metal hydroxides (NaOH or KOH), metal alkoxides (NaOR or KOR), metal amides ($NaNH_2$) or NH_3, metal cyanides (NaCN or KCN), metal carboxylate ($R'CO_2Na$), metal cyanides (NaCN or KCN), metal azides (NaN_3) and metal alkynides ($R'C{\equiv}CM$ or $R'C{\equiv}CMgX$) to generate various other functional groups. Depending on the alkyl halides and the reaction conditions, both S_N1 and S_N2 reactions can occur.

5.6.2.1 Conversion of Primary Alkyl Halides with Gilman Reagents

Coupling Reaction: Corey–House Reaction

The coupling reaction is a good synthetic way to join two alkyl groups together to produce higher alkanes. This versatile method is known as the *Corey–House reaction*. The Gilman reagent (lithium organocuprate) reacts with alkyl halide (RX) to give an alkane (R-R'), which has a higher carbon number than the starting alkyl halide. The Corey–House reaction is useful and common in organic synthesis, but this reaction is limited to a 1° alkyl halide and the alkyl groups in the Gilman reagents may be 1°, 2° or 3°.

$$R\text{–}X + R'_2\text{–CuLi} \xrightarrow[-78\ °C]{\text{Ether}} R\text{–}R' + R'\text{–Cu} + Li\text{-}X$$

X = Cl or Br

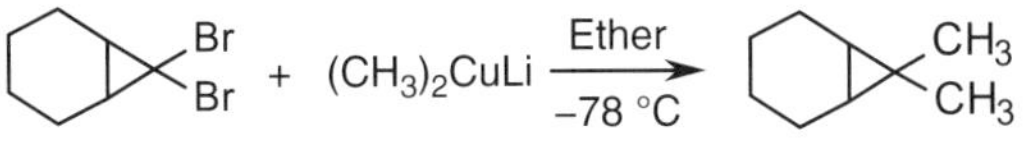

5.6.2.2 Conversion of Alkyl Halides with Metal Hydroxides

Preparation of Alcohols

Alkyl halides react with metal hydroxide (NaOH or KOH) to produce alcohols. Since hydroxide (HO⁻) is an excellent nucleophile, which conveniently reacts with ethyl chloride to yield ethanol. The reaction involves S_N2 displacement with backside attack of the hydroxide to form alcohol.

$$H_5C_2\text{-Cl} \quad + \quad Na\text{—OH} \quad \longrightarrow \quad H_5C_2\text{—OH} \quad + \quad NaCl$$

Ethyl chloride   Sodium hydoxide   Ethanol

Mechanism

Sodium hydoxide S_N2 Ethanol

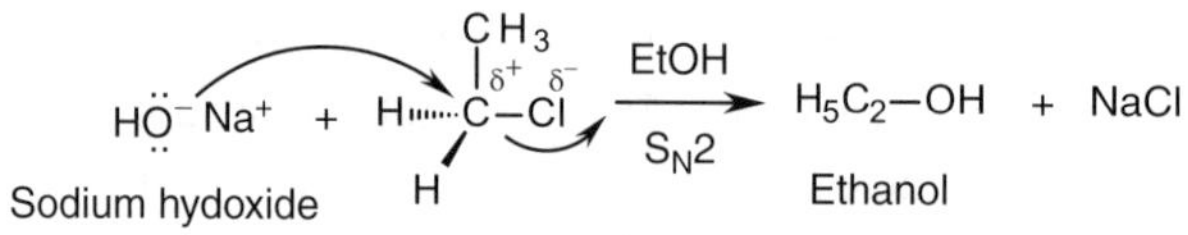

5.6.2.3 Conversion of Alkyl Halides with Metal Alkoxides

Preparation of Ethers: Williamson Ether Synthesis

Sodium or potassium alkoxides are strong bases and nucleophiles. Therefore, alkoxides (RO⁻) can react with primary alkyl halides to produce symmetrical or unsymmetrical ethers. This is known as the *Williamson ether synthesis*. The reaction is limited to primary alkyl halides. Higher alkyl halides tend to react via elimination. For example, sodium ethoxide reacts with ethyl iodide to produce diethyl ether. The reaction involves S_N2 displacement with backside attack of the alkoxide to form diethyl ether. Williamson ether synthesis is an important laboratory method for the preparation of both symmetrical and unsymmetrical ethers.

$$H_5C_2\text{—I} \quad + \quad H_5C_2\text{—ONa} \quad \xrightarrow{\text{EtOH}} \quad H_5C_2\text{—OC}_2H_5 \quad + \quad NaI$$

Ethyl iodide   Sodium ethoxide   Diethylether

Mechanism

Sodium ethoxide S_N2 Diethylether

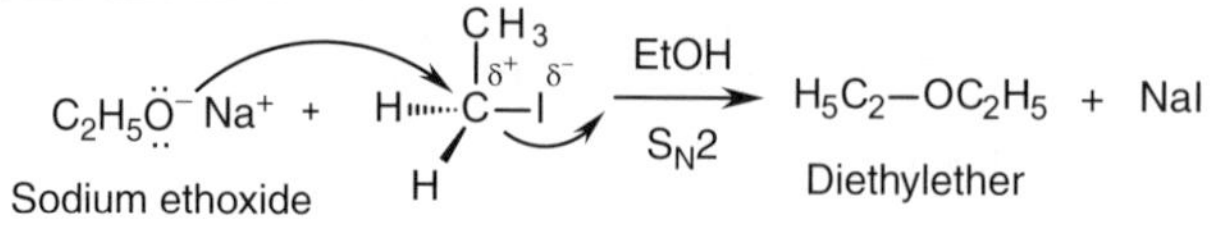

5.6.2.4 Conversion of Alkyl Halides with Sodium Amides

Preparation of Primary Amines

Alkyl halide reacts with sodium amide (NaNH$_2$) to provide a 1° amine via a S$_N$2 reaction. The reaction mechanism for the formation of a 1° amine is similar to the formation of nitrile.

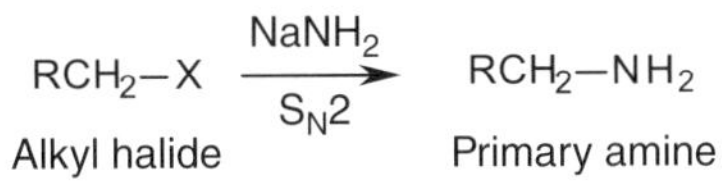

$$RCH_2-X \xrightarrow[S_N2]{NaNH_2} RCH_2-NH_2$$

Alkyl halide   Primary amine

5.6.2.5 Conversion of Alkyl Halides with Sodium Carboxylates

Preparation of Esters

Alkyl halide reacts with sodium carboxylate (R'CO$_2$Na) to give an ester via a S$_N$2 reaction. The formation of ester follows a similar mechanism to the formation of an alkyne.

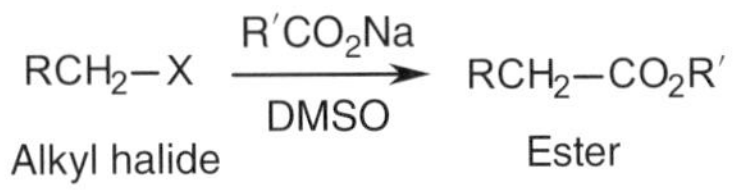

$$RCH_2-X \xrightarrow[DMSO]{R'CO_2Na} RCH_2-CO_2R'$$

Alkyl halide   Ester

5.6.2.6 Conversion of Alkyl Halides with Metal Cyanides

Preparation of Nitriles

Cyanide ion (CN$^-$) is a good nucleophile and can displace leaving groups from 1° and 2° alkyl halides. Nitriles are prepared by the treatment of alkyl halides with NaCN or KCN in DMSO. The reaction occurs rapidly at room temperature.

$$RCH_2-X + NaCN \xrightarrow{DMSO} RCH_2-CN + NaX$$

Alkyl halide   Nitrile

Mechanism

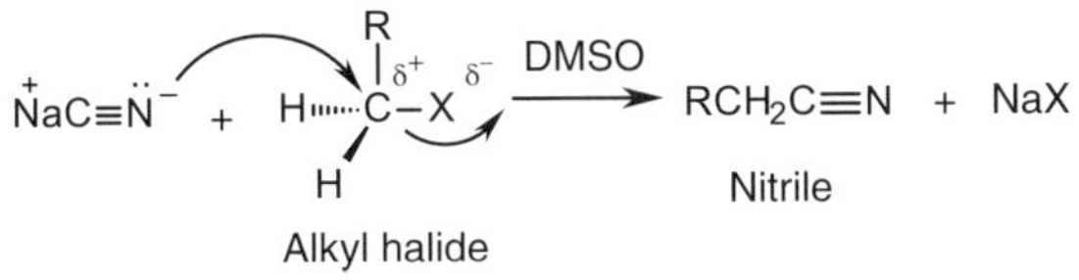

$$NaC\equiv N^- + H\text{·····}\overset{R}{\underset{H}{C}}-X \xrightarrow{DMSO} RCH_2C\equiv N + NaX$$

Alkyl halide   Nitrile

5.6.2.7 Conversion of Alkyl Halides with Metal Azides

Preparation of Alkyl Azides

The azide ion (N_3^-), a good nucleophile, can displace leaving groups from 1° and 2° alkyl halides. Alkyl azides are easily prepared from sodium or potassium azides (NaN_3 or KN_3) and alkyl halides. The reaction mechanism resembles the formation of nitrile.

$$RCH_2-X \xrightarrow[S_N2]{NaN_3} RCH_2-N_3$$

Alkyl halide   Alkyl azide

5.6.2.8 Conversion of Alkyl Halides or Tosylates with Metal Alkynides

Preparation of Internal Alkynes

The reaction of primary alkyl halides and metal alkynides ($R'C\equiv CM$ or $R'C\equiv CMgX$) yields internal alkynes. The reaction is limited to 1° alkyl halides or tosylates, since higher alkyl halides tend to react via elimination.

$$RCH_2-Y \xrightarrow[\substack{R'C\equiv CMgX \\ (M = Na, Li)}]{R'C\equiv CM \text{ or}} RCH_2C\equiv CR'$$

Y = X or OTs   Internal alkyne

Mechanism

$$R'C\equiv C{:}^- \quad + \quad H\cdots\overset{\overset{R}{|}}{\underset{\underset{H}{|}}{C}}{-}Y \longrightarrow RCH_2C\equiv CR' \quad + \quad Y{:}^-$$

Y = Cl or OTs

5.6.3 Nucleophilic Substitutions of Alcohols

Alcohols are not reactive towards nucleophilic substitution, because the hydroxyl group (—OH) is too basic to be displaced by a nucleophile. The nucleophilic substitution reaction of alcohols only occurs in the presence of an acid. The overall transformation requires the acidic conditions to replace the hydroxyl group (—OH), a poor leaving group, to a good leaving group such as H_2O. Protonation to convert the leaving group to H_2O has limited use as not all substrates or nucleophiles can be used under acidic conditions without unwanted side reactions. An alternative way is to convert the alcohol into an alkyl halide (see Section 5.6.3.2) or alkyl tosylate (see Section 5.6.3.5), which has a much better leaving group and reacts with nucleophiles without the need for an acid.

5.6.3.1 Dehydration of Alcohols: Condensation of Alcohols

Preparation of Ethers

Bimolecular dehydration is generally used for the synthesis of symmetrical ethers from unhindered 1° alcohols. Industrially diethyl ether is produced from heating ethanol at 140°C in the presence of H_2SO_4. In this reaction, ethanol is protonated in the presence of a strong acid, such as H_2SO_4. The protonated ethanol is then attacked by another molecule of ethanol to give diethyl ether. This is an acid-catalysed S_N2 type reaction. If the temperature is too high, alkene is formed via elimination. Therefore, temperature should be monitored carefully while synthesizing ether.

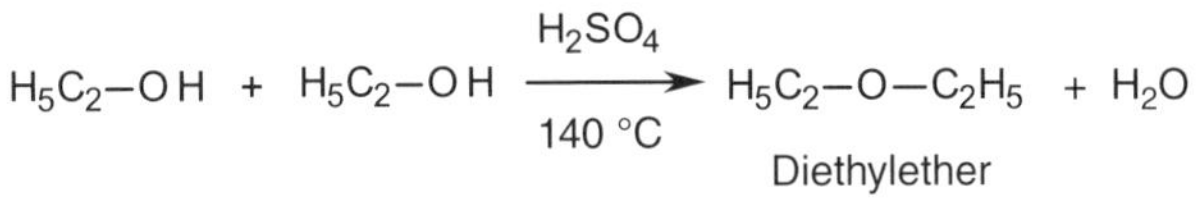

$$H_5C_2-OH \ + \ H_5C_2-OH \ \xrightarrow[140\ °C]{H_2SO_4} \ H_5C_2-O-C_2H_5 \ + \ H_2O$$

Diethylether

Mechanism

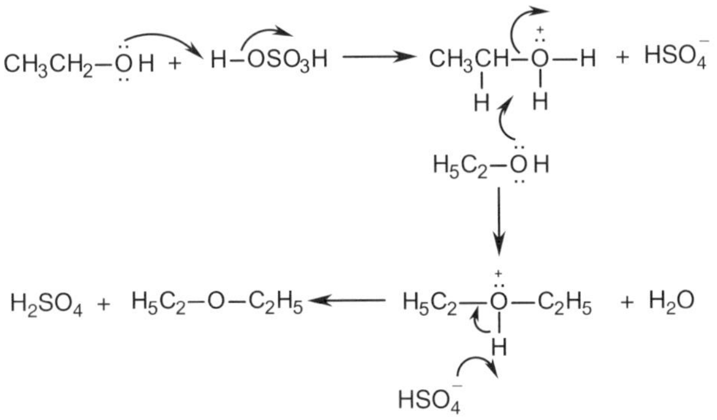

5.6.3.2 Conversion of Alcohols by Hydrogen Halides

Preparation of Alkyl Halides

The most common method for preparation of alkyl halides involves the treatment of alcohols with hydrogen halides (HX). The order of reactivity hydrogen halides parallels their acid strength; that is, $HI > HBr > HCl >> HF$. Thus, this method cannot be used for making alkyl fluorides. Among primary (1°), secondary (2°) and tertiary (3°) alcohols, the tertiary alcohols are most reactive. Primary alcohols undergo S_N2 reactions with HX. Primary alcohols with branching on the β-carbon produce rearranged products. The temperature must be kept low to avoid the formation of a E2 product.

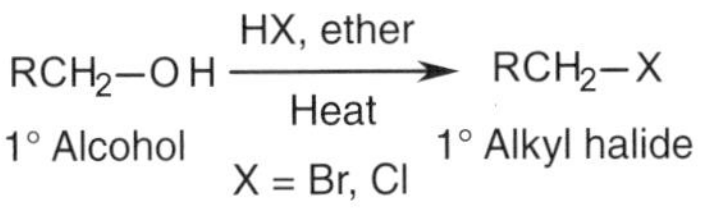

$$RCH_2-OH \ \xrightarrow[\substack{Heat \\ X = Br,\ Cl}]{HX,\ ether} \ RCH_2-X$$

1° Alcohol   1° Alkyl halide

Mechanism

$$RCH_2-\overset{\cdot\cdot}{\underset{\cdot\cdot}{O}}H \xrightleftharpoons{H^+} \underset{\underset{H}{|}}{HCR}-\overset{\overset{H}{|}}{\underset{+}{O}}-H \longrightarrow RCH_2-X + H_2O$$

$$X:^-$$

Secondary and tertiary alcohols undergo S_N1 reactions with hydrogen halides. The reaction of a HX with 3° alcohol proceeds readily at room temperature, whereas the reaction of a HX with a 2° alcohol requires heat. The reaction occurs via carbocation intermediate. Therefore, it is possible to form both substitution and elimination products. Secondary alcohols with branching on the β-carbon provide rearranged products. The temperature must be kept low to avoid the formation of a E1 product.

$$\underset{\underset{R}{|}}{\overset{\overset{R}{|}}{R-C}}-OH \xrightarrow[\substack{\text{Heat} \\ \text{X = Br, Cl}}]{\text{HX, ether}} \underset{\underset{R}{|}}{\overset{\overset{R}{|}}{R-C}}-X$$

3° Alcohol   3° Alkyl halide

Mechanism

$$\underset{\underset{R}{|}}{\overset{\overset{R}{|}}{R-C}}-\overset{\cdot\cdot}{\underset{\cdot\cdot}{O}}H \xrightleftharpoons{H^+} \underset{\underset{R}{|}\ \underset{H}{|}}{\overset{\overset{R}{|}}{R-C}}-\overset{+}{O}-H \longrightarrow \underset{\underset{R}{|}}{\overset{\overset{R}{|}}{R-C}}-X + H_2O$$

$$X:^-$$

Primary alcohol reacts with HCl in the presence of $ZnCl_2$ (a Lewis acid) to produce 1° alkyl chloride. Without the use of $ZnCl_2$, the S_N2 reaction is slow because chloride is a weaker nucleophile than bromide. The reaction rate is increased when $ZnCl_2$ is used as a catalyst. The $ZnCl_2$ coordinates to the hydroxyl oxygen and generates a better leaving group. The mixture of HCl and $ZnCl_2$ is known as a *Lucas Reagent*.

$$RCH_2-OH \xrightarrow[ZnCl_2]{HCl} RCH_2-Cl$$

1° Alcohol   1° Alkyl chloride

Mechanism

$$RCH_2-\overset{\cdot\cdot}{\underset{\cdot\cdot}{O}}H \xrightleftharpoons{ZnCl_2} \underset{\underset{H}{|}}{RCH}-\overset{\overset{H}{/}}{O}{\underset{ZnCl_2}{\overset{\cdot}{+}}} \longrightarrow R-CH_2-Cl$$

$$Cl:^-$$

                                                1° Alkyl chloride

Secondary and tertiary alcohols react via the S_N1 mechanism with the Lucas reagent. The reaction occurs via a carbocation intermediate. Thus, it is possible to form both S_N1 and E1 products. The temperature must be kept low to avoid the formation of the E1 product.

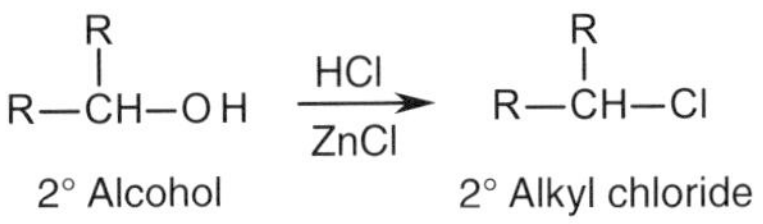

Mechanism

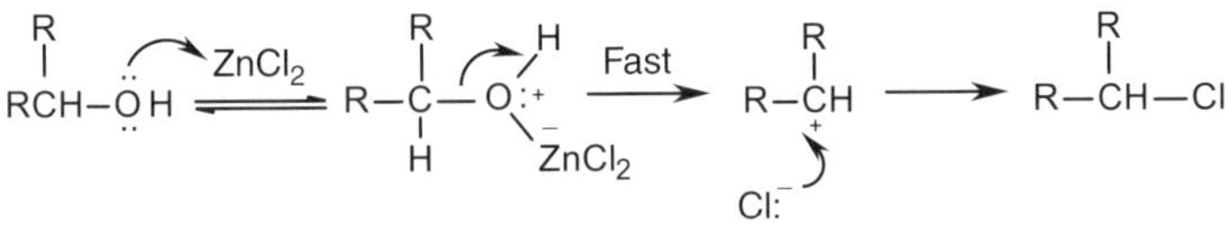

5.6.3.3 Conversion of Alcohols by Phosphorous Halides

Preparation of Alkyl Halides

Phosphorous halides react with alcohols to yield alkyl halides at low temperature (0 °C). Primary and secondary alcohols undergo S_N2 reactions with PX_3. This type of reaction does not lead to rearranged products and does not work well with 3° alcohols. PI_3 has to be generated in situ via a reaction of iodine and phosphorous.

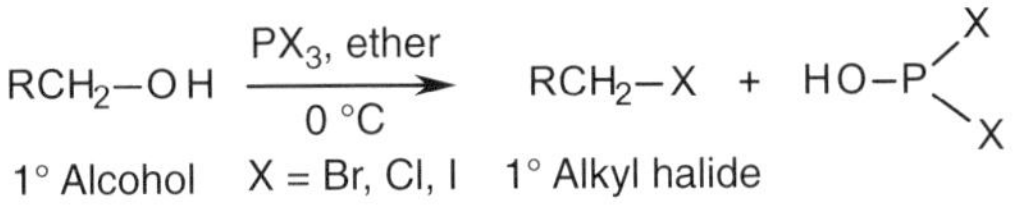

Mechanism

The hydroxyl oxygen displaces a halide, a good leaving group from the phosphorous. The halide attacks the backside of the alkyl group and displaces the positively charged oxygen, which is a good leaving group.

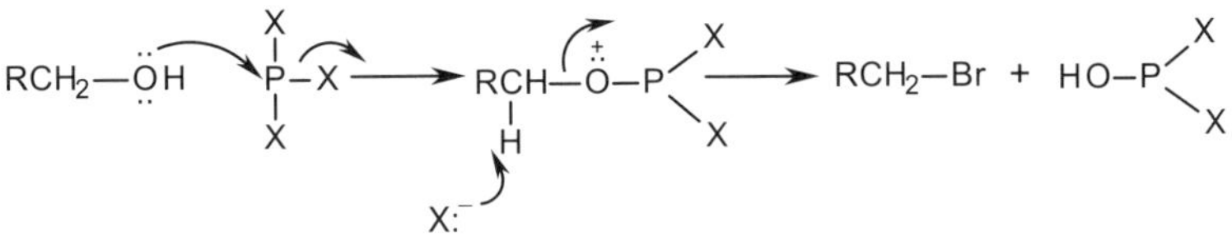

5.6.3.4 Conversion of Alcohols by Thionyl Chlorides

Preparation of Alkyl Chlorides

Thionyl chloride ($SOCl_2$) is the most widely used reagent for the conversion of 1° and 2° alcohols to corresponding alkyl chlorides. The reaction is often carried out in the presence of a base; for example, pyridine or triethylamine (Et_3N). The base catalyses the reaction and neutralizes the HCl generated during the reaction by forming pyridinium chloride ($C_5H_5NH^+Cl^-$) or triethylammonium chloride ($Et_3NH^+Cl^-$)

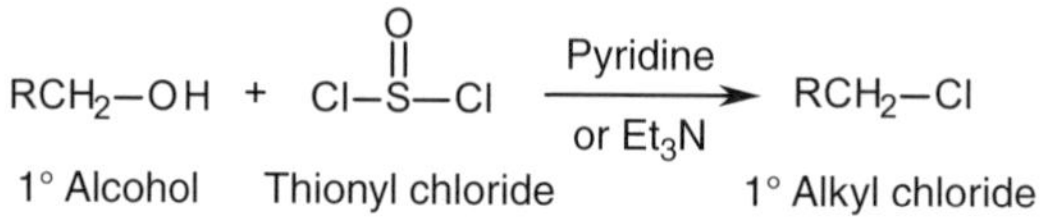

Mechanism

Thionyl chloride converts the hydroxyl group in an alcohol to a chlorosulphite leaving group that can be displaced by the chloride. Secondary or tertiary alcohols follow S_N1 reactions, whereas primary alcohols proceed via S_N2 reactions.

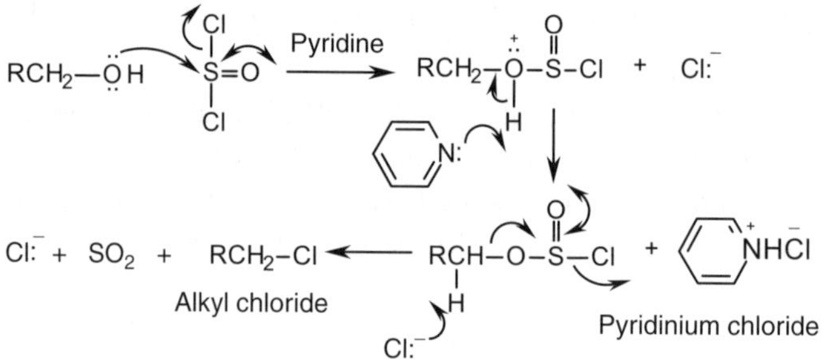

5.6.3.5 Conversion of Alcohols by Tosyl Chloride

Preparation of Tosylate Esters

Alcohol reacts with sulphonyl chloride to generate sulphonate esters via a S_N2 reaction. Tosylate ester, also called alkyl tosylate, is formed by the reaction of alcohol with *p*-toluenesulphonyl chloride (TsCl). The reaction is most commonly carried out in the presence of a base; for example, pyridine or triethylamine (Et_3N).

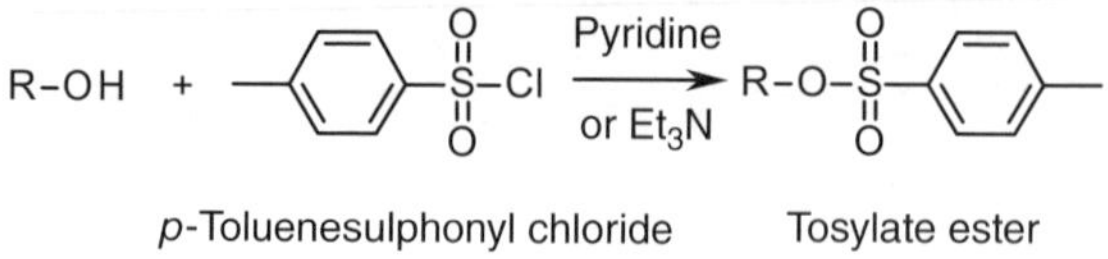

Mechanism

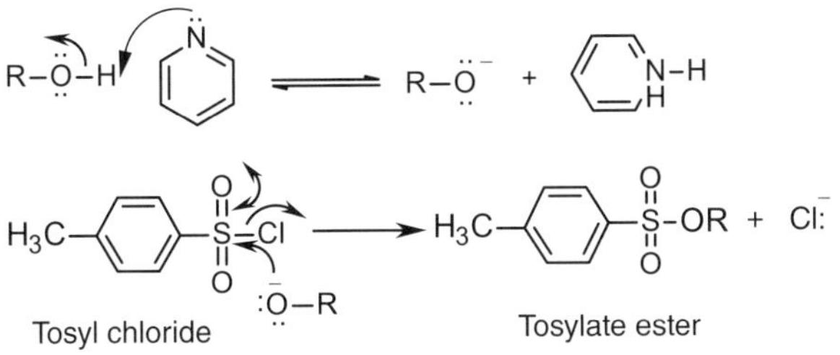

5.6.3.6 Conversion of Alcohols to Alkyl Halides via Tosylates
Preparation of Alkyl Halides
Tosylates are excellent leaving groups, and can undergo a variety of S_N2 reactions. The reaction is stereospecific, and it occurs with inversion of configuration. For example, (S)-2-butanol reacts with TsCl in pyridine to produce (S)-2-butane tosylate, which reacts readily with NaI to afford (R)-2-iodobutane via a S_N2 reaction.

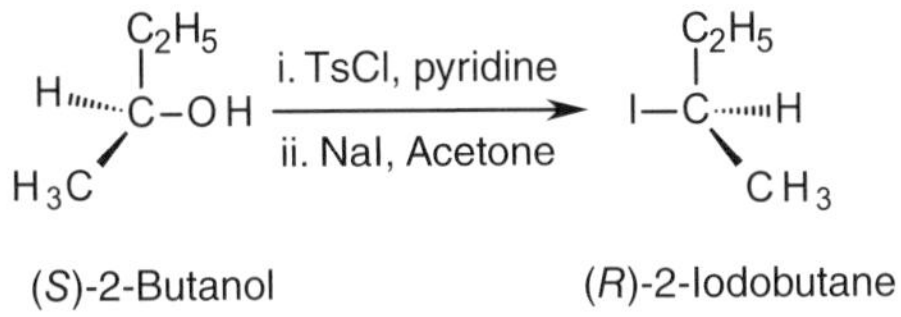

(S)-2-Butanol   (R)-2-Iodobutane

Mechanism

(S)-2-Butanol

(S)-2-Butyl tosylate

NaI
Acetone

(R)-2-Iodobutane

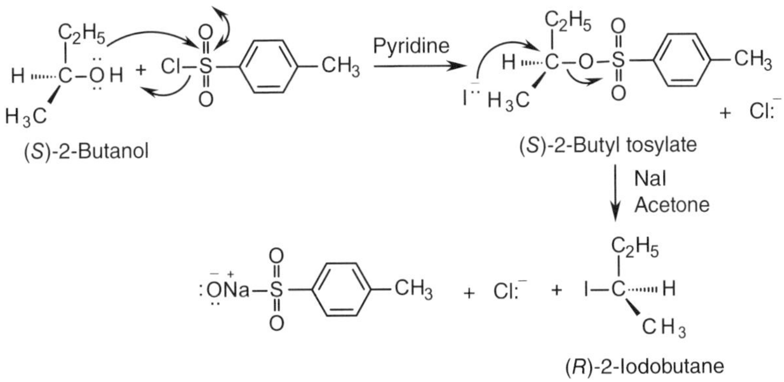

Similarly, alkyl tosylate reacts with other nucleophiles, for example H^-, X^-, HO^-, $R'O^-$, $R'-$, NH_2^- or NH_3, CN^-, N_3^- and $R'CO_2^-$, following the S_N2 reaction mechanism and produces a number of other functional groups as follows.

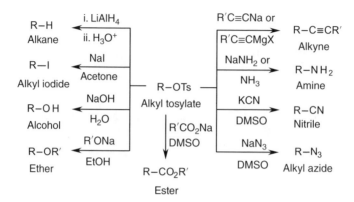

5.6.4 Nucleophilic Substitutions of Ethers and Epoxides

Ethers themselves cannot undergo nucleophilic substitution or elimination reactions, because the alkoxide anion is not a good leaving group. Thus, acid catalysis is required for the nucleophilic substitution of ethers. Ethers react with HX (usually HBr or HI) at high temperatures to form alkyl halides.

Although an epoxide and ether have the same leaving group, epoxides are more reactive than ethers due to ring strain in the three-membered ring. They undergo ring-opening reactions readily with acids as well as bases. Thus, epoxides are synthetically useful reagents and they react with a wide variety of nucleophiles. They are easily cleaved by H_2O and ROH in the presence of an acid catalyst via S_N1 reactions and by strong bases (RMgX, RLi, $NaC\equiv N$, NaN_3, $RC\equiv CM$, $RC\equiv CMgX$, $RC\equiv CLi$, $LiAlH_4$ or $NaBH_4$, NaOH or KOH and NaOR or KOR) via S_N2 reactions.

5.6.4.1 Cleavage of Ethers and Epoxides by Haloacids
5.6.4.1.1 Preparation of Alkyl Halides

Ethers can be cleaved at the ether linkage only at high temperatures using haloacids; for example, HBr or HI at high temperatures. Depending on the structure of the alkyl groups in ether, the reaction can proceed via S_N1 or S_N2. For example, methyl propylether reacts with HBr to give propyl bromide via a S_N2 reaction. Protonation of the oxygen in ether creates a good leaving group, a neutral alcohol molecule. Cleavage involves nucleophilic attack by bromide ion on the protonated ether, followed by the displacement of the weakly basic CH_3OH to produce propyl bromide.

$$CH_3CH_2CH_2-O-CH_3 \xrightarrow[\text{Heat}]{\text{HBr}} CH_3CH_2CH_2-Br + CH_3OH$$

Methyl propylether   Propyl bromide

Mechanism

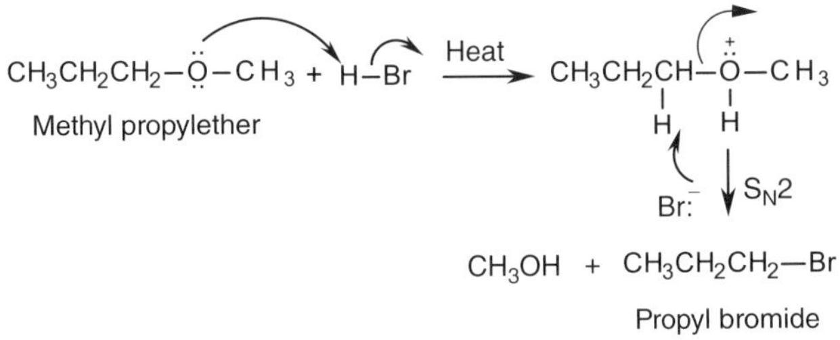

CH$_3$CH$_2$CH$_2$—Ö—CH$_3$ + H—Br $\xrightarrow{\text{Heat}}$ CH$_3$CH$_2$CH—Ö—CH$_3$

Methyl propylether

H H

Br:$^-$ ↓ S$_N$2

CH$_3$OH + CH$_3$CH$_2$CH$_2$—Br

Propyl bromide

5.6.4.1.2 *Preparation of Haloalcohols*

Ethylene oxide can be easily cleaved by HBr to yield bromoethanol. The oxygen is protonated to form a protonated ethylene oxide, which, being attacked by the halide, produces bromoethanol.

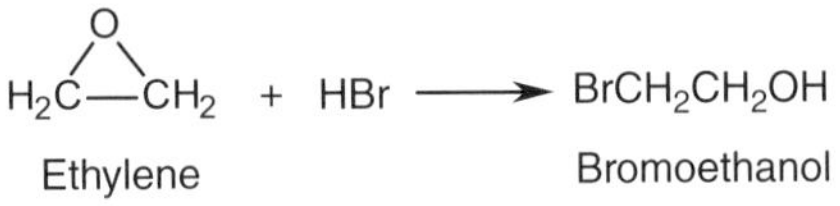

H$_2$C—CH$_2$ + HBr ⟶ BrCH$_2$CH$_2$OH

Ethylene   Bromoethanol

Mechanism

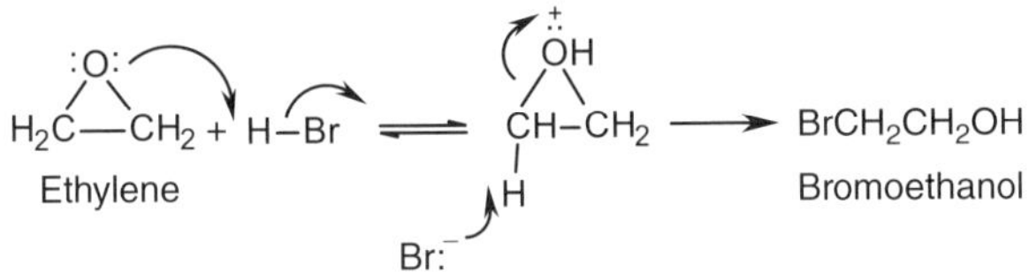

H$_2$C—CH$_2$ + H—Br ⇌ CH—CH$_2$ ⟶ BrCH$_2$CH$_2$OH

Ethylene   Bromoethanol

Br:$^-$

5.6.4.2 Acid-Catalysed Cleavage of Epoxides

Ether used as solvent in the reaction usually does not react with any reagents, but epoxide (ethylene oxide), a highly reactive cyclic ether with a ring strain, reacts with various nucleophiles. In the case of acid-catalysed unsymmetrical epoxide, the weak nucleophiles (H$_2$O and ROH) attack the most substituted carbon of the ring, and this produces 1-substituted (primary) alcohol. This reaction follows a S$_N$1 reaction.

5.6.4.2.1 *Preparation of Diols*

Acid-catalysed epoxides are easily cleaved by water. Water reacts as the nucleophile and this is referred to as *hydrolysis*. For example, hydrolysis of ethylene oxide in the presence of an acid catalyst produces 1,2-ethane diol (ethylene glycol).

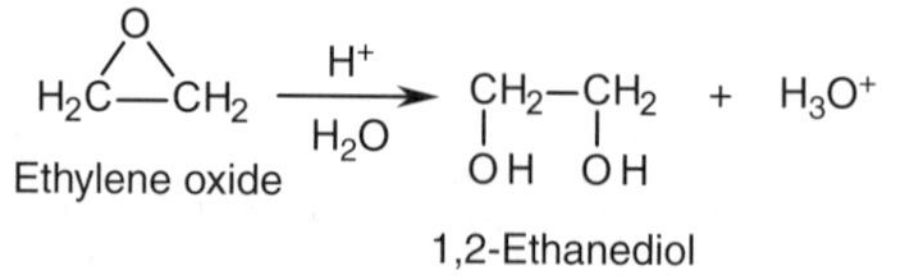

$$H_2\overset{\displaystyle O}{\overset{|}{C}}-CH_2 \xrightarrow[H_2O]{H^+} \underset{OH\ \ OH}{CH_2-CH_2} + H_3O^+$$

Ethylene oxide   1,2-Ethanediol

Mechanism

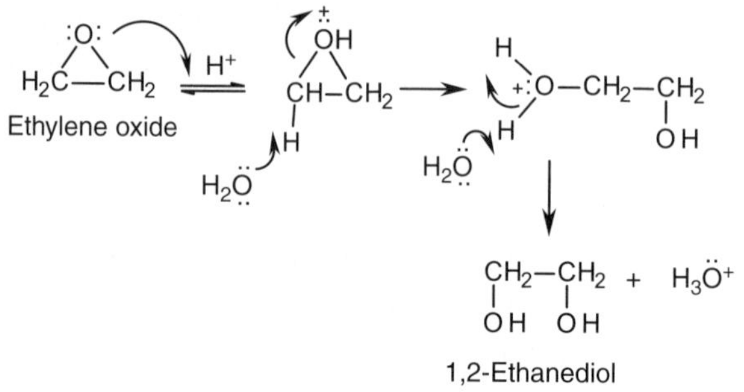

Ethylene oxide

1,2-Ethanediol

5.6.4.2.2 Preparation of Alkoxy Alcohol

Ethylene oxide is a symmetrical epoxide, which reacts with sodium methoxide to afford 2-methoxy-ethanol, after hydrolytic work-up.

Ethylene oxide   2-Methoxy-ethanol

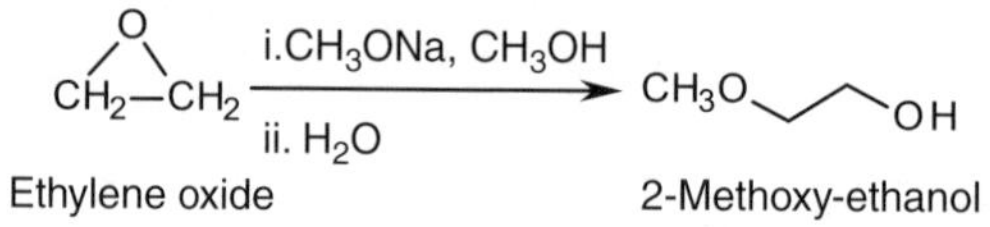

Mechanism

Epoxide is protonated in the first step, then the nucleophile, the methoxide anion, attacks the symmetrical epoxide, which ends up opening the ring. After hydrolytic work-up, it produces a primary alcohol, 2-methoxy-ethanol.

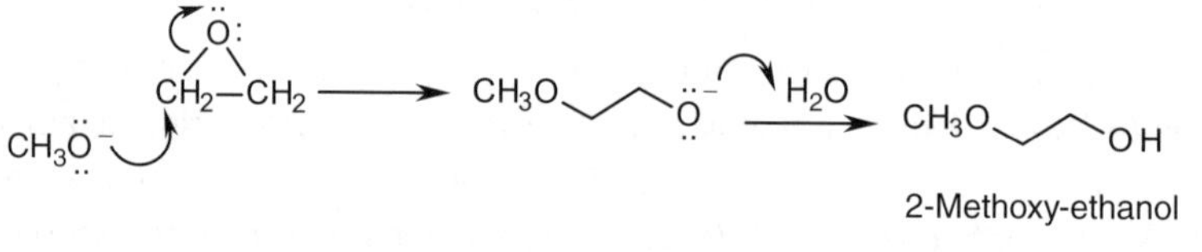

2-Methoxy-ethanol

Similarly, an acid-catalysed unsymmetrical epoxide yields a 1-substituted alcohol, resulting from the nucleophilic attack on the most substituted carbon. For example, propylene oxide reacts with CH_3OH in the presence of an acid to give 2-methoxy-1-propanol.

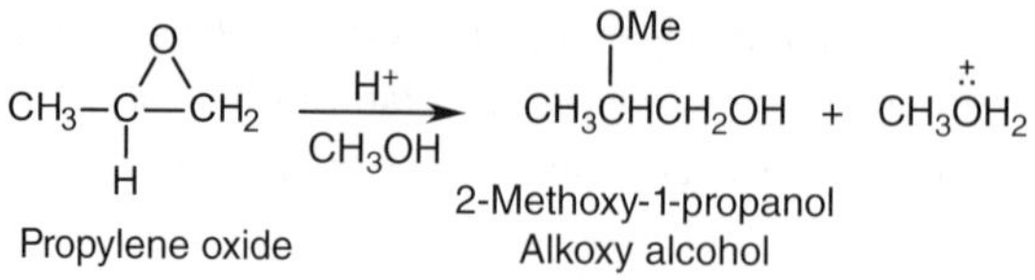

Propylene oxide   2-Methoxy-1-propanol
                     Alkoxy alcohol

Mechanism

Epoxide is protonated in the first step, then the nucleophile, a methoxide anion, attacks the most substituted carbon of the unsymmetrical epoxide, which ends up opening the ring. After hydrolytic work-up, it produces a primary alcohol, 2-methoxy-1-proanol.

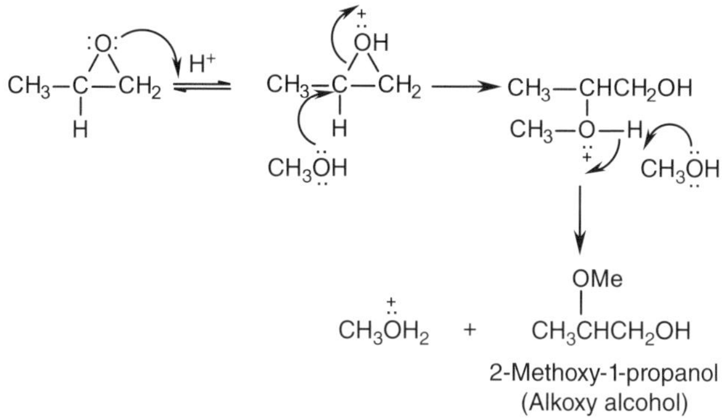

$$CH_3\overset{+}{\ddot{O}}H_2 \quad + \quad CH_3CHCH_2OH$$

2-Methoxy-1-propanol
(Alkoxy alcohol)

5.6.4.3 Base-Catalysed Cleavage of Epoxides

Preparation of Secondary Alcohols

Base-catalysed epoxide cleavage follows a S_N2 reaction in which the attack of the strong nucleophile ($Nu:^-$) takes place at the least substituted carbon of the ring. Therefore, a base-catalysed unsymmetrical epoxide forms a 2-substituted (secondary) alcohol.

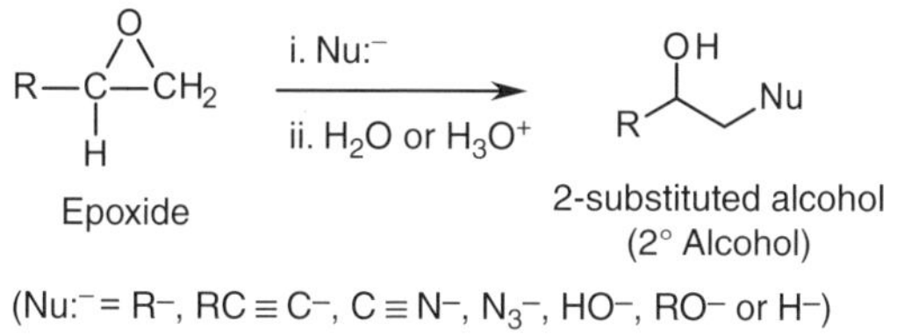

Epoxide

2-substituted alcohol
(2° Alcohol)

($Nu:^- = R^-$, $RC \equiv C^-$, $C \equiv N^-$, N_3^-, HO^-, RO^- or H^-)

Mechanism

A strong nucleophile attacks the least substituted carbon of the unsymmetrical epoxide, which ends up opening the ring. After hydrolytic work-up, it produces a secondary alcohol.

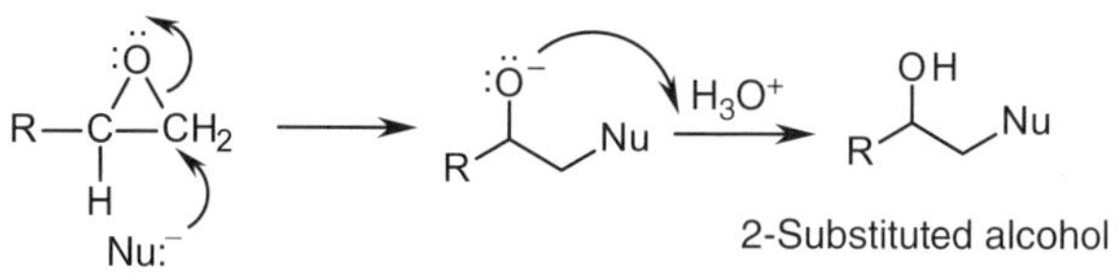

2-Substituted alcohol

Organometallic reagents (RMgX, RLi) are powerful nucleophiles. They attack epoxides at the least hindered carbon, which results in the formation of the more substituted alcohol. For example, propylene oxide is an unsymmetrical epoxide, which reacts with methyl magnesium bromide to give 2-butanol, after acidic work-up.

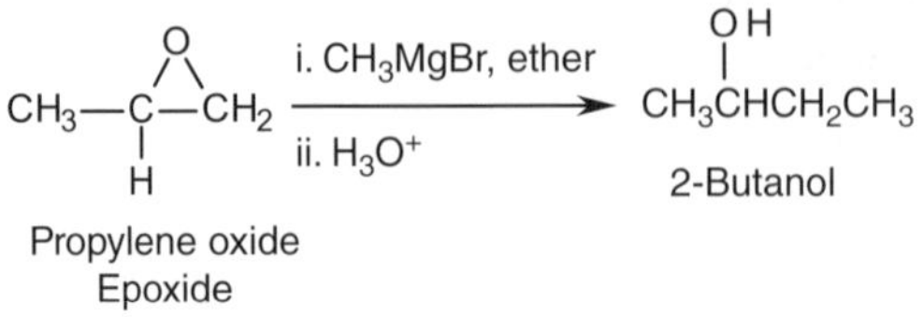

Propylene oxide
Epoxide

i. CH₃MgBr, ether

ii. H₃O⁺

2-Butanol

Mechanism

The nucleophile attacks the least substituted carbon of the unsymmetrical propylene oxide, which ends up opening the ring. After hydrolytic work-up, it yields 2-butanol.

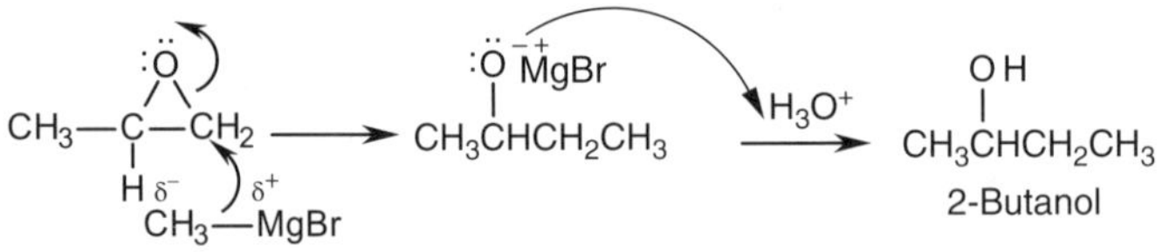

2-Butanol

5.6.5 Nucleophilic Acyl Substitutions of Carboxylic Acid Derivatives

Carboxylic acid derivatives undergo *nucleophilic acyl substitution*, where one nucleophile replaces another on the acyl carbon. Nucleophilic acyl substitution can interconvert all carboxylic acid derivatives, and the reaction mechanism varies depending on acidic or basic conditions. Nucleophiles can either be negatively charged anion (Nu:⁻) or neutral (Nu:) molecules.

General reaction mechanism

If the nucleophile is a negatively charged anion (R⁻, H⁻, HO⁻⁻, RO⁻, CN⁻), it will readily attack the carbonyl carbon and form an alkoxide tetrahedral intermediate, which in turn expels the leaving group whilse reforming the carbonyl C=O double bond.

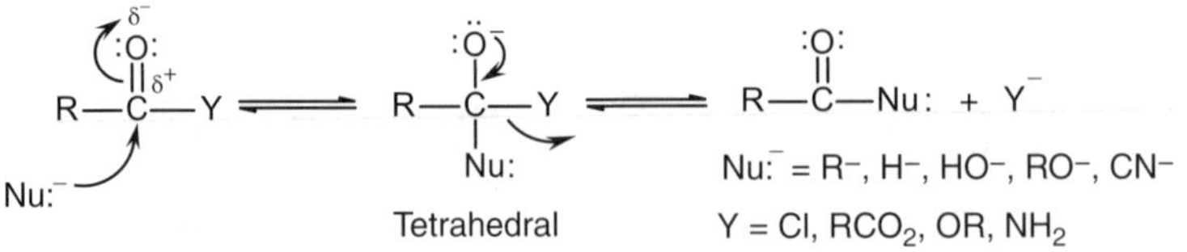

Tetrahedral

Nu:⁻ = R⁻, H⁻, HO⁻, RO⁻, CN⁻

Y = Cl, RCO₂, OR, NH₂

If the nucleophile is a neutral molecule with a lone pair of electrons (H₂O, ROH), it requires an acid catalyst for nucleophilic addition reaction to occur. Under acidic conditions, the carbonyl group becomes protonated, and thus is activated towards nucleophilic acyl substitution. Attack by a weak nucleophile generates the tetrahedral intermediate. A simultaneous deprotonation and loss of the leaving group reforms the carbonyl C=O double bond.

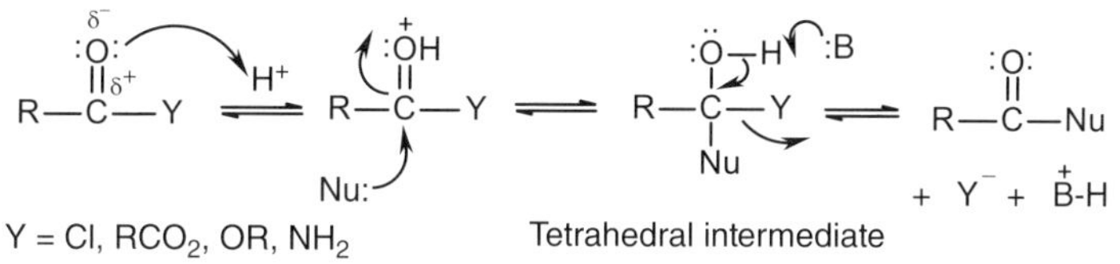

Y = Cl, RCO$_2$, OR, NH$_2$

Nu: = H$_2$O or ROH

Tetrahedral intermediate

5.6.5.1 Conversion of Carboxylic Acids by Alcohols

5.6.5.1.1 Preparation of Esters from Carboxylic Acids: Fischer Esterification

Esters are synthesized by refluxing the parent carboxylic acid and an alcohol with an acid catalyst. The equilibrium can be driven to completion by using an excess of the alcohol, or by removing the water as it forms. This is known as *Fischer esterification*.

$$R-\overset{O}{\underset{||}{C}}-OH + R'OH \xrightleftharpoons{H^+} R-\overset{O}{\underset{||}{C}}-OR' + H_3O^+$$

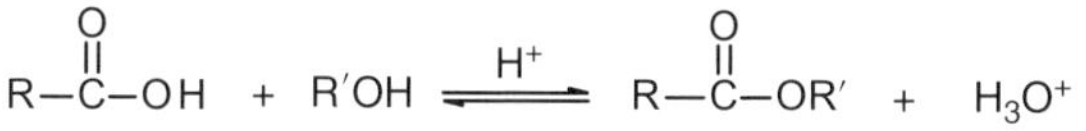

Mechanism

The carbonyl group of a carboxylic acid is not sufficiently electrophilic to be attacked by the alcohol. The acid catalyst protonates the carbonyl oxygen and activates it towards nucleophilic attack. The alcohol attacks the protonated carbonyl carbon and forms a tetrahedral intermediate. Intramolecular proton transfer converts the hydroxyl to a good leaving group as H$_2$O. A simultaneous deprotonation and loss of H$_2$O gives an ester.

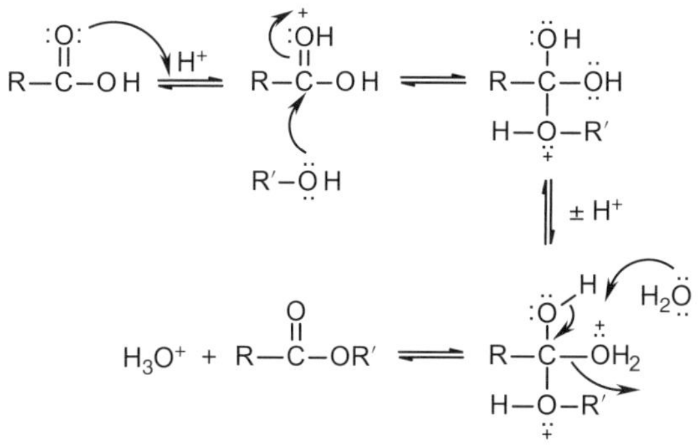

5.6.5.1.2 Preparation of Esters from Esters: Transesterification

Transesterification occurs when an ester is treated with alcohol. This reaction can be acid or base catalysed. This is where the alcohol part of the ester can be replaced with a new alcohol component. The reaction mechanism is similar to Fischer esterification.

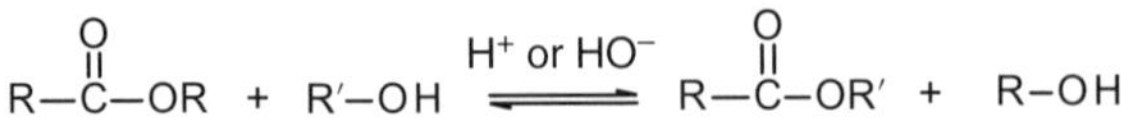

$$R-\overset{O}{\overset{\|}{C}}-OR \ + \ R'-OH \ \underset{}{\overset{H^+ \ or \ HO^-}{\rightleftharpoons}} \ R-\overset{O}{\overset{\|}{C}}-OR' \ + \ R-OH$$

5.6.5.2 Conversion of Carboxylic Acids by Thionyl or Oxalyl Chlorides

Preparation of Acid Chlorides

The best way to make acid chlorides is the reaction of a carboxylic acid with either thionyl chloride ($SOCl_2$) or oxalyl chloride ($COCl)_2$ in the presence of a base (pyridine). The mechanism of formation of acid chloride is similar to the reaction of alcohol with $SOCl_2$.

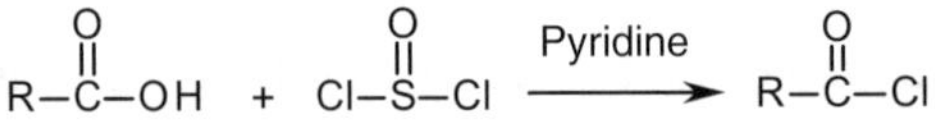

$$R-\overset{O}{\overset{\|}{C}}-OH \ + \ Cl-\overset{O}{\overset{\|}{S}}-Cl \ \xrightarrow{Pyridine} \ R-\overset{O}{\overset{\|}{C}}-Cl$$

Mechanism

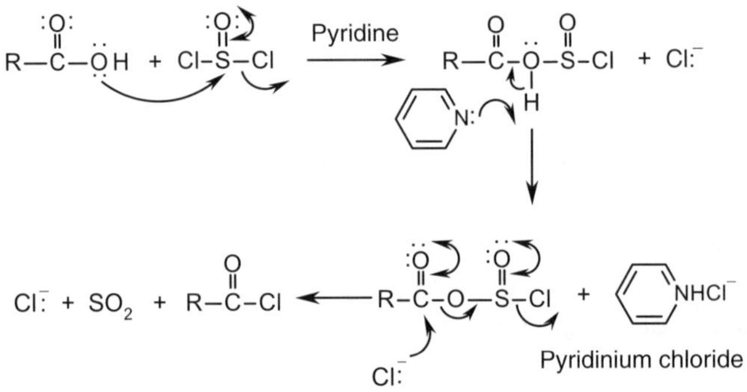

5.6.5.3 Dehydration of Carboxylic Acids

Preparation of Acid anhydrides

Acid anhydrides are prepared from dehydration of two molecules of carboxylic acids at higher temperatures. For example, acetic anhydride is produced industrially by heating acetic acid at 800 °C. Other anhydrides are difficult to synthesize directly from the corresponding carboxylic acids. Usually they are prepared from acid chloride and sodium carboxylate salt (see Section 5.6.5.6).

$$H_3C-\overset{\overset{\displaystyle O}{\|}}{C}-OH \ + \ HO-\overset{\overset{\displaystyle O}{\|}}{C}-CH_3 \ \xrightarrow{800\ °C} \ H_3C-\overset{\overset{\displaystyle O}{\|}}{C}-O-\overset{\overset{\displaystyle O}{\|}}{C}-CH_3 \ + \ H_2O$$

Acetic acids   Acetic anhydride

Mechanism

5.6.5.4 Conversion of Carboxylic Acids by Ammonia or Amines

Preparation of Amides

Ammonia, 1° and 2° amines react with carboxylic acids to afford, respectively, 1°, 2° and 3° amides through a nucleophilic acyl substitution reaction. The reaction of ammonia and a carboxylic acid initially forms a carboxylate anion and an ammonium cation. Normally, the reaction stops at this point since the carboxylate anion is a poor electrophile. However, by heating the reaction to over 100 °C, the water can be driven off as steam and amide products are formed. This is an important commercial process for the production of 1° amides.

$$R-\overset{\overset{\displaystyle O}{\|}}{C}-OH \ + \ NH_3 \longrightarrow R-\overset{\overset{\displaystyle O}{\|}}{C}-O^- + \ \overset{+}{N}H_4 \ \xrightarrow{Heat} \ R-\overset{\overset{\displaystyle O}{\|}}{C}-NH_2 \ + \ H_2O$$

5.6.5.5 Conversion of Acid Chlorides or Anhydrides by Alcohols

Preparation of Esters

Acid chlorides or acid anhydrides react with alcohols to provide esters through a nucleophilic acyl substitution. Because acid chlorides or acid anhydrides are sufficiently reactive towards weak nucleophiles, such as alcohol, no catalyst is required for this substitution reaction. Generally, the reaction is carried out in a basic solution and most commonly in pyridine (C_5H_5N) or triethylamine (Et_3N).

$$R-\overset{\overset{\displaystyle O}{\|}}{C}-Y \ + \ R'OH \ \xrightarrow[\text{or } Et_3N]{C_5H_5N} \ R-\overset{\overset{\displaystyle O}{\|}}{C}-OR'$$

Y = Cl or OCOR   Ester

Acid chloride or anhydride

Mechanism

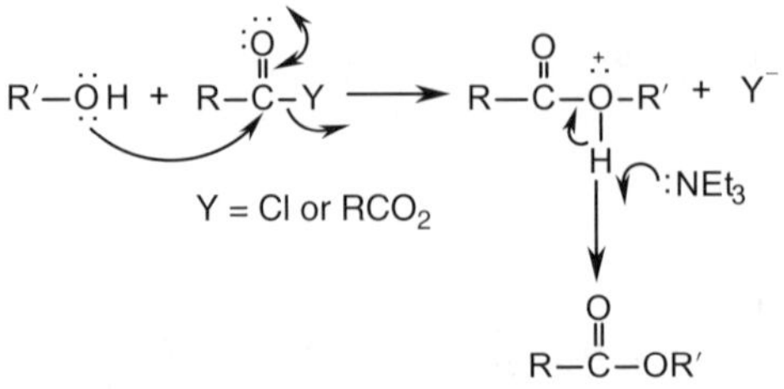

R'—ÖH + R—C̈—Y $\longrightarrow$ R—C̈—Ö—R' + Y⁻

Y = Cl or RCO₂

:NEt₃

R—C̈—OR'

5.6.5.6 Conversion of Acid Chlorides by Sodium Carboxylate

Preparation of Acid Anhydrides

Acid chloride reacts with sodium carboxylate to give acid anhydride through a nucleophilic acyl substitution reaction. Both symmetrical and unsymmetrical acid anhydrides are prepared in this way.

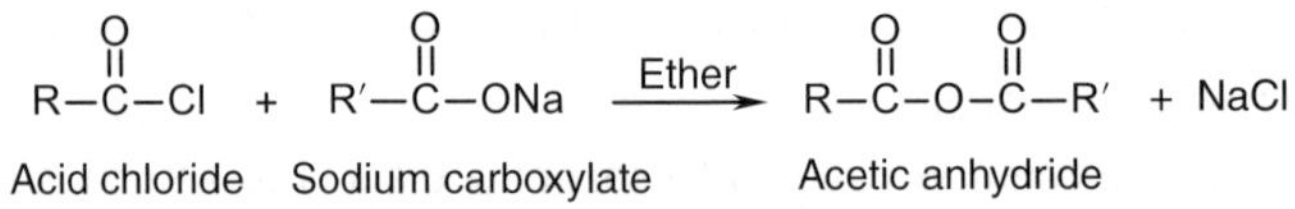

R—C̈—Cl + R'—C̈—ONa $\xrightarrow{\text{Ether}}$ R—C̈—O—C̈—R' + NaCl

Acid chloride   Sodium carboxylate   Acetic anhydride

Mechanism

R'—C̈—ONa + C=O → R—C̈—O—C̈—R' + NaCl

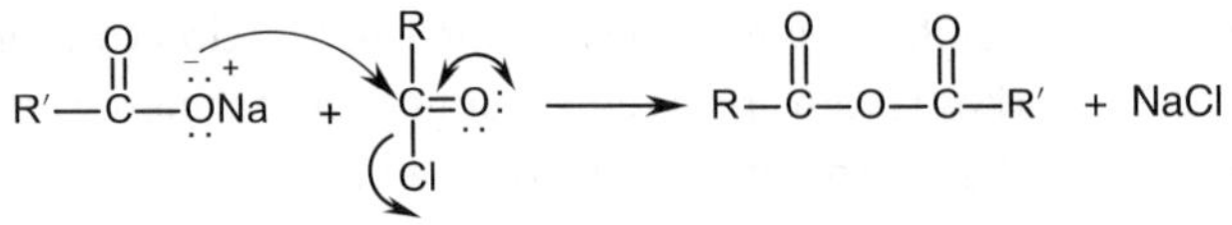

5.6.5.7 Conversion of Acid Chlorides or Anhydrides by Ammonia or Amines

Preparation of Amides

Ammonia (NH₃), 1° or 2° amines (RNH₂ or R₂NH) react with acid chlorides or anhydrides (RCOCl or RCOCO₂R) to yield 1°, 2° and 3° amides, respectively, in the presence of an excess pyridine (C₅H₅N) or triethylamine (Et₃N). In the case of an acid anhydride, 2 M equivalent of ammonia or amines is required.

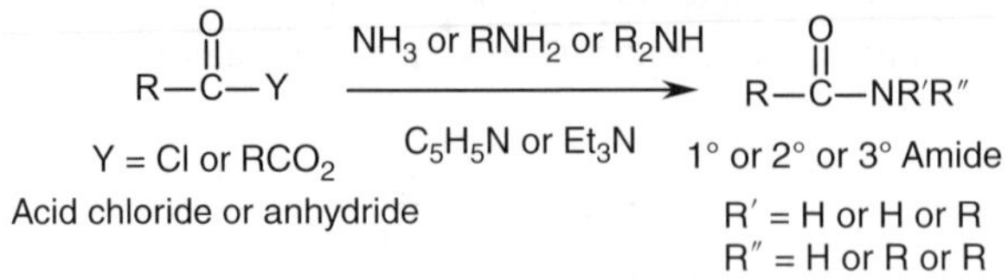

R—C̈—Y $\xrightarrow[\text{C}_5\text{H}_5\text{N or Et}_3\text{N}]{\text{NH}_3 \text{ or RNH}_2 \text{ or R}_2\text{NH}}$ R—C̈—NR'R''

Y = Cl or RCO₂   1° or 2° or 3° Amide
Acid chloride or anhydride

R' = H or H or R
R'' = H or R or R

Mechanism

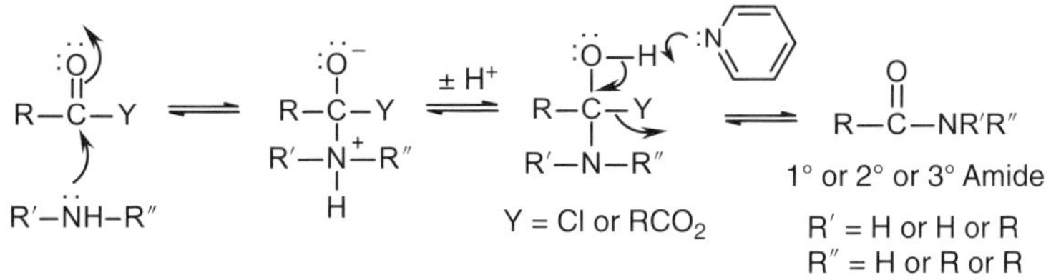

R'—NH–R"   Y = Cl or RCO₂

1° or 2° or 3° Amide

R' = H or H or R
R" = H or R or R

5.6.5.8 Conversion of Acid Chlorides and Esters by Organometallics

Preparation of Tertiary Alcohols and Ketones

Acid chloride and esters react with two equivalents of Grignard or organolithium reagents to produce a 3° alcohol. A ketone is formed by the first molar equivalent of Grignard reagent, and this immediately reacts with a second equivalent to produce the alcohol. The final product contains two identical alkyl groups at the alcohol carbon that are both derived from the Grignard reagent. This is a good route for the preparation of 3° alcohols with two identical alkyl substituents.

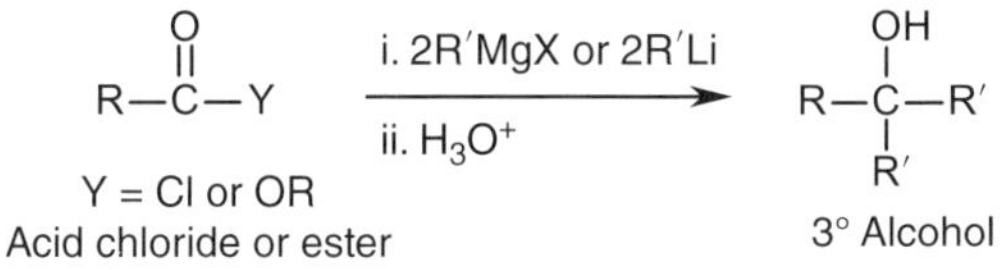

i. 2R'MgX or 2R'Li

ii. H₃O⁺

Y = Cl or OR
Acid chloride or ester

3° Alcohol

Mechanism

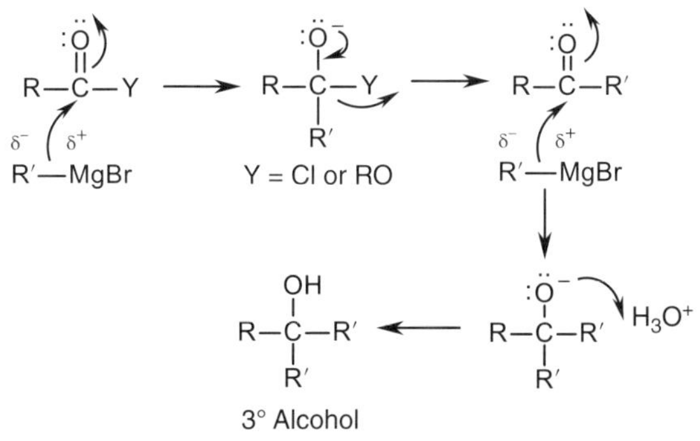

Using a weaker organometallic reagent such as Gilman reagent (R₂CuLi, organocuprate), the reaction of acid chloride can be stopped at the ketonic stage. The reaction is carried out at –78°C in ether solution and a ketone is obtained after hydrolytic work-up. Gilman reagents do not react with aldehydes, ketones, esters, amides or acid anhydrides. Thus, in the presence of other carbonyl functionalities, acid chloride selectively reacts with Gilman reagents.

$$
\underset{\text{Acid chloride}}{R-\overset{\overset{\displaystyle O}{\|}}{C}-Cl} \quad \xrightarrow[\text{ii. } H_2O]{\text{i. } R'_2CuLi,\ \text{ether}} \quad \underset{\text{Ketone}}{R-\overset{\overset{\displaystyle O}{\|}}{C}-R'}
$$

5.6.5.9 Claisen Condensation: Base-Catalysed Preparation of β-Ketoester or 1,3-Dicarbonyl Compound

When two molecules of esters undergo a condensation reaction, the reaction is called a *Claisen condensation.* Claisen condensation, like aldol condensation, requires a highly concentrated strong base. However, aqueous NaOH cannot be used in Claisen condensation because the ester can be hydrolysed by aqueous base. Therefore, most commonly used bases are non-aqueous; for example, sodium ethoxide (NaOEt) in EtOH and sodium methoxide (NaOMe) in MeOH. The product of the nucleophilic attack of an ester enolate always yields a *1,3-dicarbonyl compound.*

The enolate anion attacks the carbonyl carbon of a second molecule of ester and gives a β-ketoester (1,3-dicarbonyl compound). Therefore, Claisen condensation is a nucleophilic acyl substitution reaction. For example, two molecules of ethyl acetate condense together to form the enolate of ethyl acetoacetate, which upon addition of an acid produces ethyl acetoacetate (β-ketoester).

$$
\underset{\text{Ethyl acetates}}{H_3C-\overset{\overset{\displaystyle O}{\|}}{C}-OC_2H_5 \ + \ H_3C-\overset{\overset{\displaystyle O}{\|}}{C}-OC_2H_5} \quad \xrightarrow[\text{EtOH}]{\text{NaOEt}} \quad \underset{\substack{\text{Ethyl acetoacetate} \\ \textit{β-Ketoester}}}{H_3C-\overset{\overset{\displaystyle O}{\|}}{C}-CH_2-\overset{\overset{\displaystyle O}{\|}}{C}-OC_2H_5}
$$

As in the aldol condensation (see Section 5.4.3.12), one molecule of ester is converted to an enolate anion when an α-proton is removed by a strong base; for example, NaOEt in EtOH. It is important that one must choose the same base as the ester group (either NaOEt in EtOH or NaOMe in MeOH) to avoid getting a transesterification side product.

Base-catalysed mechanism
Removal of an α-hydrogen from the ethyl acetate by NaOEt produces an enolate anion. Nucleophilic attack of the enolate anion to the carbonyl carbon of another ethyl acetate gives an alkoxide tetrahedral intermediate. The resulting alkoxide collapses to reform the carbonyl group to produce ethyl acetoacetate and reproduces the base.

$$\underset{C_2H_5\ddot{O}^-}{\overset{:\ddot{O}:}{\underset{}{CH_2-\overset{\|}{C}-OC_2H_5}}}\;\rightleftharpoons\;\underset{\text{Enolate anion}}{H_2\bar{C}-\overset{:\ddot{O}}{\overset{\|}{C}}-OC_2H_5}$$

$$\underset{H_5C_2O-\overset{:O:}{\overset{\|}{C}}-\bar{C}H_2}{H_3C-\overset{:\ddot{O}:}{\overset{\|}{C}}-OC_2H_5}\;\longrightarrow\;H_3C-\overset{O}{\overset{\|}{C}}-CH_2-\overset{O}{\overset{\|}{C}}-OC_2H_5$$

5.6.5.10 Mixed-Claisen Condensation: Base-Catalysed Preparation of 1,3-Dicarbonyl Compound

Claisen condensation between two different ester molecules is called *mixed-Claisen condensation*. It is also possible to achieve *mixed-Claisen condensation* when one of the esters has no α-hydrogen and the other ester has a α-hydrogen, so the nucleophile is generated solely from one molecule. Thus, in the mixed-aldol condensation, the reaction takes place between an enolate and the carbonyl compound that has no α-hydrogen to produce 1,3-dicabonyl compound. For example, ethyl 2-methyl-3-oxopropanoate can be synthesized by reacting ethyl formate with ethyl propanoate using a strong base such as NaOEt in EtOH. The reaction mechanism is exactly same as the base-catalysed Claisen condensation.

$$\underset{\text{Ethyl formate}}{H-\overset{O}{\overset{\|}{C}}-OEt}\;+\;\underset{\text{Ethyl propanoate}}{CH_3CH_2-\overset{O}{\overset{\|}{C}}-OEt}\;\xrightarrow[\text{EtOH}]{\text{NaOEt}}\;\underset{CH_3}{H-\overset{O}{\overset{\|}{C}}-CH-\overset{O}{\overset{\|}{C}}-OEt}$$

Ethyl 2-methyl-3-oxopropanoate
1,3-*Dicarbonyl compound*

5.6.6 Substitution Versus Elimination

A brief comparison of key features of a substitution and an elimination reaction is presented here.

i. The strength of a base or nucleophile will dictate the order of a reaction. Thus, strong nucleophiles and bases will react more quickly and create second order kinetics.

ii. Primary halides usually undergo S_N2 with good nucleophiles, and rearrangements of 1° halides readily produce more stable cations if ionization is possible.

iii. Tertiary halides usually do not undergo S_N2 reactions, so are more likely to undergo E2 with a good base, or E1 and S_N1 otherwise.

iv. Secondary halides can react in all ways, thus they are hard to predict.

v. High temperatures always favour elimination.

vi. A moderately strong nucleophile and base will usually favour one or other type of reaction. For example, metal alkoxides (*t*-butoxide) favour elimination, bromide and iodide favour substitution.

5.7 ELECTROPHILIC SUBSTITUTIONS

Electrophilic aromatic substitution is a reaction where a hydrogen atom in an aromatic system, for example benzene, is replaced by an electrophile. There is a wide variety of electrophiles that can be introduced into a benzene ring in this way, and so electrophilic aromatic substitution is a very important method for the synthesis of substituted aromatic compounds. Some of the important electrophilic substitution reactions are halogenation, nitration, sulphonation, Friedel–Crafts alkylation and Friedel–Crafts acylation of benzene.

5.7.1 Electrophilic Substitution of Benzene

Benzene reacts with an electrophile (E^+) (usually in the presence of a Lewis acid catalyst) to form the corresponding substituted product.

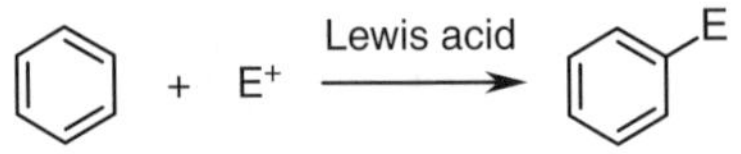

Mechanism
The electrophile takes two electrons of the six-electron π system to form a σ bond to one of the carbon atoms of the benzene ring. The arenium ion loses a proton from the carbon atom that bears the electrophile to produce the substituted benzene.

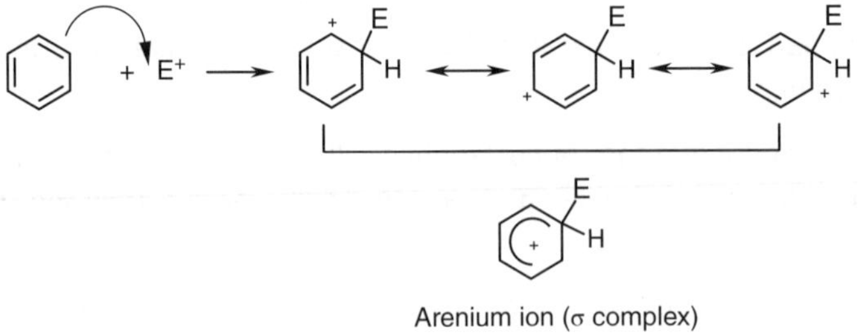

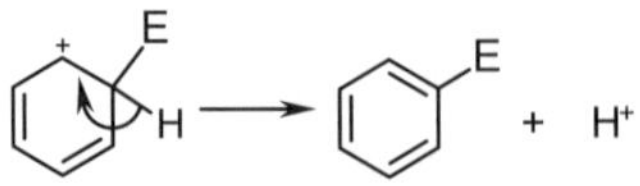

Arenium ion (σ complex)

5.7.1.1 Halogenation of Benzene

Halogen itself is not electrophilic enough to react with benzene, but the addition of a strong Lewis acid (electron pair acceptor) catalyses the reaction and leads to the substitution product. Thus, in the presence of anhydrous Lewis acid (e.g. $FeCl_3$ or $FeBr_3$), benzene reacts readily with halogens (bromine or chlorine) to produce halobenzenes (bromobenzene or chlorobenzene). Fluorine (Fl_2) reacts so rapidly with benzene that it requires special conditions and apparatus to carry out fluorination. On the other hand, iodine (I_2) is so unreactive that an oxidizing agent, for example HNO_3, has to be used to carry out iodination.

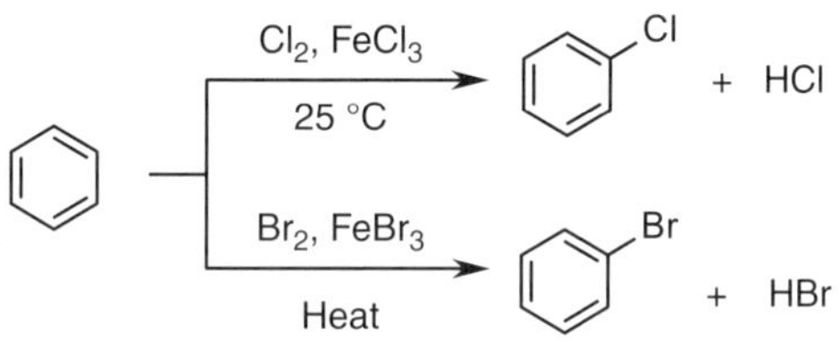

Bromination of benzene follows the same general mechanism as electrophilic aromatic substitution. The bromine molecule reacts with $FeBr_3$ by donating a pair of its electrons to it, which creates a more polar Br—Br bond.

Mechanism

Step 1: Formation of carbocation (halonium ion)

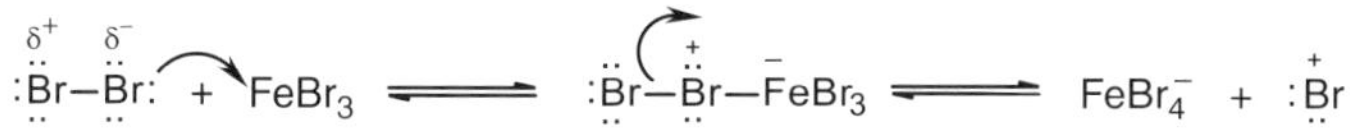

Step 2: Formation of arenium ion complex

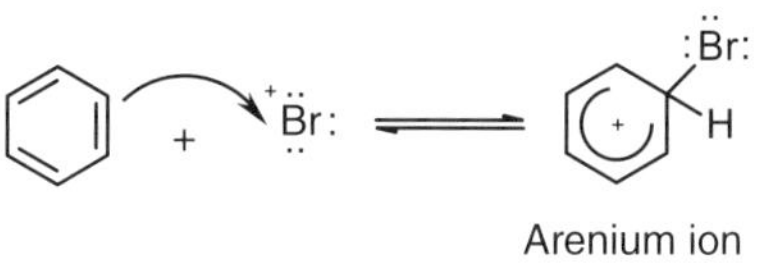

Arenium ion

Step 3: Loss of a proton from the arenium ion complex

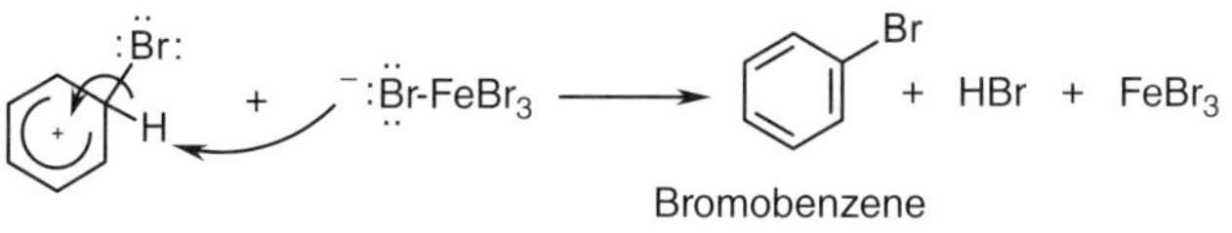

Bromobenzene

5.7.1.2 Nitration of Benzene

Benzene reacts slowly with hot concentrated nitric acid (HNO_3) to afford nitrobenzene. The reaction can be faster if a mixture of concentrated HNO_3 and concentrated sulphuric acid (H_2SO_4), which acts as a catalyst, is used. The reaction mechanism is similar to an acid-catalysed dehydration of alcohol (see Section 5.5.3). Sulphuric acid (H_2SO_4) is stronger than nitric acid (HNO_3). Thus, sulphuric acid protonates HNO_3. Loss of water (a better leaving group) from protonated HNO_3 forms a nitronium ($^+NO_2$) ion, the electrophile required for nitration. In fact, concentrated H_2SO_4 increases the rate of the reaction by increasing the concentration of the electrophiles ($^+NO_2$).

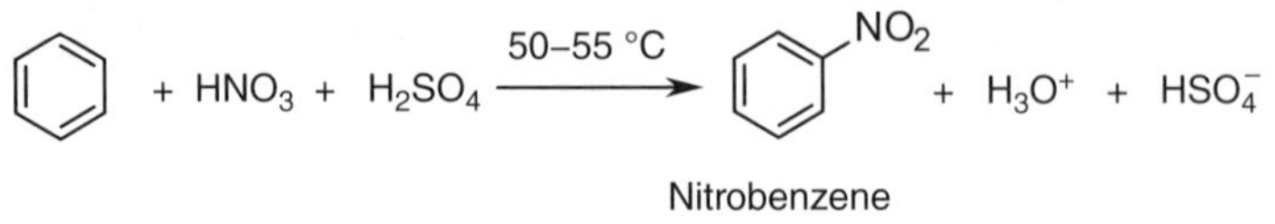

Nitrobenzene

Mechanism

Step 1: Generation of the nitronium ion ($^+NO_2$), an electrophile

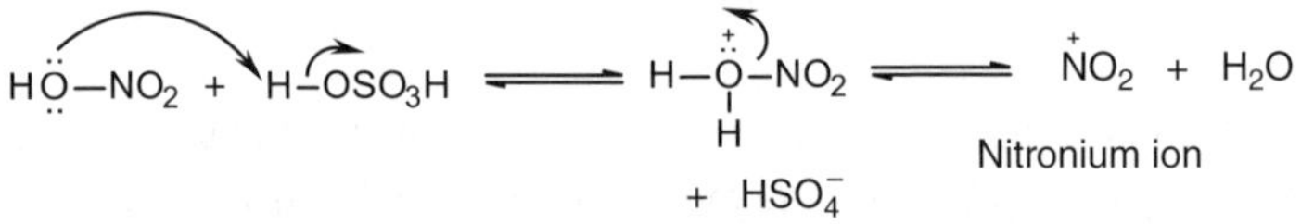

Nitronium ion

Step 2: Formation of an arenium ion complex

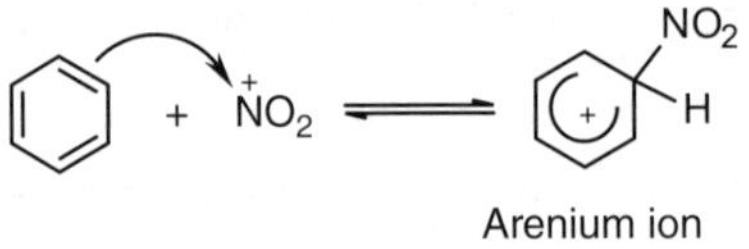

Arenium ion

Step 3: Loss of a proton from the arenium ion complex

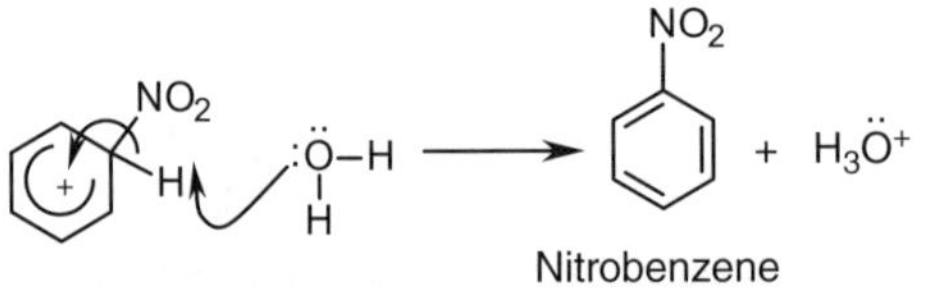

Nitrobenzene

5.7.1.3 Sulphonation of Benzene

Benzene reacts with fuming sulphuric acid at room temperature to give benzene-sulphonic acid. Fuming sulphuric acid contains added sulphur trioxide (SO_3). Sulphonation of benzene can also be carried out with concentrated H_2SO_4, but at a slower speed. In both cases, SO_3 acts as an electrophile.

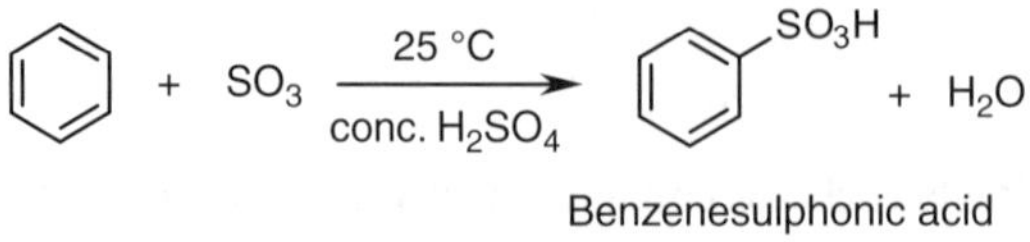

Benzenesulphonic acid

Mechanism

Step 1: Generation of SO_3, an electrophile

$$2H_2SO_4 \rightleftharpoons SO_3 + H_3O^+ + HSO_4^-$$

Step 2: Formation of arenium ion complex

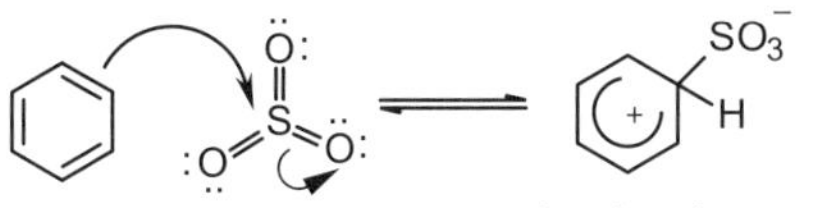

Arenium ion

Step 3: Loss of a proton from the arenium ion complex

Benzenesulphonate ion

Step 4: Protonation of the benzenesulphonate anion

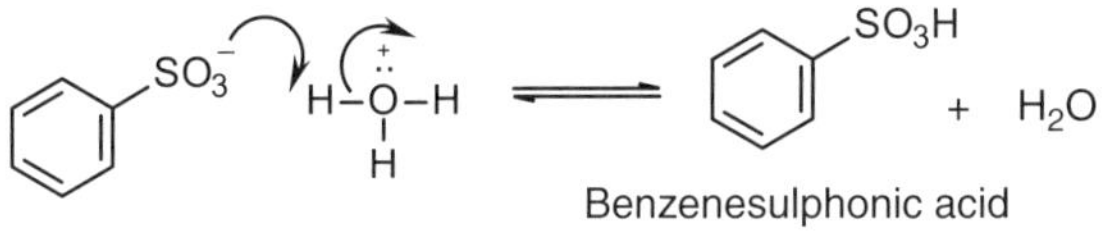

Benzenesulphonic acid

5.7.1.4 Friedel–Crafts Alkylation

First introduced by Charles Friedel and James Crafts in 1877, FC alkylation is an electrophilic aromatic substitution reaction where electrophile is a carbocation, R^+. This carbocation is generated by $AlCl_3$-catalysed ionization of alkyl halide. For example, benzene reacts with isopropylchloride in the presence of Lewis acid to produce isopropylbenzene.

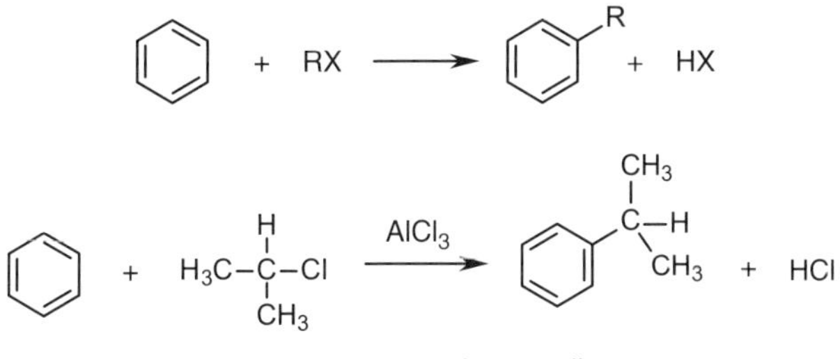

Isopropylchloride   Isopropylbenzene

Mechanism

Step 1: Formation of carbocation

Isopropylchloride   Carbocation

Step 2: Formation of arenium ion complex

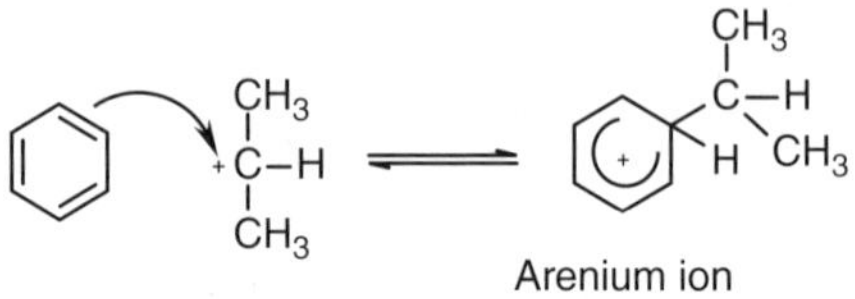

Arenium ion

Step 3: Loss of a proton from the arenium ion

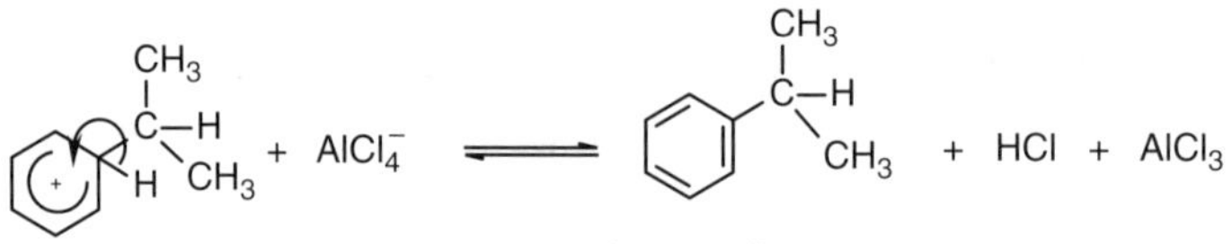

Isopropylbenzene

In the case of a 1° alkyl halide, a simple carbocation does not form. $AlCl_3$ forms a complex with a 1° alkyl halide and this complex acts as an electrophile. While this complex is not a simple carbocation, it acts as if it were and transfers a positive alkyl group to the aromatic ring.

$$\overset{\delta^+}{RCH_2} \text{------------------------} \overset{\delta^-}{Cl} : AlCl_3$$

FC alkylations are not restricted to the use of RX and $AlCl_3$. Many other pairs of reagents that form carbocations (or carbocation-like species) may be used. For example, an alkene and an acid, or an alcohol and an acid could be used.

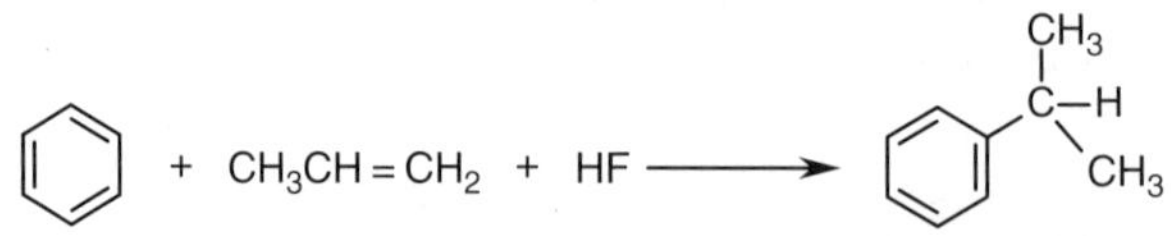

Benzene   Propene   Hydrofluoric acid   Isopropylbenzene

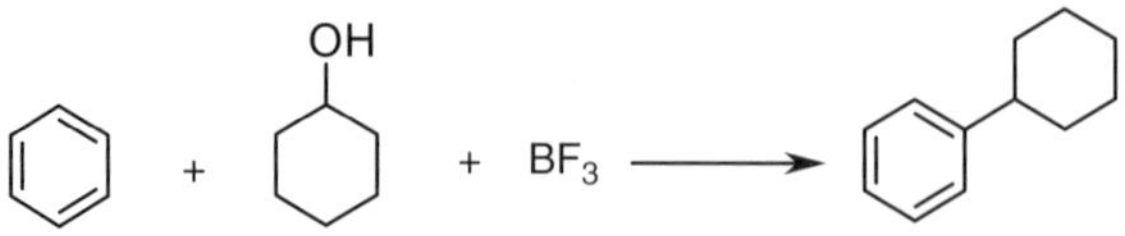

Benzene   Cyclohexanol Boron trifluoride   Cyclohexbenzene

Limitations of FC Alkylation

FC alkylations are limited to alkyl halides. Aryl or vinyl halides do not react.

FC alkylation does not occur on aromatic ring containing strong *electron withdrawing* substituents; for example, a $-NO_2$, $-CN$, $-CHO$, $-COR$, $-NH_2$, $-NHR$ or $-NR_2$ group.

Multiple substitutions often take place. Carbocation rearrangements may occur, which result in multiple products.

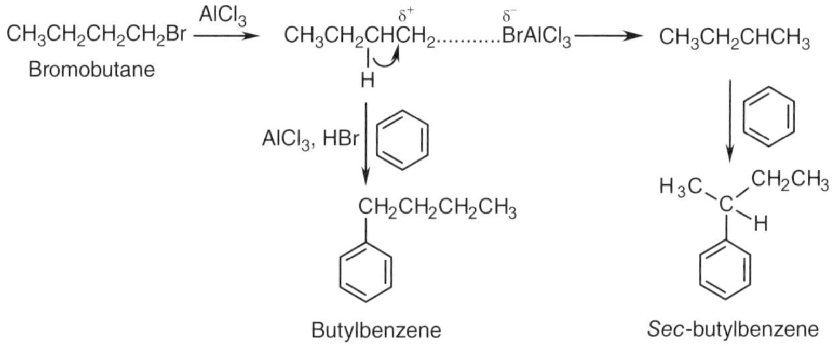

$$CH_3CH_2CH_2CH_2Br \xrightarrow{AlCl_3} CH_3CH_2\overset{\overset{\displaystyle\delta^+}{}}{C}HCH_2\ldots\ldots\overset{\displaystyle\delta^-}{BrAlCl_3} \longrightarrow CH_3CH_2CHCH_3$$

Bromobutane

AlCl₃, HBr

CH₂CH₂CH₂CH₃

Butylbenzene

H_3C, CH₂CH₃

Sec-butylbenzene

5.7.1.5 Friedel–Crafts Acylation

First introduced by Charles Friedel and James Crafts, FC acylation places an acyl group on a benzene ring. Either an acyl halide or an acid anhydride can be used for FC acylation. The acylium ion is the required electrophile, which is formed by the reaction of an acid chloride (acetyl chloride) or an acid anhydride (acetic anhydride) with a Lewis acid ($AlCl_3$).

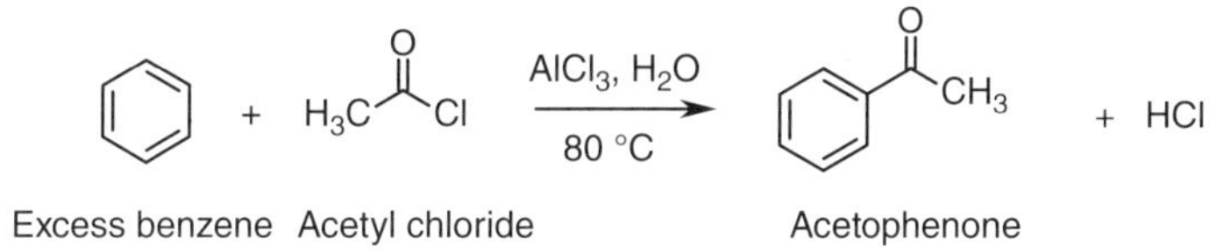

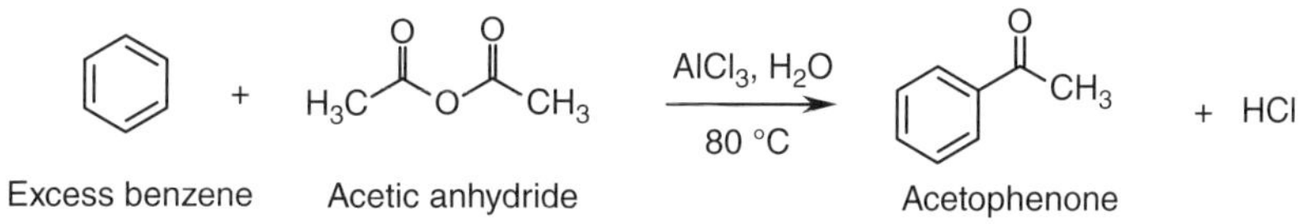

Excess benzene Acetyl chloride

$\xrightarrow[80\ °C]{AlCl_3,\ H_2O}$

Acetophenone + HCl

Excess benzene Acetic anhydride

$\xrightarrow[80\ °C]{AlCl_3,\ H_2O}$

Acetophenone + HCl

Mechanism
Step 1: Formation of acylium ion, a carbocation

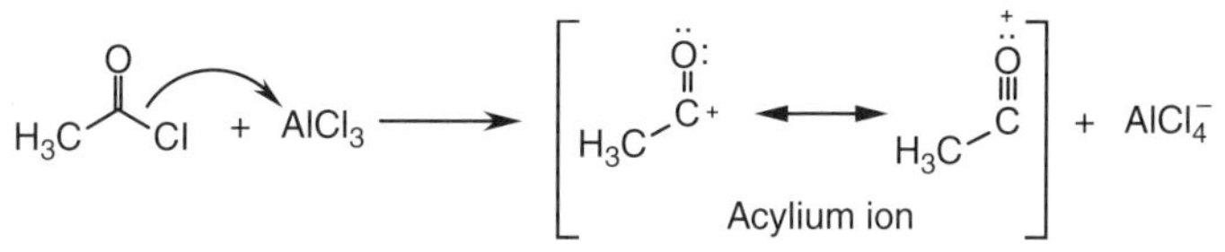

Acylium ion

Step 2: Formation of arenium ion complex

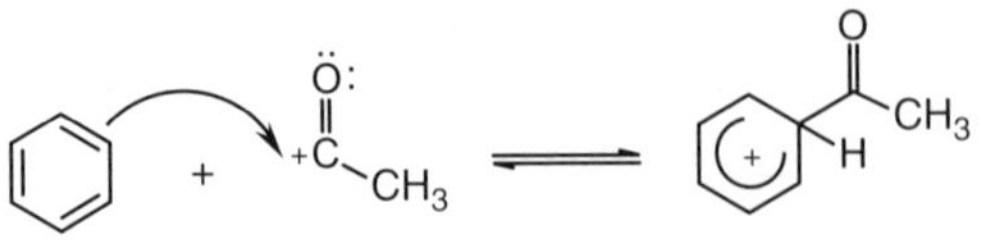

Arenium ion

Step 3: Loss of a proton from the arenium ion complex

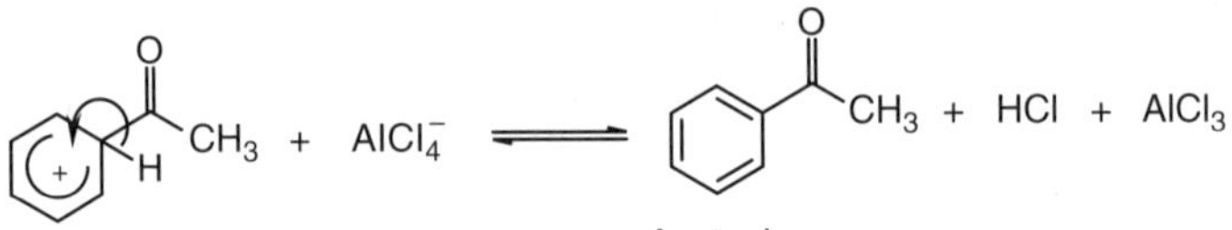

Acetophenone

5.8 HYDROLYSIS

The term *hydrolysis* comes from the word *'hydro'* meaning water and *'lysis'* meaning breakdown. A hydrolysis reaction is one in which a σ bond is cleaved by water. Usually, a hydrolysis reaction is catalysed by acid, base or hydrolysing enzyme. For example, the analgesic drug aspirin (acetyl salicylic acid) is easily hydrolysed in the presence of acid, moisture and heat to form salicylic acid.

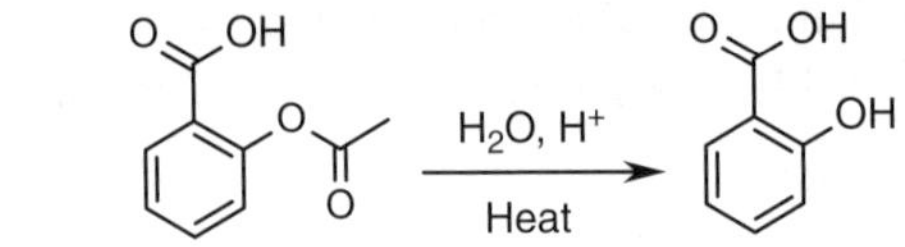

Aspirin
Acetyl salicylic acid

Salicylic acid

Glucosidase is a hydrolysing enzyme and can be used to hydrolyse various glucosides. For example, salicin, found in Willow barks, can be hydrolysed to salicyl alcohol by enzymes.

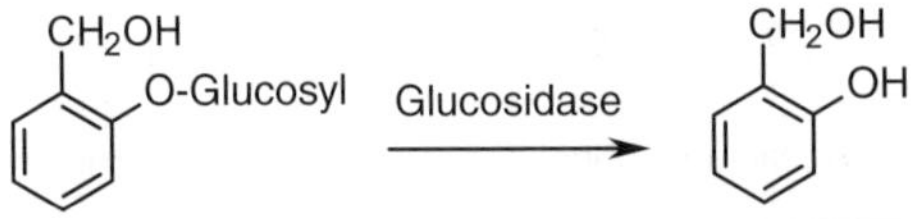

Salicin

Salicyl alcohol

5.8.1 Hydrolysis of Carboxylic Acid Derivatives

All carboxylic acid derivatives yield parent carboxylic acids on hydrolysis, catalysed either by an acid or a base. The reactivity towards hydrolysis varies greatly among the derivatives.

5.8.1.1 Hydrolysis of Acid Chlorides and Anhydrides

Preparation of Carboxylic Acids

Acid chlorides and anhydrides are highly reactive and they react with water under neutral conditions. This can be a potential problem for storage if these compounds since these compounds can be air (moisture) sensitive. Hydrolysis of these compounds can be avoided by using dry nitrogen atmospheres and anhydrous solvents and reagents. Thus, carboxylic acids are easily synthesized from acid chlorides and anhydrides by hydrolysis without any catalysts.

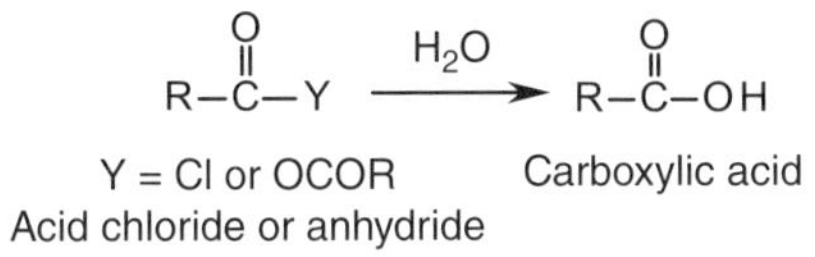
R–C(=O)–Y
H2O
R–C(=O)–OH
Y = Cl or OCOR
Acid chloride or anhydride
Carboxylic acid

Mechanism

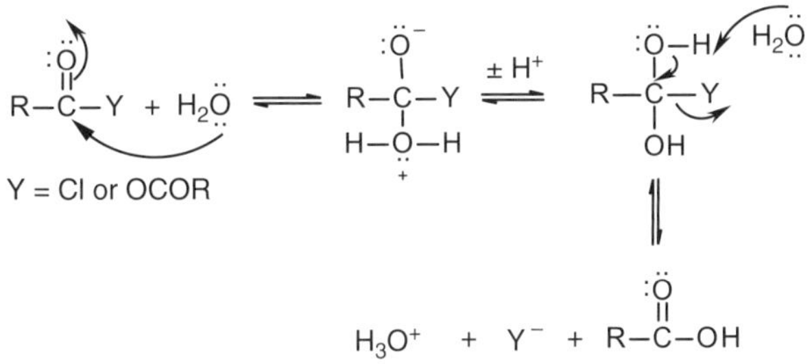
Y = Cl or OCOR
± H+
H2O
OH
H3O+ + Y− + R–C–OH

5.8.1.2 Hydrolysis of Esters

Preparation of Carboxylic Acids

The acid-catalysed hydrolysis of an ester is the reverse reaction of Fischer esterification (see Section 5.6.5.1). Addition of excess water drives the equilibrium towards the acid and alcohol formation. The base-catalysed hydrolysis is also known as *saponification*, and this does not involve the equilibrium process observed for Fischer esterification.

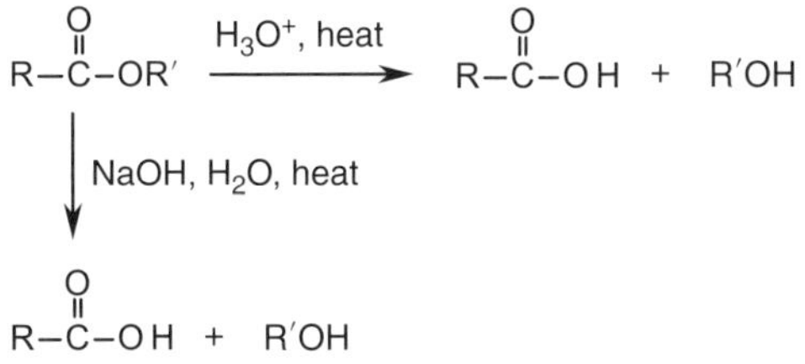
R–C(=O)–OR′
H3O+, heat
R–C(=O)–OH + R′OH
NaOH, H2O, heat
R–C(=O)–OH + R′OH

Acid-catalysed mechanism

The carbonyl group of an ester is not sufficiently electrophilic to be attacked by water. The acid catalyst protonates the carbonyl oxygen, and activates it towards nucleophilic attack. The water molecule attacks the protonated carbonyl carbon and forms a tetrahedral intermediate. Proton transfer from the hydronium ion to a second molecule of water yields an ester hydrate. Intramolecular proton transfer produces a good leaving group as an alcohol. A simultaneous deprotonation by the water and loss of alcohol affords a carboxylic acid.

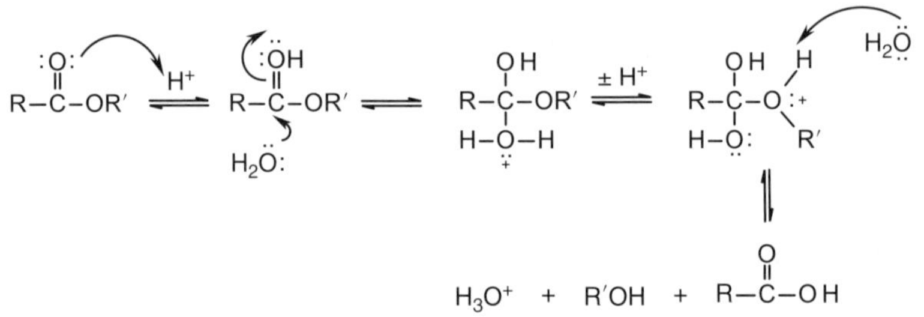

Base-catalysed mechanism

Hydroxide ion attacks the carbonyl group to generate a tetrahedral intermediate. The negatively charged oxygen can readily expel an alkoxide ion, a basic leaving group, and produce a carboxylic acid. The alkoxide ion quickly deprotonates the carboxylic acid and the resulting carboxylate ion is unable to participate in the reverse reaction. Thus, there is no equilibrium in the base-catalysed hydrolysis and the reaction goes to completion. Protonation of the carboxylate ion by addition of an aqueous acid in a separate step produces the free carboxylic acid.

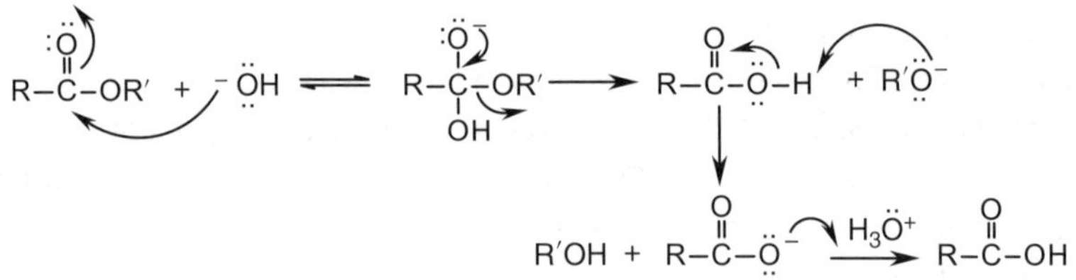

5.8.1.3 Hydrolysis of Amides

Preparation of Carboxylic Acids

Amides are the most reluctant derivatives of carboxylic acids to undergo hydrolysis. However, they can be forced to undergo hydrolysis by the use of vigorous conditions; for example, heating with 6 M HCl or 40% NaOH for prolonged periods of time.

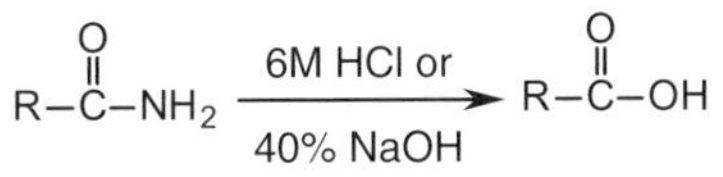

$$R-\overset{\overset{\displaystyle O}{\|}}{C}-NH_2 \quad \xrightarrow[\text{40\% NaOH}]{\text{6M HCl or}} \quad R-\overset{\overset{\displaystyle O}{\|}}{C}-OH$$

Acid-catalysed mechanism

Under acidic conditions, the hydrolysis of an amide resembles the acid-catalysed hydrolysis of an ester (see Section 5.8.1.2), with protonation of the carbonyl group providing an activated carbonyl group that undergoes nucleophilic attack by water. The intramolecular proton transfer produces a good leaving group as ammonia. Simultaneous deprotonation by water and loss of ammonia yields a carboxylic acid.

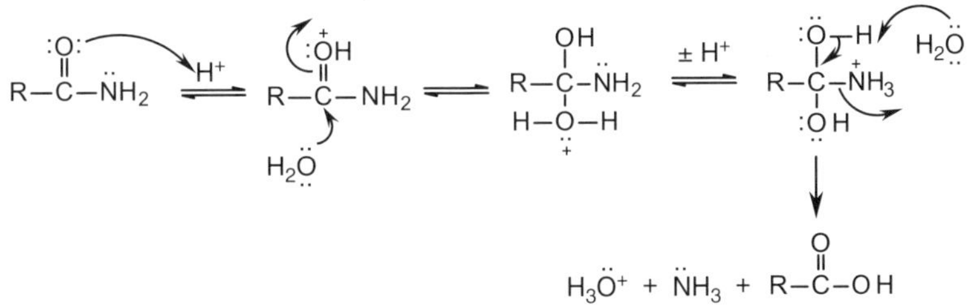

$$H_3\overset{..}{O}{}^+ \;+\; \overset{..}{N}H_3 \;+\; R-\overset{\overset{\displaystyle O}{\|}}{C}-OH$$

Base-catalysed mechanism

Hydroxide ion attacks the carbonyl, and forms a tetrahedral intermediate. The negatively charged oxygen can readily expel amide ion, a basic leaving group, and produce a carboxylic acid. The amide ion quickly deprotonates the carboxylic acid and the resulting carboxylate ion is unable to participate in the reverse reaction. Thus, there is no equilibrium in the base-catalysed hydrolysis and the reaction goes to completion. Protonation of the carboxylate by the addition of an aqueous acid in a separate step forms the free carboxylic acid.

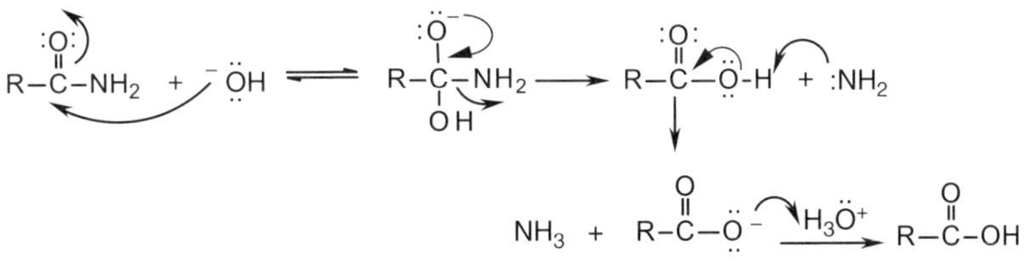

$$NH_3 \;+\; R-\overset{\overset{\displaystyle O}{\|}}{C}-\overset{..}{\underset{..}{O}}{}^- \quad \xrightarrow{H_3\overset{..}{O}{}^+} \quad R-\overset{\overset{\displaystyle O}{\|}}{C}-OH$$

5.8.1.4 Hydrolysis of Nitriles

Preparation of Primary Amides and Carboxylic Acids

Nitriles are hydrolysed to 1° amides and then to carboxylic acids either by acid or base catalysis. It is possible to stop acid hydrolysis at the amide stage by using H_2SO_4 as an acid catalyst and one mole of water per mole of nitrile. A mild basic condition (NaOH, H_2O, 50°C) only takes the hydrolysis to the amide stage and a much higher temperature (NaOH, H_2O, 200°C) is required to convert the amide to a carboxylic acid.

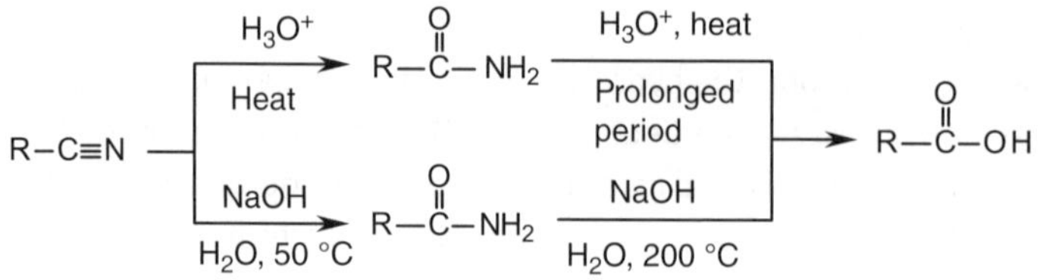

Acid-catalysed mechanism

The acid-catalysed hydrolysis of nitriles resembles the acid-catalysed hydrolysis of an amide, with protonation of the nitrogen of the cyano group activated the nucleophilic attack by water. The intramolecular proton-transfer produces a protonated imidic acid. The imidic acid tautomerizes to the more stable amide via deprotonation on oxygen and protonation on nitrogen. The acid-catalysed amide is converted to carboxylic acid in several steps, as discussed earlier for the hydrolysis of amides.

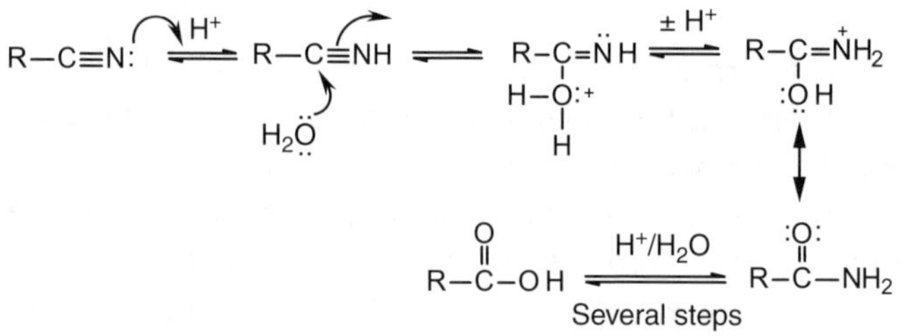

Base-catalysed mechanism

The hydroxide ion attacks the nitrile carbon, followed by the protonation on the unstable nitrogen anion to generate an imidic acid. The imidic acid tautomerizes to the more stable amide via deprotonation on oxygen and protonation on nitrogen. The base-catalysed amide is converted to carboxylic acid in several steps, as discussed earlier for the hydrolysis of amides.

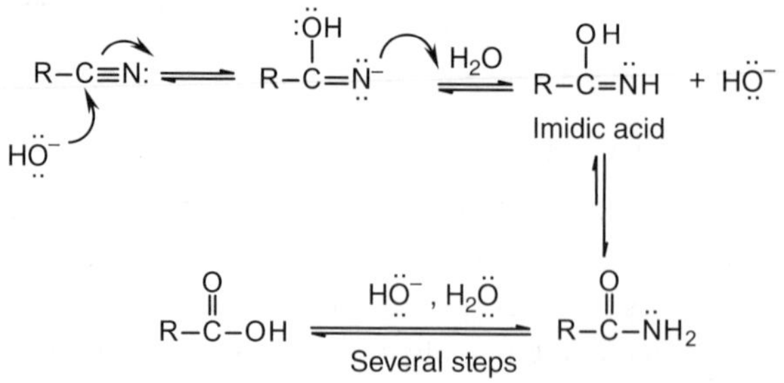

5.9 OXIDATION–REDUCTION REACTIONS

Oxidation is a loss of electrons and reduction is a gain of electrons. However, in the context of organic chemistry, *oxidation* means the loss of hydrogen, the addition of oxygen or the addition of halogen. A general symbol for oxidation is [O]. Thus, oxidation can also be defined as a reaction that increases the content of any element more electronegative than carbon. *Reduction* is the addition of hydrogen, the loss of oxygen or the loss of halogen. A general symbol for reduction is [H]. The conversion of ethanol to acetaldehyde and acetaldehyde to acetic acid are oxidation reactions, and the reverse reactions are reduction reactions.

Ethanol   Acetaldehyde   Acetic acid

5.9.1 Oxidizing and Reducing Agents

Oxidizing agents are reagents that seek electrons and are electron-deficient species; for example, chromic acid (H_2CrO_4), potassium permanganate ($KMnO_4$), osmium tetroxide (OsO_4) and nitric acid (HNO_3). Therefore, oxidizing agents are classified as *electrophiles*. In the process of gaining electrons, oxidizing agents become reduced. Oxidation results in an increase in the number of C—O bonds or a decrease in the number of C—H bonds.

On the other hand, *reducing agents* are reagents that give up electrons and are electron-rich species; for example, sodium borohydride ($NaBH_4$) and lithium aluminiumhydride ($LiAlH_4$). Therefore, reducing agents are classified as *nucleophiles*. In the process of giving up electrons, reducing agents become oxidized. Reduction results in an increase in the number of C—H bonds or a decrease in the number of C—O bonds.

5.9.2 Oxidation of Alkenes

5.9.2.1 Preparation of Epoxides

Alkenes undergo a number of oxidation reactions in which the C=C is oxidized. The simplest epoxide, ethylene oxide, is prepared by catalytic oxidation of ethylene with Ag at high temperatures (250 °C).

Ethylene oxide

Alkenes are also oxidized to epoxides by peracid or peroxy acid (RCO_3H); for example, peroxybenzoic acid ($C_6H_5CO_3H$). A peroxyacid contains an extra oxygen atom compared to carboxylic acid and this extra oxygen is added to the double bond of an alkene to give an epoxide. For example, cyclohexene reacts with peroxybenzoic acid to produce cyclohexane oxide.

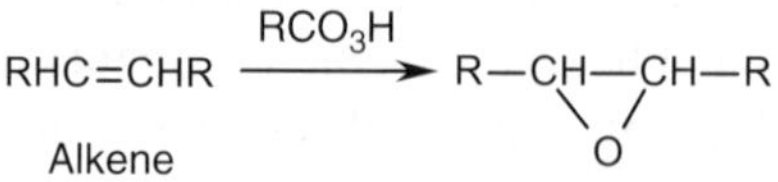

$$\text{RHC=CHR} \xrightarrow{\text{RCO}_3\text{H}} \text{R-CH-CH-R}$$

Alkene   Epoxide

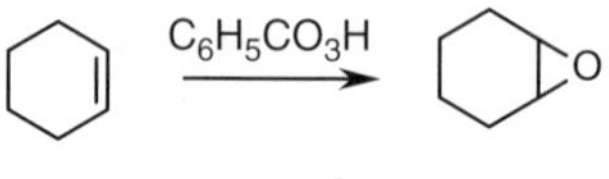

Cyclohexene   Cyclohexane oxide

The addition of oxygen to an alkene is stereospecific (see Section 5.4.2.13). Therefore, a *cis*-alkene provides a *cis*-epoxide and a *trans*-alkene affords a *trans*-epoxide.

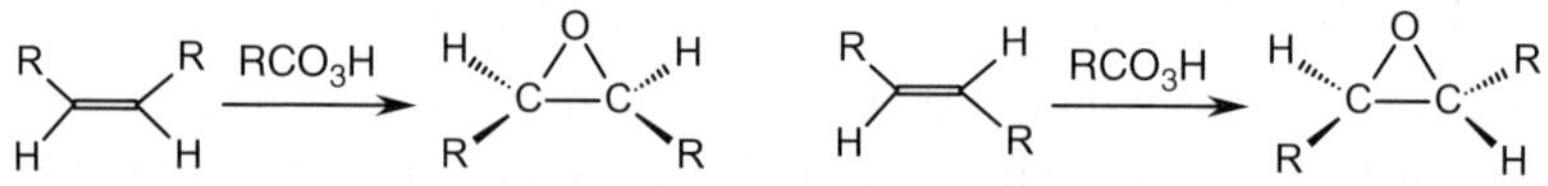

cis-Alkene   *cis*-Epoxide   *trans*-Alkene   *trans*-Epoxide

5.9.2.2 Preparation of Carboxylic Acids or Ketones

Reaction of an alkene with hot basic potassium permanganate ($KMnO_4$) results in cleavage of the double bond and formation of highly oxidized carbons. Therefore, unsubstituted carbon atoms become CO_2, monosubstituted carbon atoms become carboxylates and disubstituted carbon atoms become ketones. This can be used as a chemical test (known as the *Baeyer test*) for alkenes and alkynes in which the purple colour of the $KMnO_4$ disappears and a brown MnO_2 residue is formed.

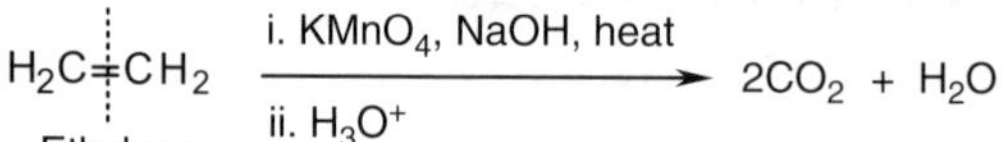

$$\text{H}_2\text{C}\text{=}\text{CH}_2 \xrightarrow[\text{ii. H}_3\text{O}^+]{\text{i. KMnO}_4,\ \text{NaOH, heat}} 2\text{CO}_2 + \text{H}_2\text{O}$$

Ethylene

$$\text{CH}_3\text{CH=CHCH}_3 \xrightarrow[\text{H}_2\text{O, heat}]{\text{KMnO}_4,\ \text{NaOH}} 2\text{CH}_3\text{-C-O}^- \xrightarrow{\text{H}_3\text{O}^+} 2\text{H}_3\text{C-C-OH}$$

(*cis* or *trans*)-2-Butene   Acetates   Acetic acids

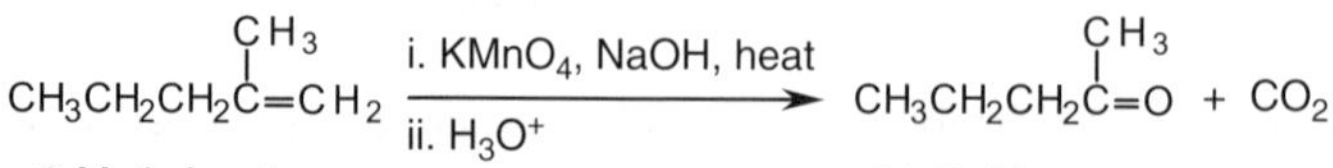

$$\text{CH}_3\text{CH}_2\text{CH}_2\overset{\overset{\text{CH}_3}{|}}{\text{C}}\text{=CH}_2 \xrightarrow[\text{ii. H}_3\text{O}^+]{\text{i. KMnO}_4,\ \text{NaOH, heat}} \text{CH}_3\text{CH}_2\text{CH}_2\overset{\overset{\text{CH}_3}{|}}{\text{C}}\text{=O} + \text{CO}_2$$

2-Methylpentene   Methyl butanone

5.9.3 Oxidation of Alkynes

Preparation of Diketones or Carboxylic Acids

Alkynes are oxidized to diketones by cold, dilute and basic potassium permanganate ($KMnO_4$).

$$R-C{\equiv}C-R' \xrightarrow[\text{ii. NaOH, } H_2O_2]{\text{i. Cold } KMnO_4} R-\overset{\overset{O}{\|}}{C}-\overset{\overset{O}{\|}}{C}-R'$$

Alkyne → Diketone

When the reaction condition is too warm or basic, the oxidation proceeds further to generate two carboxylate anions, which, on acidification, yield two carboxylic acids.

$$R-C{\equiv}C-R' \xrightarrow[\text{KOH, heat}]{KMnO_4} R-CO_2^- + {}^-O_2C-R' \xrightarrow{H_3O^+} R-CO_2H + R'-CO_2H$$

Carboxylates

Unsubstituted carbon atoms are oxidized to CO_2 and monosubstituted carbon atoms to carboxylic acids. Therefore, oxidation of 1-butyne with hot basic $KMnO_4$ followed by an acidification produces propionic acid and carbon dioxide.

$$C_2H_5C{\equiv}CH \xrightarrow[\text{ii } H_3O^+]{\text{i. } KMnO_4,\ KOH,\ \text{heat}} C_2H_5-\overset{\overset{O}{\|}}{C}-OH + CO_2$$

1-Butyne → Propionic acid

5.9.4 Hydroxylation of Alkenes

When hydroxyl groups of the diol are added to the same side of the double bond, this is as known as *syn*-addition. When hydroxyl groups are added to the opposite side of the double bond, the process is called *anti*-addition. Thus, *syn*-diol or *anti*-diol can be selectively prepared using appropriate reaction conditions.

5.9.4.1 *syn*-Hydroxylation of Alkenes

Preparation of *syn*-Diols

Hydroxylation of alkenes is the most important method for the synthesis of 1,2-diols (also called glycol). Alkenes react with cold, dilute and basic $KMnO_4$ or osmium tetroxide (OsO_4) and hydrogen peroxide to give *cis*-1,2-diols. The products are always *syn*-diols, since the reaction occurs with *syn*-addition.

Cyclopentene → (i. Cold $KMnO_4$; ii. NaOH, H_2O) → *cis*-1,2-Cyclopentane diol (A meso compound)

$H_2C{=}CH_2$ Ethene → (i. OsO_4, Pyridine; ii. H_2O_2) → *syn*-1,2-Ethanediol Ethylene glycol

5.9.4.2 *anti*-Hydroxylation of Alkenes

Preparation of *anti*-Diols

Alkenes react with peroxyacids (RCO_3H) followed by hydrolysis leading to the formation of *trans*-1,2-diols. The products are always *anti*-diols, since the reaction occurs with *anti*-addition.

Cyclopentene

trans-1,2-Cyclopentane diol
(A racemic mixture)

Propene

anti-1,2-Propanediol
Propylene glycol

5.9.5 Oxidative Cleavage of *syn*-Diols

Preparation of Ketones and Aldehydes

The treatment of an alkene to *syn*-hydroxylation, followed by a periodic acid (HIO_4) cleavage is an alternative to the *ozonolysis reduction*, followed by a reductive work-up. *Syn*-Diols are oxidized to aldehydes and ketones by periodic acid (HIO_4). This oxidation reaction divides the reactant into two pieces, thus it is called an oxidative cleavage. The reaction involves the formation of a cyclic periodate ester intermediate, which cleaves to generate two carbonyl compounds.

Alkene

syn-Diol

Ketone

Aldehyde

5.9.6 Ozonolysis of Alkenes

Alkenes can be cleaved by ozone followed by oxidative or reductive work-up to generate carbonyl compounds. The products derived from an ozonolysis reaction depend on the reaction conditions. If ozonolysis is followed by reductive work-up (Zn/H_2O), the products are aldehydes and/or ketones. Unsubstituted carbon atoms are oxidized to formaldehyde, monosubstituted carbon atoms to aldehydes and disubstituted carbon atoms to ketones. When ozonolysis is followed by oxidative work-up ($H_2O_2/NaOH$), the products are carboxylic acids and/or ketones. Unsubstituted carbon atoms are oxidized to formic acids, monosubstituted carbon atoms to carboxylic acids and disubstituted carbon atoms to ketones.

5.9.6.1 Preparation of Aldehydes and Ketones

Alkenes are directly oxidized to aldehydes and/or ketones by ozone (O_3) at very low temperatures (−78 °C) in methylene chloride, followed by reductive work-up.

For example, 2-methyl-2-butene reacts with O_3, followed by a reductive work-up to yield acetone and acetaldehyde. This reducing agent prevents aldehyde from oxidation to carboxylic acid.

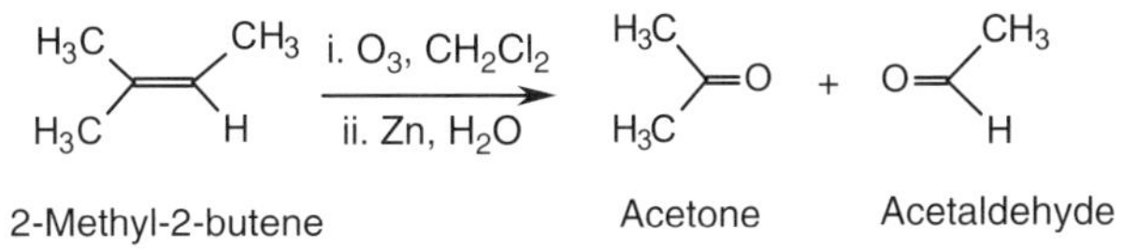

2-Methyl-2-butene   Acetone   Acetaldehyde

5.9.6.2 Preparation of Carboxylic Acids and Ketones

Alkenes are oxidized to carboxylic acids and/or ketones by ozone (O_3) at very low temperatures $(-78\,°C)$ in methylene chloride (CH_2Cl_2), followed by oxidative work-up. For example, 2-methyl-2-butene reacts with O_3, followed by an oxidative work-up to produce acetone and acetic acid.

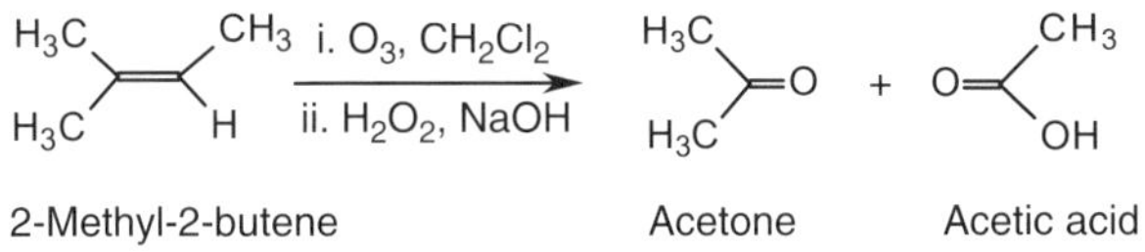

2-Methyl-2-butene   Acetone   Acetic acid

5.9.7 Ozonolysis of Alkynes

Preparation of Carboxylic Acids

Ozonolysis of alkynes followed by hydrolysis gives similar products to those obtained from permanganate oxidation. This reaction does not require oxidative or reductive work-up. Unsubstituted carbon atoms are oxidized to CO_2 and mono-substituted carbon atoms to carboxylic acids. For example, ozonolysis of 1-butyene followed by a hydrolysis forms propionic acid and carbon dioxide gas.

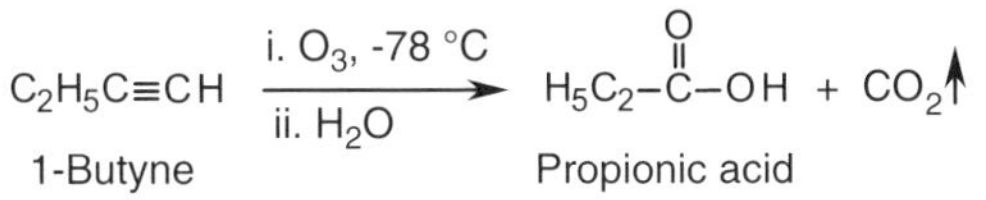

1-Butyne   Propionic acid

5.9.8 Oxidation of Alcohols

5.9.8.1 Preparation of Carboxylic Acids: Oxidation of Primary Alcohols

Primary alcohols are oxidized either to aldehydes or carboxylic acids, depending on the oxidizing reagents and conditions used. Primary alcohols are oxidized to carboxylic acids using a variety of aqueous oxidizing agents, including potassium permanganate ($KMnO_4$ in basic solution), chromic acid (H_2CrO_4 in aqueous acid) and Jones' reagent (CrO_3 in acetone). Potassium permanganate is most commonly used for oxidation of a 1° alcohol to a carboxylic acid. The reaction is generally carried out in an aqueous basic solution. A brown precipitate of MnO_2 indicates that the oxidation has taken place.

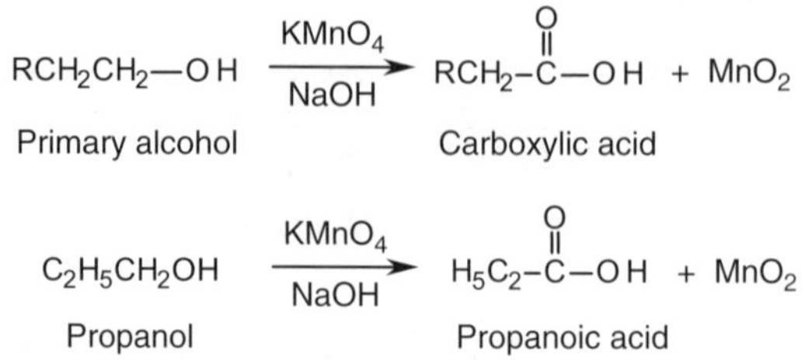

$$RCH_2CH_2\text{—}OH \xrightarrow[\text{NaOH}]{KMnO_4} RCH_2\text{-}\overset{\displaystyle O}{\overset{\displaystyle \|}{C}}\text{-}OH + MnO_2$$

Primary alcohol   Carboxylic acid

$$C_2H_5CH_2OH \xrightarrow[\text{NaOH}]{KMnO_4} H_5C_2\text{-}\overset{\displaystyle O}{\overset{\displaystyle \|}{C}}\text{-}OH + MnO_2$$

Propanol   Propanoic acid

Chromic acid is synthesized in situ by the reaction of sodium dichromate ($Na_2Cr_2O_7$) or chromic trioxide (CrO_3), sulphuric acid and water.

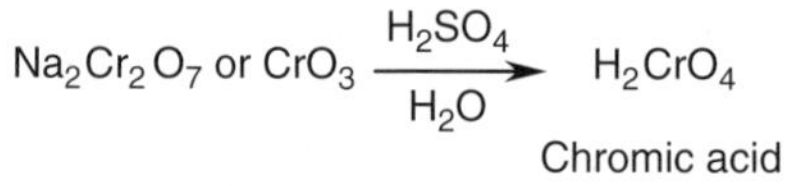

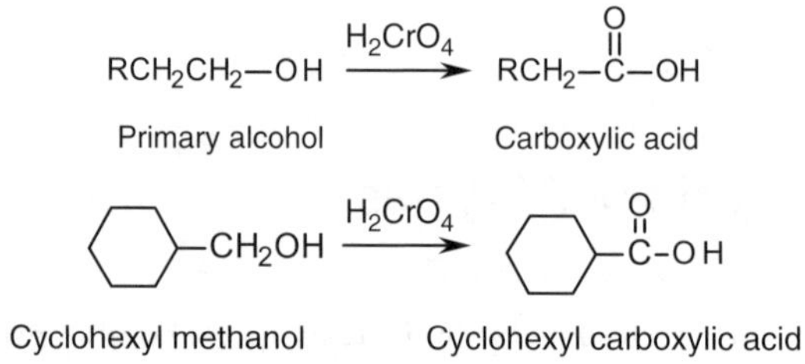

$$Na_2Cr_2O_7 \text{ or } CrO_3 \xrightarrow[\text{H}_2\text{O}]{H_2SO_4} H_2CrO_4$$

Chromic acid

$$RCH_2CH_2\text{—}OH \xrightarrow{H_2CrO_4} RCH_2\text{-}\overset{\displaystyle O}{\overset{\displaystyle \|}{C}}\text{-}OH$$

Primary alcohol   Carboxylic acid

Cyclohexyl methanol   Cyclohexyl carboxylic acid

5.9.8.2 Preparation of Aldehydes: Selective Oxidation of Primary Alcohols

A convenient reagent that selectively oxidizes primary alcohols to aldehyde is anhydrous pyridinium chlorochromate or abbreviated to PCC ($C_5H_5NH^+CrO_3Cl^-$). It is made from chromium trioxide and pyridine under acidic conditions in dry dichloromethane (CH_2Cl_2).

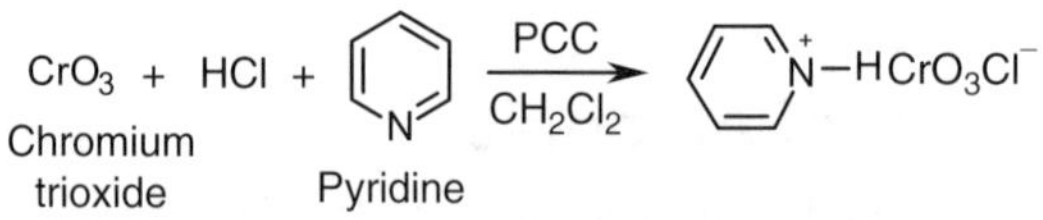

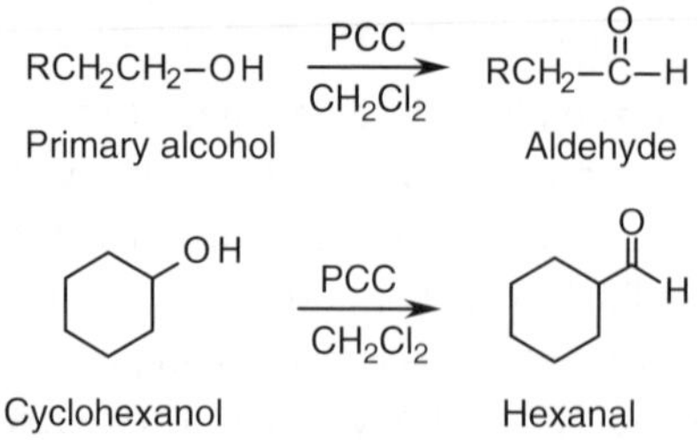

$$CrO_3 + HCl + \text{Pyridine} \xrightarrow[\text{CH}_2\text{Cl}_2]{PCC} \text{N}\text{—}HCrO_3Cl$$

Chromium trioxide   Pyridine

$$RCH_2CH_2\text{—}OH \xrightarrow[\text{CH}_2\text{Cl}_2]{PCC} RCH_2\text{-}\overset{\displaystyle O}{\overset{\displaystyle \|}{C}}\text{-}H$$

Primary alcohol   Aldehyde

Cyclohexanol   Hexanal

5.9.8.3 Preparation of Ketones: Oxidation of Secondary Alcohols

Any oxidizing reagents, including chromic acid (H_2CrO_4), Jones' reagent or PCC, can be used to oxidize 2° alcohols to ketones. However, the most common reagent used for oxidation of 2° alcohols is chromic acid (H_2CrO_4).

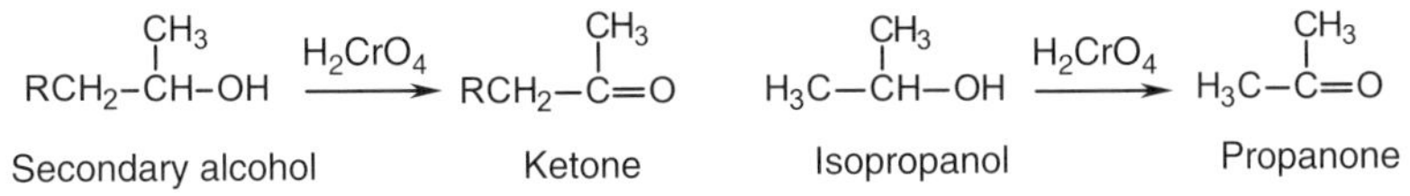

Secondary alcohol   Ketone   Isopropanol   Propanone

Mechanism

Chromic acid reacts with isopropanol to afford a chromate ester intermediate. An elimination reaction occurs by removal of a hydrogen atom from the alcohol carbon and departure of the chromium group with a pair of electrons. The Cr is reduced from Cr(VI) to Cr(IV) and the alcohol is oxidized.

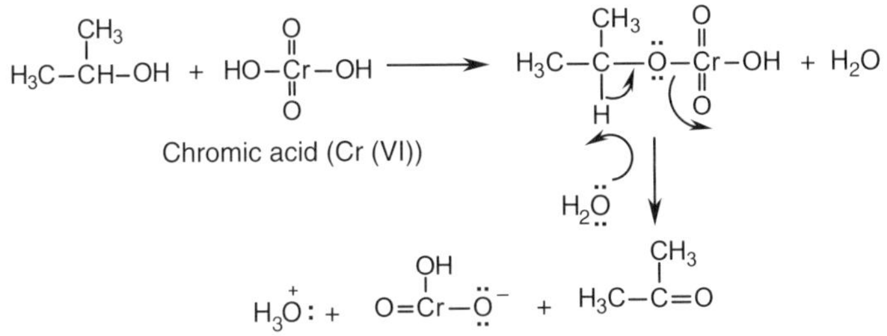

Chromic acid (Cr (VI))

5.9.9 Oxidation of Aldehydes and Ketones

Preparation of Carboxylic Acids: Oxidation of Aldehydes

Any aqueous oxidizing reagent, for example chromic acid (CrO_3 in aqueous acid), Jones' reagent (CrO_3 in acetone) and $KMnO_4$ in basic solution, can oxidize aldehydes to carboxylic acids.

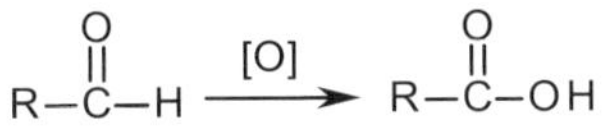

Aldehydes can also be oxidized selectively in the presence of other functional groups using silver (I) oxide (Ag_2O) in aqueous ammonium hydroxide (*Tollen's reagent*). Since ketones have no H on the carbonyl carbon, they do not undergo this oxidation reaction.

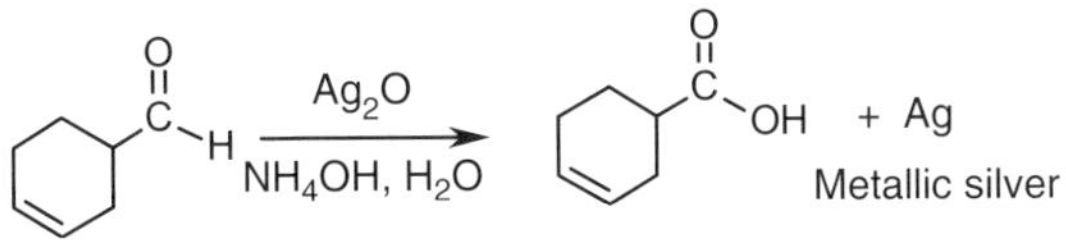

Metallic silver

5.9.10 Baeyer–Villiger Oxidation of Aldehydes or Ketones

Preparation of Carboxylic Acids or Esters: Peroxyacids Oxidation

Aldehyde reacts with peroxyacid (RCO_3H) to yield carboxylic acid. Most oxidizing reagents do not react with ketones. However, ketone reacts with peroxyacid (RCO_3H) to give an ester. Cyclic ketones afford lactones (cyclic esters). This reaction is known as *Baeyer--Villiger oxidation*, which can simply be defined as the oxidative cleavage of a C—C bond adjacent to a carbonyl group, which converts ketones to esters and cyclic ketones to lactones. A peroxyacid contains an extra oxygen atom than a carboxylic acid. This extra oxygen is inserted between the carbonyl carbon and R group (R = H in an aldehyde, and R = alkyl group in a ketone).

Aldehyde	Peroxyacid	Carboxylic acid

Ketone	Peroxyacid	Ester

5.9.11 Reduction of Alkyl Halides

Preparation of Alkanes

Lithium aluminium hydride ($LiAlH_4$), a strong reducing agent, reduces alkyl halides to alkanes. Essentially, a hydride ion (H^-) acts as a nucleophile displacing the halide. A combination of metal and acid, usually Zn with acetic acid (AcOH), can also be used to reduce alkyl halides to alkanes.

$$CH_3CH_2CH_2Br \xrightarrow[\text{LiAlH}_4,\ \text{THF}]{\text{Zn, AcOH or}} CH_3CH_2CH_3$$

Propyl bromide → Propane

5.9.12 Reduction of Organometallics

Preparation of Alkanes

Organometallics are generally strong nucleophiles and bases. They react with weak acids, for example, water, alcohol, carboxylic acid and amine, to become protonated and to produce hydrocarbons. Thus, small amounts of water or moisture can destroy organometallic compounds. For example, ethylmagnesium bromide or ethyllithium reacts with water to form ethane. It is a convenient way to reduce an alkyl halide to an alkane via Grignard and organolithium synthesis.

$$CH_3CH_2-Br \xrightarrow[\text{Dry ether}]{Mg} CH_3CH_2-MgBr \xrightarrow{H_2O} CH_3CH_3 + Mg(OH)Br$$

Ethyl bromide   Ethyl magnesium   Ethane
bromide

$$CH_3CH_2-Br \xrightarrow[\text{Ether}]{2Li} CH_3CH_2-Li + LiBr \xrightarrow{H_2O} CH_3CH_3 + LiOH$$

Ethyl bromide   Ethyllithium   Ethane

5.9.13 Reduction of Alcohols via Tosylates

Preparation of Alkanes

Generally, an alcohol cannot be reduced directly to an alkane in one step, because the −OH group is a poor leaving group so hydride displacement is not a good option.

$$R-OH \xrightarrow[\quad\quad]{LiAlH_4} \text{✗} \quad R-H$$

Alcohol   Alkane

However, the hydroxyl group can easily be converted to an excellent leaving group, and that allows the reaction to proceed with ease. One such conversion involves tosyl chloride and the formation of a highly reactive tosylate ester. Generally, tosylate esters are superior leaving groups as they can form stable anions. They undergo substitution and elimination reactions easily and often more reactive than alkyl halides. For example, cyclopentanol will not be reduced by LiAlH$_4$, unless it is converted to cyclopentyl tosylate and the corresponding tosylate ester is conveniently reduced to cyclopentane.

Cyclopentanol   Cyclopentyl tosylate   Cyclopentane

5.9.14 Reduction of Aldehydes and Ketones

Aldehydes and ketones are reduced to 1° and 2° alcohols, respectively, by hydrogenation with metal catalysts (Raney nickel, Pd—C and PtO$_2$). They are also reduced to alcohols relatively easily with a mild reducing agent, for example NaBH$_4$, or powerful reducing agent; for example, LiAlH$_4$. The key step in the reduction is the reaction of hydride with the carbonyl carbon.

5.9.14.1 Preparation of Alcohols: Catalytic Reduction

Catalytic hydrogenation using H$_2$ and a catalyst reduces aldehydes and ketones to 1° and 2° alcohols, respectively. The most common catalyst for these hydrogenations is Raney nickel, although PtO_2 and Pd—C can also be used. The C=C double bonds are reduced quicker than C=O double bonds. Therefore, it is not possible to reduce C=O selectively in the presence of a C=C without reducing both by this method.

$$H_2C=CHCH_2CH_2-\overset{\overset{\displaystyle O}{\|}}{C}-H \xrightarrow[\text{Raney Ni}]{H_2} CH_3CH_2CH_2CH_2CH_2OH$$

Pentanol

$$H_2C=CHCH_2-\overset{\overset{\displaystyle O}{\|}}{C}-CH_3 \xrightarrow[\text{Raney Ni}]{H_2} CH_3CH_2CH_2-\overset{\overset{\displaystyle CH_3}{|}}{C}HOH$$

2-Pentanol

5.9.14.2 Preparation of Alcohols: Hydride Reduction

The most useful reagents for reducing aldehydes and ketones are the metal hydride reagents. Complex hydrides are the source of hydride ions and the two most commonly used reagents are $NaBH_4$ and $LiAlH_4$. Lithium aluminium hydride is extremely reactive with water and must be used in an anhydrous solvent; for example, dry THF.

$$\overset{+}{NaH}-\overset{\overset{\displaystyle H}{|}}{\underset{\displaystyle H}{B}}-H \qquad \overset{+}{LiH}-\overset{\overset{\displaystyle H}{|}}{\underset{\displaystyle H}{Al}}-H$$

$$R-\overset{\overset{\displaystyle O}{\|}}{C}-Y \xrightarrow[\text{ii. } H_2O \text{ or } H_3O^+]{\text{i. } NaBH_4 \text{ or } LiAlH_4} R-\overset{\overset{\displaystyle OH}{|}}{\underset{\displaystyle H}{C}}-Y$$

Y = H or R

$1°$ or $2°$ Alcohol

Mechanism

Hydride ions attack carbonyl groups, generating alkoxide ions, and protonation furnishes alcohols. The net result of adding H⁻ from $NaBH_4$ or $LiAlH_4$ and H⁺ from aqueous acids is the addition of the elements of H_2 to the carbonyl π bond.

$$R-\overset{\overset{\displaystyle :O^{\delta-}}{\|^{\delta+}}}{\underset{\displaystyle H^-}{C}}-Y \longrightarrow R-\overset{\overset{\displaystyle :O^-}{|}}{\underset{\displaystyle H}{C}}-Y \xrightarrow{H_3O^+} R-\overset{\overset{\displaystyle OH}{|}}{\underset{\displaystyle H}{C}}-Y$$

Y = H or R

Selectivity of hydride reduction

Sodium borohydride is a more selective and milder reagent than $LiAlH_4$. It cannot reduce esters or carboxylic acids, whereas $LiAlH_4$ reduces esters and carboxylic acids to $1°$ alcohols. These hydride sources do not reduce alkene double bonds. Therefore, when a compound contains both a C=O group and a C=C bond, selective reduction of one functional group can be achieved by choosing an appropriate reagent.

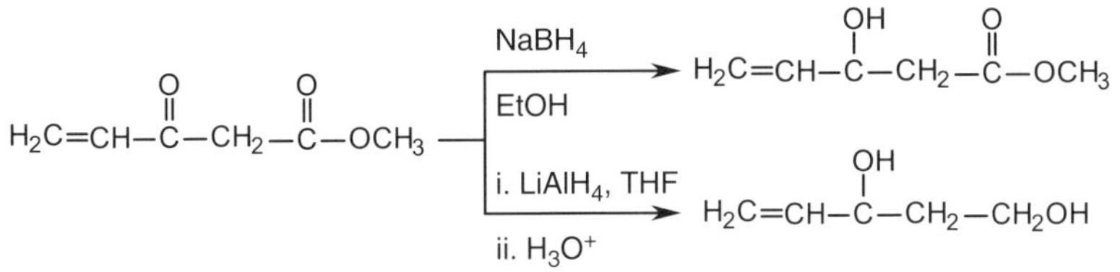

Stereochemistry of hydride reduction

Hydride converts a planar sp^2-hybridized carbonyl carbon to a tetrahedral sp^3-hybridized carbon. Thus, hydride reduction of an achiral ketone with $LiAlH_4$ or $NaBH_4$ gives a racemic mixture of alcohol when a new stereocentre is formed.

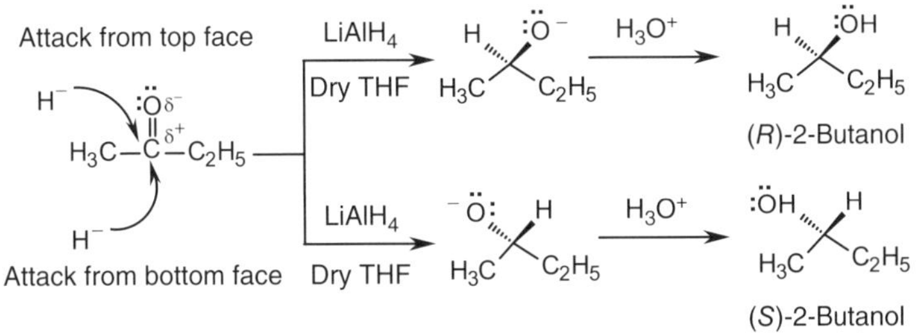

5.9.15 Clemmensen Reduction

Preparation of Alkanes

This method is used for the reduction of acyl benzenes to alkyl benzenes, but it also reduces aldehydes and ketones to alkanes.

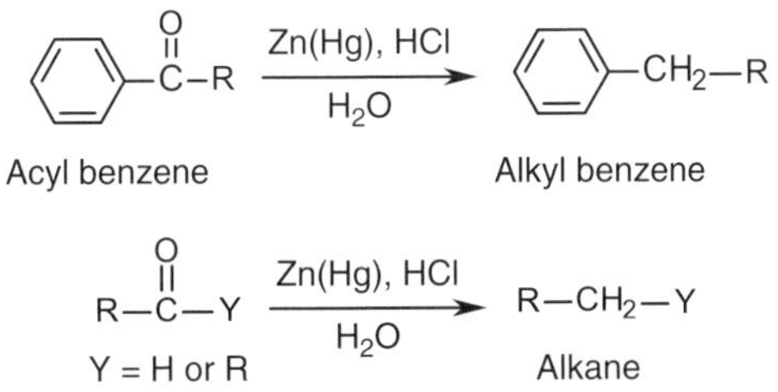

Acyl benzene   Alkyl benzene

Y = H or R   Alkane

Sometimes, the acidic conditions used in Clemmensen reduction are unsuitable for certain molecules. In these cases, *Wolff–Kishner reduction* is employed, which occurs under basic conditions.

5.9.16 Wolff–Kishner Reduction

Preparation of Alkanes

This method reduces acyl benzenes as well as aldehydes and ketones, but does not reduce alkenes, alkynes or carboxylic acids. Hydrazine reacts with aldehyde or ketone to yield hydrazone (see Section 5.4.3.8), which is then treated with a strong base (NaOH or KOH) to generate an alkane.

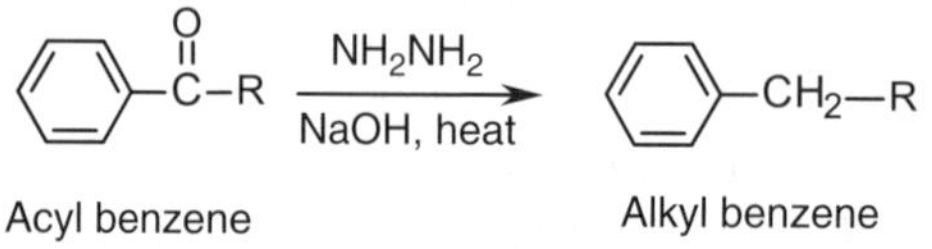

$$\text{Acyl benzene} \xrightarrow[\text{NaOH, heat}]{\text{NH}_2\text{NH}_2} \text{Alkyl benzene}$$

Acyl benzene → Alkyl benzene

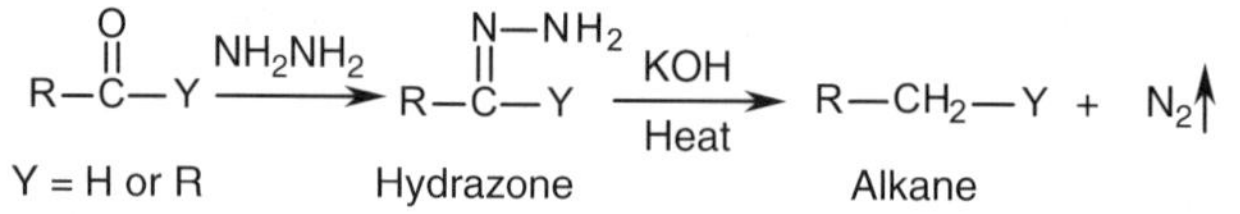

$$R-\overset{O}{\underset{||}{C}}-Y \xrightarrow{\text{NH}_2\text{NH}_2} R-\overset{N-NH_2}{\underset{||}{C}}-Y \xrightarrow[\text{Heat}]{\text{KOH}} R-CH_2-Y + N_2\uparrow$$

Y = H or R   Hydrazone   Alkane

Mechanism

The aqueous base deprotonates the hydrazone and the anion produced is resonance stabilized. The carbanion picks up a proton from water and another deprotonation by the aqueous base generates an intermediate, which is set up to eliminate a molecule of nitrogen (N_2) and to provide a new carbanion. This carbanion is quickly protonated by water, giving the final reduced product as an alkane.

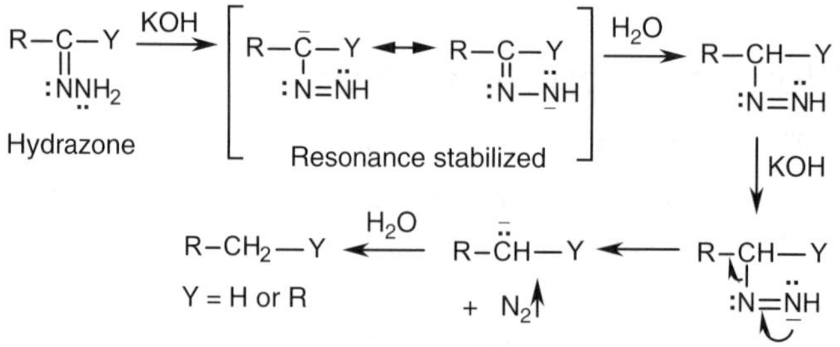

5.9.17 Reduction of Acid Chlorides

5.9.17.1 Preparation of Primary Alcohols: Catalytic or Hydride Reductions

Acid chlorides are easier to reduce than carboxylic acids and other carboxylic acid derivatives. They are reduced conveniently all the way to 1° alcohols by metal hydride reagents ($NaBH_4$ or $LiAlH_4$), as well as by catalytic hydrogenation (H_2/Pd–C).

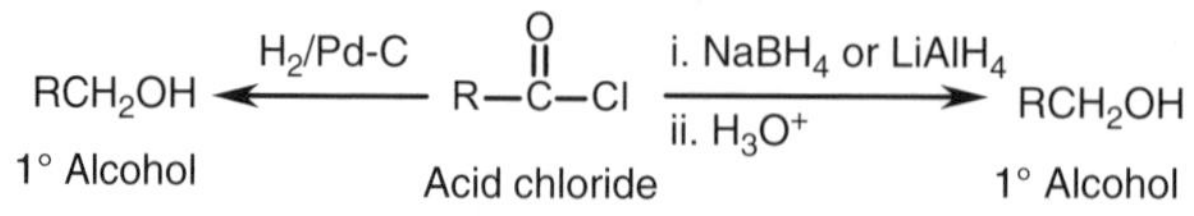

$$RCH_2OH \xleftarrow{H_2/\text{Pd-C}} R-\overset{O}{\underset{||}{C}}-Cl \xrightarrow[\text{ii. } H_3O^+]{\text{i. NaBH}_4 \text{ or LiAlH}_4} RCH_2OH$$

1° Alcohol   Acid chloride   1° Alcohol

5.9.17.2 Preparation of Aldehydes: Partial and Selective Reduction of Acid Chlorides

Sterically bulky reducing agents, for example lithium tri-t-butoxyaluminium hydride, can partially and selectively reduce acid chlorides to aldehydes at very low temperatures (−78 °C). Lithium tri-t-butoxyaluminium hydride, LiAlH(O-t-Bu)$_3$, has three electronegative oxygen atoms bonded to aluminium, which makes this reagent less nucleophilic than LiAlH$_4$.

$$\overset{+}{Li}\ (CH_3)_3CO\overset{\overset{\displaystyle H}{|}}{\underset{\underset{\displaystyle OC(CH_3)_3}{|}}{\overset{-}{Al}}}-OC(CH_3)_3$$

Lithium tri-*tert*-butoxyaluminium hydride

$$R-\overset{\overset{\displaystyle O}{||}}{C}-Cl \xrightarrow[\text{ii. } H_3O^+]{\text{i. LiAlH(O-}t\text{-Bu)}_3,\ -78\ ^\circ C} R-\overset{\overset{\displaystyle O}{||}}{C}-H$$

Aldehyde

5.9.18 Reduction of Esters

5.9.18.1 Preparation of Primary Alcohols: Hydride Reduction of Esters

Esters are harder to reduce than acid chlorides, aldehydes and ketones. They cannot be reduced with milder reducing agents, for example NaBH$_4$, or by catalytic hydrogenation. Only LiAlH$_4$ can reduce esters. Esters react with LiAlH$_4$ generating aldehydes, which react further to give 1° alcohols.

$$R-\overset{\overset{\displaystyle O}{||}}{C}-OR \xrightarrow[\text{ii. } H_3O^+]{\text{i. LiAlH}_4} RCH_2OH$$

1° Alcohol

5.9.18.2 Preparation of Aldehydes: Partial and Selective Reduction of Esters

Sterically bulky reducing agents, such as diisobutylaluminium hydride (DIBAH), can partially and selectively reduce esters to aldehydes. The reaction is carried out at very low temperatures (−78 °C) in toluene. DIBAH has two bulky isobutyl groups, which make this reagent less reactive than LiAlH$_4$.

$$(CH_3)_2CHCH_2-\overset{\overset{\displaystyle H}{|}}{Al}-CH_2CH(CH_3)_2$$

Diisobutylaluminium hydride (DIBAH)

$$R-\overset{\overset{\displaystyle O}{||}}{C}-OR \xrightarrow[\text{ii. } H_2O]{\text{i. DIBAH, -78 }^\circ C} R-\overset{\overset{\displaystyle O}{||}}{C}-H$$

Aldehyde

5.9.19 Hydride Reduction of Carboxylic Acids

Preparation of Primary Alcohols

Carboxylic acids are considerably less reactive than acid chlorides, aldehydes and ketones towards reduction. They cannot be reduced by catalytic hydrogenation or sodium borohydride ($NaBH_4$) reduction. This requires the use of a powerful reducing agent; for example, $LiAlH_4$. The reaction needs two hydrides (H^-) from $LiAlH_4$, since the reaction proceeds through an aldehyde, but it cannot be stopped at that stage. Aldehydes are more easily reduced than carboxylic acids and $LiAlH_4$ reduces all the way back to 1° alcohol.

$$R-\overset{\overset{\displaystyle O}{\|}}{C}-OH \xrightarrow[\text{ii. } H_3O^+]{\text{i. } LiAlH_4} RCH_2OH$$

1° Alcohol

5.9.20 Reduction of Oximes or Imine Derivatives

Preparation of Amines: Catalytic Hydrogenation or Hydride Reductions

The most general method for synthesizing amines involves the reduction of oximes and imine derivatives prepared from aldehydes or ketones (see Section 5.4.3.8). By catalytic hydrogenation or by $LiAlH_4$ reduction, while 1° amines are synthesized from oxime or unsubstituted imine, 2° amines are obtained from substituted imine. Unsubstituted imines are relatively unstable and are reduced in situ.

$$\underset{\substack{\text{Oxime} \\ Y = H \text{ or } R}}{R-\overset{\overset{\displaystyle N-OH}{\|}}{C}-Y} \text{ or } \underset{\substack{\text{Imine} \\ Y = H \text{ or } R}}{R-\overset{\overset{\displaystyle NH}{\|}}{C}-Y} \xrightarrow[LiAlH_4]{H_2/Pd\text{-}C \text{ or }} \underset{\substack{1° \text{ amine}}}{R-\overset{\overset{\displaystyle NH_2}{|}}{C}H-Y}$$

$$\underset{\substack{\text{Imine} \\ Y = H \text{ or } R}}{R-\overset{\overset{\displaystyle N-R'}{\|}}{C}-Y} \xrightarrow[LiAlH_4]{H_2/Pd\text{-}C \text{ or }} \underset{\substack{2° \text{ amine}}}{R-\overset{\overset{\displaystyle NHR'}{|}}{C}H-Y}$$

Tertiary amines are made from iminium salts by catalytic hydrogenation or by $LiAlH_4$ reduction. The iminium salts are usually unstable, and so are reduced as they are formed by a reducing agent already in the reaction mixture. A mild reducing agent, for example sodium cyanoborohydride ($NaBH_3CN$), can also be used.

$$\underset{\substack{\text{Iminium salt} \\ Y = H \text{ or } R}}{R-\overset{\overset{\displaystyle \overset{+}{N}R'_2}{\|}}{C}-Y} \xrightarrow[NaBH_3CN]{H_2/Pd\text{-}C \text{ or }} \underset{\substack{3° \text{ amine}}}{R-\overset{\overset{\displaystyle NR'_2}{|}}{C}H-Y}$$

5.9.21 Reduction of Amides, Azides and Nitriles

5.9.21.1 Preparation of Primary Amines

Primary amides, alkyl azides and alkyl nitriles are reduced to primary amines by catalytic hydrogenation (H_2/Pd—C or H_2/Pt—C) or LiAlH$_4$ reduction. They are less reactive towards NaBH$_4$. Unlike LiAlH$_4$ reduction of all other carboxylic acid derivatives, which affords 1° alcohols, the LiAlH$_4$ reduction of amides, azides and nitriles yields amines. Acid is not used in the work-up step, since amines are basic. Thus, hydrolytic work-up is employed to prepare amines. When the nitrile group is reduced, an NH$_2$ and an extra CH$_2$ are introduced into the molecule.

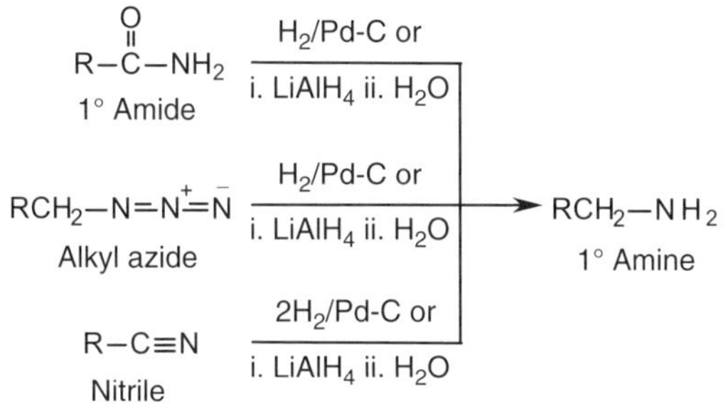

5.9.21.2 Preparation of Secondary and Tertiary Amines: Reduction of Secondary and Tertiary Amides

Secondary and tertiary amides are reduced to corresponding amines by catalytic hydrogenation (H_2/Pd—C or H_2/Pt—C) or LiAlH$_4$ reduction.

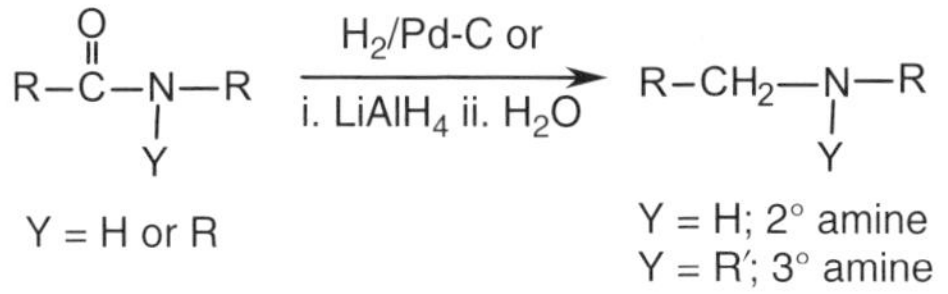

5.9.21.3 Preparation of Aldehydes: Selective Reduction of Nitriles

Reduction of a nitrile with a less powerful reducing reagent, for example DIBAH, can selectively produce aldehyde. The reaction is carried out at very low temperatures (−78 °C) in toluene.

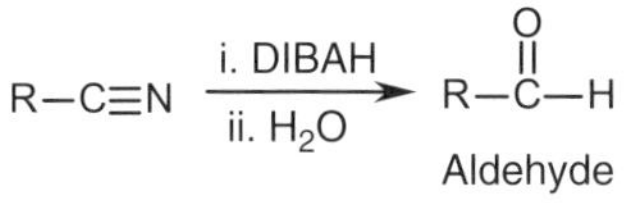

5.9.22 Reductive Amination of Aldehydes and Ketones

Preparation of Primary, Secondary and Tertiary Amines

Reductive amination is a versatile preparation of alkyl amines. This reaction involves the conversion of a carbonyl group to an amine via imine and enamine intermediate. Ammonia, primary and secondary amines react with aldehydes or ketones to form imines or enamines (see Sections 5.4.3.8 and 5.4.3.9), which are then conveniently reduced either by sodium cyanoborohydride ($NaBH_3CN$) in acetic acid (CH_3CO_2H) or catalytic hydrogenation to afford 1°, 2° and 3° amines, respectively.

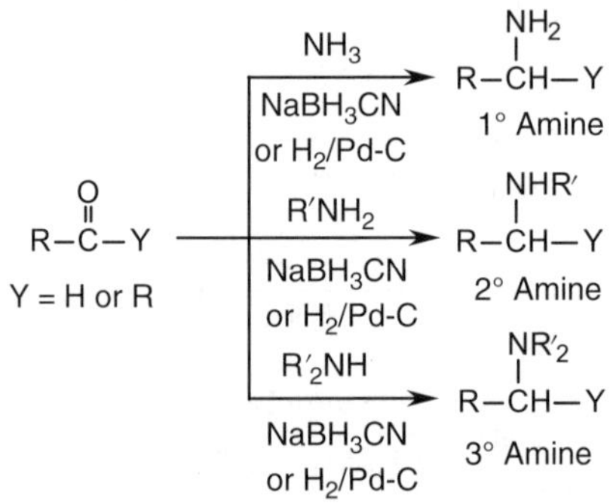

5.10 PERICYCLIC REACTIONS

Pericyclic reactions are concerted reactions that take place in a single step without any intermediates and involve a cyclic redistribution of bonding electrons. The concerted nature of these reactions gives fine stereochemical control over the generation of the product. The best-known examples of this reaction are the Diels–Alder reaction (cycloaddition) and sigmatropic rearrangement. Note that in a *sigmatropic rearrangement*, one bond is broken while another bond is formed across a π system (see Section 5.10.4).

5.10.1 Diels–Alder Reaction

In the *Diels–Alder reaction*, a conjugated diene reacts with an α,β-unsaturated carbonyl compound, generally called a *dienophile*. A dienophile is a reactant that loves a diene. The most reactive dienophiles usually have a carbonyl group, but it may also have another electron-withdrawing group, for example, a cyano, nitro, haloalkene or sulphone group conjugated with a carbon–carbon double bond.

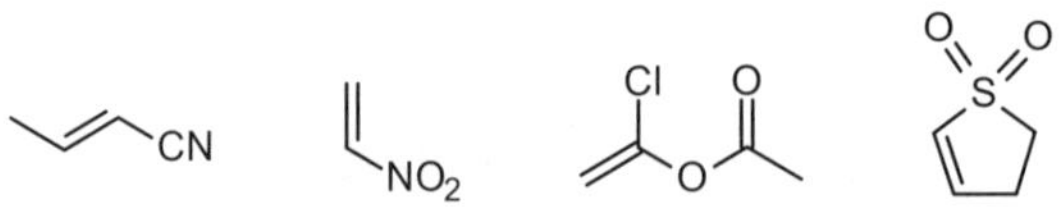

Dienophiles other than the carbonyl group directly linked to the conjugated system

The Diels–Alder reaction is in fact a [4+2] cycloaddition reaction, where C-1 and C-4 of the conjugated diene system become attached to the double-bonded carbons of the dienophile to form a six-membered ring. For example, 1,3-butadiene reacts with maleic anhydride to give tetrahydrophthalic anhydride on heating.

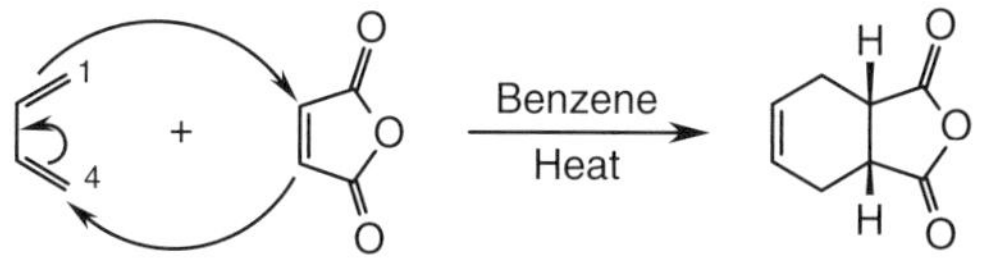

Benzene
Heat

1, 3-Butadiene   Maleic anhydride   Tetrahydrophthalic anhydride
A conjugated diene   A dienophile   95%

Different types of cyclic compounds can be prepared just by varying the structures of the conjugated diene and the dienophile. Compounds containing carbon–carbon triple bonds can be used as dienophiles to provide compounds with two bonds as shown next.

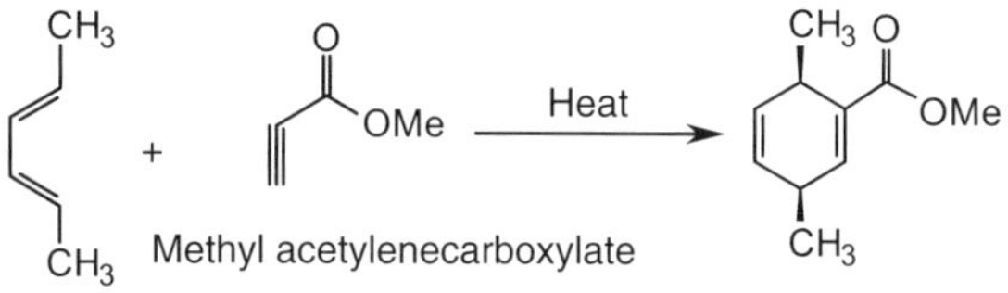

Heat

1,4-Dimethyl-1, 3-butadiene   Methyl acetylenecarboxylate

Methyl cis-3,6-dimethyl-
1,4-cyclohexadiene-1-carboxylate

In the case of a cyclic conjugated diene, the Diels–Alder reaction yields a bridged bicyclic compound. A bridged bicyclic compound contains two rings that share two nonadjacent carbons. For example, cyclopentadiene reacts with ethylene to produce norbornene.

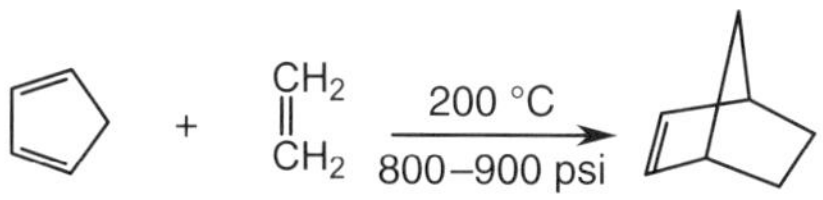

200 °C
800–900 psi

Cyclopentadiene Ethylene   Norbornene

Cycloaddition is used extensively in the synthesis of chiral natural products and pharmaceutical agents because the reaction can determine the relative configuration of up to four chiral centres in a single reaction.

5.10.2 Essential Structural Features for Dienes and Dienophiles

In the Diels–Alder reaction, the conjugated diene can be cyclic or acyclic and it may contain different types of substituents. A conjugated diene can exist in two different conformations, an s-*cis* and an s-*trans*. The 's' stands for single bond or σ-bond; for example, s-*cis* means the double bonds are *cis* about the single bond. In the Diels–Alder reaction, the conjugated diene has to be in an s-*cis* conformation. A conjugated diene that is permanently in an s-*trans* conformation cannot undergo this reaction. This s-*cis* feature must also be present in conjugated cyclic dienes for Diels–Alder reaction. Note that in a normal demand Diels–Alder reaction, a *dienophile* has an electron-withdrawing group in conjugation with the alkene, and in an inverse-demand scenario, a dienophile is conjugated with an electron-donating group.

s-*cis* conformation   s-*trans* conformation

s-*cis* conformation   s-*trans* conformation
Undergoes Diels–Alder reaction   Does not undergo Diels–Alder reaction

Cyclic conjugated dienes that are s-*cis* conformation, for example cyclopentadiene and 1,3-cyclohexadiene, are highly reactive in Diels–Alder reactions. In fact, cyclopentadiene is reactive both as a diene and as a dienophile and forms dicyclopentadiene at room temperature. When dicyclopentadiene is heated to 170 °C, a reverse Diels–Alder reaction takes place and reforms the cyclopentadiene.

Diene   Dienophile   Dicyclopentadiene

5.10.3 Stereochemistry of the Diels–Alder Reaction

The Diels–Alder reaction is stereospecific. The stereochemistry of the dienophile is retained in the product, that is, *cis* and *trans* dienophiles yield different diastereoisomers in the product. For example, freshly distilled cyclopentadiene, with an s-*cis* configuration, reacts with maleic anhydride to give *cis*-norbornene-5,6-endo-dicarboxylic anhydride.

Cyclopentadiene Maleic anhydride   *cis*-norbornene-5,6-
                                        endo-dicarboxylic anhydride

There are two possible configurations, *endo* and *exo*, for bridged bicyclic compounds resulting from the reaction of a cyclic diene and cyclic dienophile. A substituent on a bridge is *endo* if it is closer to the longer of the two other bridges, and it
is *exo* if it is closer to the shorter bridge. Most of these reactions result in an *endo*
product. However, if this reaction is reversible and thermodynamically controlled,
the *exo* product is formed.

The *exo* product   Furan   Maleic anhydride   The *endo* product
(more stable)   (less stable)

5.10.4 Sigmatropic Rearrangements

Sigmatropic rearrangements are unimolecular processes and involve the movement
of a σ bond with the simultaneous rearrangement of the π-system. In this rearrangement reaction, a σ bond is broken in the reactant, a new σ bond is formed in
the product and the π bonds rearrange. However, the number of π bonds does not
change; that is, the reactant and the product possess the same number of π bonds.
Sigmatropic reactions are usually uncatalysed, although Lewis acid catalysts are
sometimes used. Sigmatropic rearrangement plays an important role in the biosynthesis of vitamin D, a fat-soluble vitamin, in our body. In fact, vitamin D is a family
of compounds that are biosynthetically derived from cholesterol. For example,
vitamin D_2 (found in plants and better known as ergocalciferol or calciferol) and
vitamin D_3 (found in animal tissues and often referred to as cholecalciferol).

Bond broken   Heat   New bond formed

This reaction can occur through hydrogen shift, alkyl shift (Cope rearrangement) or Claisen rearrangement.

5.10.5 Hydrogen Shift

A sigmatropic rearrangement involves the migration of a σ bond from one position
to another with a simultaneous shift of the π bonds. For example, a hydrogen atom
and its σ bond can be migrated in (*Z*)-1,3-pentadiene. This is known as *hydrogen
shift*. Hydrogen shifts occur at 4n + 1 positions in a *suprafacial* fashion. It can also
take place at 4n + 3 positions in an *antarafacial* fashion. *Antarafacial* means that
opposite faces are involved, whereas it is *suprafacial* when both changes occur at

the same face. Many sigmatropic rearrangements and Diels–Alder reactions can be either suprafacial or antarafacial, and this dictates the stereochemistry. Anta-rafacial hydrogen shifts are observed in the conversion of lumisterol to vitamin D.

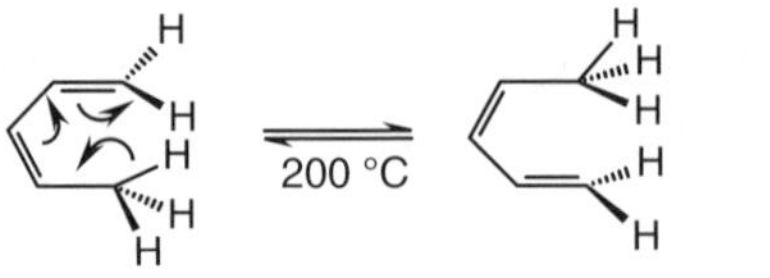

(Z)-1,3-pentadiene   (Z)-1,3-pentadiene

5.10.6 Alkyl Shift: Cope Rearrangement

In addition to the migration of hydrogen atoms in sigmatropic rearrangements, *alkyl shifts* also take place. A large number of such reactions occur with a migra-tion of a carbon atom and a σ bond, but do not have ionic intermediates. More specifically, these reactions involve methyl shifts at 4n + 3 positions in a suprafacial fashion with inversion of stereochemistry.

Alkyl shift is evident in the *Cope rearrangement*. A Cope rearrangement is a [3,3] sigmatropic rearrangement of a 1,5-diene. This reaction leads to the formation of a six-membered ring transition state. As [3,3] sigmatropic rearrangements involve three pairs of electrons, they take place by a suprafacial pathway under thermal conditions.

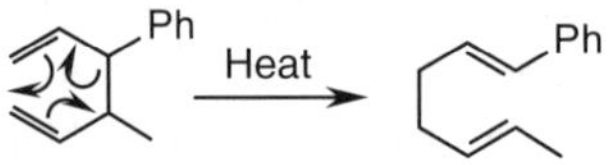

5.10.7 Claisen Rearrangement

Sigmatropic rearrangements involving the cleavage of a σ bond at an oxygen atom are called *Claisen rearrangements*. A Claisen rearrangement is [3,3] sigma-tropic rearrangement of an allyl vinyl ether to afford γ, δ-unsaturated carbonyl compound. Like the Cope rearrangement, this reaction also forms a six-mem-bered ring transition state. This reaction is exothermic and occurs by a suprafacial pathway under thermal conditions.

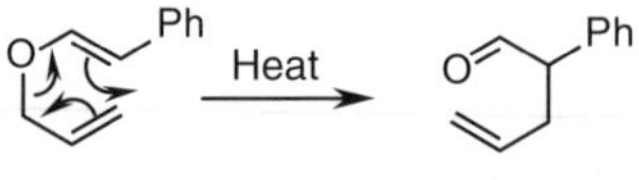

Claisen rearrangement plays an important part in the biosynthesis of several natural products. For example, the chorismate ion is rearranged to a prephenate ion by Claisen rearrangement, which is catalysed by the enzyme chorismate mutase. This prephenate ion is a key intermediate in the shikimic acid pathway for the biosynthesis of phenylalanine, tyrosine and many other biologically important natural products.

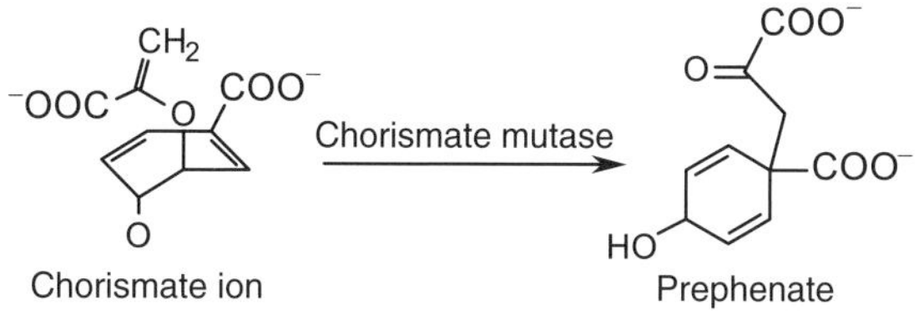

Chorismate ion →(Chorismate mutase)→ Prephenate

Chapter 6
Heterocyclic Compounds

Learning objectives

After completing this chapter, students should be able to

- provide an overview of heterocyclic compounds and their derivatives;
- describe various reactions of heterocyclic compounds.

6.1 HETEROCYCLIC COMPOUNDS AND THEIR DERIVATIVES

Cyclic compounds that have one or more of the atoms other than carbon, for example N, O, or S (heteroatoms), in their rings are called *heterocyclic compounds* or *heterocycles*; for example, pyridine, tetrahydrofuran, thiophene and so on. Heterocyclic chemistry is an important branch of organic chemistry as almost two-thirds of organic compounds are heterocyclic compounds and many drug molecules, for example artemisinin, taxol, vincristine and vinblastine, belong to this class. Heterocycles also include many of the biochemical material essential to life; for example, nucleic acids.

Chemistry for Pharmacy Students: General, Organic and Natural Product Chemistry,
Second Edition. Lutfun Nahar and Satyajit Sarker.
© 2019 John Wiley & Sons Ltd. Published 2019 by John Wiley & Sons Ltd.

Pyridine
N is the hetero-atom

Tetrahydrofuran
O is the hetero-atom

Thiophene
S is the hetero-atom

Among the heterocyclic compounds, there are aromatic, for example pyridine, as well as non-aromatic, for example tetrahydrofuran (THF), compounds. Similarly, there are saturated (e.g. tetrahydrofuran) and unsaturated (e.g. pyridine) hetero- cyclic compounds. Heterocycles also differ in their ring sizes; for example, pyridine has a six-member ring, whereas tetrahydrofuran is a five-membered oxygen-con- taining heterocyclic compound.

6.1.1 Medicinal Importance of Heterocyclic Compounds

Heterocyclic compounds play important roles in medicine and biological sys- tems. As mentioned previously, a great majority of important drugs and natural products, for example caffeine, nicotine, morphine, penicillins and cephalo- sporins, are heterocyclic compounds. The purine and pyrimidine bases, two nitrogenous heterocyclic compounds, are structural units of RNA and DNA. Serotonin, a neurotransmitter found in our body, is responsible for various bodily functions.

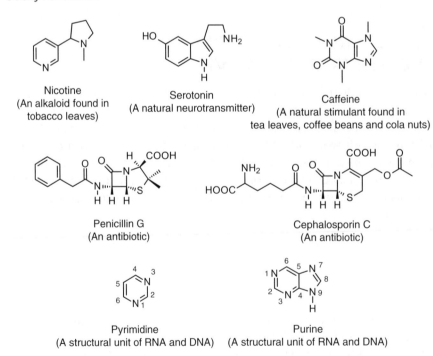

Nicotine
(An alkaloid found in tobacco leaves)

Serotonin
(A natural neurotransmitter)

Caffeine
(A natural stimulant found in tea leaves, coffee beans and cola nuts)

Penicillin G
(An antibiotic)

Cephalosporin C
(An antibiotic)

Pyrimidine
(A structural unit of RNA and DNA)

Purine
(A structural unit of RNA and DNA)

6.1.2 Nomenclature of Heterocyclic Compounds

Most of the heterocycles are known by their trivial names; for example, pyridine, indole, quinoline, thiophene and so on. However, there are some general rules to be followed in a heterocycle, especially in the use of suffixes to indicate the ring size, saturation or unsaturation as shown in the following table. For example, from the name, pyrid*ine*, where the suffix is *ine*, one can understand that this heterocyclic compound contains nitrogen, has a six-membered ring system and is unsaturated.

	Ring with nitrogen		Ring without nitrogen	
Ring size	Maximum unsaturation	Saturation	Maximum unsaturation	Saturation
3	irine	iridine	irene	irane
4	ete	etidine	ete	etane
5	ole	olidine	ole	olane
6	ine	—	ine	ane
7	epine	—	epine	epane
8	ocine	—	ocine	ocane
9	onine	—	onine	onane
10	ecine	—	ecine	ecane

Monocyclic heterocycles containing 3–10 members and one or more hetero-atoms are named systematically by using a prefix or prefixes to indicate the nature of the heteroatoms as presented in the following table. For example, *thia*cyclobutane contains the heteroatom sulphur (S).

Element	Prefix	Element	Prefix	Element	Prefix
O	oxa	P	phospha	Ge	germa
S	thia	As	arsa	Sn	stanna
Se	selena	Sb	stiba	Pb	plumba
Te	tellura	Bi	bisma	B	bora
N	aza	Si	sila	Hg	mercura

Two or more identical heteroatoms are indicated by use of the multiplying prefixes *di-*, *tri-* or *tetra-*. When more than one distinct heteroatom is present, the appropriate prefixes are cited in the name in descending order of group number in the periodic table; for example, *oxa-* takes precedence over *aza-*. If both lie within the same group of the periodic table (see Section 2.2), then the order is determined by increasing atomic number; for example, *oxa-* precedes *thia-*.

In unsaturated heterocycles, if the double bonds can be arranged in more than one way, their positions are defined by indicating the N or C atoms (which are not multiple bonded, and consequently carry an 'extra' hydrogen atom), by 1H-, 2H- and so on. For example, 1H-azepine and 2H-azepine.

1H-Azepine   2H-Azepine

Important aromatic heterocycles that contain a single heteroatom include pyridine, quinoline, isoquinoline, pyrrole, thiophene, furan and indole.

Pyridine   Quinoline   Isoquinoline

Pyrrole   Thiophene   Furan   Indole

Derivatives of these heterocyclic compounds are named in the same way as other compounds, by adding the name of the substituent, in most cases as a prefix to the name of the heterocycle, and a number to indicate its position on the ring system; for example, 2-methylpyridine, 5-methylindole and 3-phenylthiophene.

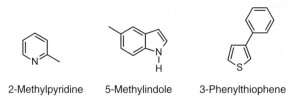

2-Methylpyridine   5-Methylindole   3-Phenylthiophene

Heterocyclic aromatic compounds can also have two or more heteroatoms. If one of the heteroatoms is a nitrogen atom and the compound has a five-membered system, their names all end in *azole* and the rest of the name indicates other heteroatoms. For example, pyrazole and imidazole are two isomeric heterocycles that contain two nitrogen atoms in the ring, thiazole has a sulphur atom and a nitrogen atom in the ring, and oxazole contains an oxygen atom and a nitrogen atom. In imidazole and oxazole, two heteroatoms are separated by a carbon atom, whereas in their isomers, pyrazole and isoxazole, the heteroatoms are directly linked to each other. The six-membered aromatic heterocycles with two nitrogen atoms can exist in three isomeric forms, the most important being pyrimidine.

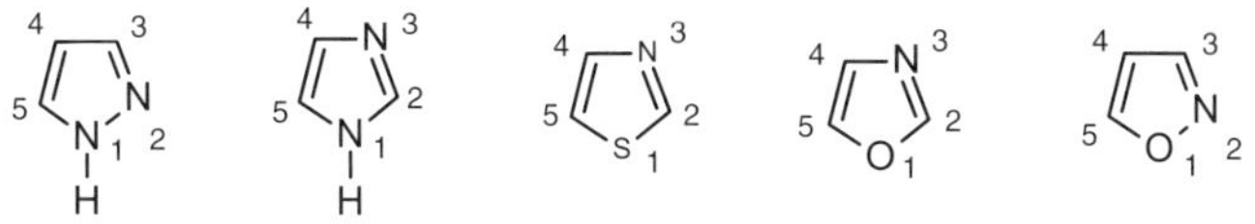

| Pyrazole | Imidazole | Thiazole | Oxazole | Isoxazole |

There are a number of fully saturated non-aromatic heterocycles. For example, pyrrolidine, tetrahydrofuran, isoxazolidine and piperidine are fully saturated derivatives of pyrrole, furan, isoxazole and pyridine, respectively. Partially saturated derivatives, for example 2-pyroline, 2-isoxazoline, 1,4-dihydropyridine, are also known.

Saturated	Partially saturated	Fully saturated
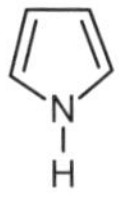 Pyrrole	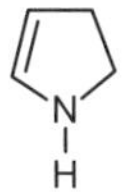 2-Pyrroline	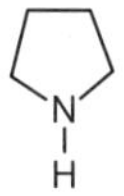 Pyrrolidine
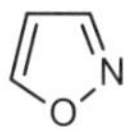 Isoxazole	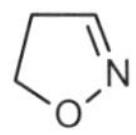 2-Isoxazoline	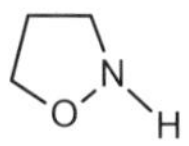 Isoxazolidine
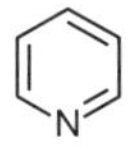 Pyridine	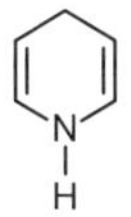 1,4-Dihydropyridine	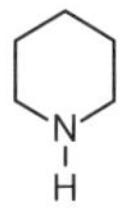 Piperidine

6.1.3 Physical Properties of Heterocyclic Compounds

A large number of structurally diverse compounds belong to the class, *heterocycles*. This makes it extremely difficult to generalize the physical properties of these compounds, because they vary significantly depending on the saturation-unsaturation status, aromatic-nonaromatic behaviour, ring sizes, type and number of heteroatoms present. Saturated heterocycles, known as *alicyclic heterocycles*, containing five or more atoms have physical and chemical properties typical of acyclic compounds that contain the same heteroatoms. These compounds undergo the same reactions as their open-chain analogues. On the other hand, aromatic heterocycles display characteristic and often complex reactivity. However, aromatic heterocycles show general patterns of reactivity associated with certain 'molecular fragments' such that the reactivity of a given heterocycle can be anticipated. Physical and chemical properties of selected important heterocyclic compounds are discussed under each compound sub-heading.

6.2 PYRROLE, FURAN AND THIOPHENE: UNSATURATED HETEROCYCLES

Pyrrole is a nitrogen-containing unsaturated five-membered heterocyclic aromatic compound. It shows aromaticity by delocalization of lone pair of electrons from nitrogen. In pyrrole, there are four π electrons, two short of the Hückel criteria for aromaticity. The nitrogen atom is sp^2-hybridized, formally containing a lone pair of electrons in the *p* orbital at right angles to the ring. However, the system delocalizes and pushes the lone pair of electrons into the ring to complete the sextet required for aromaticity. The nonbonding electrons on the nitrogen atom become a part of the aromatic sextet. A small number of simple pyrroles occur in nature. However, biologically more significant natural pyrroles are rather less simple; they are tetrameric pyrrole derivatives, known as porphyrins; for example, chlorophyll-a and haeme.

Furan, also known as furane and furfuran, is an oxygen-containing five-membered aromatic heterocyclic compound that is usually produced when wood, especially pine wood, is distilled. The highly electronegative oxygen holds on the electron density tightly. Although it has a lone pair of electrons, these electrons cannot delocalize easily and so the system is generally considered to be almost non-aromatic or weakly aromatic. Furan and methylfurans in food, especially in ready-to-eat jarred or canned foods, can cause long-term liver damage. Furan and the related compounds 2- and 3-methylfurans are chemical contaminants that naturally form during heated food processing, including cooking. These substances have always been present in cooked or heated foods.

Thiophene is a sulphur-containing five-membered unsaturated heterocycle. The lone pair of electrons of the sulphur are in the 3s orbital and are less able to interact with the π electrons of the double bonds. Therefore, thiophene is considered weakly aromatic. Acetylenic thiophenes are found in some higher plant species; for example, in the roots of *Echinops ellenbeckii* of the family Asteraceae. Acetylenic thiophenes are known to possess various biological properties; for example, antimalarial properties. Generally, the thiophene ring is present in many important agrochemical and pharmaceutical products.

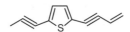

Acetylenic thiophene

6.2.1 Physical Properties of Pyrrole, Furan and Thiophene

Pyrrole is a weakly basic compound. However, as the nonbonding electrons on the nitrogen atom are part of the aromatic sextet and no longer available for protonation, it has an extremely low basicity ($pK_a = \sim$15). Pyrrole accepts a proton on one of the carbon atoms adjacent to the nitrogen atom, whereas the proton on the nitrogen atom can be removed by hydroxide ion yielding its conjugate base.

Conjugate acid
$pK_a = -3.80$

Pyrrole
$pK_a = \sim$15

Conjugate base

Salts containing the pyrrole anion can easily be prepared in this way. The pair of nonbonding electrons on N in pyrrole is much less available for protonation than the pair on ammonia. Thus, pyrrole is much less basic than NH_3 ($pK_a = 36$); that is, much stronger acid than NH_3.

Furan and thiophene, both are clear and colourless liquids at room temperature. While furan is extremely volatile and highly flammable with a boiling point close to room temperature (31.4 °C), the boiling point of thiophene is 84 °C. Thiophene possesses a mildly pleasant odour.

6.2.2 Preparations of Pyrrole, Furan and Thiophene

A general way of synthesizing heterocyclic compounds is by cyclization of a dicarbonyl or diketo compound using a nucleophilic reagent that introduces the desired heteroatom.

6.2.2.1 Paal–Knorr Synthesis

This is a useful and a straightforward method for the synthesis of five-membered heterocyclic compounds; for example, pyrrole, furan and thiophene. In fact, the Paal–Knorr Pyrrole Synthesis is the condensation of a 1,4-dicarbonyl compound with an excess of a primary amine or ammonia to give a pyrrole. However, necessary precursors, for example dicarbonyl compounds, are not readily available. Ammonia, primary amines, hydroxylamines or hydrazines are used as the nitrogen components for the synthesis of pyrrole. The Paal–Knorr synthesis was initially reported independently by German chemists Carl Paal and Ludwig Knorr in 1884 as a method for the preparation of furans and was adapted for pyrroles and thiophenes. However, its mechanism was not fully understood until it was elucidated in 1990s.

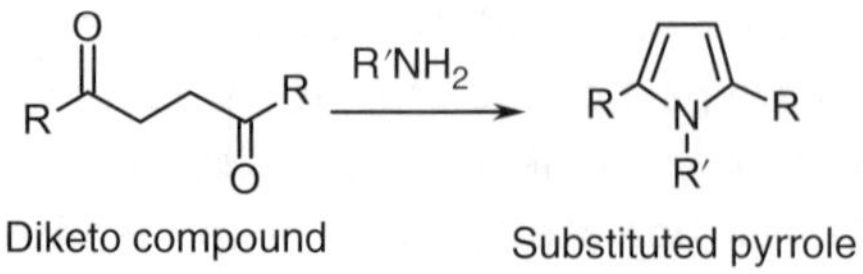

Diketo compound   Substituted pyrrole

Paal–Knorr synthesis can be used to synthesize furan and thiophene ring systems. A simple dehydration of a diketo (1,4-dicarbonyl) compound provides furan system, whereas thiophene or substituted thiophenes can be prepared by treating 1,4-dicarbonyl compounds with hydrogen sulphide (H_2S) and hydrochloric acid (HCl).

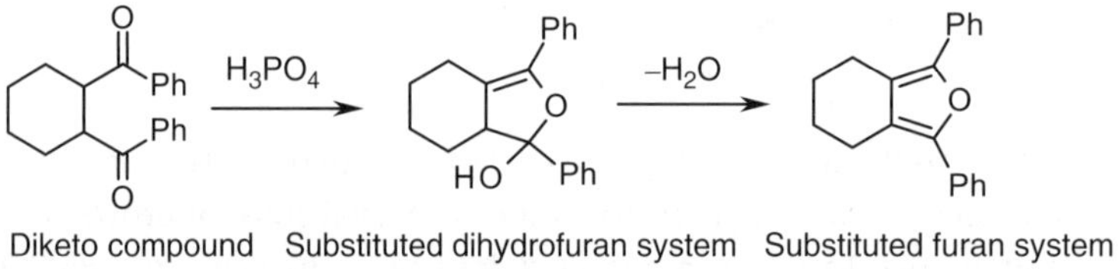

Diketo compound   Substituted dihydrofuran system   Substituted furan system

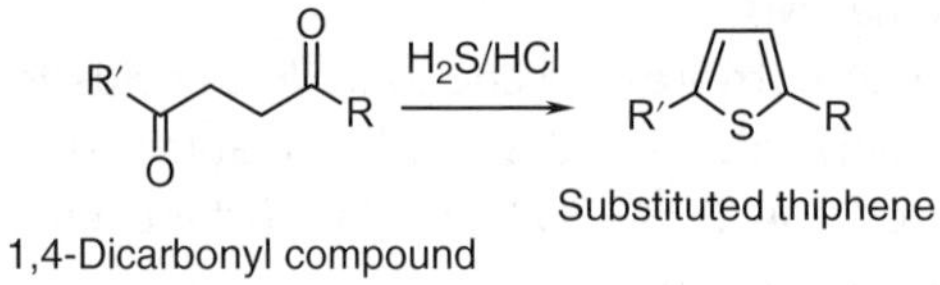

1,4-Dicarbonyl compound   Substituted thiphene

6.2.2.2 Commercial Preparation of Pyrrole, Furan and Thiophene

Pyrrole is obtained commercially from coal tar or by treating furan with NH_3 over an alumina catalyst at 400 °C.

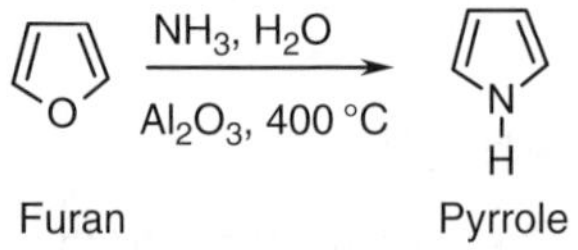

Furan   Pyrrole

Furan is synthesized by decarbonylation of furfural (furfuraldehyde), which itself can be prepared by acidic dehydration of the pentose sugars found in oat hulls, corncobs and rice hulls.

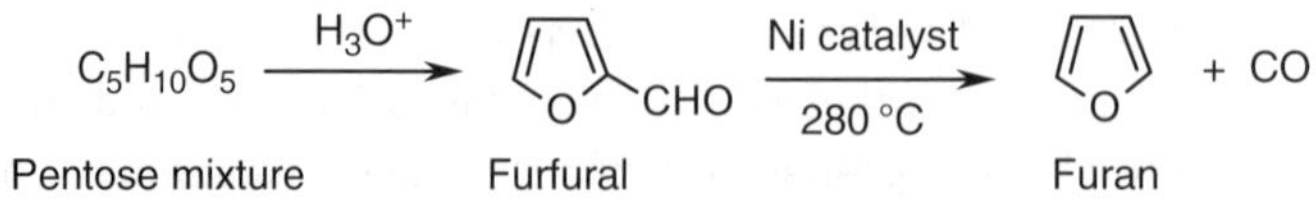

Pentose mixture   Furfural   Furan

Thiophene is found in small amounts in coal tar, and commercially it is prepared from the cyclization of butane or butadiene with sulphur at 600 °C.

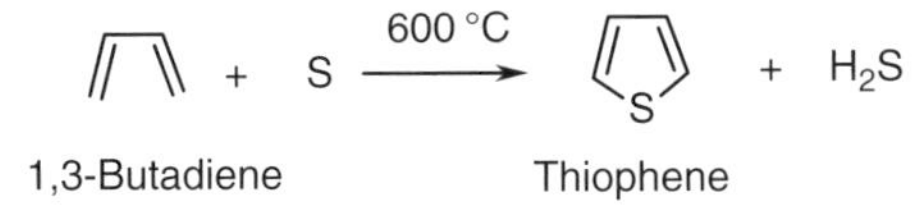

1,3-Butadiene   Thiophene

6.2.2.3 Hantzsch Synthesis

Hantzsch synthesis, reported in 1881 by Arthur Rudolf Hantzsch involves a reaction of an α-haloketone with a β-ketoester and NH_3 or a primary amine yielding substituted pyrrole.

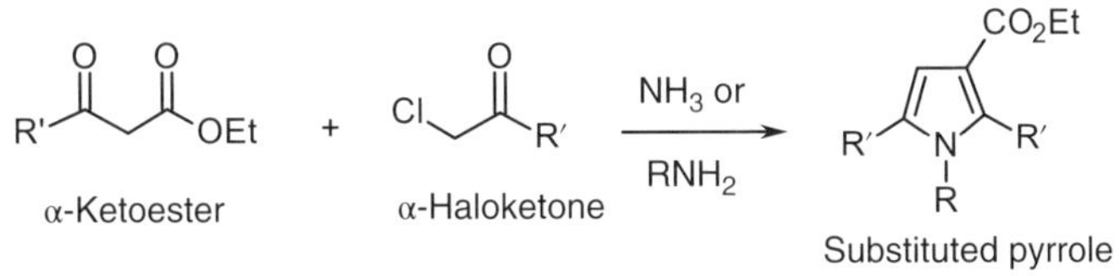

α-Ketoester   α-Haloketone   Substituted pyrrole

Substituted furan can be prepared by using *Feist–Benary synthesis*, which is similar to the *Hantzsch synthesis* of pyrrole ring. This synthetic route was introduced by Franz Feist and Erich Benary. In this reaction, α-haloketone reacts with 1,3-dicarbonyl compound (β-ketoester) in presence of pyridine to form substituted furan.

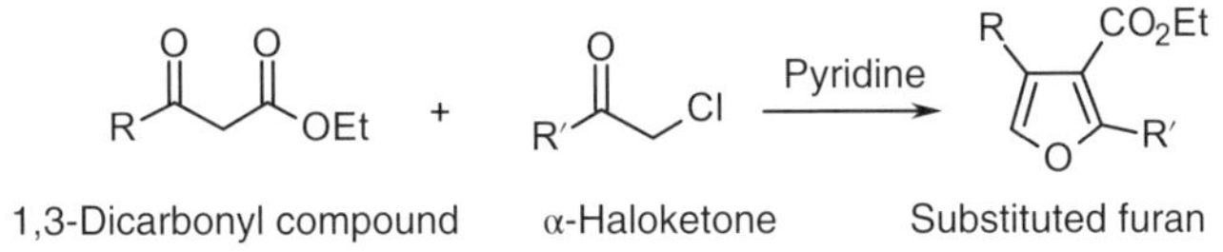

1,3-Dicarbonyl compound   α-Haloketone   Substituted furan

6.2.3 Reactions of Pyrrole, Furan and Thiophene

Pyrrole, furan and thiophene undergo electrophilic substitution reactions. However, the reactivity of reaction varies significantly among these heterocycles. The ease of electrophilic substitution is usually: furan > pyrrole > thiophene > benzene. Clearly, all three heterocycles are more reactive than benzene towards electrophilic substitution. Electrophilic substitution generally occurs at C-2; that is, the position next to the heteroatom.

6.2.3.1 Vilsmeier Reaction

The *Vilsmeier reaction*, also known as the Vilsmeier–Haack reaction, first introduced by Anton Vilsmeier and Albrecht Haack, is the chemical reaction of a substituted amide with phosphorus oxychloride and an electron-rich arene to produce an aryl aldehyde or ketone. Following this principle, formylation of pyrrole, furan or thiophene is carried out using a combination of phosphorous oxychloride ($POCl_3$) and N,N-dimethylformamide (DMF). This reaction proceeds by formation of the electrophilic Vilsmeier complex, followed by electrophilic substitution of the heterocycle. The formyl group is generated in the hydrolytic workup.

Pyrrole   i. $POCl_3$, DMF   ii. H_2O 2-Formylpyrrole

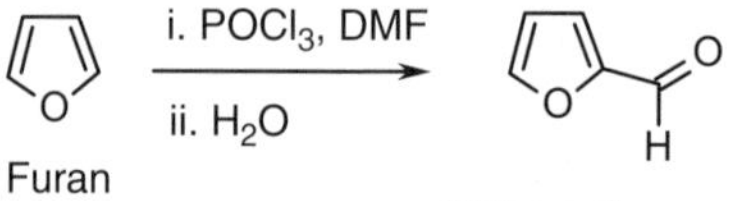

Furan   i. $POCl_3$, DMF   ii. H_2O 2-Formylfuran

6.2.3.2 Mannich Reaction

Pyrrole and alkyl substituted furan undergo *Mannich reaction*, which is an organic reaction that comprises an amino alkylation of an acidic proton placed next to a carbonyl group by formaldehyde and a primary or secondary amine or ammonia. Thiophene also undergoes this reaction, but instead of acetic acid HCl is used.

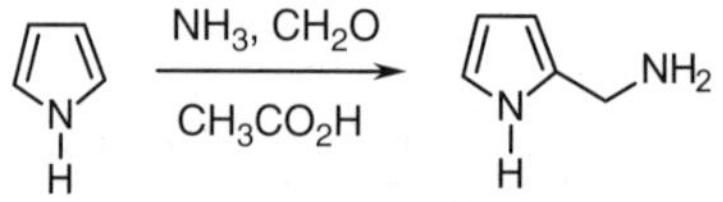

Pyrrole NH_3, CH_2O / CH_3CO_2H Pyrrole-2-methanamine

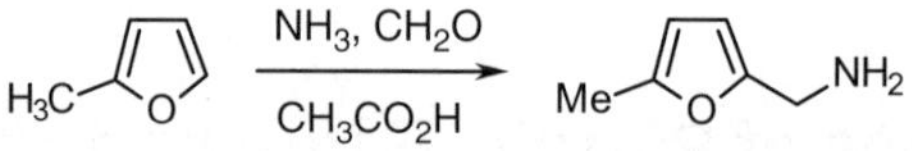

Methyl furan NH_3, CH_2O / CH_3CO_2H (5-Methyl-2-furyl)methanamine

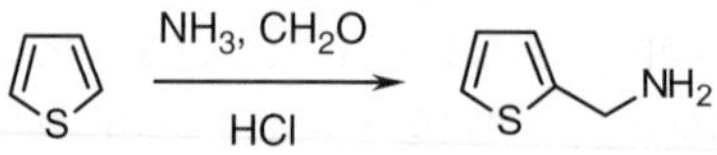

Thiophene NH_3, CH_2O / HCl   2-Thienylmethanamine

6.2.3.3 Sulphonation

Pyrrole, furan and thiophene undergo sulphonation with the pyridine-sulphur trioxide complex ($C_5H_5N^+SO_3^-$) as shown here.

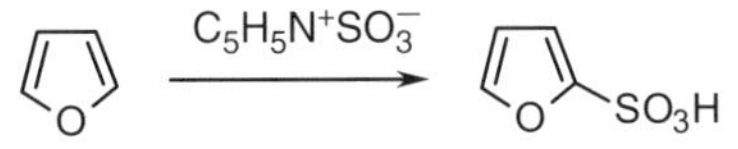

$C_5H_5N^+SO_3^-$

Furan → 2-Sulphonylfuran (SO$_3$H)

$C_5H_5N^+SO_3^-$

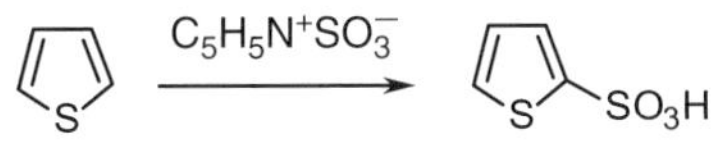

Thiophene → 2-Sulphonylthiophene (SO$_3$H)

6.2.3.4 Nitration

Instead of a mixture of nitric acid and sulphuric acid, nitration of these three hetero-cycles is carried out with acetyl nitrate (formed from nitric acid and acetic anhydride). Nitration is in place mainly at one of the carbon atoms next to the heteroatom.

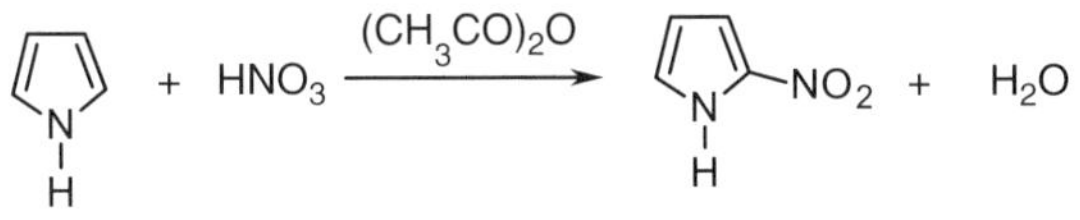

Pyrrole + HNO_3 $\xrightarrow{(CH_3CO)_2O}$ 2-Nitropyrrol (NO_2) + H_2O

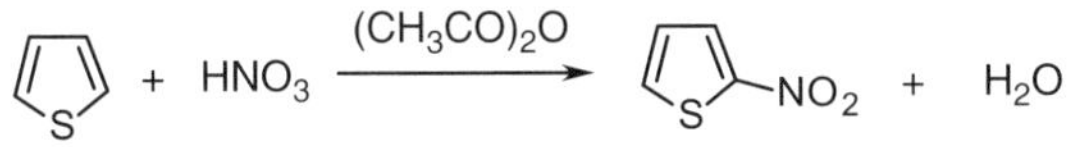

Thiophene + HNO_3 $\xrightarrow{(CH_3CO)_2O}$ 2-Nitrothiophene (NO_2) + H_2O

6.2.3.5 Bromination

The five-membered aromatic heterocycles are more reactive towards electrophiles than benzene and the reactivity is similar to that of phenol. These compounds undergo electrophilic bromination. However, reaction rates vary considerably and for pyrrole, furan and thiophene the rates are 5.6×10^8, 1.2×10^2 and 1.00, respectively. While unsubstituted five-membered aromatic heterocycles provide a mixture of bromo derivatives, for example bromothiophenes, substituted het-erocycles yield a single product.

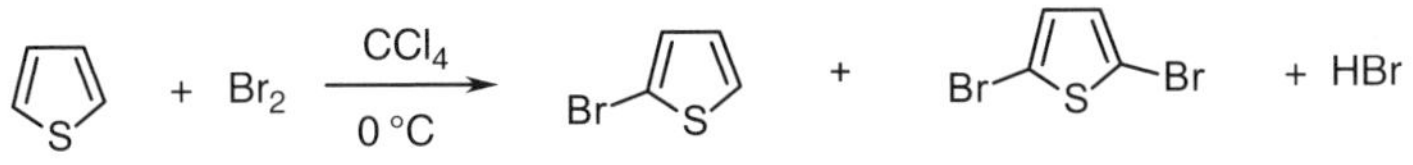

Thiophene + Br_2 $\xrightarrow[0\,°C]{CCl_4}$ 2-Bromothiophene + 2,5-Dibromothiophene + HBr

Pyrrole-COOCH$_3$ $\xrightarrow{Br_2}$ Br-pyrrole-COOCH$_3$

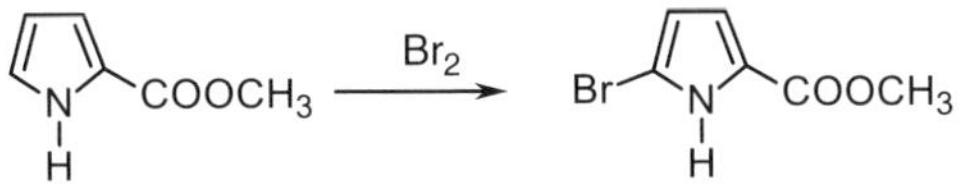

6.2.3.6 FC Acylation and Alkylation

As pyrroles and furans are not stable in presence of Lewis acids, which are necessary for FC alkylations and acylations, only thiophine, which is stable in Lewis acids, can undergo these reactions. Thiophene reacts with benzoyl chloride in the presence of aluminium chloride to afford phenyl 2-thienyl ketone.

Thiophene   Benzoyl chloride   Phenyl 2-thienyl ketone

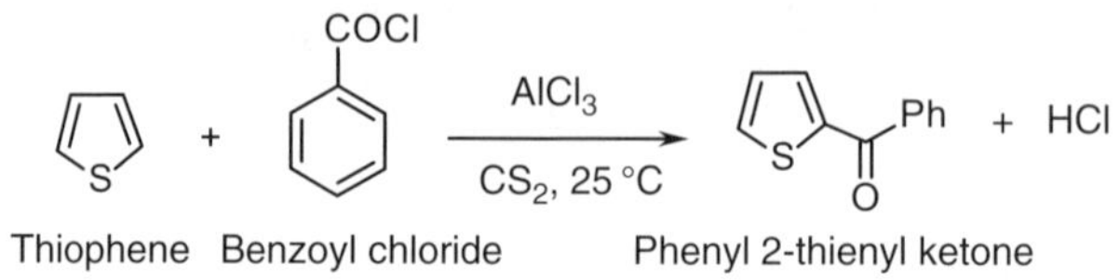

Thiophene derivative reacts with ethylbromide in the presence of a Lewis acid to bring in 3-ethyl substituent on the ring of the thiophene derivative.

Thiophene derivative   3-Ethyl thiophene derivative

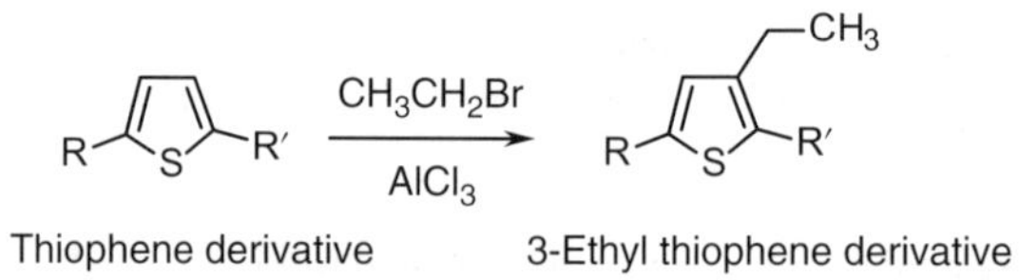

6.2.3.7 Ring Opening of Substituted Furan

Furan may be regarded as a cyclic hemiacetal that has been dehydrated and is hydrolysed back to a dicarbonyl compound when heated with dilute mineral acid (H_2SO_4).

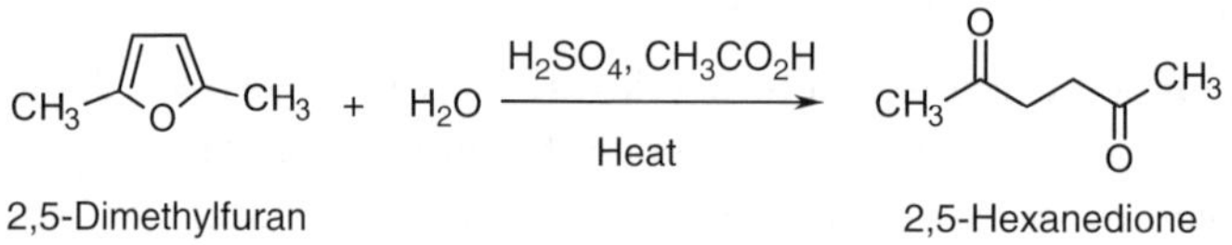

2,5-Dimethylfuran   2,5-Hexanedione

6.2.3.8 Addition Reaction of Furan

Furan reacts with bromine by 1,4-addition reactions, not electrophilic substitution. When this reaction is carried out in methanol (MeOH), the isolated product is formed by solvolysis of the intermediate dibromide. Note that *solvolysis* is a type of nucleophilic substitution (S_N1)/(S_N2) or elimination, where the nucleophile is a solvent molecule.

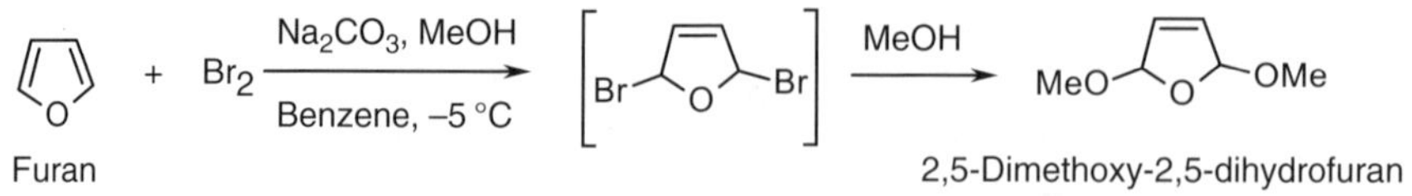

Furan   2,5-Dimethoxy-2,5-dihydrofuran

6.2.3.9 Catalytic Hydrogenation of Furan

Catalytic hydrogenation of furan with a palladium (Pd) catalyst gives tetrahydrofuran (THF), which is a clear, low-viscosity liquid with a diethyl ether-like smell.

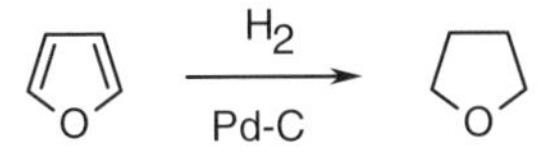

Furan   Tetrahydrofuran
THF

6.3 PYRIDINE

Pyridine (C_5H_5N) is a nitrogen-containing unsaturated six-membered heterocyclic aromatic compound. It is similar to benzene and conforms to the Hückel's rule for aromaticity. Pyridine, a tertiary amine, has a lone pair of electrons instead of a hydrogen atom, but the six π electrons are essentially the same as benzene. It was discovered in 1849 by the Scottish chemist Thomas Anderson as one of the constituents of bone oil. A number of drug molecules possess pyridine or modified pyridine skeleton in their structures; for example, the antihypertensive drug, amlodipine, and the antifungal drug, pyridotriazine.

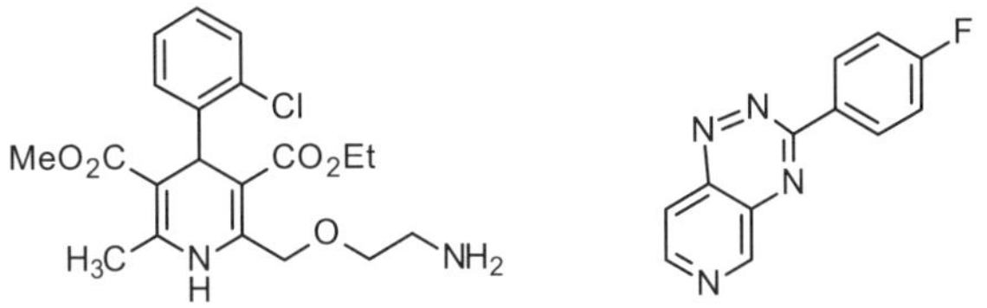

Amlodipine
(An antihypertensive agent)

Pyridotriazine
(An antifungal drug)

6.3.1 Physical Properties of Pyridine

Pyridine is a liquid (bp: 115 °C) with an unpleasant smell. It is a polar aprotic solvent and is miscible with both water and organic solvents. The dipole moment of pyridine is 1.57 D. Pyridine is an excellent donor ligand in metal complexes. It is highly aromatic and moderately basic in nature with a pK_a 5.23; that is, a stronger base than pyrrole but weaker than alkylamines. The lone pair of electrons on the nitrogen atom in pyridine is available for bonding without interfering with its aromaticity. Protonation of pyridine results in a pyridinium ion ($pK_a = 5.16$), which is a stronger acid than a typical ammonium ion (e.g. piperinium ion, $pK_a = 11.12$), because the acidic hydrogen of a pyridinium ion is attached to a sp^2-hybridized nitrogen that is more electronegative than an sp^3-hybridized nitrogen.

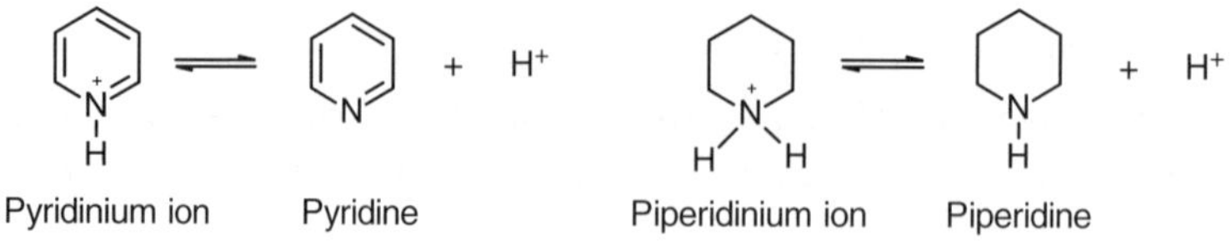

Pyridinium ion   Pyridine   Piperidinium ion   Piperidine

6.3.2 Preparations of Pyridine

Among the methods available for the preparation of a pyridine system, *Hantzsch synthesis* is probably the most important and widely used synthetic route. However, a pyridine ring can be synthesized from the reaction between pentan-2,4-di-one and ammonium acetate. Cyclization of 1,5-diketones is also considered as a convenient method for the synthesis of corresponding pyridine derivatives. Commercially, pyridine is obtained from the distillation of coal tar.

6.3.2.1 Hantzsch Synthesis

The reaction of 1,3-dicarbonyl compounds with aldehydes and NH_3 provides a 1,4-dihydropyridine, which can be aromatized by oxidation with nitric acid or nitric oxide. Instead of NH_3, primary amine can be used to give 1-substituted 1,4-dihydropyridines.

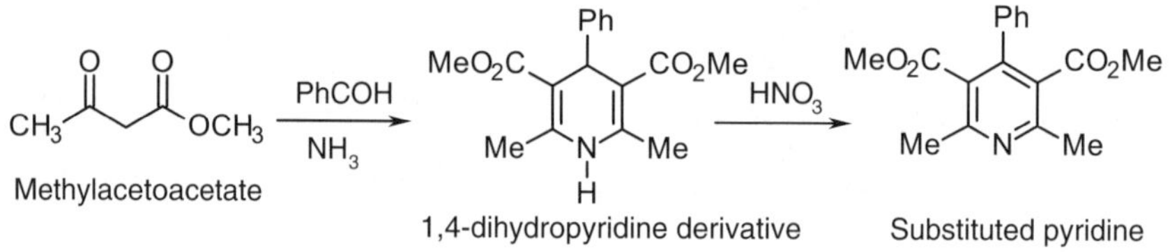

Methylacetoacetate   1,4-dihydropyridine derivative   Substituted pyridine

6.3.2.2 Cyclization of 1,5-Diketones

The reaction between 1,5-diketones and NH_3 affords dihydropyridine systems, which can easily be oxidized to pyridines.

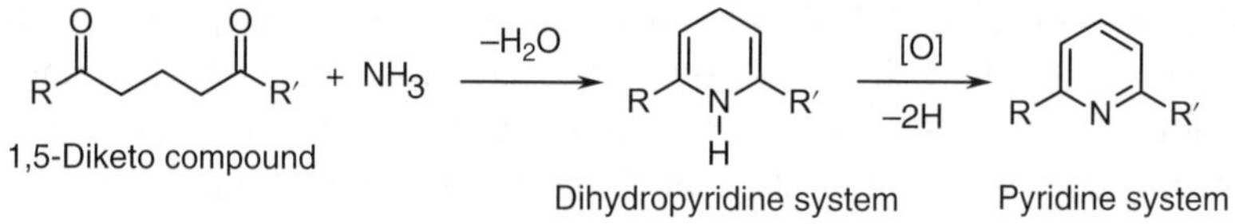

1,5-Diketo compound   Dihydropyridine system   Pyridine system

6.3.3 Reactions of Pyridine

Generally, reactions that are characteristic of benzene may proceed with pyridine either at more complicated conditions, with low yield or both. Because of the decreased electron density in the aromatic system, electrophilic substitutions are not favoured in pyridine in usual reaction conditions, but pyridine and its derivatives favour addition of nucleophiles at the electron-rich nitrogen atom. The nucleophilic addition at the nitrogen atom creates further deactivation of the aromatic properties and seriously hinders electrophilic substitution. On the contrary, free radical and nucleophilic substitutions take place more conveniently in pyridine than in benzene.

6.3.3.1 Electrophilic Substitutions

Pyridine's electron-withdrawing nitrogen causes the ring carbons to have significantly less electron density than the ring carbons of benzene. Thus, pyridine is less reactive than benzene towards electrophilic aromatic substitution. However, pyridine undergoes some electrophilic substitution reactions under drastic conditions, for example high temperature, and the yields of these reactions are usually quite low. The main substitution takes place at C-3.

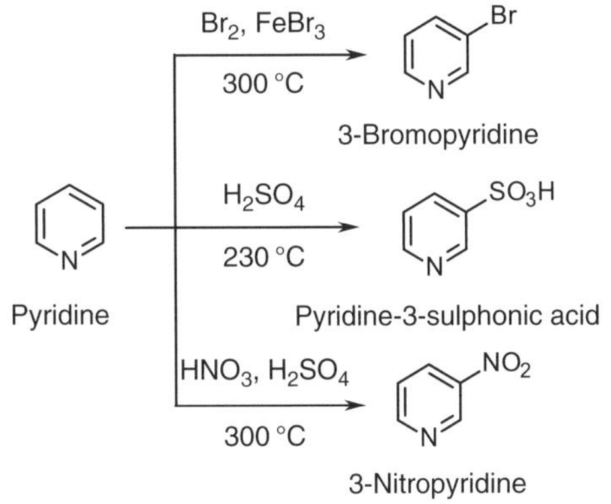

6.3.3.2 Nucleophilic Aromatic Substitutions

Pyridine is more reactive than benzene towards nucleophilic aromatic substitutions because of the presence of electron-withdrawing nitrogen in the ring. Nucleophilic aromatic substitutions of pyridine occur at C-2 (or C-6) and C-4 positions.

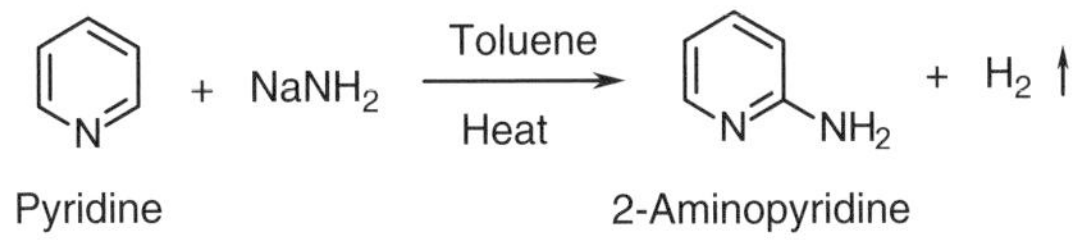

These nucleophilic substitution reactions are rather facile, when better leaving groups, for example halide ions, are present. Reaction occurs by addition of the nucleophile to the C=N bond, followed by loss of halide ion from the anion intermediate.

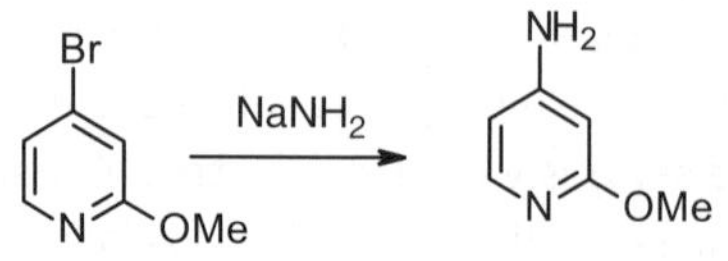

4-Bromo-2-methoxypyridine   4-Amino-2-methoxypyridine

6.3.3.3 Reactions as an Amine

Pyridine is a tertiary amine and undergoes reactions characteristic to tertiary amines. For example, pyridine undergoes S_N2 reactions with alkyl halides and it reacts with hydrogen peroxide to form an N-oxide.

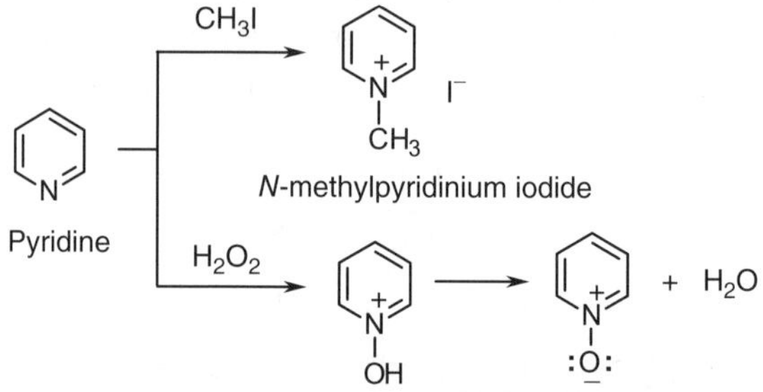

N-methylpyridinium iodide

Pyridine

Pyridine-N-oxide

6.4 OXAZOLE, IMIDAZOLE AND THIAZOLE

Oxazole, imidazole and thiazole systems contain a five-membered ring and two hetero-atoms, one of which is a nitrogen atom. The hetero-atoms are separated by a carbon atom in the ring. The second heteroatoms are oxygen, nitrogen and sulphur for oxazole, imidazole and thiazole systems, respectively. *Azoles* are a class of five-membered heterocyclic compounds containing a nitrogen atom and at least one other non-carbon atom, such as O, N or S, as part of the ring.

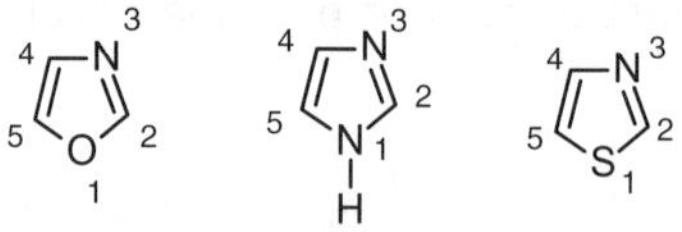

Oxazol   Imidazol   Thiazole

Oxazole, imidazol and thiazole are isomeric with the 1,2-azoles; for example, isoxazole, pyrazole and isothiazole (see Section 6.5). The aromatic characters of the oxazole, imidazole and thiazole systems arise from delocalization of a lone pair of electrons from the second hetero-atom.

Histamine, an important mediator of inflammation, gastric acid secretion and other allergic manifestations, contain an imidazole ring system. Thiamine (also known as vitamin B$_1$), an essential vitamin, possesses a quaternized thiazole ring and is found in many foods including yeast, cereal grains, beans, nuts and meat.

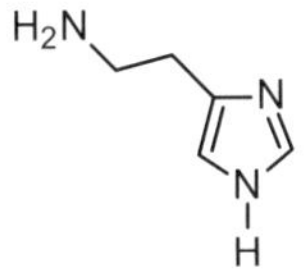

Histamine

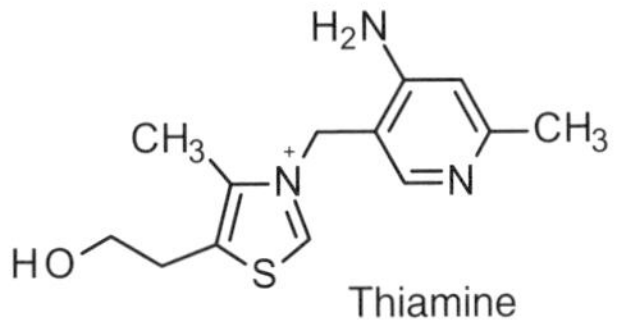

Thiamine
Vitamin B$_1$

Apart from some plant and fungal secondary metabolites, the occurrence of an oxazole ring system in nature is rather limited. However, oxaprozin is a synthetic *non-steroidal anti-inflammatory drug* (NSAID) that contains an oxazole ring system. It may be worth-mentioning here that oxaprozin works by reducing the levels of *prostaglandins* (chemicals responsible for pain, fever and inflammation), by blocking the enzyme cyclooxygenase that produces prostaglandin. The most common side effects of oxaprozin include a rash, ringing in the ears, headaches, dizziness, drowsiness, abdominal pain, nausea, diarrhoea, constipation, heartburn and shortness of breath.

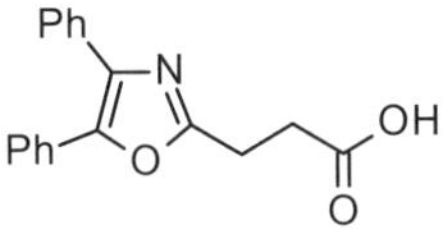

Oxaprozin
(Non-steroidal anti-inflammatory drug)

6.4.1 Physical Properties of Oxazole, Imidazole and Thiazole

Among these 1,3-azoles, imidazole is the most basic compound. The increased basicity of imidazole can be accounted for from the greater electron-releasing ability of two nitrogen atoms relative to a nitrogen atom and a heteroatom of higher electronegativity. Some of the physical properties of these compounds are presented here.

1,3-Azoles	pK_a	bp (°C)	Solubility in H$_2$O	Physical state
Oxazole	0.8	69–70	Sparingly soluble	Clear to pale yellow liquid
Imidazole	7.0	255–256	Soluble	Clear to pale yellow crystalline flake
Thiazole	2.5	116–118	Sparingly soluble	Clear to pale yellow liquid

6.4.2 Preparations of Oxazole, Imidazole and Thiazole

6.4.2.1 Preparation of Oxazole

Cyclocondensation of amides, through dehydration, leads to the formation of corresponding oxazoles. This synthesis is known as *Robinson–Gabriel synthesis*, named after Sir Robert Robinson and Siegmund Gabriel, who described the reaction in 1909 and 1910, respectively. Simply, Robinson–Gabriel synthesis involves the intramolecular reaction of a 2-acylamino-ketone followed by a dehydration to afford an oxazole. A cyclodehydrating agent is needed to catalyse the reaction. Several acids or acid anhydrides, for example phosphoric acid, phosphorous oxychloride, phosgene and thionyl chloride, can bring about this dehydration.

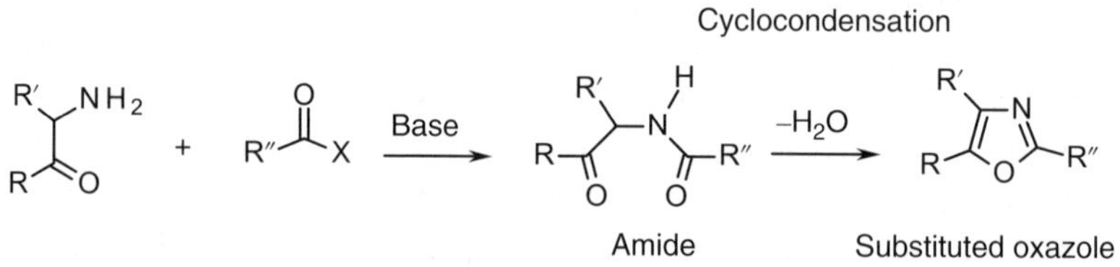

Substituted oxazole

6.4.2.2 Preparation of Imidazole

The condensation of a 1,2-dicarbonyl compound with ammonium acetate and an aldehyde results in the formation of an imidazole skeleton.

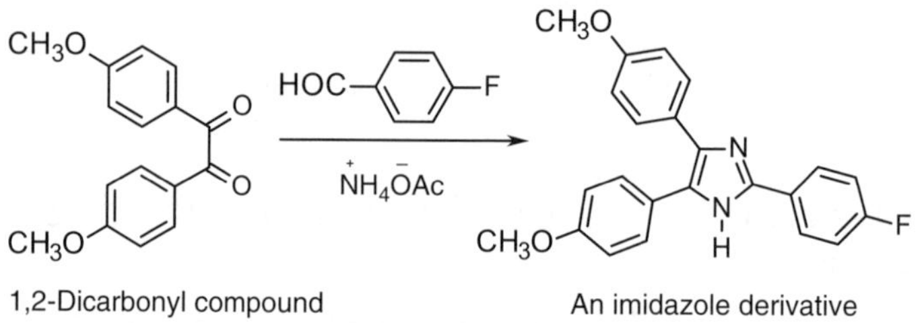

1,2-Dicarbonyl compound   An imidazole derivative

6.4.2.3 Preparation of Thiazole

Hantzsch synthesis can be applied to prepare thiazole system from thioamides. The reaction involves an initial nucleophilic attack by sulphur followed by a cyclocondensation.

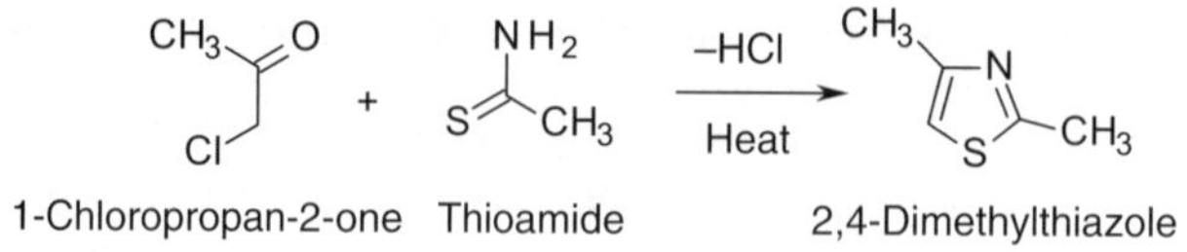

1-Chloropropan-2-one   Thioamide   2,4-Dimethylthiazole

A modification of this method involves the use of thiourea instead of a thioamide, as shown here.

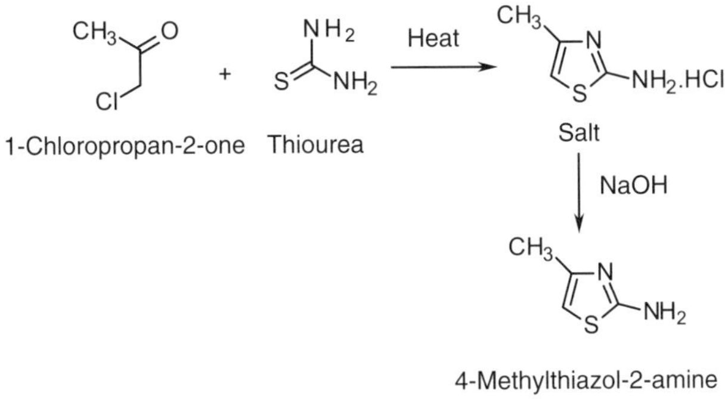

1-Chloropropan-2-one   Thiourea

Salt

4-Methylthiazol-2-amine

6.4.3 Reactions of Oxazole, Imidazole and Thiazole

The presence of the pyridine-like nitrogen deactivates the 1,3-azoles towards electrophilic attack and increases their affinity towards nucleophilic attack.

6.4.3.1 Electrophilic Substitutions

Although oxazole, imidazole and thiazoles are not very reactive towards aromatic electrophilic substitution reactions; the presence of any electron-donating group on the ring can facilitate electrophilic substitution. For example, 2-methoxythiazole is more reactive than thiazole itself. Some examples of electrophilic substitutions of oxazole, imidazole and thiazoles and their derivatives are presented here.

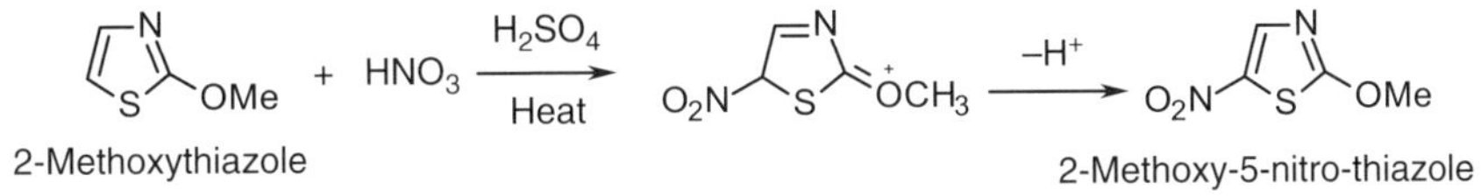

2-Methoxythiazole

2-Methoxy-5-nitro-thiazole

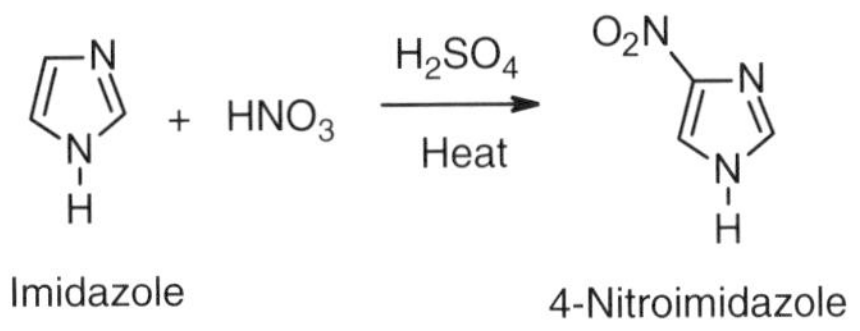

Imidazole

4-Nitroimidazole

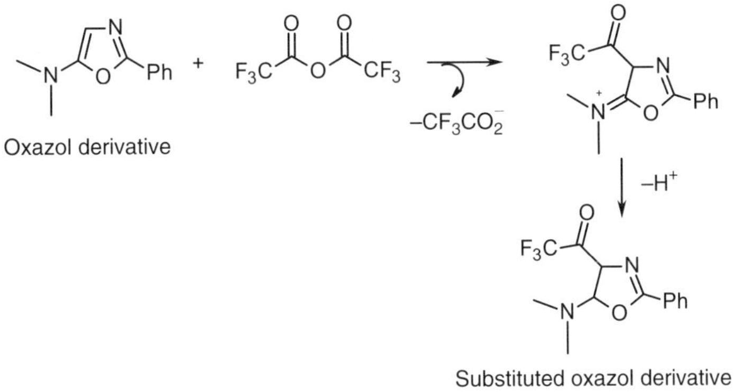

Oxazol derivative

Substituted oxazol derivative

6.4.3.2 Nucleophilic Aromatic Substitutions

1,3-Azoles are more reactive than pyrrole, furan or thiophene towards nucleophilic attack. Some examples of nucleophilic aromatic substitutions of oxazole, imidazole and thiazoles and their derivatives are given next. In the reaction with imidazole, the presence of a nitro group in the reactant can activate the reaction because that group can act as an electron-acceptor.

Imidazole   3-Bromo-2-nitro-thiophene   1-(2-Nitro-3-thienyl)imidazole

No activation is required for 2-halo-1,3-azoles, which can undergo nucleophilic aromatic substitutions quite easily.

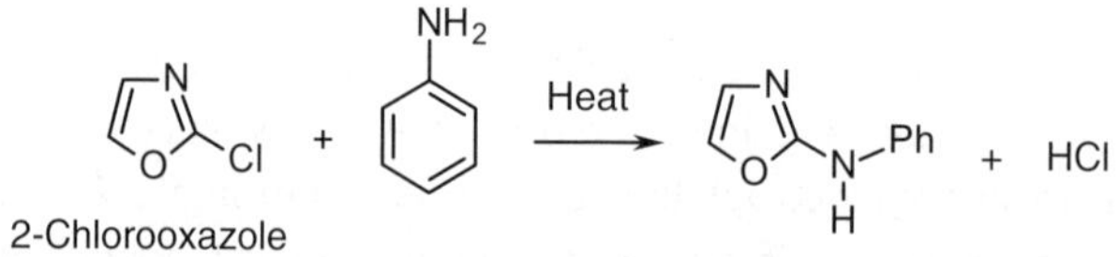

2-Chlorooxazole

N-Phenyloxazol-2-amine

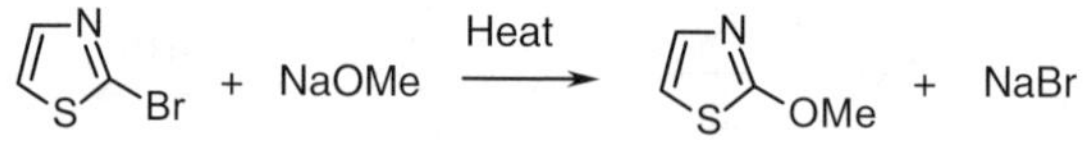

2-Bromothiazole   2-Methoxythiazole

6.5 ISOXAZOLE, PYRAZOLE AND ISOTHIAZOLE

Isoxazole, pyrazole and isothiazole constitute the 1,2-azoles of heterocycles that contain two heteroatoms, one of which is a nitrogen atom. The second heteroatom is oxygen, nitrogen or sulphur, respectively, for isoxazole, pyrazole and isothiazole. The aromaticity of these compounds is due to the delocalization of a lone pair of electrons from the second heteroatom to complete the aromatic sextet.

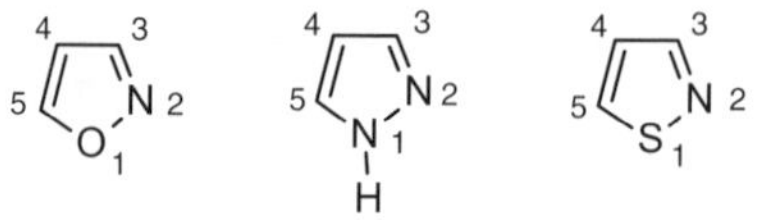

Isoxazole   Pyrazole   Isothiazole

The 1,2-azoles of heterocycles is important in medicine. For example, leflunomide (an anti-arthritis drug), cycloserine (an antibiotic) and gaboxadol (an antidepressant), all contain a substituted isoxazole system.

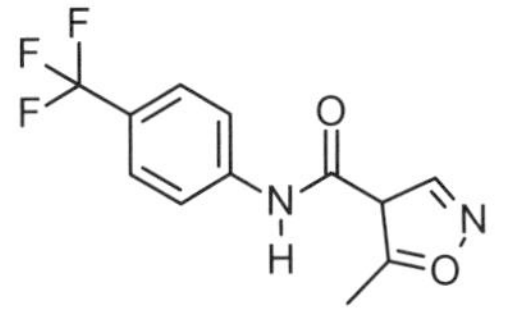

Leflunomide
(Used to treat rheumatoid arthritis)

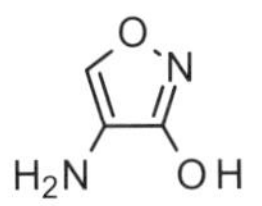

Cycloserine
(An antibacterial drug)

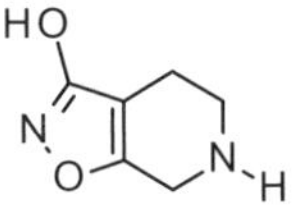

Gaboxadol
(An antidepressant drug)

Similarly, among the notable drugs containing a pyrazole ring are celecoxib (Celebrex), an NSAID and the anabolic steroid stanozolol. The pyrazole ring system is also found in several pesticides as fungicides, insecticides and herbicides, including chlorfenapyr, fenpyroximate, fipronil, tebufenpyrad, tolfenpyrad and tralopyril.

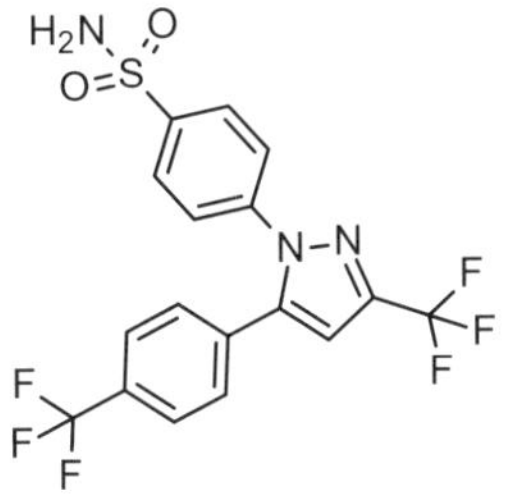

Celecoxib
(A non-steoirdal anti-inflammatory drug)

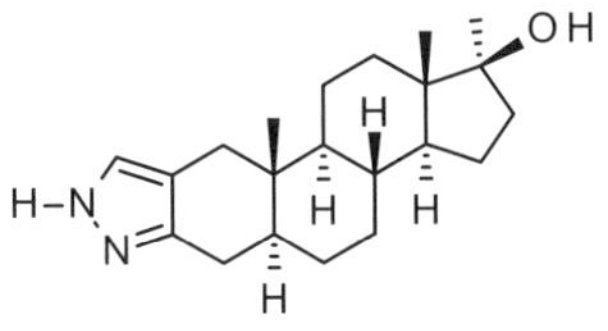

Stanozolol
(An anabolic steroid)

Ziprasidone (trade name: Geodon; an antipsychotic drug) and perospirone (trade name: Lullan; also an antipsychotic drug) are examples of drugs that contain an isothiazole ring system. Both these drugs are used in the treatment of schizophrenia as well as mania. Ziprasidone is also used off-label for depression, bipolar maintenance and post-traumatic stress disorder (PTSD).

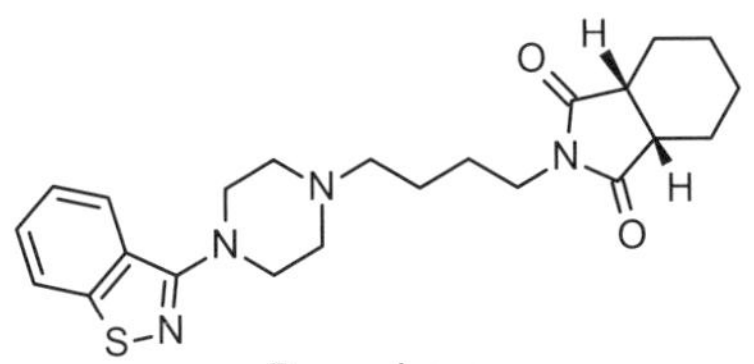

Perospirone
(An antipsychotic drug)

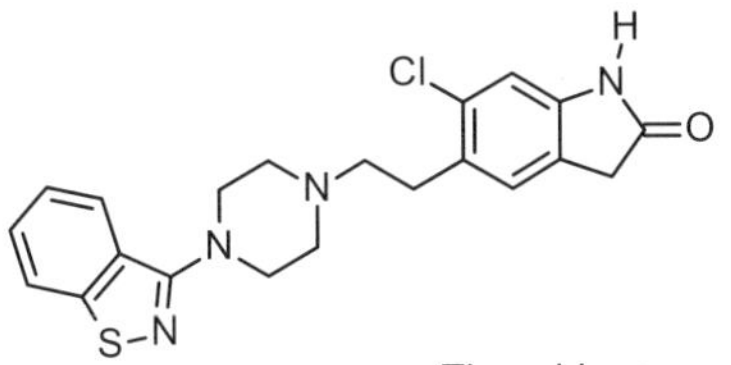

Ziprasidone
(An antipsychotic drug)

6.5.1 Physical Properties of Isoxazole, Pyrazole and Isothiazole

The 1,2-azoles are basic compounds because of the lone pair of electrons on the nitrogen atom, which is available for protonation. However, these compounds are much less basic than their isomers, 1,3-azoles, owing to the electron withdrawing effect of the adjacent heteroatom. Some of the physical properties of these compounds are as follows.

1,2-Azoles	pK_a	bp (°C)	mp (°C)	Physical state
Isoxazole	−2.97	95	—	Liquid
Pyrazole	2.52	186–188	60–70	Solid
Isothiazole	—	114	—	Liquid

6.5.2 Preparations of Isoxazole, Pyrazole and Isothiazole

6.5.2.1 Isoxazole and Pyrazole Synthesis

While 1,3-diketones undergo condensation with hydroxylamine to produce isoxazoles, with hydrazine they provide corresponding pyrazoles.

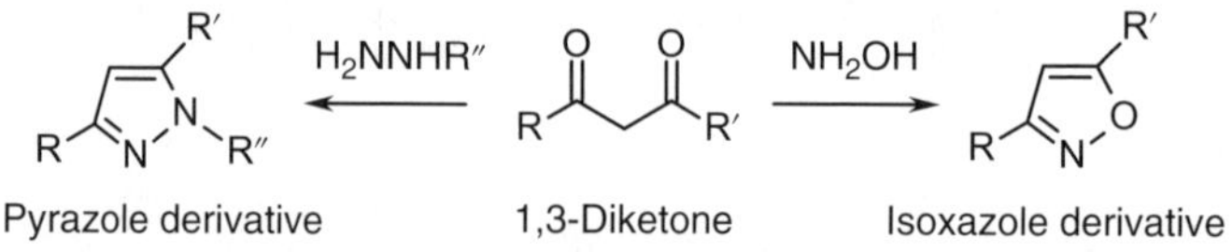

Pyrazole derivative   1,3-Diketone   Isoxazole derivative

6.5.2.2 Isothiazole Synthesis

Isothiazole can be prepared conveniently from thioamide in the following way.

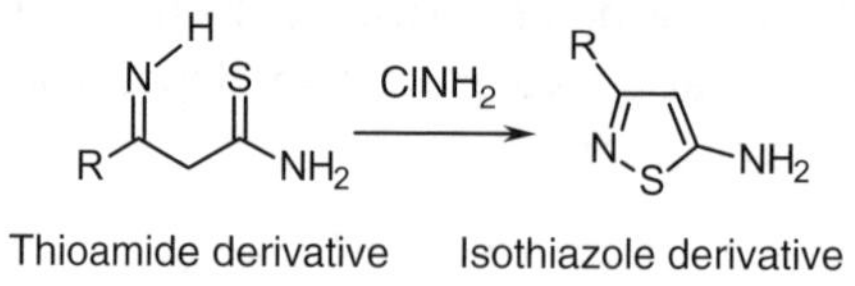

Thioamide derivative   Isothiazole derivative

6.5.3 Reactions of Isoxazole, Pyrazole and Isothiazole

Like 1,3-azoles, due to the presence of a pyridine-like nitrogen atom in the ring, 1,2-azoles are also much less reactive towards electrophilic substitutions than furan, pyrrole or thiophene. However, 1,2-azoles undergo electrophilic substitutions under appropriate reaction conditions and the main substitution takes place at the C-4 position; for example, bromination of 1,2-azoles. Nitration and sulphonation of 1,2-azoles can also be carried out, but only under vigorous reaction conditions.

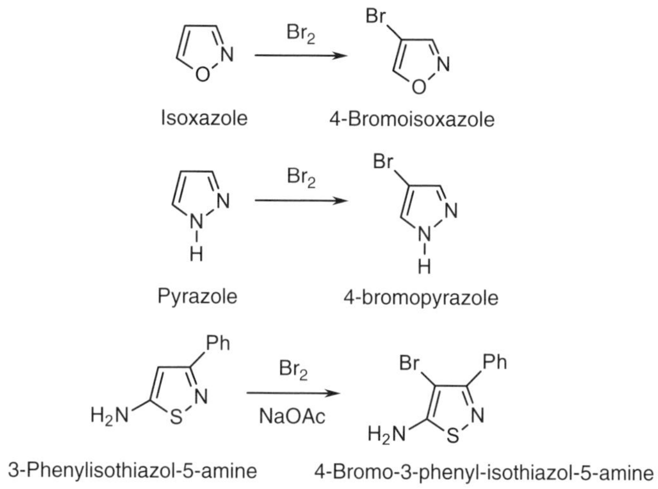

Br2
Br
N
O
Isoxazole
4-Bromoisoxazole
Br2
Br
N
N
H
Pyrazole
4-bromopyrazole
Ph
Br2
NaOAc
Br
Ph
H2N
S
N
3-Phenylisothiazol-5-amine
4-Bromo-3-phenyl-isothiazol-5-amine

6.6 PYRIMIDINE

Pyrimidine is a six-membered aromatic heterocyclic compound that contains two nitrogen atoms, separated by a carbon atom, in the ring. Nucleic acids, DNA and RNA, contain substituted purines and pyrimidines. Cytosine, uracil, thymine and alloxan are just a few of the biologically significant modified pyrimidine compounds, the first three being the components of the nucleic acids.

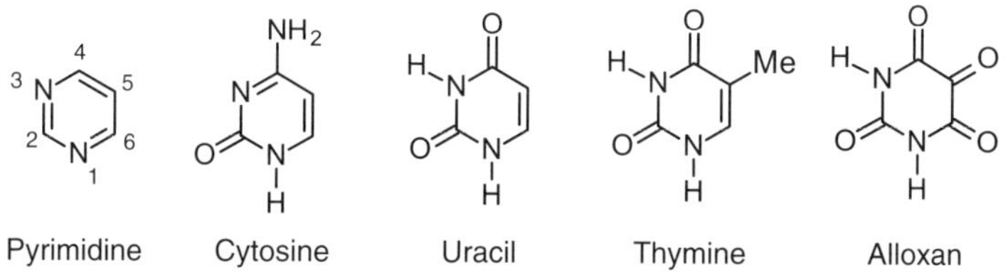

NH2
O
Me
N
H
Pyrimidine
Cytosine
Uracil
Thymine
Alloxan

A number of drug molecules contain modified pyrimidine skeleton, the best-known examples being the anticancer drug, 5-fluorouracil, which is structurally similar to thymine, the antiviral drug, AZT, currently being used in the treatment of AIDS and phenobarbital, a well-known sedative.

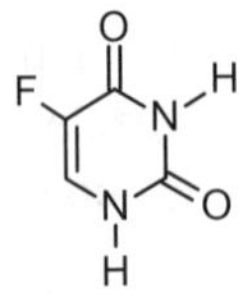

5-Fluorouracil
(An anticancer drug)

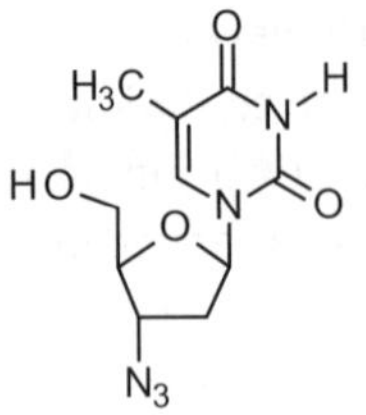

AZT
(An antiviral drug)

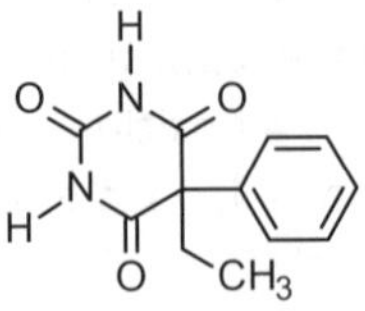

Phenobarbital
(A sedative)

Two positional isomers of pyrimidine are pyridiazine and pyrazine, which only differ structurally from of pyrimidine in terms of the position of the nitrogen atoms in the ring. These three heterocycles together with their derivatives are known as *diazines*.

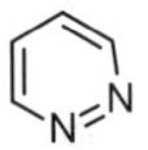

Pyridiazine 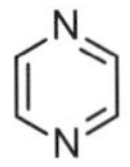   Pyrazine

6.6.1 Physical Properties of Pyrimidine

Pyrimidine is a weaker base than pyridine because of the presence of the second nitrogen. Its conjugate acid is a much stronger acid ($pK_a = 1.0$). The pK_a values of the N-1 hydrogen in uracil, thymine and cytosine are 9.5, 9.8 and 12.1, respectively. Pyrimidine is a hygroscopic solid (bp: 123–124 °C, mp: 20–22 °C) and soluble in water.

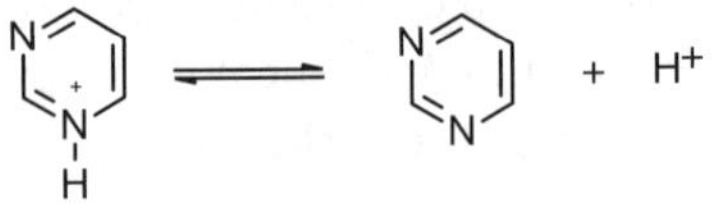

Conjugate acid of pyrimidine   Pyrimidine

6.6.2 Preparations of Pyrimidine

The combination of *bis*-electrophilic and *bis*-nucleophilic components is the basis of general pyrimidine synthesis. A reaction between an amidine (urea or thiourea or guanidine) and a 1,3-diketo compound affords corresponding pyrimidine systems. These reactions are usually facilitated by acid or base catalysis.

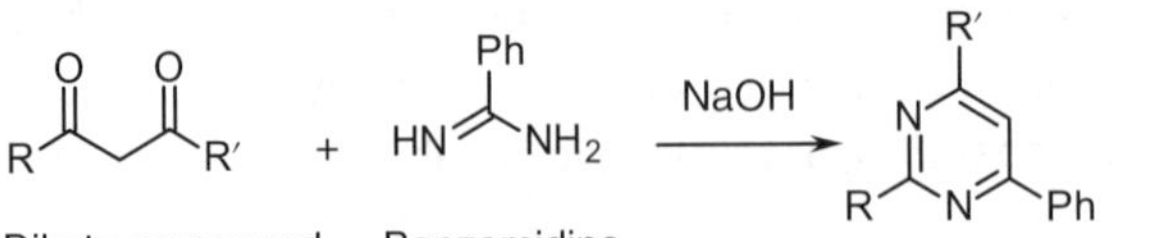

1,3-Diketo compound   Benzamidine

Pyrimidine derivative

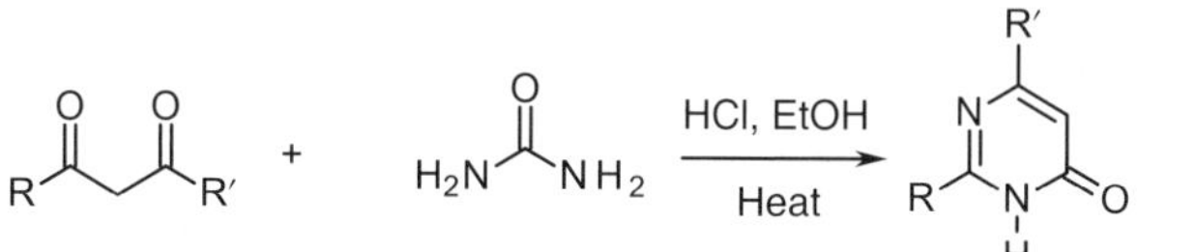

1,3-Diketo compound + Urea → Pyrimidine derivative

(HCl, EtOH, Heat)

6.6.3 Reactions of Pyrimidine

6.6.3.1 Electrophilic Aromatic Substitutions

The chemistry of pyrimidine is similar to that of pyridine with the notable exception that the second nitrogen in the aromatic ring makes it less reactive towards electrophilic substitutions. For example, nitration can only be carried out when there are two ring-activating substituents present on the pyrimidine ring (e.g. 2,4-dihydroxypyrimidine or uracil). The most activated position towards electrophilic substitution is C-5.

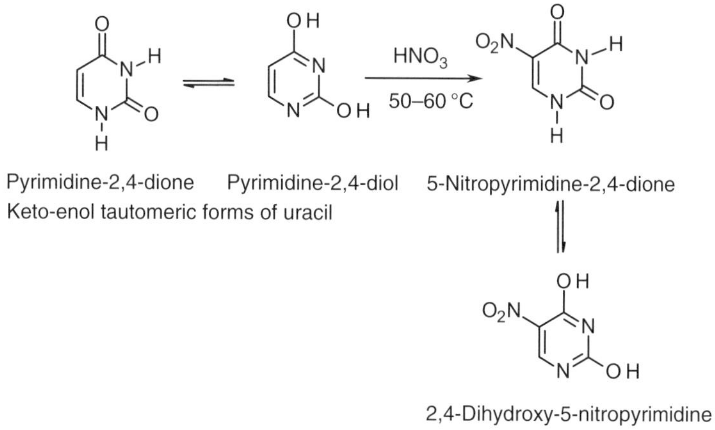

Pyrimidine-2,4-dione   Pyrimidine-2,4-diol   5-Nitropyrimidine-2,4-dione
Keto-enol tautomeric forms of uracil

2,4-Dihydroxy-5-nitropyrimidine

6.6.3.2 Nucleophilic Aromatic Substitutions

Pyrimidine is more reactive than pyridine towards nucleophilic aromatic substitution, again due to the presence of the second electron-withdrawing nitrogen in the pyrimidine ring. Leaving groups at C-2, C-4 or C-6 positions of pyrimidine can be displaced by nucleophiles.

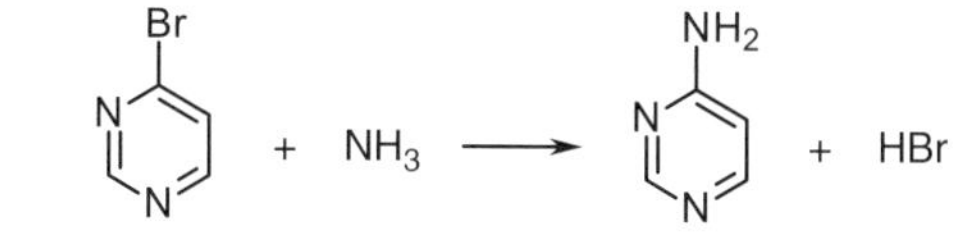

4-Bromopyrimidine + NH$_3$ → 4-Aminopyrimidine + HBr

2-Chloropyrimidine + NaOMe → 4-Methoxypyrimidine + NaCl

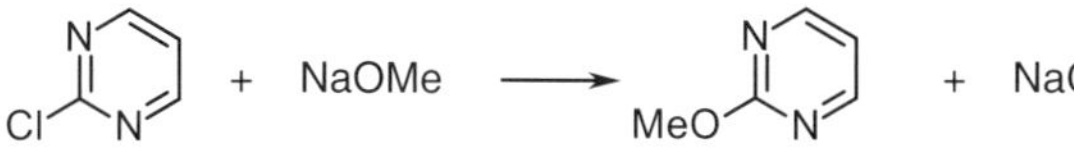

6.7 PURINE

Purine contains a pyrimidine ring fused with an imidazole nucleus. Guanine and adenine are two purine bases that are found in nucleic acids, DNA and RNA.

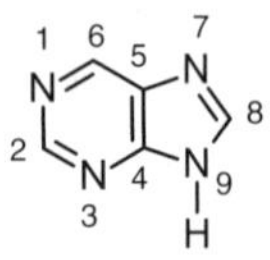

Purine 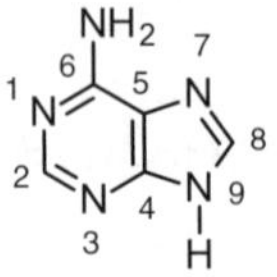   Adenine 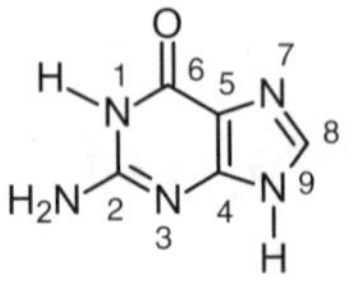   Guanine

Several purine derivatives are found in nature; for example, xanthine, hypo-xanthine and uric acid. The pharmacologically important (central nervous system (CNS)-stimulant) xanthine alkaloids, for example caffeine, theobromine and the-ophylline, are found in tea leaves, coffee beans and coco. The actual biosynthesis of purines involves construction of a pyrimidine ring onto a pre-formed imid-azole system.

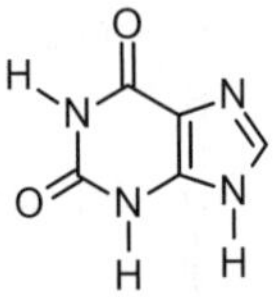

Xanthine 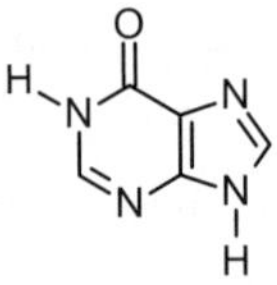   Hypoxanthine 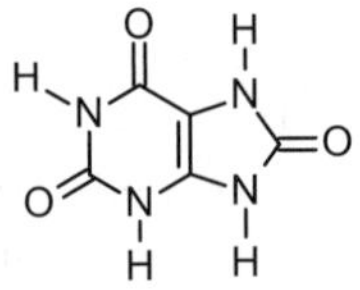   Uric acid

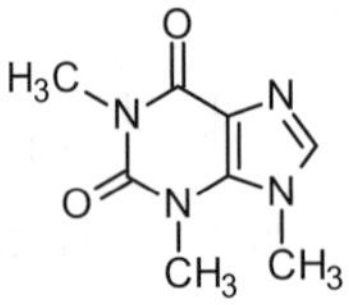

Caffeine 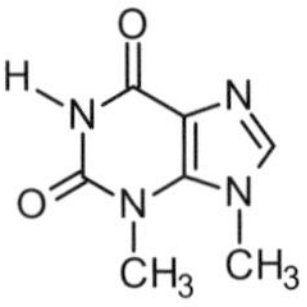   Theobromine 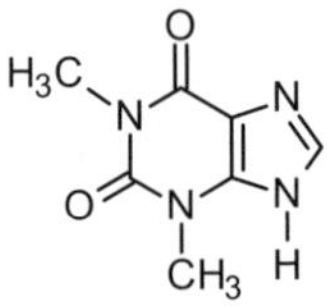   Theophylline

The purine and pyrimidine bases play an important role in the metabolic processes of cells through their involvement in the regulation of protein synthesis. Thus, several synthetic analogues of these compounds are used to interrupt with the cancer cell growth. One of such examples is an adenine mimic, 6-mercaptopu-rine, which is a well-known anticancer drug.

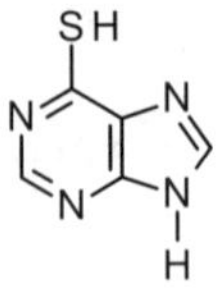

6-Mercaptopurine
(An anticancer drug)

6.7.1 Physical Properties of Purine

Purine is a basic crystalline solid (mp: 214 °C). As it consists of a pyrimidine ring fused to an imidazole ring, it possesses the properties of the both rings. The electron-donating imidazole ring makes the protonated pyrimidine part less acidic ($pK_a = 2.5$) than unsubstituted protonated pyrimidine ($pK_a = 1.0$). On the other hand, the electron-withdrawing pyrimidine ring makes the hydrogen of *N*-9 ($pK_a = 8.9$) more acidic than the corresponding *N*-1 hydrogen of imidazole ($pK_a = 14.4$).

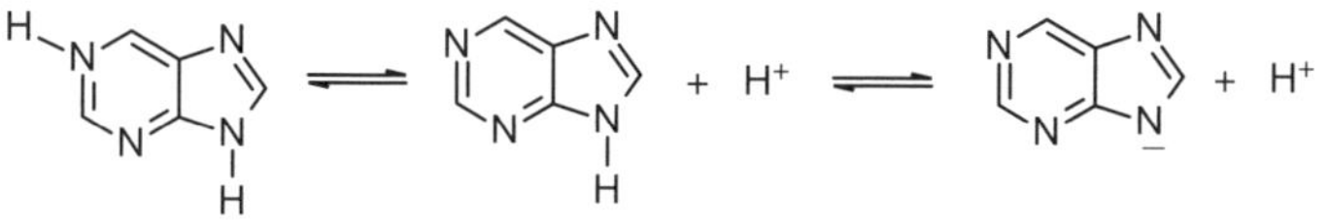

6.7.2 Preparations of Purine

The major site of purine synthesis is in the liver. Synthesis of the purine nucleotides begins with phosphoribosyl pyrophosphate (PRPP) and leads to the first fully formed nucleotide, inosine 5′-monophosphate (IMP).

6.7.3 Reactions of Purine

6.7.3.1 Nucleophilic Substitutions

Aminopurines react with dilute nitrous acid to provide the corresponding hydroxy compounds.

Adenine Purin-6-ol

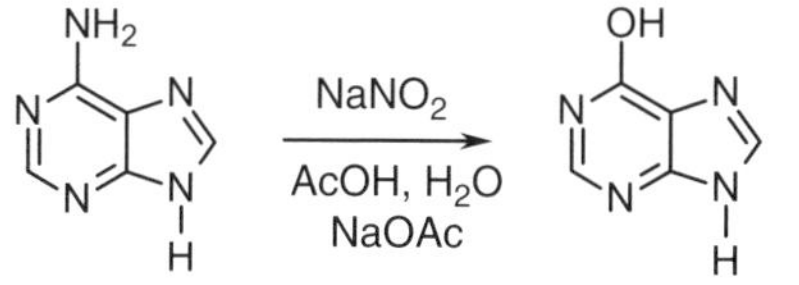

6.7.3.2 Deamination of Aminopurines

Adenine undergoes deamination to give hypoxanthine and guanine is deaminated to xanthine.

Adenine Hypoxanthine Guanine Xanthine

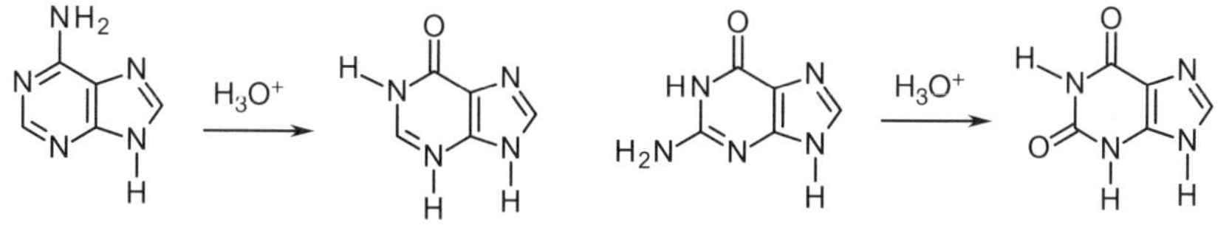

6.7.3.3 Oxidation of Xanthine and Hypoxanthine

Xanthine and hypoxanthine can be oxidized enzymatically with xanthine oxidase to yield uric acid.

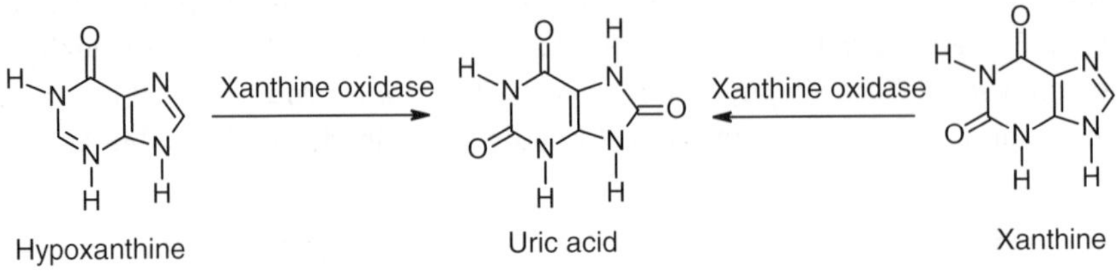

Hypoxanthine   Xanthine oxidase   Uric acid   Xanthine oxidase   Xanthine

6.8 QUINOLINE AND ISOQUINOLINE

Quinoline and isoquinoline, known as benzopyridines, are two isomeric heterocyclic compounds that have two rings, a benzene and a pyridine ring, fused together. In quinoline, this fusion is at C2/C3, whereas in isoquinoline, this is at C-3/C4 of the pyridine ring. Like benzene and pyridine, these benzopyridines are also aromatic in nature.

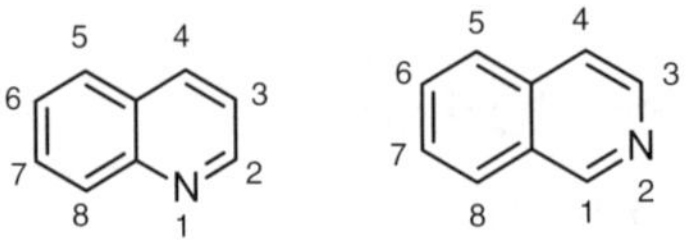

Quinoline   Isoquinoline

A number of naturally occurring pharmacologically active alkaloids possess a quinoline and isoquinoline skeleton. For example, papaverine, from *Papaver somniferum*, is an isoquinoline alkaloid, and quinine from *Cinchona* barks is a quinoline alkaloid that has antimalarial properties.

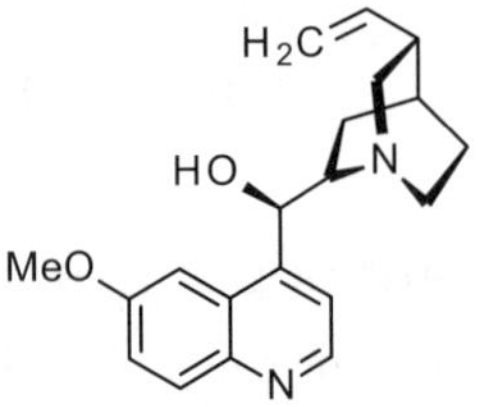

Quinine
(An antimalarial drug)

6.8.1 Physical Properties of Quinoline and Isoquinoline

Quinoline and isoquinoline are basic in nature. Like pyridine, the nitrogen atom of quinoline and isoquinoline is protonated under the usual acidic conditions. The conjugate acids of quinoline and isoquinoline have similar pK_a values (4.85 and 5.14, respectively) to that of the conjugate acid of pyridine.

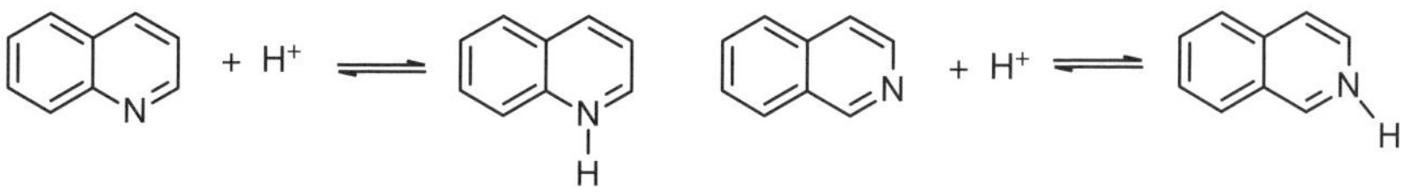

| Quinoline | Conjugate acid of quinoline | Isoquinoline | Conjugate acid of isoquinoline |

Quinoline, when exposed to light, first forms a yellow liquid and then slowly a brown liquid. It is slightly soluble in water but dissolves readily in many organic solvents. Isoquinoline crystallizes to platelets, is sparingly soluble in water but dissolves well in ethanol, acetone, diethyl ether, carbon disulphide (CS_2) and other common organic solvents. It is also soluble in dilute acids as the protonated derivative. Some other physical properties of these compounds are shown next.

Name	bp (°C)	mp (°C)	Physical state
Quinoline	238	−15.0	A colourless hygroscopic liquid with a strong odour
Isoquinoline	242	26–28	A colourless hygroscopic liquid at room temperature with a penetrating, unpleasant odour.

6.8.2 Preparations of Quinoline and Isoquinoline

6.8.2.1 Quinoline Synthesis: Skraup Synthesis

The *Skraup synthesis*, named after the Czech chemist Zdenko Hans Skraup, is used to synthesize quinoline skeleton by heating aniline with glycerol, using sulphuric acid as a catalyst and dehydrating agent. Ferrous sulphate is often added as a moderator, as the reaction can be violently exothermic. Note that an exothermic reaction is a chemical reaction that releases energy by light or heat, and is the opposite to an endothermic reaction. The most likely mechanism of this synthesis is that glycerol is dehydrated to acrolein, which undergoes conjugate addition to the aniline. This intermediate is then cyclized, oxidized and dehydrated to produce the quinoline system.

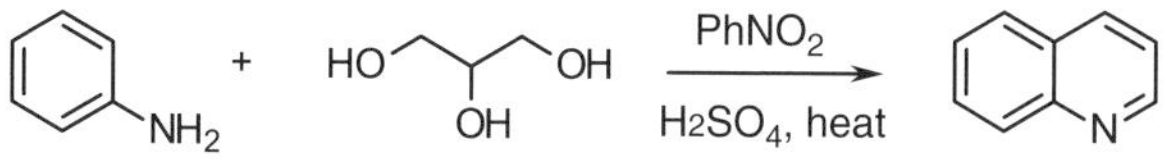

| Aniline | Glycerol | | Quinoline |

6.8.2.2 Quinoline Synthesis: Friedländer Synthesis

A modified version of the *Friedländer synthesis*, named after the German chemist Paul Friedländer, utilizing a 2-nitroaryl carbonyl compound, is sometimes used to synthesize a quinoline skeleton. *Friedländer synthesis* itself is somewhat complicated, because of the difficulty in preparing the necessary 2-aminoaryl carbonyl compounds.

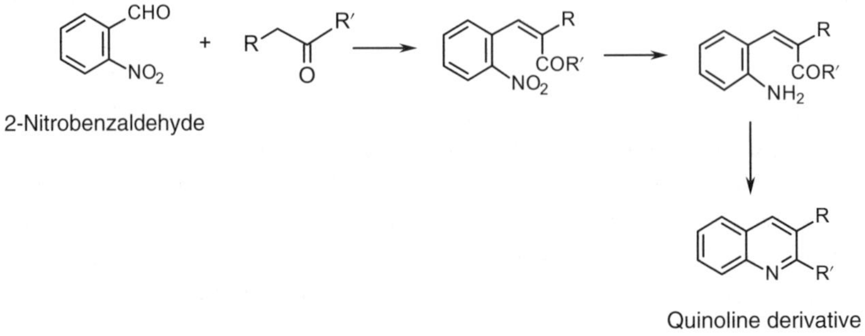

2-Nitrobenzaldehyde

Quinoline derivative

6.8.2.3 Isoquinoline Synthesis: Bischler–Napieralski Synthesis

The *Bischler–Napieralski synthesis*, first introduced in 1893 by August Bischler and Bernard Napieralski, is used to prepare isoquinolines. First β-phenylethylamine is acylated and then cyclodehydrated using phosphoryl chloride, phosphorous pentoxide or other Lewis acids to yield dihydroisoquinoline, which can be aromatized by dehydrogenation with palladium. For example, the synthesis of papaverine, a pharmacologically active isoquinoline alkaloid.

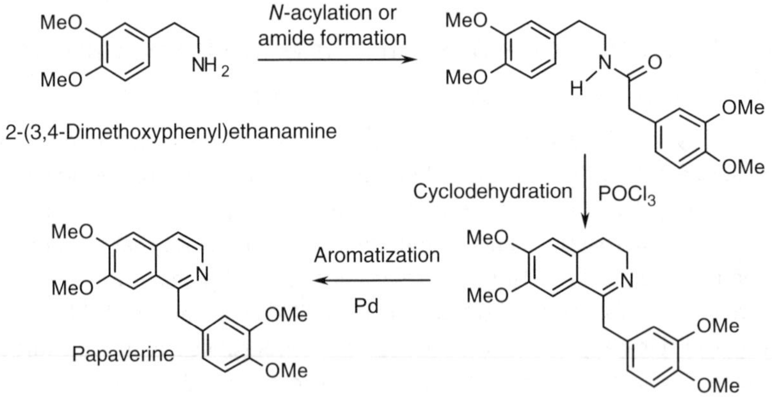

2-(3,4-Dimethoxyphenyl)ethanamine

Papaverine

6.8.2.4 Isoquinoline Synthesis: Pictet–Spengler Synthesis

The *Pictet–Spengler synthesis*, first introduced by Amé Pictet and Theodor Spengler in 1911, is another method of preparing isoquinolines. β-Phenylethylamine reacts with an aldehyde to form an imine, which undergoes acid-catalysed cyclization resulting in the synthesis of tetrahydroisoquinoline system. Again, tetrahydroiso-quinoline can be aromatized by palladium dehydrogenation to synthesize an iso-quinoline system.

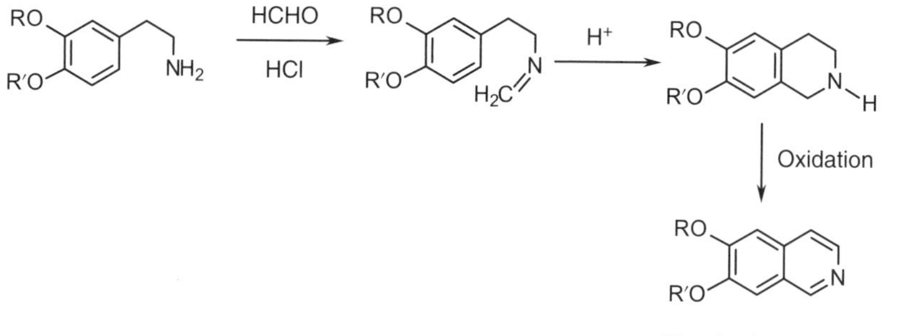

Tetrahydroisoquinoline

6.8.3 Reactions of Quinoline and Isoquinoline

6.8.3.1 Electrophilic Aromatic Substitutions

Quinoline and isoquinoline undergo electrophilic aromatic substitution on the benzene ring, because a benzene ring is more reactive than a pyridine ring towards such a reaction. Substitution generally occurs at C-5 and C-8; for example, bromi-nation of quinoline and isoquinoline.

Quinoline   5-Bromoquinoline 8-Bromoquinoline

Quinoline   5-Bromoisoquinoline 8-Bromoisoquinoline

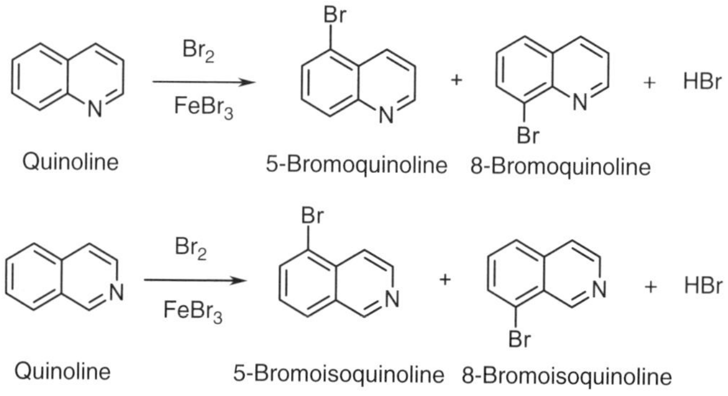

6.8.3.2 Nucleophilic Substitutions

Nucleophilic substitutions in quinoline and isoquinoline occur on the pyridine ring, because a pyridine ring is more reactive than a benzene ring towards such a reac-tion. While this substitution takes place at C-2 and C-4 in quinoline, isoquinoline undergoes nucleophilic substitution only at C-1.

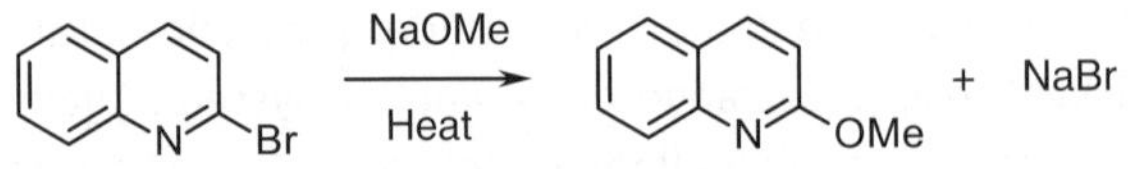

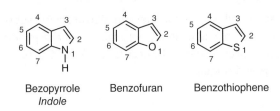

2-Bromoquinoline → NaOMe, Heat → 2-Methoxyquinoline + NaBr

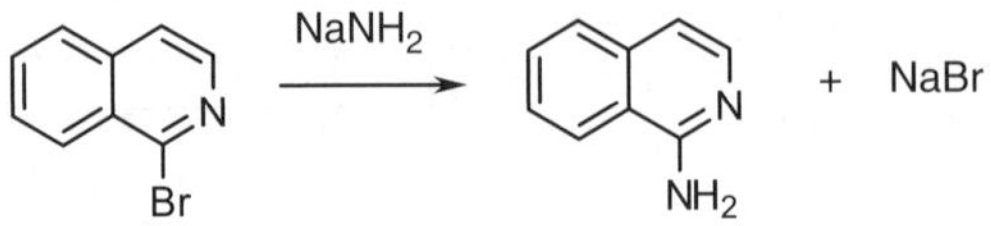

Bromoisoquinoline → NaNH₂ → Isoquinolinamine + NaBr

6.9 INDOLE

Indole contains a benzene ring fused with a pyrrole ring at C-2/C-3 and can be described as benzopyrrole. Indole is a 10 π electron aromatic system achieved from the delocalization of the lone pair of electrons on the nitrogen atom. Benzofuran and benzothiophene are quite similar to benzopyrrole and indole, with different heteroatoms, oxygen and sulphur, respectively.

Bezopyrrole
Indole

Benzofuran

Benzothiophene

Indole group of compounds is one of the most prevalent groups of alkaloids found in nature. A number of important pharmacologically active medicinal products and potential drug candidates contain an indole system. For example, serotonin, a well-known neurotransmitter, and tryptophan, one of the naturally occurring essential amino acids, both contain an indole skeleton.

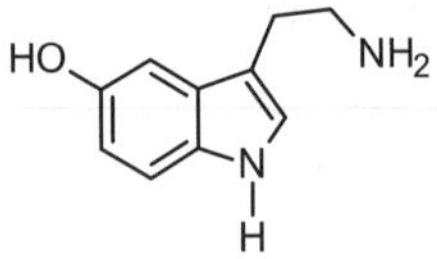

5-Hydroxytrptamine
Serotonin
(A neurotransmitter)

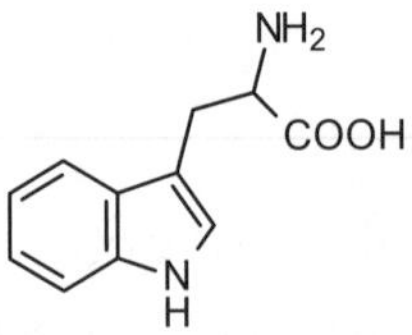

Tryptophan
(An essential amino acid)

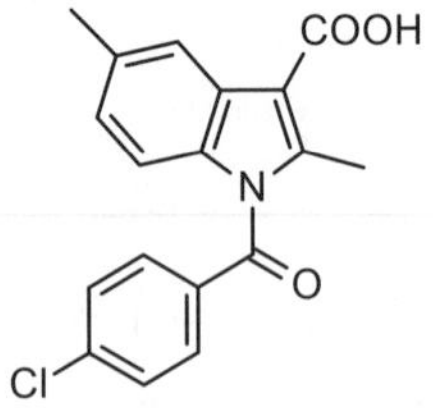

Indomethacin
(An anti-arthritic drug)

A dimeric indole alkaloid, vincristine, from *Vinca rosea*, is a well-known anti-cancer and antitumour drug (mainly used against non-Hodgkin lymphoma) and belongs to this indole heterocycle group of compounds. Both indomethacine, a synthetic indol-3-acetic acid derivative used in the treatment of rheumatoid arthritis and lysergic acid diethylamide (LSD), used in the treatment of alcoholism pain and cluster headaches, have an indole skeleton. Examples of some of the other well-known drugs that contain an indole skeleton and their applications are summarized in the following table.

Indole drugs	Applications	Indole drugs	Applications
Apaziquone	Anticancer	Indalpine	Antidepressant
Arbidol	Antiviral	Mitragynine	Opoid agonist
Atevirdine	Anti-HIV	Mitraphylline	Anticancer
Binedaline	Antidepressant	Oxypertine	Antipsychotic
Cediranib	Anticancer	Peridopril	Antihypertensive
Delavirdine	Anti-HIV	Reserpine	Antihypertensive
Doleasetron	Anti-emetic	Zafirlukast	Anti-asthmatic

6.9.1 Physical Properties of Indole

Indole is a weakly basic compound. The conjugate acid of indole is a strong acid with a pK_a value −2.4. Indole is a white solid (bp: 253–254 °C, mp: 52–54 °C) at room temperature and possesses an intense faecal smell. However, at low concentrations, it has a flowery smell. Indole is slightly soluble in water, but well soluble in organic solvents; for example, ethanol, ether and benzene.

6.9.2 Preparations of Indole

6.9.2.1 Fischer Indole Synthesis

Cyclization of arylhydrazones by heating with an acid or Lewis acid catalyst yields an indole system. The most commonly used catalyst is $ZnCl_2$. The disadvantage of this reaction is that unsymmetrical ketones give mixtures of indoles if R′ also has an α-methylene group.

Phenylhydrazine   Indole derivative

6.9.2.2 Leimgruber–Batcho Synthesis

Aminomethylenation of *o*-nitrotoluene followed by hydrogenation affords an indole. In fact, the first step is the formation of an enamine followed by reductive cyclization. Usually, the reductive cyclization is effected by Raney nickel and hydrazine. Willy Leimgruber and Andrew Batcho synthesized a large number of indoles and employed several reduction conditions with and without the isolation of the nitroenamine.

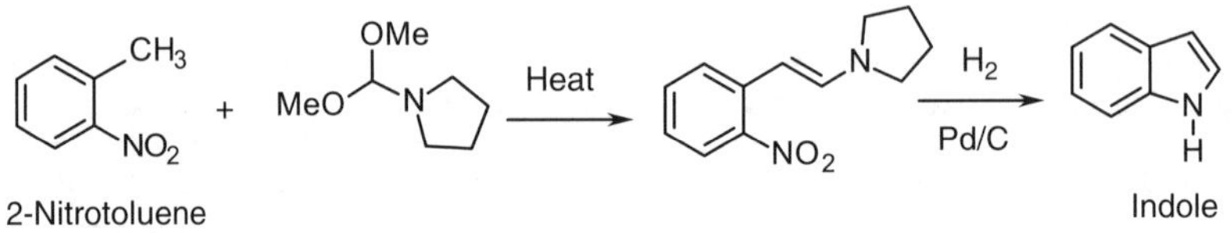

2-Nitrotoluene   Indole

6.9.3 Reactions of Indole

6.9.3.1 Electrophilic Aromatic Substitution

Electrophilic aromatic substitution of indole occurs on the five-membered pyrrole ring, because it is more reactive towards such reactions than a benzene ring. As an electron-rich heterocycle, indole undergoes electrophilic aromatic substitution primarily at C-3; for example, bromination of indole.

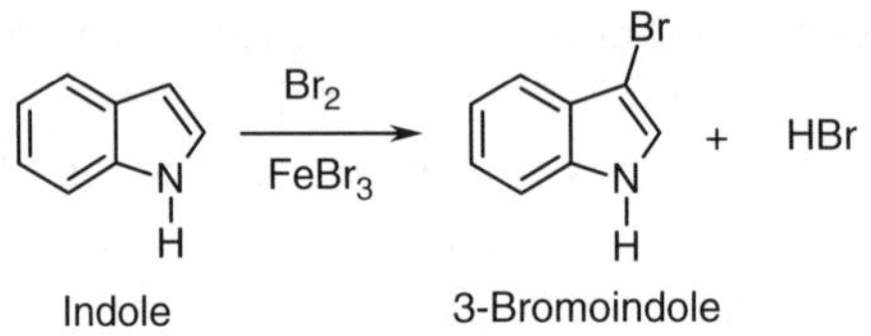

Indole   3-Bromoindole

6.9.3.2 Mannich Reaction

This is another example of electrophilic aromatic substitution where indole can produce an aminomethyl derivative.

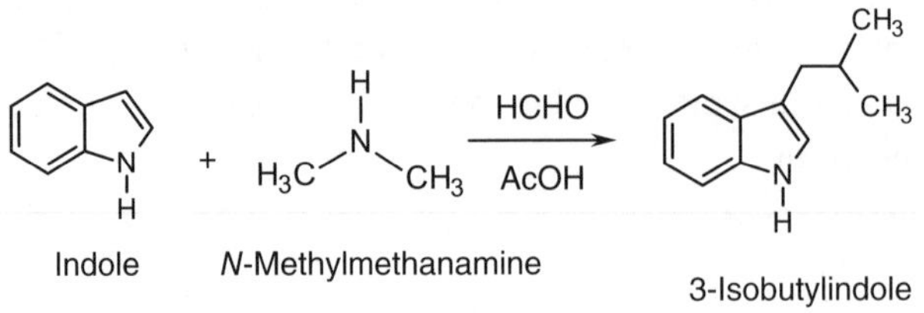

Indole   *N*-Methylmethanamine

3-Isobutylindole

6.9.3.3 Vilsmeier Reaction

Using the Vilsmeier reaction, an aldehyde group can be brought in at C-3 of an indole skeleton.

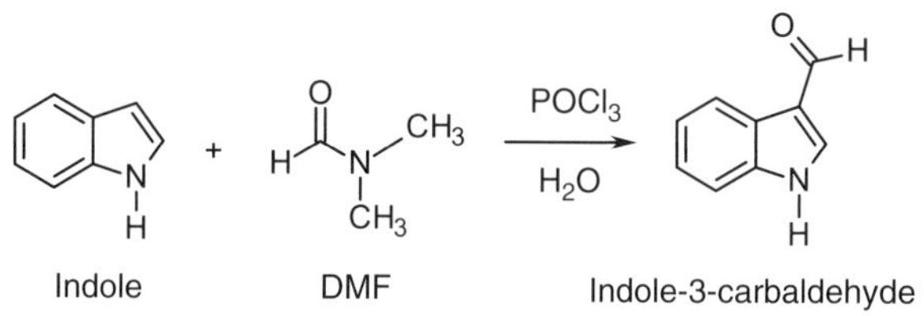

Indole   DMF   Indole-3-carbaldehyde

6.9.4 Test for Indole

Indole is a component of the amino acid tryptophan, which can be broken down by the bacterial enzyme tryptophanase. When tryptophan is broken down, the presence of indole can be detected through the use of *Kovacs' reagent*. Kovacs' reagent, which is yellow, reacts with indole and forms a red colour on the surface of the test tube. Kovac's reagent is prepared by dissolving 10 g of p-aminobenzaldehyde in 150 ml of isoamylalcohol and then slowly adding 50 m of concentrated HCl.

Chapter 7
Nucleic Acids

Learning objectives

After completing this chapter, students should be able to

- provide an overview of heterocyclic aromatic chemistry;
- classify amino acids, describe the properties of amino acids and discuss the formation of peptides;
- explain the fundamentals of the chemistry of nucleic acids.

7.1 NUCLEIC ACIDS

The nucleic acids, deoxyribonucleic acid (DNA) and ribonucleic acid (RNA), are the chemical carriers of a cell's genetic information. They are biopolymers made of *nucleotides* joined together to form a long chain. These biopolymers are often found associated with proteins and in this form they are called *nucleoproteins*. Each nucleotide comprises a *nucleoside* bonded to a phosphate group and each nucleoside is composed of an aldopentose sugar, ribose or 2-deoxyribose, linked to a heterocyclic purine or pyrimidine base (see Section 6.7).

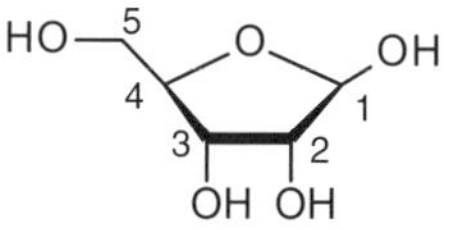

Ribose

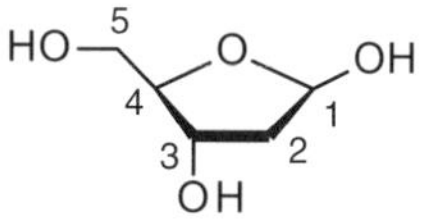

2-Deoxyribose

Chemistry for Pharmacy Students: General, Organic and Natural Product Chemistry,
Second Edition. Lutfun Nahar and Satyajit Sarker.
© 2019 John Wiley & Sons Ltd. Published 2019 by John Wiley & Sons Ltd.

The sugar component in RNA is *ribose*, whereas in DNA, it is *2-dexoyribose*. In deoxyribonucleotide, the heterocyclic bases are purine bases, adenine and guanine, and pyrimidine bases, cytosine and thymine. In ribonucleotide, adenine, guanine, and cytosine are present, but not thymine, which is replaced by uracil, another pyrimidine base.

In the nucleotides, while the heterocyclic base is linked to C-1 of the sugar through an *N*-glycosidic β-linkage, the phosphoric acid is bonded by a phosphate ester linkage to C-5. When the sugar is a part of a nucleoside, the numbering of sugars starts with 1′, that is, C-1 becomes C-1′. For example, 2′-deoxyadenosine 5′-phosphate and uridine 5′-phosphate.

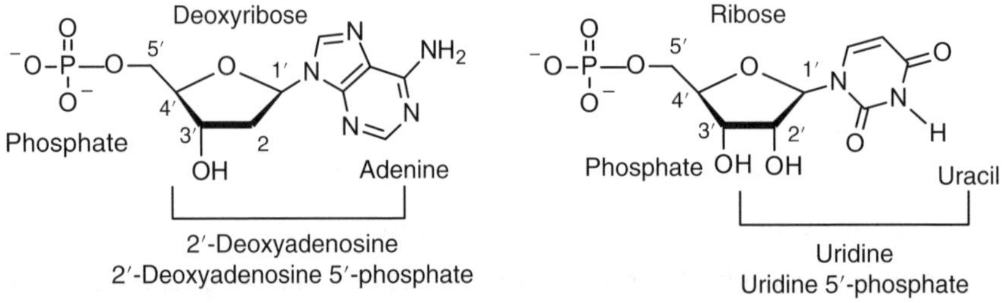

2′-Deoxyadenosine
2′-Deoxyadenosine 5′-phosphate

Uridine
Uridine 5′-phosphate

Despite being structurally similar, DNA and RNA differ in size and their functions within a cell. The molecular weights of DNA, found in the nucleus of cells, can be up to 150 billion and length of up to 12 cm, whereas the molecular weight of RNA, found in outside the cell nucleus, can only be up to 35 000.

Deoxyribonucleic acid (DNA)	
Name of the nucleotide	Composition
2′-Deoxyadenosine 5′-phosphate	Adenine + Deoxyribose + Phosphate Nucleoside is 2′-deoxyadenosine, composed of adenine and deoxyribose
2′-Deoxyguanosine 5′-phosphate	Guanine + Deoxyribose + Phosphate Nucleoside is 2′-deoxyguanosine, composed of guanine and deoxyribose
2′-Deoxycytidine 5′-phosphate	Cytosine + Deoxyribose + Phosphate Nucleoside is 2′-deoxycytidine, composed of cytosine and deoxyribose
2′-Deoxythymidine 5′-phosphate	Thymine + Deoxyribose + Phosphate Nucleoside is 2′-deoxythymidine, composed of thymine and deoxyribose

Adenosine 5′-phosphate	Adenine + Ribose + Phosphate Nucleoside is adenosine, composed of adenine and ribose
Guanosine 5′-phosphate	Guanine + Ribose + Phosphate Nucleoside is guanosine, composed of guanine and ribose
Cytidine 5′-phosphate	Cytosine + Ribose + Phosphate Nucleoside is cytidine, composed of cytosine and ribose
Uridine 5′-phosphate	Uracil + Ribose + Phosphate Nucleoside is uridine, composed of uracil and ribose

7.1.1 Synthesis of Nucleosides and Nucleotides

A reaction between a suitably protected ribose or 2-deoxyribose and an appropriate purine or pyrimidine base yields a nucleoside. For example, guanosine can be synthesized from a protected ribofuranosyl chloride and a chloromercurieguanine.

2,3,5-Tri-*O*-acetyl-
β-D-ribofuranosylchloride   Chloromercurieguanine

Guanosine

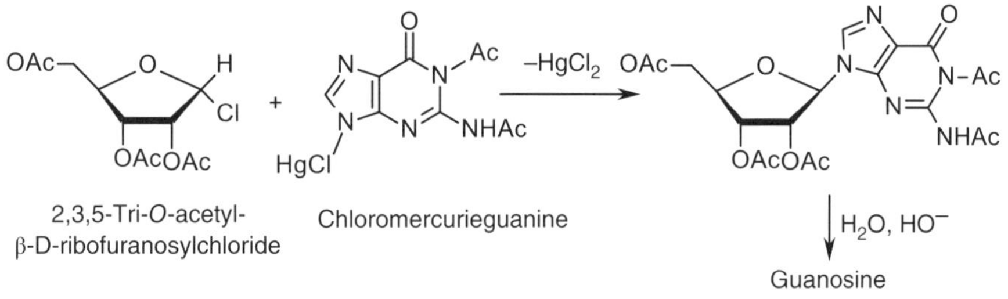

Nucleosides can also be prepared through the formation of the heterocyclic base on a protected ribosylamine derivative.

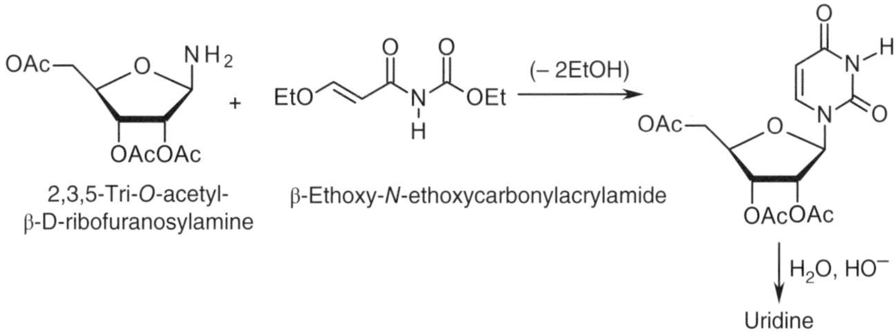

2,3,5-Tri-*O*-acetyl-
β-D-ribofuranosylamine   β-Ethoxy-*N*-ethoxycarbonylacrylamide

Uridine

Phosphorylation of nucleosides produces corresponding nucleotides. Phosphorylating agents, for example dibenzylphosphochloridate, are used in this reaction. To perform phosphorylation at C-5′, the other two hydroxyl functionalities at C-2′ and C-3′ have to be protected, usually with an isopropylidene group. At the final step, this protecting group can be removed by mild acid-catalysed hydrolysis, and a hydrogenolysis cleaves the benzylphosphate bonds. Note that hydrogenolysis is a chemical reaction whereby a carbon-carbon or carbon-heteroatom single bond is cleaved or undergoes lysis (breakdown) by hydrogen. Usually hydrogenolysis is conducted catalytically using hydrogen gas.

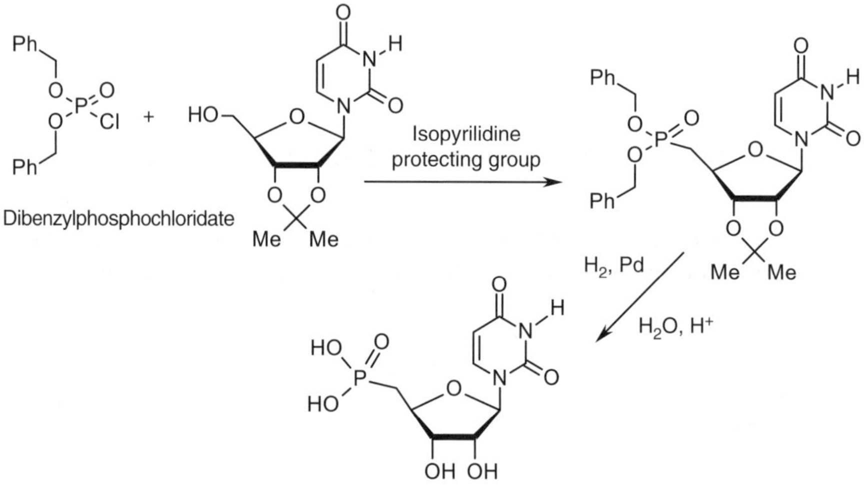

Dibenzylphosphochloridate

Isopyrilidine protecting group

H₂, Pd

H₂O, H⁺

Uridine 5′-phosphate
(A nucleotide)

7.1.2 Structure of Nucleic Acids

7.1.2.1 Primary Structure

The *primary structure* of the nucleic acid is formed by the covalent backbone consisting of ribo/deoxyribo nucleotides linked to each other by phosphodiester bonds. Nucleotides join together in DNA and RNA by forming a phosphate ester bond between the 5′-phosphate group on one nucleotide and the 3′-hydroxyl group on the sugar (ribose or 2′-deoxyribose) of another nucleotide. In the nucleic acids, these phosphate ester links provide the nucleic acids with a long unbranched chain with a 'backbone' of sugar and phosphate units with heterocyclic bases sticking out from the chain at regular intervals. One end of the nucleic acid polymer has a free hydroxyl at C-3′ (the *3′-end*), and the other end has a phosphate at C-5′ (the *5′-end*).

The structure of nucleic acids depends on the sequence of individual nucleotides. The actual base sequences for many nucleic acids from various species are available to date. Instead of writing the full name of each nucleotide, abbreviations are used; for example, A for adenine, T for thymidine, G for guanosine and C for cytidine. Thus, a typical DNA sequence might be presented as TAGGCT.

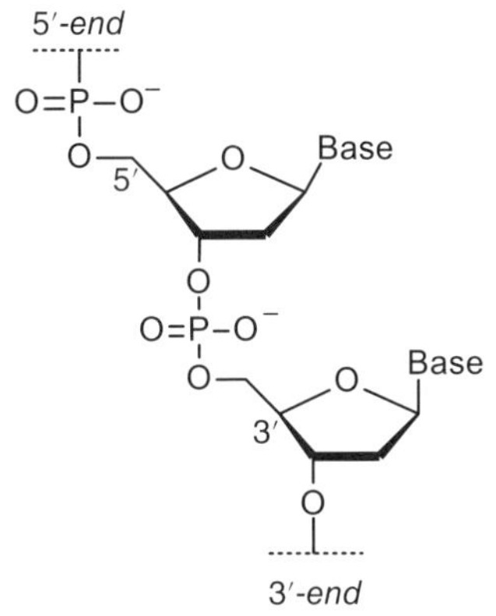

Generalized structure of DNA

7.1.2.2 Secondary Structure: Base Pairing

The base sequence along the chain of a DNA contains the genetic information. Samples of DNA isolated from different tissues of the same species have the same proportions of heterocyclic bases, but the samples from different species often have different proportions of bases. For example, human DNA comprises about 30% adenine, 30% thymine, 20% guanine and 20% cytosine, while the bacterium *Staphylococcus aureus* contains about 30% adenine, 30% thymine, 20% guanine and 20% cytosine. In these examples, it is clear that the bases in DNA occur in pairs. Adenine and thymine are usually present in equal amounts, so are cytosine and guanine. In late 1940s, Erwin Chargaff, an Austro-Hungarian biochemist, pointed out these regularities and summarized as follows:

i. The total mole percentage of purines is approximately equal to that of the pyrimidines, that is $(\%G + \%A)/(\%C + \%T) \cong 1$.

ii. The mole percentage of adenine is nearly equal to that of thymine, that is $\%A/\%T \cong 1$, and same is true for guanine and cytosine, that is $\%G/\%C \cong 1$.

To provide explanations for some of these earlier findings, the *secondary structure* of DNA was first proposed by James Watson and Francis Crick in 1953, and was verified shortly thereafter through X-ray crystallographic analysis by Maurice Wilkins. According to the Watson–Crick model, DNA consists of two polynucleotide strands coiled around each other in a *double helix* like the handrails on a spiral staircase. The two strands run in opposite directions and are held together by hydrogen bonds between specific pairs of bases. Adenine and thymine form strong hydrogen bonds to each other, but not to cystine or guanine. Similarly, cystine and guanine form strong hydrogen bonds to each other, but not to adenine or thymine. The discovery of the double helix, the twisted-ladder structure of deoxyribonucleic acid (DNA), by James Watson and Francis Crick, marked a milestone in the history of science and gave birth of modern molecular biology, which is largely concerned with understanding how genes control the chemical processes within cells.

Guanine   Cytosine   Adenine   Thymine

Hydrogen bonding between base pairs of the DNA double helix

The base pairs are on the inside of the helix and the sugar-phosphate backbone is on the outside. The pitch of the helix is such that 10 successive nucleotide pairs form one complete turn in 34 Å (the repeat distance). The exterior width of the spiral is about 20 Å, and the internal distance between 1′-positions of ribose units on opposite chains is about 11 Å.

The two strands of DNA double helix are not identical, but complementary to each other in a way such that, whenever cytosine occurs in one strand, a guanine occurs opposite in the other strand and the same situation is true for adenine and thymine. This complementary pairing of bases explains why A and T are always found in equal amounts, as are C and G. It is this complementary behaviour of the two strands that explains how a DNA molecule replicates itself at the time of cell division and thereby passes on the genetic information to each other of the two daughter cells.

The two strands of the double helix coil in such a way that two types of 'grooves' are formed, a major groove 1.2 nm wide and a minor groove 600 pm wide. A number of flat, polycyclic molecules fit sideways into the groove between the strands and intercalate, or insert themselves between the stacked base pairs. Many cancer causing and cancer preventing agents exert their actions through intercalating with DNA.

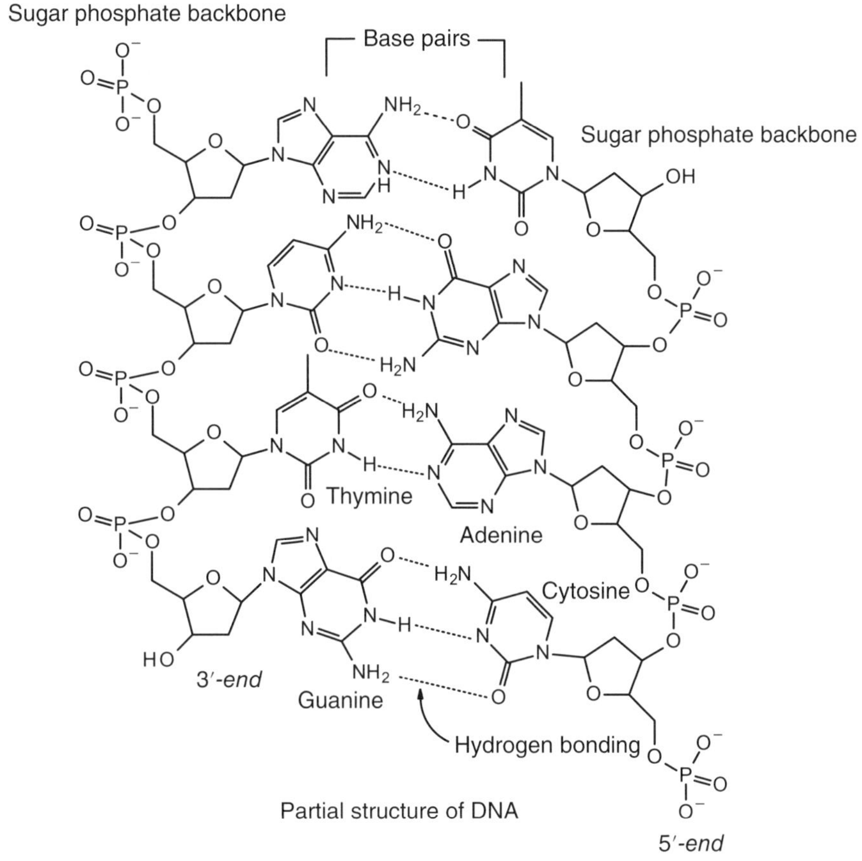

Partial structure of DNA

While the sugar-phosphate backbone of DNA is completely regular, the sequence of heterocyclic base pairs along the backbone can be of different permutations. It is the precise sequence of base pairs that carries the genetic information.

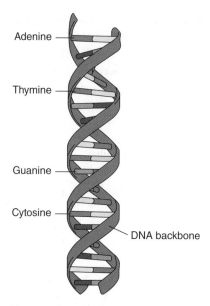

Adenine

Thymine

Guanine

Cytosine

DNA backbone

Heterocyclic base pairs backbone of DNA

7.1.3 Nucleic Acids and Heredity

The genetic information of an organism is stored as a sequence of deoxyribonu-
cleotides strung together in the DNA chain. Three fundamental processes are
involved in the transfer of this stored genetic information.

Replication. This process creates the identical copies of DNA, so that
information can be preserved and handed down to offspring.

Transcription. This process reads the stored genetic information, brings it out
of the nucleus to ribosomes where protein synthesis occurs.

Translation. In this process, the genetic messages are decoded and used to
build proteins.

7.1.3.1 Replication of DNA

Replication of DNA is an enzymatic process that starts with the partial unwinding
of the double helix. Just before the cell division, the double strand begins to
unwind. As the strands separate and bases are exposed, new nucleotides line up
on each strand in a complementary fashion, A to T, and C to G. Two new strands
now begin to grow, which are complementary to their old template strands. Two
new identical DNA double helices are generated in this way, and these two new
molecules can then be passed on, one to each daughter cell. As each of the new

DNA molecules contains one strand of old DNA and one new, the process is called as *semiconservative replication.*

Addition of new nucleotide units to the growing chain occurs in the 5′ to C′ direction and is catalysed by the enzyme DNA polymerase. The most important step is the addition of a 5′-mononucleoside triphosphate to the free 3′-hydroxyl group of the growing chain as the 3′-hydroxyl attacks the triphosphate and expels a diphosphate leaving group.

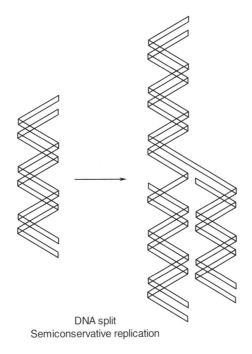

DNA split
Semiconservative replication

7.1.3.2 Transcription: Synthesis of RNA

Transcription starts with the process by which the genetic information is transcribed onto a form of RNA, called mRNA. Ribonucleic acid, RNA, is structurally similar to DNA with the exceptions that its nucleotides contain ribose instead of a 2′-deoxyribose and the base thymine is replaced by uracil. There are three major types of RNA depending on their specific functions. However, all three types of RNA are much smaller than DNA and they are single- rather than double-stranded.

i.   *Messenger RNA (mRNA)* carries genetic information from DNA to ribosomes where protein synthesis occurs.

ii. *Ribosomal RNA (rRNA),* complexed with proteins (nucleoproteins), provides the physical make up of ribosomes.

iii. *Transfer RNA (tRNA)* transports amino acids to the ribosomes for protein synthesis.

Protein synthesis takes place in the cell nucleus with the synthesis of mRNA. Part of the DNA double helix unwinds adequately to expose a portion corresponding to at least one *gene* on a single chain. Ribonucleotides, present in the cell nucleus, assemble along the exposed DNA chain by pairing with the bases of DNA in a similar fashion to that observed in DNA base pairing. However, in RNA uracil replaces thymine. The ribonucleotide units of mRNA are joined into a chain by the enzyme *RNA polymerase.* Once the mRNA is synthesized, it moves into the cytoplasm where it acts as a template for protein synthesis. Unlike what is seen in DNA replication where both strands are copied, only one of the two DNA strands is transcribed into mRNA. The strand that contains the gene is called the *coding* or *sense strand.* The strand that gets transcribed is known as the *template* or *antisense strand.* As the template strand and the coding strand are complementary, and as the template strand and the RNA molecule are also complementary, the RNA molecule produced during transcription is a copy of the coding strand, with the only exception that the RNA molecule contains a U everywhere the DNA coding strand has a T.

Ribosomes are small granular bodies scattered throughout cytoplasm, and this is the place where protein synthesis starts. rRNA itself does not directly govern protein synthesis. A number of ribosomes get attached to a chain of mRNA and form *polysome,* along which with mRNA acting as the template, protein synthesis occurs. One of the major functions of rRNA is to bind the ribosomes to the mRNA chain.

tRNA is the smallest of all three types of RNA mentioned here, and consequently much more soluble than mRNA and rRNA. This is why sometimes tRNA is also called *soluble RNA.* tRNA transports amino acids, building blocks of protein synthesis, to specific areas of the mRNA of the polysome. tRNAs are composed of a small number of nucleotide units (70–90 units) folded into several loops or arms through base pairing along the chain.

7.1.3.3 Translation: RNA and Protein Biosynthesis

Translation is the process by which mRNA directs protein synthesis. In this process, the message carried by mRNA is read by tRNA. Each mRNA is divided into codons, ribonucleotide triplets that are recognized by small amino acid-carrying molecules of tRNA, which deliver the appropriate amino acids needed for protein synthesis.

RNA directs biosynthesis of various peptides and proteins essential for any living organisms. Protein biosynthesis seems to be catalysed by mRNA rather than protein-based enzymes and occurs on the ribosome. On the ribosome, mRNA acts as a template to pass on the genetic information that it has transcribed from DNA. The specific ribonucleotide sequence in mRNA forms an 'instruction' or *codon* that determines the order in which different amino acid residues are to be joined. Each 'instruction' or codon along the mRNA chain comprises a sequence of three ribonucleotides that is specific for a given amino acid. For example, the codon U-U-C on mRNA directs incorporation of the amino acid phenylalanine into the growing protein.

7.1.4 DNA Fingerprinting

DNA fingerprinting, also known as DNA-typing, is a method of identification that compares fragments of DNA. This technique was first developed by Alec Jeffreys in 1984 and was originally used to detect the presence of genetic diseases. With the exception of identical twins, the complete DNA of each individual is unique. This technique was first used in 1985 to solve immigration and paternity cases, and also to identify identical twins.

In 1984, it was discovered that human genes contain short, repeating sequences of noncoding DNA, called *short tandem repeats* (STRs). The STR loci are slightly different for every individual except identical twins. By sequencing these loci, a unique pattern for each individual can be obtained. On the basis of this fundamental discovery, the technique of DNA fingerprinting was developed.

The DNA fingerprinting technique has now been applied almost routinely in all modern forensic laboratories to solve various crimes. When a DNA sample is collected from a crime scene, for example from blood, hair, skin or semen, the sample is subjected to cleavage with restriction endonucleases to cut out fragments containing the STR loci. The fragments are then amplified using the polymerase chain reaction (PCR), and the sequence of the fragments are determined. If the DNA profile from a known individual and that obtained from the DNA from the crime scene matches, the probability is approximately 82 billion to 1 that the DNA is from the same person.

The DNA of a father and offspring are related, but not completely identical. Thus, in paternity cases, DNA fingerprinting technique comes very handy and the identity of the father can be established with a probability of 100 000 to 1.

7.2 AMINO ACIDS AND PEPTIDES

Amino acids, as the name implies, contain both an amino and a carboxylic acid group and are the building blocks of proteins. Twenty different amino acids are used to synthesize proteins and these are alanine (ala, A), arginine (arg, R), asparagine (asn, N), aspartic acid (asp, D), cysteine (cys, C), glutamine (gln, Q), glutamic acid (glu, E), glycine (gly, G), histidine (his, H), isoleucine (ile, I), leucine (leu, L), lysine (lys, K), methionine (met, M), phenylalanine (phe, F), proline (pro, P), serine (ser, S), threonine (thr, T), tryptophan (trp, W), tyrosine (tyr, Y) and valine (val, V). The shape and other properties of each protein are dictated by the precise sequence of amino acids in it. Most amino acids are optically active and almost all the 20 naturally occurring amino acids that comprise proteins are of the L-form. While the (R) and (S) system can be used to describe the absolute stereochemistry of amino acids, conventionally the D and L system is more popular for amino acids.

Aliphatic amino acids

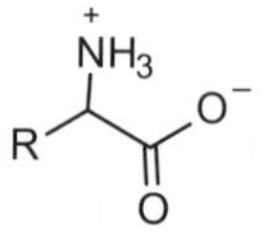

Alanine R=Me
Glycine R=H
Leucine R=CH$_2$CH(CH$_3$)$_2$
Valine R=CH(CH$_3$)$_2$

Isoleucine R =

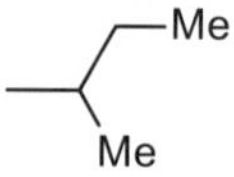

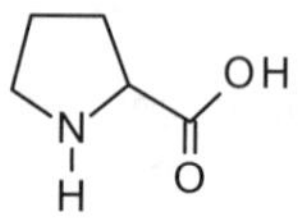

Proline
(A cyclic amino acid)

Aromatic amino acids

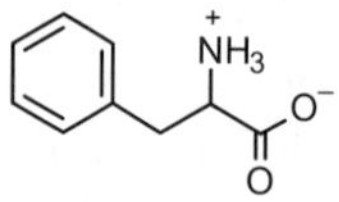

Phenylalanine

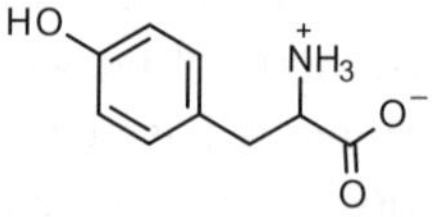

Tyrosine

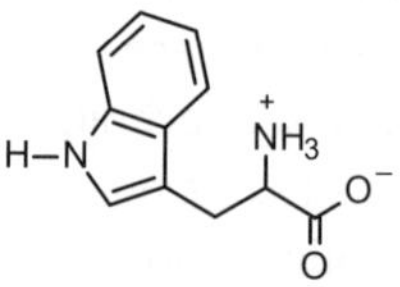

Tryptophan

Acidic amino acids

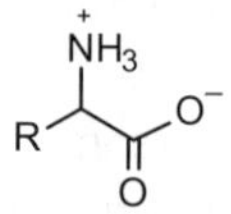

Aspartic acid R = CH$_2$COOH
Glutamic acid R = CH$_2$CH$_2$COOH

Basic amino acids

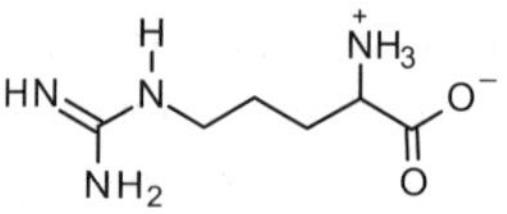

Arginine

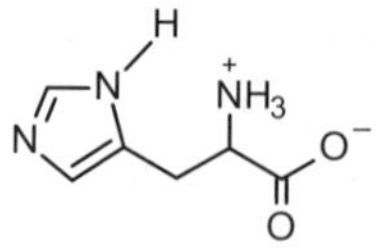

Histidine

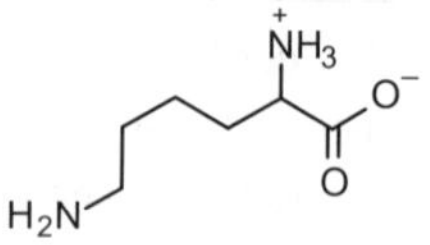

Lysine

Hydroxylic amino acids

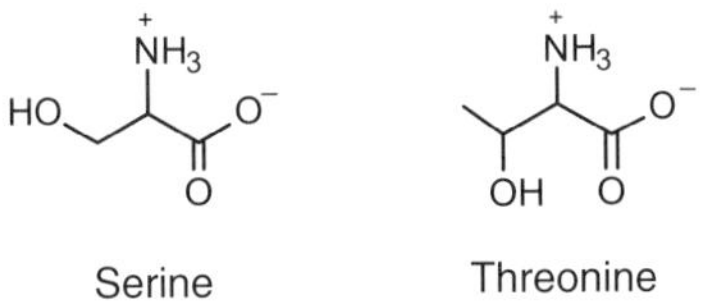
Serine
Threonine

Sulphur-containing amino acids

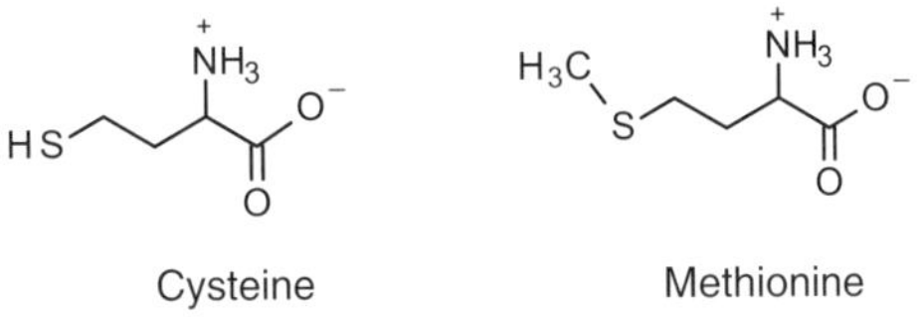
Cysteine
Methionine

Amidic amino acids

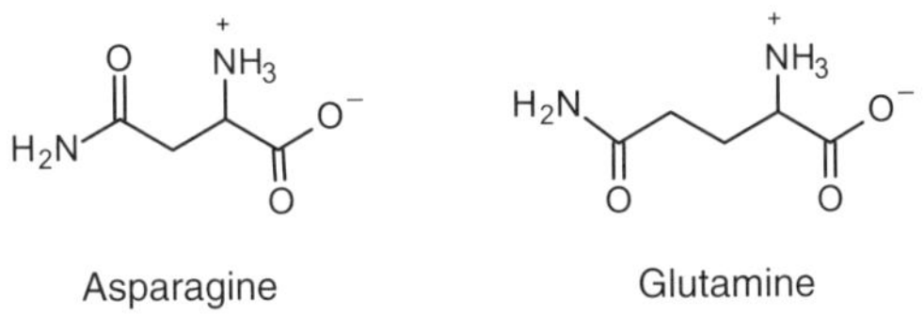
Asparagine
Glutamine

Peptides are biologically important polymers in which α-amino acids are joined into chains through amide linkages called *peptide bonds*. A peptide bond is formed from the amino group (—NH_2) of one amino acid and the carboxylic acid group (—COOH) of another. The term *peptide bond* implies the existence of the peptide group, which is commonly written in text as —CONH—. Two molecules (amino acids) linked by a peptide bond form a *dipeptide*. A chain of molecules linked by peptide bonds is called a *polypeptide*. Proteins are large peptides. A protein is made up of one or more polypeptide chains, each of which consists of amino acids. Instead of writing out complex formulae, sequences of amino acids are commonly written using three- or one-letter codes; for example, ala-val-lys (three-letter) or AVK (one-letter). The ends of a peptide are labelled as the amino end or amino terminus, and the carboxy end or carboxy terminus.

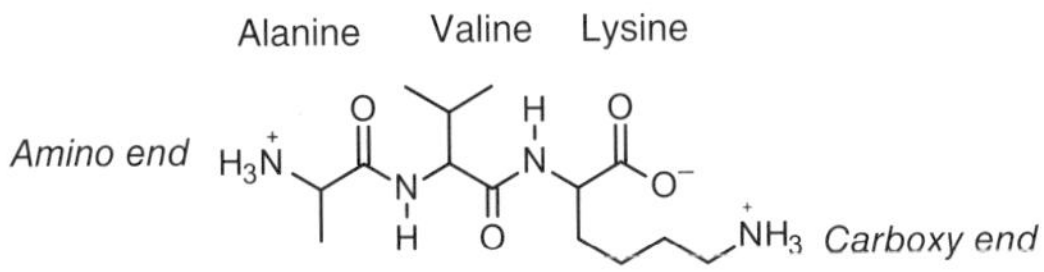
Alanine
Valine
Lysine
Amino end
Carboxy end

Alanylvalyllysine
(ala-val-lys or AVK)

Large peptides of biological significance are known by their trivial names; for example, insulin is an important peptide composed of 51 amino acid residues.

7.2.1 Fundamental Structural Features of an Amino acid

Each amino acid consists of a carbon atom to which is attached a hydrogen atom an amino group (NH_2), a carboxyl group (—COOH) and one of 20 different R groups. It is the structure of the R group (side chain) that determines the identity of an amino acid and its special properties. The side chains (R group), depending on the functional groups, can be aliphatic, aromatic, acidic, basic, hydroxylic, sulphur-containing and amidic (containing amide group). However, proline has an unusual ring structure where the side chain is bonded at its terminus to the main chain nitrogen.

Alanine
(An amino acid)

The zwitterionic
structure of alanine

An amino acid, with an overall charge of zero, can contain two groups of opposite charge in the same molecule. Molecules containing oppositely charged groups are known as *zwitterions*. For amino acids, a *zwitterionic* structure (see Section 1.4) is possible because the basic amino group can accept a proton and the acidic carboxylic group can donate a proton.

7.2.2 Essential Amino Acids

All living organisms can synthesize amino acids. However, many higher animals are deficient in their ability to synthesize all of the amino acids they need for their proteins. Thus, these higher animals require certain amino acids as a part of their diet. Human beings also must include adequate amounts of eight different amino acids in their diet, which they cannot synthesize in their body. These are known as *essential* amino acids. The eight essential amino acids are: valine, leucine, isoleucine, phenylalanine, tryptophan, threonine, methionine and lysine. Sometimes, arginine and histidine are also included in the category of essential amino acids.

7.2.3 Glucogenic and Ketogenic Amino Acids

The carbon skeletons of the amino acids can be used to provide metabolic energy. Several amino acids can be classified as *glucogenic* and *ketogenic* because of their degradation products.

Amino acids that are converted to glucose or glycogen are called glucogenic amino acids. Alanine, arginine, asparagine, cysteine, glutamine, glycine, histidine, hydroxyproline, methionine, proline, serine and valine are *glucogenic amino acids*. Glucogenic amino acids give rise to a net production of pyruvate or the TCA cycle, such as α-ketoglutarate or oxaloacetate, all of which are precursors to glucose via gluconeogenesis.

Amino acids that yield ketone bodies (acetylCoA or acetoacetylCoA, neither of which can bring about net glucose production) are called ketogenic amino acids. Leucine and lysine are *ketogenic amino acids.* Some amino acids, for example threonine, isoleucine, phenylalanine, tyrosine and tryptophan, can be both ketogenic and glycogenic.

7.2.4 Amino Acids in Human Body

All human tissues are capable of synthesizing the nonessential amino acids, amino acid remodelling and conversion of non-amino acid carbon skeletons into amino acids and other derivatives that contain nitrogen. However, the liver is the major site of metabolism of nitrogenous compounds in the body. Dietary proteins are the primary source of essential amino acids (or nitrogen). Digestion of dietary proteins produces amino acids, which are absorbed through epithelial cells and enter the blood. Various cells take up these amino acids that enter the cellular pools.

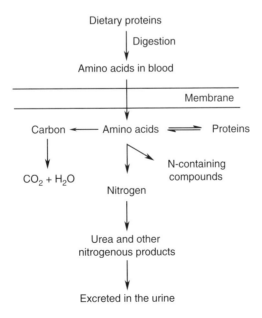

In our body, amino acids are used for the synthesis of proteins and other nitro-gen-containing compounds, or they are oxidized to afford energy. Cellular proteins, hormones (thyroxine, epinephrine and insulin), neurotransmitters, creatine phosphate, the haeme of haemoglobin, cytochrome, melanin (skin pigment) and nucleic acid bases (purine and pyrimidine) are the examples in the amino acid-derived nitrogen-containing biologically important group of compounds found in humans.

7.2.5 Acid–Base Properties of Amino Acids

The neutral forms of amino acids are *zwitterions* (see Section 1.4). This is why amino acids are insoluble in apolar aprotic solvents, for example ether, but most non-protonated amines and unionized carboxylic acids dissolve in ether. For the same reason, amino acids usually have high melting points, for example, the melting point of glycine is 262 °C, and large dipole moments. The high melting points and greater water solubility than in ether are salt-like characters, not the characters of uncharged organic molecules. This salt-like characteristic is found in all zwitterionic compounds. Water is the best solvent for most amino acids because it solvates ionic groups much as it solvates the ions of a salt. A large dipole moment is characteristic of zwitterionic compounds that contain great deal of separated charge. The pK_a values for amino acids are also typical of zwitterionic forms of neutral molecules. Peptides can also exist as zwitterions, that is at pH values near 7, amino groups are protonated and carboxylic acid groups are ionized.

7.2.6 Isoelectric Points of Amino Acids and Peptides

Isoelectric point (pI) or *isoelectric pH* is the pH at which a molecule carries no net electrical charge; that is, zero charge. It is an important measure of the acidity or basicity of an amino acid. To have a sharp *isoelectric point*, a molecule must be amphoteric, that is, it must have both acidic and basic functional groups, as found in amino acids. For an amino acid with only one amino and one carboxylic acid group, the pI can be calculated from the pK_as of this molecule.

$$pI = \frac{pK_{a1} + pK_{a2}}{2}$$

For amino acids with more than two ionizable groups, for example lysine, the same formula is used but the two pK_as used are those of the two groups that lose and gain a charge from the neutral form of the amino acid.

The process that separates proteins according to their isoelectric point is called *isoelectric focusing*. At a pH below the pI, proteins carry a net positive charge, whereas above the p*I* they carry a net negative charge. Applying this principle, *gel electrophoretic methods* have been developed to separate proteins. The pH of an electrophoretic gel is determined by the buffer used for that gel. If the pH of the buffer is above the p*I* of the protein being run, the protein will migrate to the positive pole (negative charge is attracted to a positive pole). Similarly, if the pH of the buffer is below the p*I* of the protein being run, the protein will migrate to the negative pole of the gel (positive charge is attracted to the negative pole). If the protein is run with a buffer pH that is equal to the p*I*, it will not migrate at all. This also applies for individual amino acids.

Chapter 8
Natural Product Chemistry

Learning objectives

After completing this chapter, students should be able to

- describe the origins, chemistry, biosynthesis and pharmaceutical importance of various classes of natural products including alkaloids, carbohydrates, glycosides, iridoids and secoiridoids, phenolics, steroids and terpenoids.

8.1 INTRODUCTION TO NATURAL PRODUCTS

8.1.1 Natural Products

Natural products are simply the products of natural origins; that is, plants, microbes and animals. Natural products can be:

i.   an entire organism (e.g. a plant, an animal or a microorganism);

ii. a part of an organism (e.g. leaves or flowers of a plant, or an isolated animal organ);

iii. an extract of an organism or part of an organism (e.g. a water extract of tea leaves);

iv. an exudate (e.g. resins, oleoresins, balsams and gums);

v.   or pure compounds (e.g. alkaloids, coumarins, flavonoids, lignans, steroids and terpenoids) isolated from plants, animals or microorganisms.

Chemistry for Pharmacy Students: General, Organic and Natural Product Chemistry, Second Edition. Lutfun Nahar and Satyajit Sarker.
© 2019 John Wiley & Sons Ltd. Published 2019 by John Wiley & Sons Ltd.

However, in practice, the term 'natural products' refers to secondary metabolites. *Secondary metabolites* are small organic molecules (mol. wt. <1500 amu) biosynthesized by an organism, but are not directly involved in the fundamental functions of an organism; for example, usual growth, development and reproduction. As the secondary metabolites do not seem to have any primary functions in any organisms, for example plants, and because many of them possess protective roles, especially against herbivores and pathogenic microbes, it is widely assumed that secondary metabolites have evolved because of their protective value. It has been observed that the production and release of secondary metabolites in the environment may increase when the organism (e.g. plant) is attacked by herbivores or pathogens (commonly referred to as 'external stimuli').

Secondary metabolites are different from *primary metabolites*, which are directly involved in primary physiological functions (or intrinsic functions) in an organism; for example, growth, development and reproduction. Some common examples of primary metabolites may include lactic acid and certain amino acids. However, the distinction between a primary and secondary metabolite can sometime be difficult to ascertain, and there are some metabolites that can be categorized as primary or secondary or both. For example, plant growth regulators, such as brassinosteroids, salicylic acid, jasmonates, systemin, polyamines, nitric oxide and signal peptides, may be classified as both primary and secondary metabolites due to their role in plant growth and development. Besides, some of them are intermediates between primary and secondary metabolism.

8.1.2 Natural Products in Medicine

The use of natural products, especially higher plants, for healing is as ancient and universal as medicine itself. Palaeoanthropological work at the cave site of Shanidar, located in the Zagros Mountains of Kurdistan in Iraq, suggested that more than 60 000 years ago, the primitive humans, Neanderthals, might have known about the medicinal properties of various plants, as evidenced by pollen deposits in one of the graves at the site, and also from the DNA analysis of their dental deposits. The more precise therapeutic use of plants certainly goes back to the Sumerian and the Akkadian civilizations about the third millennium BC. In fact, one of the earliest records of natural products were depicted on clay tablets in cuneiform from Mesopotamia (2600 BC) that documented *Cupressus sempervirens* oils and *Commiphora myrrha* (myrrh), which are still in use to treat coughs, colds and inflammation. More than 1000 plants and plant-derived substances could be found on those tablets. Hippocrates (ca. 460–377 BC), traditionally considered to be the father of medicine, was one of the ancient Greek physicians and authors. He described medicinal natural products of plant and animal origins, and listed approximately 400 different plant species for medicinal purposes.

A chronological list of contributions from various authors and researchers in documenting medicinal plants is presented in the following table.

Time line	Contributor	Details
2900 BC	Ebers Papyrus	An Egyptian pharmaceutical record that lists over 700 plant-based drugs ranging from gargles, pills and infusions, to ointments
1100 BC	The Chinese Materia Medica	Listed various medicinal plants and their uses
300 BC	Theophrastus	Described various medicinal herbs
215 BC	Wu Shi Er Bing Fang	Plant-based prescriptions for 52 ailments
200 BC	Shennong Herbal/Wugushen	365 Plant medicines
659 AD	The Tang Herbal	850 Plant-based drugs
40–90 AD	Pedanius Dioscorides	Compiled *De Materia Medica*, which described the dosage and efficacy of about 600 plant-derived medicines and laid the foundations of pharmacology in Europe
129–200 AD	Galen	Recorded 540 plant-derived medicines and demonstrated that herbal extracts contain not only beneficial components, but also harmful ingredients

Natural products have been an integral part of the ancient traditional medicine systems; for example, Chinese, Ayurvedic and Egyptian. Even now, continuous traditions of natural-product therapy exist throughout the world, especially in Asia where numerous minerals, animal substances and plants are still in common use. According to the World Health Organization (WHO), more than 3.5 billion people in the developing world depend on plant-based traditional medicines (also known as or complementary medicine or phytotherapy) for their primary health care. Over 100 million Europeans are currently traditional and complementary medicine (T&CM) users, with 20% of them regularly using T&CM and the same percentage preferring health care that includes T&CM. There are many more T&CM users in Africa, Asia, Australia and North America.

In China alone, over 7000 plant species are utilized as medicinal agents. The output of Chinese Materia Medica products was estimated to amount to US$83.1 billion in 2012, an increase of more than 20% from the year 2011. The number of traditional Chinese medicine (TCM) visits in China was 907 million in 2009, which accounts for 18% of all medical visits; the number of TCM inpatients was 13.6 million or 16% of the total in all hospitals surveyed.

In the Republic of Korea, annual expenditures on traditional medicine were US$4.4 billion in 2004, rising to US$7.4 billion in 2009 and out-of-pocket spending for medicinal natural products in the USA was US$14.8 billion in 2008. About 80% of the population in Singapore and the Republic of Korea, where the conventional health-care system is quite well established, uses traditional medicine still today. In Saudi Arabia, individuals appeared to pay US$560 per annum for traditional and complementary medicines.

Nature has been an important source of therapeutic agents for thousands of years. An impressive number of modern top-selling drugs have been derived from natural sources. Anticancer drugs vincristine and vinblastine from *Vinca rosea* (also known as *Catharanthus roseus*; common name: Periwinkle), narcotic analgesic morphine from *Papaver somniferum*, antimalarial drug artemisinin from *Artemisia annua* (the 2015 Nobel Prize in Physiology and Medicine was awarded to the Chinese scientist Youyou Tu for her key contributions to the discovery of artemisinin), the anticancer drug Taxol® from *Taxus brevifolia* and antibiotic penicillins from *Penicillium* ssp. are just a few examples.

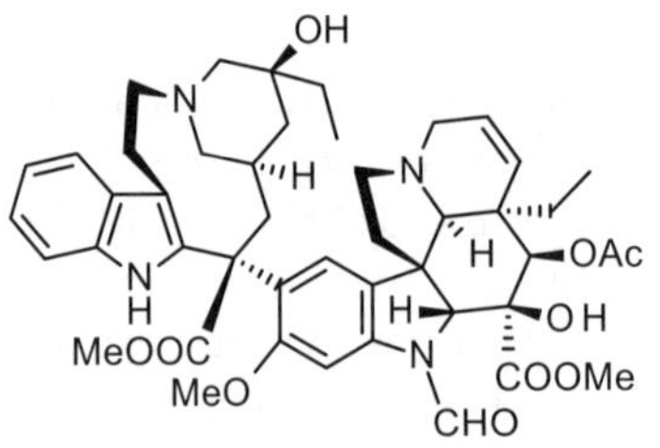

Vincristine

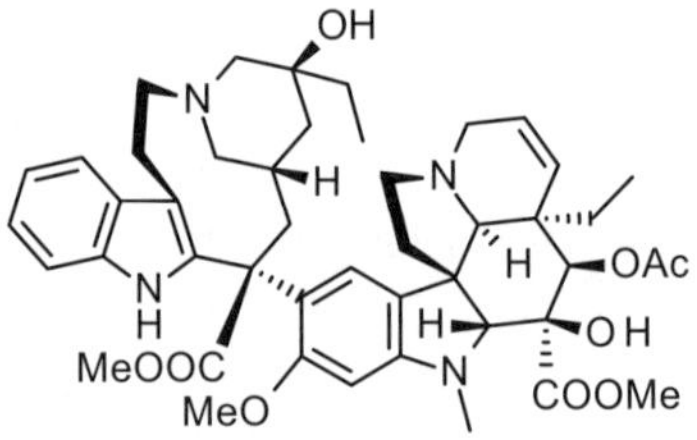

Vinblastine

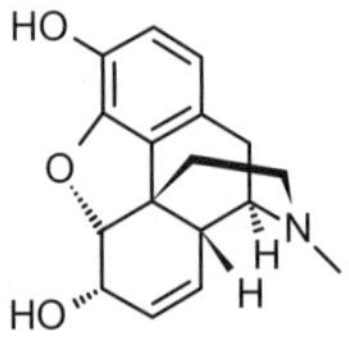

Morphine

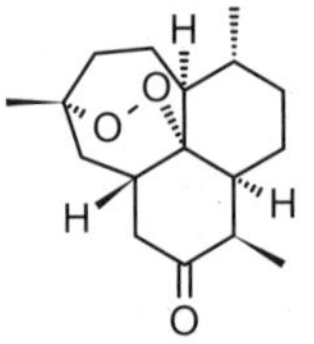

Artemisinin

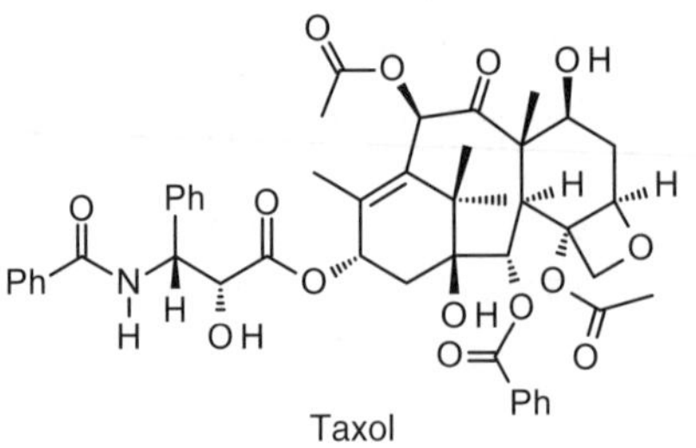

Taxol

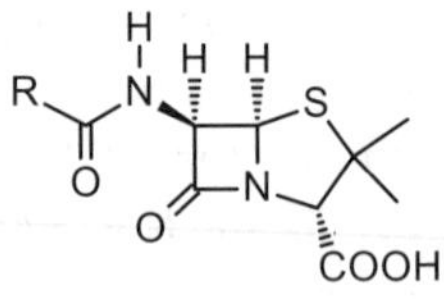

Penicillin G   R =

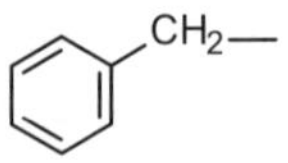

Penicillin V   R =

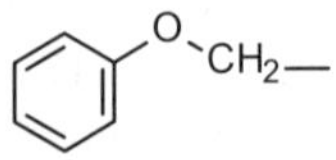

Apart from the natural product derived modern medicines, as outlined earlier in this section, natural products are also used directly in the 'natural' pharmaceutical industry that is growing rapidly in Europe and North America, as well as in the traditional medicine programmes being incorporated into the primary health care systems in Singapore, Republic of Korea, Mexico, The People's Republic of China, Nigeria and other countries.

8.1.3 Drug Discovery and Natural Products

Although *drug discovery* may be considered a recent concept that evolved from modern science during the twentieth century, in reality, the concept of drug discovery dates back many centuries and has its origins in nature. Time and time again, humans have turned to nature for cures and discovered unique drug molecules. Thus, the term *natural products* has become synonymous to the concept of drug discovery. In modern drug discovery and development processes, natural products play an important role at the early stage of 'lead' discovery, that is, discovery of the active (determined by various bioassays) natural molecule, which or its structural analogues could be an ideal drug candidate.

Natural products have been a wellspring of drugs and drug leads. Compounds derived from natural products have made a big impact on the pharmaceutical industry and will continue to do so in the years to come, because of unique chemical diversity that exists in natural products. Natural products remain the best sources of drugs and drug leads and this remains true today, despite the fact that many pharmaceutical companies have somewhat ignored natural products research in favour of high-throughput screening (HTS) of combinatorial libraries. In the area of cancer drug development, during 1940–2014, of the 175 approved small molecules, 131 (75%) were other than synthetic, with 85 (49%) being either natural products or directly derived from natural products. From 1981 to date, 79 (80%) out of 99 small molecule anticancer drugs are natural product-based/inspired, with 53 (54%) being either natural products or derived therefrom.

In other therapeutic areas, the contribution of natural product structures is quite remarkable, for example the anti-infective area is dominated by natural products and their structures. About 78% of antibacterial agents and 74% of anticancer drug candidates are natural products or structural analogues of natural products. Of the 1010 new chemical entities (NCEs) for pharmaceutical research approved between January 1981 and June 2006, 43 were unaltered natural products and a further 232 were second generation natural products derivatives, which are primarily semi-synthetic natural product analogues created to improve properties such as solubility and pharmacokinetics. About 61% of the 877 small molecule NCEs introduced as drugs worldwide during 1981–2002 could be traced back to or were developed from natural products. These include natural products (6%), natural product derivatives (27%), synthetic compounds with natural product derived pharmacophores (5%) and synthetic compounds inspired by a natural product; that is, a natural product mimic (23%). In 2000, approximately 60% of all drugs in clinical trials for the multiplicity of cancers were of natural origins. In 2001, eight (simvastatin, pravastatin, amoxicillin, clavulanic acid, clarithromycin, azithromycin, ceftriaxone, cyclosporine and paclitaxel) of 30 top selling medicines were natural products or derived from natural products, and these eight drugs together totalled US$16 billion in sales. Among the 20 approved small molecule NCEs in 2010, half of them were natural products.

Despite the outstanding record and statistics regarding the success of natural products in drug discovery, 'natural product drug discovery' has been neglected by many big pharmaceutical companies in the recent past. The declining popularity of natural products as a source of new drugs began in 1990s due to some practical reasons, for example the apparent lack of compatibility of natural products with the modern HTS programmes, where a significant degrees of automation, robotics and computations are used, the complexity in the isolation and identification of natural products, and the cost and time involved in natural products 'lead' discovery process. Complexity in the chemistry of natural products, especially in the case of novel structural types, also became the rate-limiting step in drug discovery programmes. Despite being neglected by the pharmaceutical companies, attempts to discover a new drug 'lead' from natural sources has never stopped, but continued especially in academia and some semi-academic research organizations where more traditional approaches to natural product drug discovery have been applied.

Neglected for years, natural product drug discovery has drawn attention and immense interest again, and is on the verge of a comeback in the main stream of drug discovery ventures. In recent years, a significant revival of interests in natural products as a potential source for new medicines has been observed among academics as well as several pharmaceutical companies. This extraordinary comeback of natural products in drug discovery research is mainly due to the following factors.

i.   Combinatorial chemistry's promise to fill drug development pipelines with *de novo* synthetic small molecule drug candidates has been somewhat unsuccessful.

ii. The practical difficulties of natural product drug discovery have been mostly overcome by the advances in separation and identification technologies, and in the speed and sensitivity of structure elucidation process.

iii. Various natural products libraries are now available for HTS for new lead discovery.

iv. Natural products still offer the unique and incomparable chemical diversity and remain the main source of NCEs.

v. Moreover, only a small fraction of the world's biodiversity has ever been explored for bioactivity to date. For example, there are at least 250000 species of higher plants that exist on this planet, but merely 5–10% of these terrestrial plants have been investigated so far.

vi. In addition, re-investigation of previously investigated plants has continued to offer new bioactive compounds that have the potential to be developed as drugs.

vii. New natural product sources, for example marine sponges, are being explored.

While several biologically active compounds have been found in marine organisms, for example antimicrobial compound cephalosporin C (*Cephalosporium acremonium* and *Streptomyces* spp.), and antiviral compounds like avarol and avarone from marine sponges, for example *Dysidea avara*, research in this area is still in its infancy.

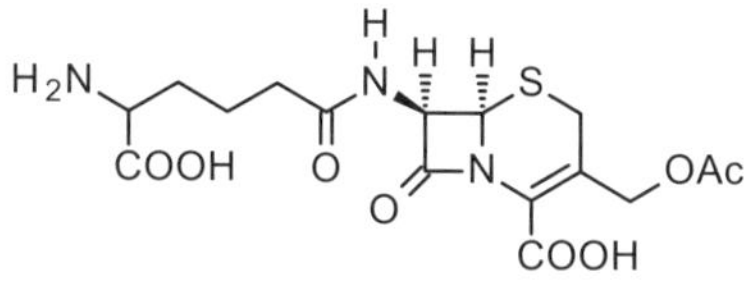

Cephalosporin C

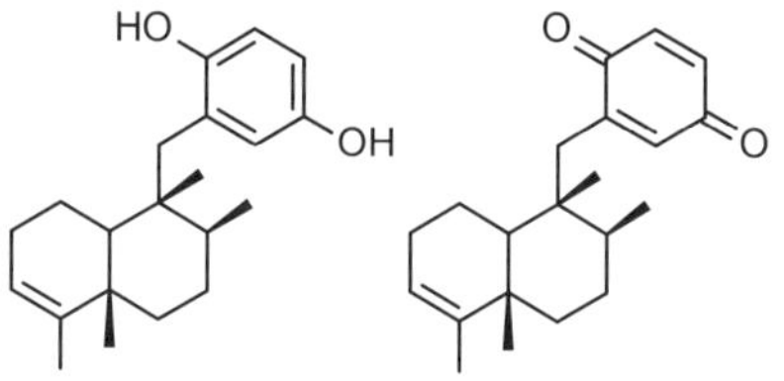

Avarol   Avarone

Now let us have a look at the summary of the traditional as well as the modern drug discovery processes involving natural products.

8.1.3.1 Natural Product Drug Discovery: The Traditional Way

In the traditional, rather more academic method of drug discovery from natural products, drug targets are exposed to crude natural products extracts and, in the case of a hit, that is, any evidence of activity, the extract is fractionated and the active compound is isolated and identified. Every step of fractionation and isolation is usually guided by bioassays and the process is called *bioassay-guided isolation*. The following scheme presents an overview of a bioassay-guided traditional natural product drug discovery process.

Sometimes, straightforward natural products isolation route, irrespective of bioactivity, is also applied, which results in the isolation of a number of natural compounds (*small compound library*) suitable for undergoing any bioactivity screening. However, this process can be slow, inefficient and labour intensive, and it does not guarantee that a 'lead' from screening would be chemically workable or even patentable.

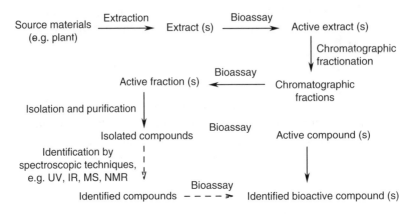

8.1.3.2 Natural Product Drug Discovery: The Modern Processes

Modern drug discovery approaches involve HTS, where applying automation and robotics aided by computational tools, hundreds of molecules can be screened using several assays within a short time and with very little amounts of compounds. In order to incorporate natural products in the modern HTS programmes, a natural products library (a collection of *dereplicated* natural products) is needed. *Dereplication* is the process by which one can eliminate recurrence or re-isolation of the same or similar compounds from various extracts or fractions. A number of hyphenated techniques are used for dereplication, for example LC-PDA (liquid chromatography-photo-diode-array detector), LC–MS (liquid chromatography-mass spectroscopy) and LC–NMR (liquid chromatography-nuclear magnetic resonance spectroscopy). Note that a hyphenated technique originates from the coupling of a separation technique, generally a chromatographic technique, and an on-line detection technology, normally one or more spectroscopic techniques; for example, LC–MS.

While in the recent past, it was extremely difficult, time consuming and labour intensive to build such a library from purified natural products, with the advent of newer and improved technologies related to separation, isolation and identification of natural products, and introduction of various computational tools and robotics, the situation has improved remarkably. The following scheme shows an outline of the modern natural product drug discovery process. Now, it is possible to build a 'high quality', 'chemically diverse' and 'dereplicated' natural products library that is suitable for any modern HTS programmes. Natural products libraries can also be of crude extracts, chromatographic fractions or semi-purified compounds. However, the best result can be guaranteed from a fully identified pure natural products library as it provides scientists with the opportunity to handle the 'lead' rapidly for further developmental work, for example total or partial synthesis, dealing with formulation factors, *in vivo* assays and pre-clinical and clinical trials.

To continue to exploit natural sources for potential drug candidates, the focus must be on exploiting newer approaches for natural product drug discovery. These approaches include the application of genomic tools, seeking novel sources of organisms from the environment, new screening technologies, for example *in silico* screening (also known as virtual screening), and improved processes of sample preparation for screening samples. Particularly, the increasing popularity in *in silico* screening of natural products, due to the introduction of innovative computational tools in processing structurally complex natural products to predict their macromolecular targets and possible functions, has added a new dimension to the natural product drug discovery process. A classic example of a natural-product-inspired drug as a result of *in silico* screening (*in silico* pharmacophore modelling) is the cholesterol lowering drug rosuvastatin, which was developed as a synthetically accessible mimetic of the natural product mevastatin, a secondary metabolite from *Penicillium citrinum*.

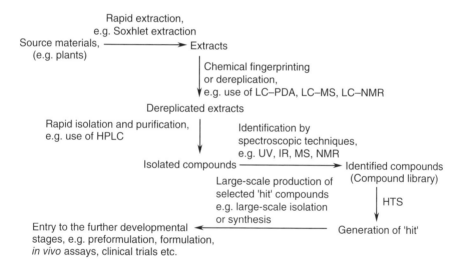

It is noteworthy that these two compounds possess a consensus pharmacophore but have different chemical structures. *In silico* screening of natural products often successfully forecasts the role that natural product-derived fragments and fragment-like natural products may play in the next generation of new drug discovery and development. In addition, the recent efforts in the synthesis of diversity-oriented combinatorial libraries based on natural product-like compounds is an attempt to enhance the productivity of synthetic chemical libraries.

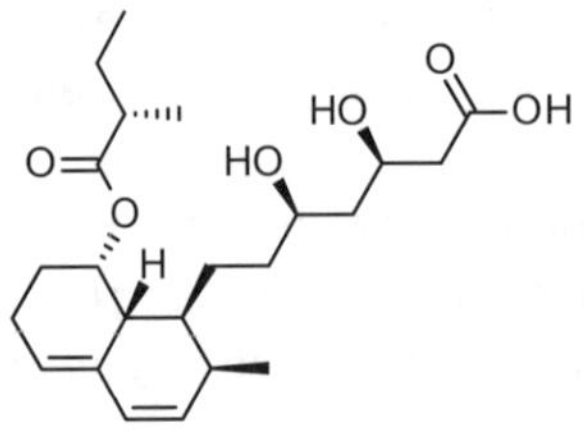

Mevastatin

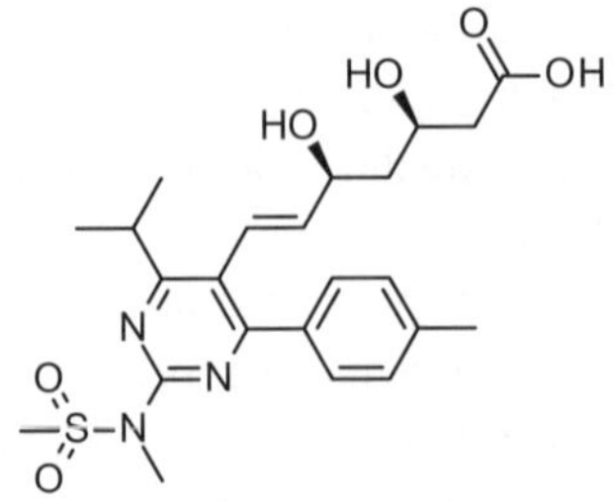

Rosuvastatin

8.2 ALKALOIDS

Alkaloids are a large, naturally occurring group of pharmacologically active nitrogen-containing secondary metabolites of plants, microbial or animal origin. In most alkaloids, the nitrogen atom is a part of the ring. Alkaloids are biosynthetically derived from amino acids. The name 'alkaloid' derives from the word 'alkaline', which means a water-soluble base. A number of natural alkaloids and their derivatives have been developed as drugs to treat various diseases; for example, morphine, taxol, ephedrine, coniine, quinine and reserpine. In fact, morphine was the very first drug ever to be purified and developed from a plant origin (*P. somniferum*). However, some alkaloids are powerful poisons. It is a well-known fact that, in 399 BC, the famous Greek philosopher Socrates was condemned to death by drinking a cup of hemlock, a cocktail of poisonous alkaloids (mainly coniine) from *Coniium maculatum* (Apiaceae).

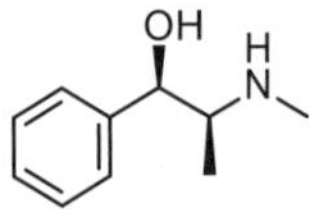

(1R, 2S)-(−)-Ephedrine, a bronchodilator from *Ephedra sinica* (Ephedraceae)

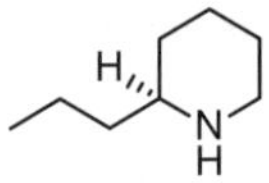

(S)-(+)-Coniine, a poison, from *Coniium maculatum* (Apiaceae)

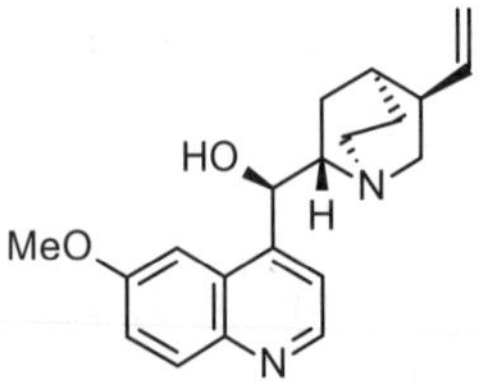

Quinine, an antimalarial drug, from *Cinchona succirubra*

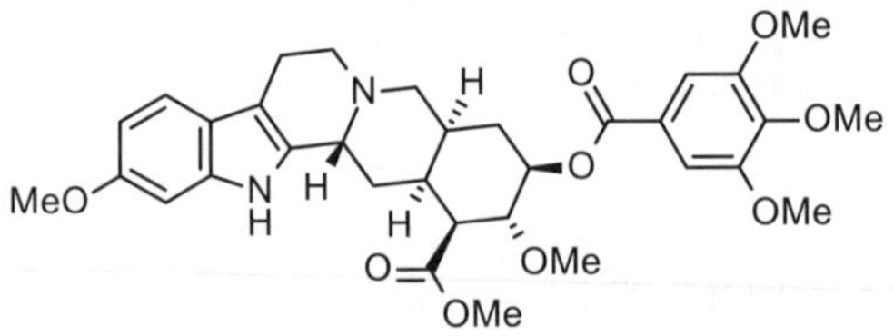

Reserpine, an antihypertensive and antipsychotic drug, from *Rauwolfia serpentina* (Apocynaceae)

To date, over 3000 different types of alkaloids have been identified from 4000 plant species. Certain plant families, for example Amaryllidaceae, Papaveraceae,

Ranunculaceae and Solanaceae, are particularly rich in alkaloids. The Rutaceae family is also known to produce quinoline and furoquinoline type alkaloids. A few alkaloids are also produced by organisms other than plants; for example, the New World beaver (*Castor canadensis*), poison-dart frogs (*Phyllobates*), ergot and a few other fungi.

8.2.1 Properties of Alkaloids

Alkaloids are basic compounds. They form water-soluble salts with mineral acids. In fact, one or more nitrogen atoms that are present in an alkaloid, typically as 1°, 2° or 3° amines, contribute to the basicity of the alkaloid. The degree of basicity varies considerably depending on the structure of the molecule and presence and location of the functional groups. Most alkaloids are colourless crystalline solids and bitter in taste. However, only a few, for example coniine and nicotine, are liquids and a few are coloured, for example berberine. Most alkaloids are generally optically active. The vast majority of free alkaloidal bases are generally insoluble or very sparingly soluble in water, but are fairly soluble in organic solvents; for example, either, chloroform, *n*-hexane, benzene and petroleum ether. However, alkaloidal salts are freely soluble in water and either insoluble or sparingly soluble in organic solvents.

8.2.2 Classification of Alkaloids

Alkaloids are generally classified according to the amino acid that provides both the nitrogen atom and the fundamental alkaloidal skeleton. However, alkaloids can also be grouped together on the basis of their generic structural similarities. The following table shows different major types of alkaloids, their generic skeleton and specific examples.

Class/Structural types	Generic structure	Examples
Aporphine *(Tyrosine derived)*	 Aporphine R = Me Noraporphine R = H	Boldine
Betaines *(Serine derived)*	 Betaine	Choline, muscarine and neurine

Class/Structural types	Generic structure	Examples
Imidazole *(Histidine derived)*	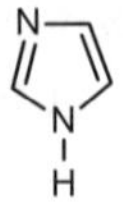 Imidazole	Pilocarpine
Indole *(Tryptophan derived)*	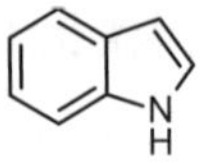 Indole	
Tryptamines	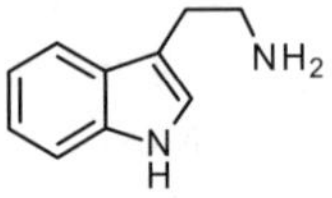 Tryptamine	Moschamine, moschamindole, psilocybin and serotonin
Ergolines	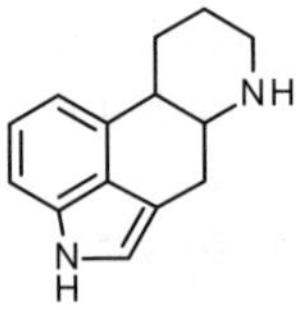 Ergoline	Ergine, ergotamine and lysergic acid
β-Carbolines	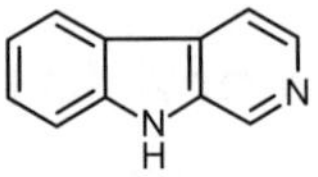 β-Carboline	Emetin, harmine and reserpine
Indolizidine *(L-Lysine derived)*	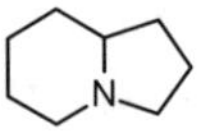 Indolizidine	Swainsonine and castanospermine
Isoquinoline *(Tyrosine derived)*	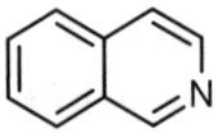 Isoquinoline	Codeine, berberine, morphine, papaverine, sanguinarine and thebaine
Macrocyclic spermines and spermidines *(Ornithine derived)*	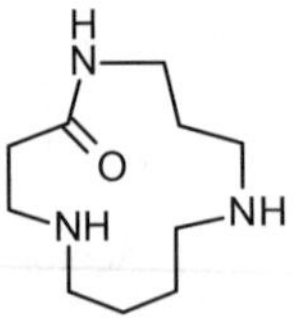 Spermine	Celabenzine

Class/Structural types	Generic structure	Examples
Norlupinane *(Lysine derived)*	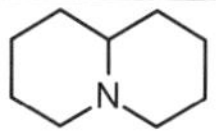 Norlupinane	Cytisine and lupanine
Phenethylamine *(Phenylalanine derived)*	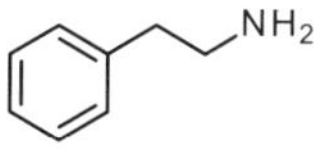 Phenylethylamine	Ephedrine and mescaline
Purine *(Xanthosine derived)*	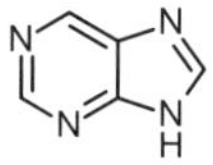 Purine	Caffeine, theo-bromine and theophylline
Pyridine and piperidine *(Nicotinic acid derived)*	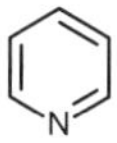 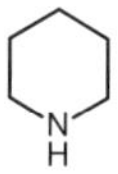 Pyridine Piperidine	Arecoline, coniine, nicotine, piperine and trigonelline
Pyrrole and pyrrolidine *(Ornithine derived)*	 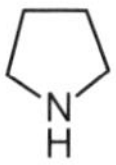 Pyrrole Pyrrolidine	Hygrine, cusco-hygrine and nicotine
Pyrrolizidine *(Putrescine derived)*	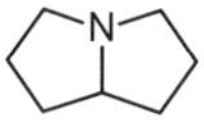 Pyrrolizidine	Echimidine and symphitine
Quinoline *(Tryptophan/anthranilic acid derived)*	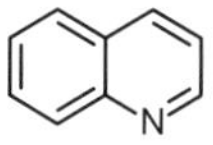 Quinoline	Cinchinine, bru-cine, quinine and quinidine
Terpenoidal/Steroidal *(Cholesterol derived)*	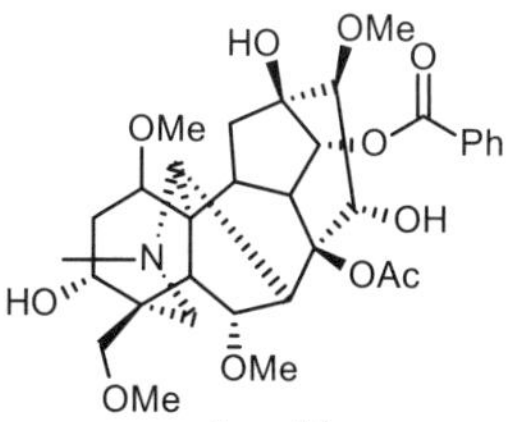 Aconitine	Aconitine

Class/Structural types	Generic structure	Examples
Terpenoidal Steroidal	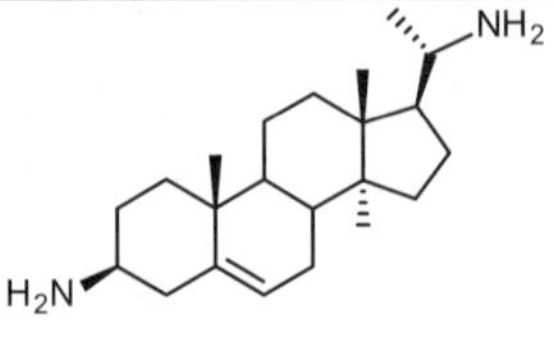 Steroidal alkaloid	Batrachotoxin, conanine, irehdi-amine A, solanine, samandarine and tomatillidine
Tropane *(Ornithine derived)*	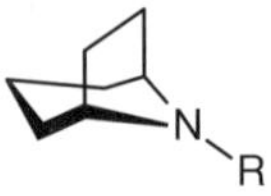 Tropane R = Me Nortropane R = H	Atropine, cocaine, ecgonine, hyoscine and scopalamine

8.2.2.1 Pyridine and Piperidine Alkaloids

Alkaloids, for example arecoline, coniine, nicotine, piperine and trigonelline, possess a pyridine or modified pyridine heterocyclic ring system (e.g. piperidine ring).

8.2.2.1.1 Piperine

Piperine, a component of black pepper (*Piper nigrum*), discovered by Hans Christian Ørsted in 1819, is a yellow crystalline solid (molecular formula: $C_{17}H_{19}NO_3$). It is present in various traditional medicine preparations and is also used as an insecticide. Piperine has various effects on human drug metabolizing enzymes and is marketed under the brand name Bioperine® as an adjunct for increasing bioavailability of various dietary supplements, especially curcumin, one of the active ingredients of turmeric (*Curcuma longa*).

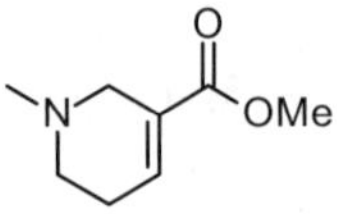

Arecoline 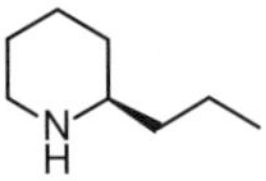   Coniine

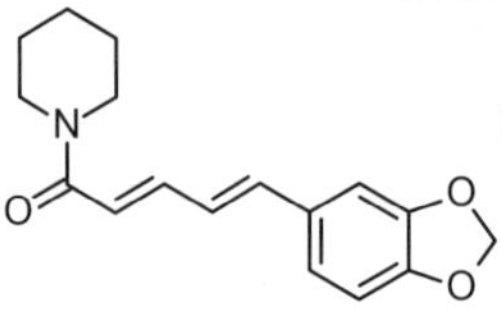

Piperine 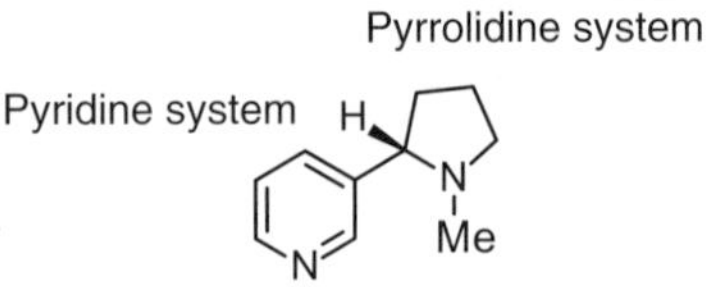   Nicotine

8.2.2.1.2 Coniine

Coniine (molecular formula: $C_8H_{17}N$) is a poisonous alkaloid found in hemlock poison and the 'Yellow Pitcher Plant' (*Sarracenia flava*). Chemically, it is known as (*S*)-2-propylpiperidine. Coniine contributes to the foul smell of hemlock (*C. maculatum*). It is a neurotoxin that causes respiratory paralysis and is toxic to all classes of live-stock and humans. Coniine has an important place in organic chemistry history as it is the first of the important classes of alkaloids synthesized by the German chemist Albert Ladenburg in 1886.

8.2.2.1.3 Nicotine

Nicotine (molecular formula: $C_{10}H_{14}N_2$) is the major pharmacologically active component of tobacco, *Nicotiana tabacum*, and is also found extensively in other species of the family Solanaceae; for example, tomato, potato, aubergine and green pepper. Nicotine is a hygroscopic oily liquid and miscible with water in its base as well as its salt form. It possesses two nitrogenous ring systems; one is pyridine but the other is a pyrrolidine. Thus, this alkaloid can be classified either under pyridine or pyrrolidine.

Nicotine is a potent nerve poison and is included in many insecticide preparations. In lower concentrations, nicotine is a potent parasympathomimetic stimulant, that is, it increases activity, alertness and memory, and this stimulant property is one of the main factors that contribute to the dependence-forming properties of tobacco smoking. It increases heart-rate and blood pressure and reduces appetite. In higher doses, nicotine acts as a depressant. In large doses, it can cause nausea and vomiting. The main symptoms of the withdrawal of nicotine intake include irritability, headaches, anxiety, cognitive disturbances and sleep disruption. The primary therapeutic use of nicotine is in the treatment of nicotine dependence, where controlled levels of nicotine are given to individuals through gums, dermal patches, lozenges, electronic cigarettes or nasal sprays.

8.2.2.2 Pyrrole and Pyrrolidine Alkaloids

These alkaloids contain pyrrole or a modified pyrrole (e.g. pyrrolidine) ring system. The simplest example of this class containing a pyrrolidine skeleton is nicotine. A pyrrolidine ring is the central structure of the amino acids proline and hydroxy-proline. These alkaloids are also found in many drug preparations, for example procyclidine hydrochloride, which is an anticholinergic drug mainly used for the treatment of drug-induced Parkinsonism, akathisia and acute dystonia, and is also used to reduce side effects associated with antipsychotic treatment for schizo-phrenia. The Chinese herbal medicine Tai Zi Shen or Prince Ginseng (*Pseudostellaria heterophylla*), used as a paediatric or geriatric tonic, contains the pyrrole alkaloid, 3-furfuryl-pyrrole-2-carboxylate.

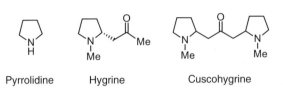

Pyrrolidine   Hygrine   Cuscohygrine

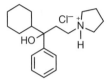   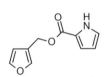

Procyclidine hydrochloride
(An anticholinergic drug)

3-Furfuryl-pyrrole-2-carboxylate

8.2.2.2.1 Hygrine

Hygrine (molecular formula: $C_8H_{15}NO$) was first isolated from coca leaves (*Erythroxylum coca*) by the German chemist Carl Liebermann in 1889. It is a thick yellow oil with a pungent taste and odour.

8.2.2.2.2 Cuscohygrine

Cuscohygrine (molecular formula: $C_{13}H_{24}N_2O$) is a symmetrical dimeric pyrrolidine alkaloid found in coca, and also in many species of the Solanaceae; for example, *Atropa belladonna*, *Datura innoxia* and *Datura stramonium*. It is an oil but is soluble in water. This alkaloid was also first isolated by Carl Liebermann in 1889.

8.2.2.3 Tropane Alkaloids

This is a group of alkaloids that possess a 8-methyl-8-aza-bicyclo[1,2,3]octane or tropane skeleton; for example, atropine, cocaine and scopolamine. Tropane alkaloids occur mainly in plants from the families Erythroxylaceae and Solanaceae. 8-Aza-bicyclo[1,2,3]octane, that is, tropane without the 8-methyl group, is known as *nortropane*.

Atropine   Tropane R = Me   Cocaine
              Nortropane R = H

Scopolamine or Hyoscine
(A medication for motion sickness)

8.2.2.3.1 Atropine

Atropine (molecular formula: $C_{17}H_{23}NO_3$) is a tropane alkaloid, first isolated from 'deadly nightshade' (*A. belladonna*) in the 1930s, and also found in many other plants of the Solanaceae family. Atropine is a racemic mixture of D-hyoscyamine and L-hyoscyamine. However, the most of its pharmacological properties are due to its L-isomer and because of its binding to muscarinic acetylcholine receptors. Atropine is a *competitive antagonist* of the muscarinic acetylcholine receptors. The main medicinal use of atropine is as an ophthalmic drug. Usually, a salt of atropine, for example atropine sulphate, is used in pharmaceutical preparations. It is used as a cycloplegic to paralyse accommodation temporarily, and as a mydriatic to dilate the pupils. It is contraindicated in patients predisposed to narrow angle glaucoma. Injections of atropine are used in the treatment of bradycardia (an extremely low heart rate), asystole and pulseless electrical activity in cardiac arrest. It is also used as an antidote for poisoning by organophosphate insecticides and nerve gases. The major adverse effects of atropine include ventricular fibrillation, tachycardia, nausea, blurred vision, loss of balance and photophobia. It also produces confusion and hallucination in elderly patients. Overdoses of atropine can be fatal. The antidote for atropine poisoning is physostigmine or pilocarpine.

Physostigmine or pilocarpine
(An antidote for atropine poisoning)

8.2.2.3.2 Cocaine

Cocaine, also known as 'coke' (molecular formula: $C_{17}H_{21}NO_4$), is a white crystalline tropane alkaloid found mainly in the coca plant (*Erythroxylum coca*). It was first isolated by the German chemist Friedrich Gaedcke in 1855. Cocaine is a potent central nervous system (CNS) stimulant and appetite suppressant. For its euphoretic effect, cocaine is often used recreationally, and it is one of the most common drugs of abuse and addiction. It is used as a topical anaesthetic in the eye, throat and nose surgery. Note that possession, cultivation and distribution of cocaine is illegal for nonmedicinal and nongovernment sanctioned purposes in virtually all over the world. The side effects of cocaine include twitching, paranoia and impotence, which usually increase with frequent usage. With excessive dosage it creates hallucinations, paranoid delusions, tachycardia, itching and formication. Cocaine overdose leads to tachyarrhythmias and elevated blood pressure, and can be fatal. Severe cardiac adverse events, particularly sudden cardiac death, become a serious risk at high doses due to cocaine's blocking effect on cardiac sodium channels.

8.2.2.4 Quinoline Alkaloids

The chemistry of quinoline heterocycle has already been discussed in Chapter 7. Any alkaloids that possess a quinoline, that is, a 1-azanaphthalene, 1-benzazine or benzo[b]pyridine, skeleton is known as a quinoline alkaloid; for example, quinine. Quinoline itself is a colourless hygroscopic liquid with strong odour and is slightly soluble in water, but readily miscible with organic solvents. Quinoline is toxic. Short-term exposure to the vapour of quinoline causes irritation of the nose, eyes and throat, dizziness and nausea. It may also cause liver damage.

Quinoline

8.2.2.4.1 Quinine

Quinine (molecular formula: $C_{20}H_{24}N_2O_2$) is a white crystalline quinoline alkaloid first isolated from Cinchona bark (*Cinchona succirubra*) in 1820 by Pierre Joseph Pelletier and Joseph Caventou and is well-known as an antimalarial drug. Quinine is extremely bitter and also possesses antipyretic, analgesic and anti-inflammatory properties. Although quinine used to be the drug of choice for the treatment of malaria caused by *Plasmodium falciparum*, as of 2006 it is no longer recommended by the WHO (World Health Organization) as a first-line treatment for malaria, and it should be used only when artemisinins are not available. Quinine can be used to treat nocturnal leg cramps, lupus and arthritis. However, its use in the treatment of leg cramps is discouraged and has become less common because of a Food and Drug Administration warning, which states that this practice is associated with life-threatening side effects. In fact, it has a narrow therapeutic window. Quinine is an extremely basic compound and is available in its salt forms; for example, sulphate, hydrochloride and gluconate.

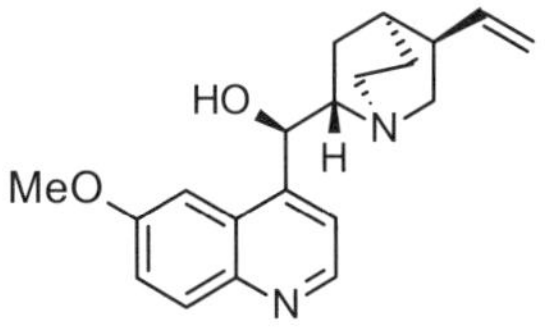

Quinine

Despite being a wonder drug against malaria, quinine in therapeutic doses can cause various side effects, for example nausea, vomiting and cinchonism, and in some patients, pulmonary oedema. It may also cause paralysis if accidentally injected into a nerve. An overdose of quinine may have fatal consequences. Non-medicinal uses of quinine include its uses in small quantities as a flavouring agent in tonic water and bitter lemon. For its bitterness, quinine is used in the gin and tonic cocktail and as an aperitif.

8.2.2.4.2 Quinidine

Quinidine (molecular formula: $C_{20}H_{24}N_2O_2$) is a stereoisomer of quinine found in Cinchona bark. Chemically, it is known as (2-ethenyl-4-azabicyclo[2.2.2]oct-5-yl)-(6-methoxyquinolin-4-yl)-methanol or 6′-methoxycinchonan-9-ol. It is used as a Class 1 anti-arrhythmic agent to prevent ventricular arrhythmias. Intravenous injection of quinidine is offered in the treatment of *P. falciparum* malaria. Among the adverse effects, quinidine induces thrombocytopenia (low platelet counts) and may lead to thrombocytic purpurea. Like quinine, it may also cause cinchonism.

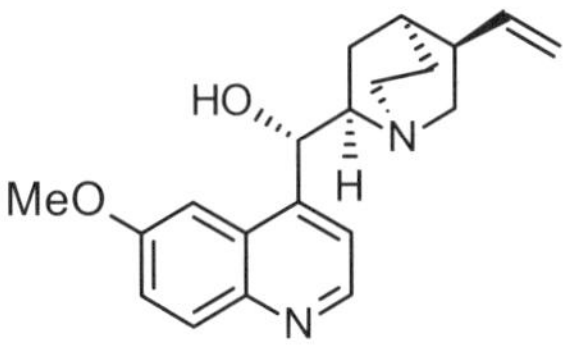

Quinidine

8.2.2.5 Isoquinoline Alkaloids

Isoquinoline is in fact an isomer of quinoline and chemically known as benzo[c]pyridine or 2-benzanine. Both quinoline and isoquinoline are benzopyridines, which are composed of a benzene ring fused with a pyridine ring. Any alkaloid that possesses an isoquinoline skeleton is known as an isoquinoline alkaloid; for example, papaverine and morphine. The isoquinoline backbone is biosynthesized from the aromatic amino acid tyrosine.

Isoquinoline itself is a colourless hygroscopic liquid at room temperature. It has an unpleasant odour. It is slightly soluble in water, but very soluble in ethanol, acetone, ether and other common organic solvents. It is a weak base with a pK_a of 5.14. Isoquinoline alkaloids play an important part in medicine. A number of these alkaloids are available as drugs. Some examples of isoquinoline derivatives with medicinal values are summarized in the following table. In addition to their medicinal uses, isoquinolines are used in the manufacture of dyes, paints, insecticides and as a solvent for the extraction of resins.

Isoquinoline alkaloids	Medicinal uses
Dimethisoquin	Anaesthetic
Quinapril	Antihypertensive agent
2,2'-Hexadecamethylenediisoquinolinium dichloride	Topical antifungal agent
Papaverine	Vasodilator
Morphine	Narcotic analgesic

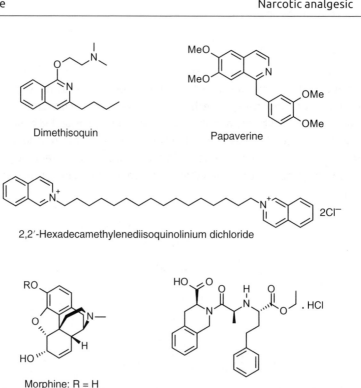

Dimethisoquin

Papaverine

2,2'-Hexadecamethylenediisoquinolinium dichloride

Morphine: R = H
Codeine: R = OMe

Quinapril hydrochloride

8.2.2.5.1 Papaverine

Papaverine (molecular formula: $C_{20}H_{21}NO_4$) is an isoquinoline alkaloid first isolated from poppy seeds (*P. somniferum*, Family: Papaveraceae) by Georg Merck in 1848. This alkaloid is used mainly in the treatment of visceral spasm, vasospasm and erectile dysfunction. It is also used as a cerebral and coronary vasodilator, and in the treatment of acute mesenteric ischemia. Papaverine may be used as a smooth muscle relaxant in microsurgery. In pharmaceutical preparations, it is used in its salt form; for example, hydrochloride, codecarboxylate, adenylate and teprosylate. The usual side effects of papaverine treatment include polymorphic ventricular tachycardia, constipation, increased transaminase levels, hyperbiliru-binemia and vertigo. Uncommon side effects may include flushing of the face, excessive sweating, cutaneous eruption, arterial hypotension, tachycardia, loss of appetite, jaundice, eosinophilia, thrombocytopenia, mixed hepatitis, headache, allergic reaction, chronic active hepatitis and paradoxical aggravation of cerebral vasospasm.

8.2.2.5.2 Morphine

Morphine (molecular formula: $C_{17}H_{19}NO_3$), a habit forming Class A analgesic drug (narcotic analgesic), is the major bioactive constituent of opium poppy seeds and was first isolated between 1803 and 1805 by Friedrich Sertürner. In fact, it is believed to be the first isolation of a biologically active ingredient from a plant. It is a modified isoquinoline alkaloid. Like other opium constitu-ents (opiates), for example heroin, morphine acts directly on the CNS to relieve pain. Morphine is used for the treatment of post-surgical pain and chronic pain (e.g. cancer pain), as an adjunct to general anaesthesia and an antitussive for severe coughs. It is frequently used for pain from myocardial infarction and during labour. Side effects of morphine treatment generally include impair-ment of mental performance, decreased respiratory effort and low blood pressure, euphoria, drowsiness, loss of appetite, constipation, lethargy and blurred vision. A large overdose may cause asphyxia and death by respiratory depression.

8.2.2.6 Phenylethylamines

Phenylethylamine, a neurotransmitter or neuromodulator, is a monoamine. It is a primary amine and the amino group is attached to a benzene ring through a two-carbon or ethyl group. Although the nitrogen is not a part of the ring, phen-ylethylamine and its derivatives are classified as alkaloids. Phenylethylamine itself is a colourless liquid at room temperature, and soluble in water, ethanol and ether. It has a fishy odour. Phenylethylamine moiety can be found in various complex ring systems; for example, the ergoline system in lysergic acid diethyl-amide (LSD) or the morphinan system in morphine. It is biosynthesized by several plants and some animals, including humans. Certain fungi and bacteria also afford phenylethylamines. Several alkaloids of this class are used as neurotransmitters, stimulants (e.g. ephedrine, cathinone and amphetamine), hallucinogens (e.g. mes-caline), bronchodilaotrs (e.g. ephedrine and salbutamol),and antidepressants (e.g. bupropion).

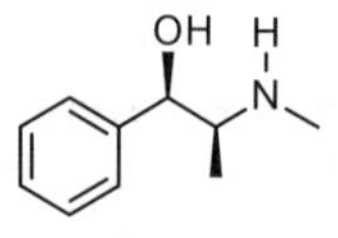

Ephedrine
(A constituent of *Ephedra sinica*)

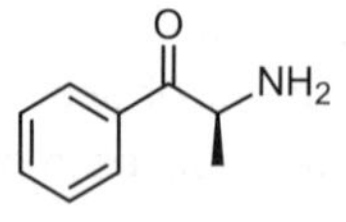

Cathinone
(A constituent of *Catha edulis*)

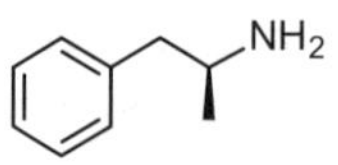

Amphetamine
(Commercially known as Speed)

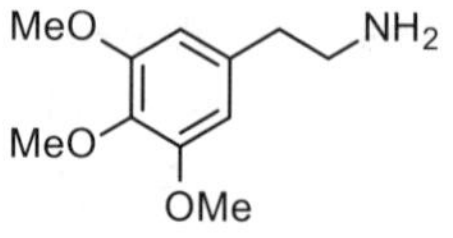

Mescaline
(A hallucinogen from the cactus
Lophophora williamsii)

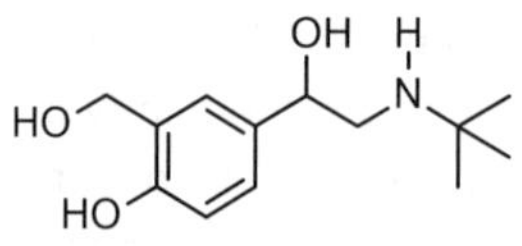

Salbutamol
(A bronchodilator)

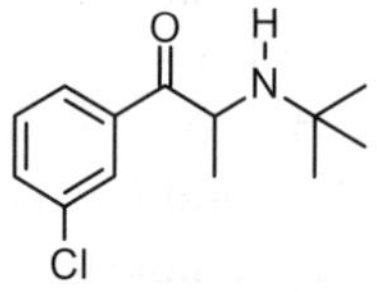

Bupropion
Known as Wellbutrin
(An antidepressant)

8.2.2.6.1 *Ephedrine*

Ephedrine (molecular formula: $C_{10}H_{15}NO$) is a sympathomimetic phenylethylamine alkaloid, which was first isolated from *Ephedra sinica* in 1885. Ephedrine in its natural form has long been used in TCM for the treatment of asthma and bronchitis and was documented during the Han dynasty (206 BC to 220 AD). Nowadays, ephedrine is prescribed for increasing blood pressure and as a bronchodilator. It is well known for promoting modest short-term weight loss. It should not be used together with certain antidepressants, for example norepinephrine-dopamine reuptake inhibitors (NDRIs) like bupropion, as this increases the risk of symptoms due to excessive serum levels of norepinephrine. It is also contraindicated in closed-angle glaucoma, phaeochromocytoma and general anaesthesia. Appropriate caution must be taken in the use of ephedrine in patients with inadequate fluid replacement, impaired adrenal function, hypoxia, acidosis, hypertension, diabetes and cardiovascular diseases.

8.2.2.7 Indole Alkaloids

Indole chemistry has already been discussed in Chapter 7. This is one of the major groups of naturally occurring bioactive alkaloids containing well over 4000 different compounds, and can be classified into three main categories, tryptamine and its derivatives, ergoline and its derivatives, and β-carboline and its derivatives. The first indole alkaloid, strychnine, was isolated by Pierre Joseph Pelletier and Joseph Bienaimé Caventou in 1818 from the plants of the *Strychnos* genus; for example *S. nux-vomica*.

Indole   Tryptamine

β-Carboline
Ergoline   R, R′ and R″ are various substituents

The pharmacological properties of some indole alkaloids has long been known. For example, Aztecs used the psilocybin mushrooms that contain the alkaloids psilocybin and psilocin; *Rauwolfia serpentina,* which contains reserpine, was a common medicine in India around 1000 BC; Africans used the roots of the *Iboga,* which contain ibogaine, as a stimulant. Many modern drugs, for example reserpine, possess an indole structure.

8.2.2.7.1 Tryptamine Derivatives

Tryptamine, chemically known as 3-(2-aminoethyl)-indole, is a monoamine alkaloid found widespread in plants, fungi and animals. Biosynthetically tryptamine derives from the amino acid tryptophan. Tryptamine acts as the precursor of many other indole alkaloids, many of which are neurotransmitters and psychedelic drugs. Substitutions to the tryptamine skeleton provide a group of compounds collectively known as *tryptamines*; for example, serotonin, an important neurotransmitter, is the 5-hydroxy derivative of tryptamine, and melatonin, a hormone found in all living creatures, is actually 5-methoxy-N-acetyltryptamine. Some of the pharmacologically active natural tryptamines are psilocybin (4-phosphoryloxy-N,N-dimethyltryptamine) from 'magic mushrooms' (*Psilocybe cubensis* and *Psilocybe semilanceata*), DMT (N,N-dimethyltryptamine) from a number of plants and DET (N,N-diethyltryptamine), an orally active hallucinogenic drug and psychedelic compound of moderate duration. Many synthetic tryptamines, for example sumatriptan (5-methylamino sulphonyl-N,N-dimethyltryptamine), a drug used for the treatment of migraine, are also available.

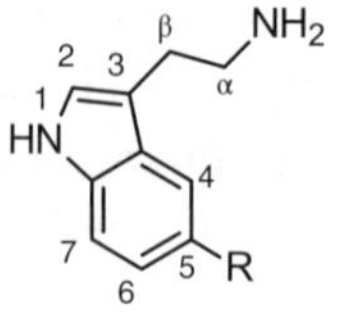

Tryptamine R = H
Serotonin R = OH

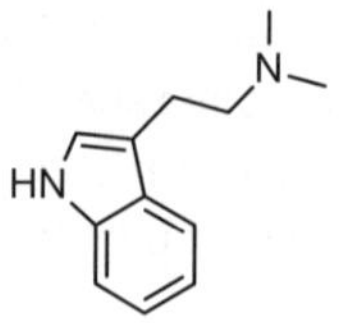

N,N-Dimethyltryptamine (DMT)

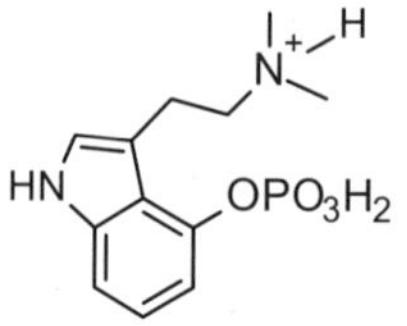

Psilocybin
(A psychedelic tryptamine alkaloid)

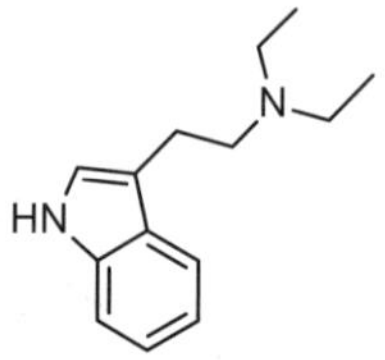

N,N-Diethyltryptamine (DET)
(A hallucinogenic drug)

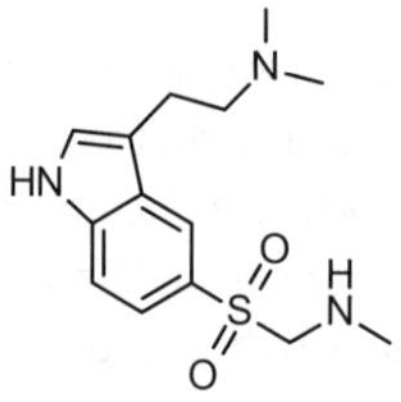

Sumatriptan
(A drug for migraine treatment)

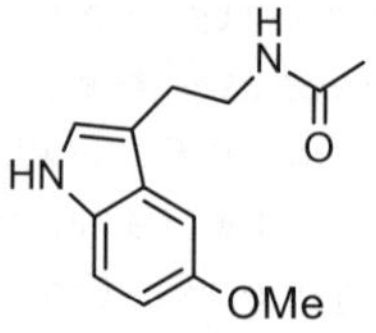

Melatonin
(A hormone regulates sleep)

8.2.2.7.2 Ergolines

Alkaloids that contain an ergoline skeleton as shown next are called ergoline alkaloids, and some of them are psychedelic drugs; for example, LSD. A number of ergoline derivatives are used clinically as a vasoconstrictor (e.g. 5-HT 1 agonists ergotamine), in the treatment of migraine and Parkinson's disease, and some are implicated to the disease ergotism. Ergoline alkaloids were first isolated from ergot, a fungus that infects grain and causes ergotism.

With the isolation of ergotamine by Arthur Stoll in 1918, the first therapeutic use of isolated ergoline alkaloids began. Although ergoline alkaloids are found in lower fungi, some species of flowering plants are also known to biosynthesize these alkaloids. For example, the Mexican species *Turbina corymbosa* and *Ipomoea tricolor* of the family Convolvulaceae produce ergolines.

Ergine (molecular formula: $C_{16}H_{17}N_3O$) is the amide of D-lysergic acid, and commonly known as LSA (D-lysergic acid amide) or LA-111 (D-lysergamide). It is an ergoline alkaloid that occurs in various species of the Convolvulaceae, and in some species of fungus. *Argyreia nervosa* (Hawaiian baby woodrose), *Ipomoea violacea* (morning glory) and *Rivea corymbosa* (ololiuhqui) are three major sources of this alkaloid.

Ergoline

D-Lysergic acid: R = CO_2H
(Found in ergot fungus)

Ergine: R = $CONH_2$
(Occurs in various species of vines of the Convolvulaceae)

LSD: R = $CON(C_2H_5)_2$
(A psychedelic drug)

D-Lysergic acid diethylamide (molecular formula: $C_{20}H_{25}N_3O$), also known as LSD or LSD-25, is a semi-synthetic psychedelic drug, synthesized from the natural precursor lysergic acid found in ergot, a grain fungus. It is a colourless, odourless and mildly bitter compound. LSD produces altered experience of senses, emotions, memories, time and awareness for 8–14 hours. Moreover, LSD may cause visual effects; for example, moving geometric patterns, 'trails' behind moving objects and brilliant colours. The most significant adverse effect is impairment of mental functioning while intoxicated; it may trigger panic attacks or a feeling of extreme anxiety.

8.2.2.7.3 β-Carbolines

Alkaloids that possess a 9*H*-pyrid-[3,4-b]-indole skeleton are called β-carboline alkaloids (also known as norharmane), which are found in several plants and animals. The structure of β-carboline is similar to that of tryptamine, with the ethylamine chain re-connected to the indole ring via an extra carbon atom, to generate a three-membered ring structure. The biosynthesis of β-carbolin alkaloids follows a similar pathway to tryptamine. The β-carbolines, for example harmine, harmaline and tetrahydroharmine, play an important role in the pharmacology of the psychedelic brew ayahuasca. Some β-carbolines, notably tryptoline and pinoline, are formed naturally in the human body.

β-Carboline: R = R′ = H
Harmine: R = Me, R′ = OMe
Harmane: R = Me, R′ = H

Pinoline: R = R″ = H, R′ = OMe
Tetrahydroharmine: R = Me, R″ = OMe
Tryptoline: R = R′ = R″ = H

Harmaline

The major sources of β-carboline alkaloids with their medicinal or pharmacological properties are summarized here.

β-Cabboline alkaloids	Natural sources	Medicinal or pharmacological properties
Harmine and harmaline	Seeds of 'harmal' (*Peganum harmala*) and *Banisteriopsis caapi*	CNS-stimulant, acts by inhibiting the metabolism of serotonin and other monoamines

8.2.2.8 Purine Alkaloids

Alkaloids that contain a purine skeleton (see Section 6.7), which is a heterocyclic aromatic organic compound that consists of a pyrimidine ring fused to an imidazole ring, are commonly known as purine alkaloids; for example, caffeine and theobromine. We have already learnt that two of the bases in nucleic acids, adenine and guanine, are purines.

Purine R = H
Adenine R = NH2

Guanine

Caffeine R = Me
Theobromine R = H

8.2.2.8.1 Caffeine

Caffeine (molecular formula: $C_8H_{10}N_4O_2$) is a xanthine (purine) alkaloid, found mainly in tea leaves (*Camellia sinensis*) and coffee beans (*Coffea arabica*). Chemically, caffeine is known as 1,3,7-trimethyl-1*H*-purine-2,6(3*H*,7*H*)-dione. Caffeine is sometimes called guaranine when found in guarana (*Paullinia cupana*), mateine when found in mate (*Ilex paraguariensis*) and theine when found in tea. Caffeine is found in a number of other plants where it acts as a natural pesticide. Its form is odourless white needles or powder. Apart from its presence in tea and coffee that we drink regularly, caffeine is also an ingredient of a number of soft drinks. Caffeine is a potent CNS and metabolic stimulant, and is used both recreationally and medically to reduce physical fatigue and to restore mental alertness. It stimulates the CNS first at the higher levels, resulting in increased alertness and wakefulness, faster and clearer flow of thought, increased focus and better general body coordination, and later at the spinal cord level at higher doses. It is used in combination with a number of pain killers. Caffeine is also used with ergotamine in the treatment of migraine and cluster headaches as well as to overcome the drowsiness caused by antihistamines. Side effects of caffeine include increased blood pressure, vasoconstriction and chronic arterial stiffness.

8.2.2.9 Terpenoidal Alkaloids

The diverse group of natural products classed as terpenoids and steroids includes some compounds that contain nitrogen and are called terpenoidal alkaloids. Biosynthetically, they can be set apart from all other alkaloids because their carbon skeletons are derived from mevalonic acid, while most other alkaloids have skeletons constructed largely of amino acid residues. To date, a plethora of such alkaloids, based on mono-, sesqui-, di- and tri-terpenoidal and steroidal skeletons, have been isolated from various plant species.

8.2.2.9.1 Aconite Alkaloids

Aconitine (molecular formula $C_{34}H_{47}NO_{11}$) is an example of an aconite alkaloid. It is soluble in organic solvents, for example $CHCl_3$, C_6H_6, slightly soluble in alcohol or ether, but insoluble in water. It is an extremely toxic substance obtained from the plants of the genus *Aconitum* (family: Ranunculaceae), commonly known as 'aconite' or 'monkshood'. It is a neurotoxin, and used for creating models of cardiac arrhythmia. The use of aconitine as a poison can be traced back to the Egyptian history; Cleopatra VII of Egypt poisoned her brother Ptolemy XIV with aconitine to put her son on the throne. In China, aconitine is used in small doses as an analgesic and as a blood coagulant.

Aconitine

8.2.2.9.2 Steroidal Alkaloids

These alkaloids have a core steroidal skeleton as part of the molecule; for example, solanine. Several structural varieties exist in steroidal alkaloids. The following discussion focuses on just a few selected steroidal alkaloids. Steroidal alkaloids can be classified into two major classes: the *Solanum* type (e.g. solanidine and solanine) and the *Veratrum* type (e.g. veratramine).

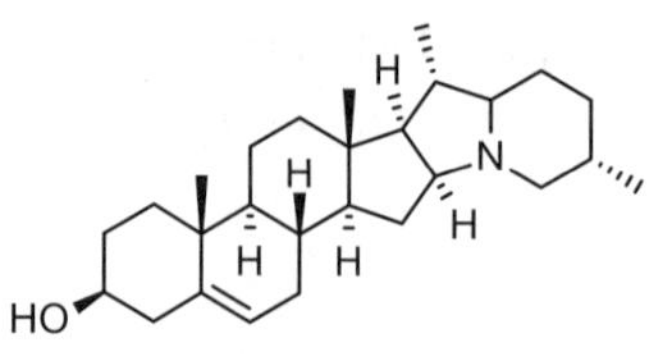

Solanidine

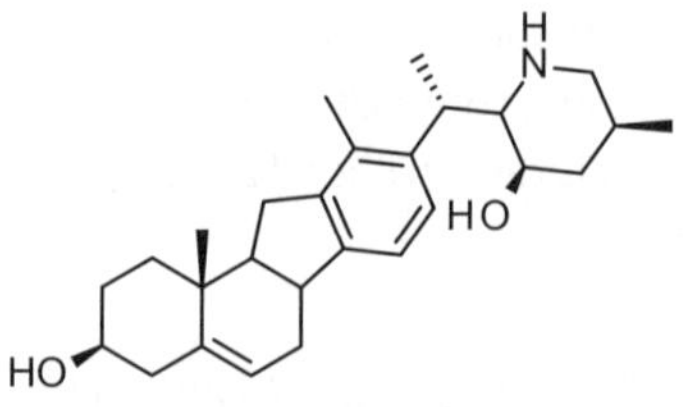

Veratramine

Solanine is a poisonous steroidal alkaloid, also known as glycoalkaloid, found in the nightshade family (Solanaceae) and within the genus *Solanum* (*Solanum tuberosum*, *Solanum lycopersicum* and *Solanum melongena*). It was first isolated by H. S. Rooke in 1820 from the berries of *Solanum nigrum* (the European black nightshade). It is extremely toxic, even in small quantities. Solanine has both fungicidal and pesticidal properties and it is one of the plant's natural defences.

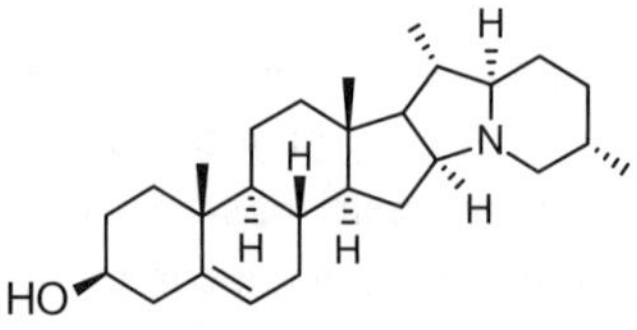

Solanine

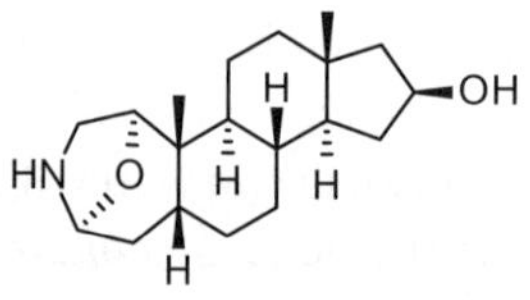

Samandarin

Solanine hydrochloride has been used as a commercial pesticide. It has sedative and anticonvulsant properties, and has sometimes been used for the treatment of asthma, as well as for coughs and common colds. However, gastrointestinal and neurological disorders result from solanine poisoning. Symptoms include nausea, diarrhoea, vomiting, stomach cramps, burning of the throat, headaches and dizziness. Other adverse reactions, in more severe cases, include hallucinations, loss of sensation and paralysis, fever, jaundice, dilated pupils and hypothermia. Solanine overdose can be fatal.

Samandarin (molecular formula: $C_{19}H_{31}NO_2$) is the major steroidal alkaloid of the skin glands of the fire salamander frog (*Salamandra salamandra*) and is extremely toxic. The toxicities of samandarin include muscle convulsions, raised blood pressure and hyperventilation.

Veratramine is a hypotensive steroidal alkaloid found in the rhizomes of *Veratrum nigrum*, commonly known as black false hellebore. It may decrease the heart rate, and creates muscular rigidity and clonic convulsions.

8.2.2.10 Betaines

Alkaloids that contain the betaine (*N,N,N*-trimethylglycine or TMG) skeleton are included in this class; for example, muscarine. Betaine itself is used to treat high homocysteine levels, and sometimes as a mood enhancer.

Betaine

Muscarine

8.2.2.10.1 Muscarine

Muscarine (molecular formula $C_9H_{20}^+NO_2$), first isolated by the German chemists Oswald Schmiedeberg and Richard Koppe in 1869 from fly agaric *Amanita muscaria*, occurs in certain mushrooms, especially in the species of the genera *Inocybe* and *Clitocybe*. Mushrooms from the genera *Entoloma* and *Mycena* also have muscarine, which can be dangerous if ingested. Muscarine can be found in harmless trace amounts in the genera *Boletus*, *Hygrocybe*, *Lactarius* and *Russula*. It is a parasympathomimetic substance. It causes profound activation of the peripheral parasympathetic nervous system, which may result in convulsions and death. Muscarine mimics the action of the neurotransmitter acetylcholine at the muscarinic acetylcholine receptors.

8.2.2.11 Macrocyclic Alkaloids

This group of alkaloids possess a macrocycle and, in most cases, nitrogen is a part of the ring system. The macrocyclic spermine group of alkaloids is one such example. These polyamine alkaloids are found in a number of plant families; for example, Acanthaceae, Ephedraceae, Leguminosae (Fabaceae) and Scrophulariaceae. They possess various biological properties. For example, budmunchiamines L4 and L5, two antimalarial spermine alkaloids isolated from *Albizia adinocephala* (Leguminosae). Many of these alkaloids can also be described as macrocyclic peptide alkaloids as they form cyclopeptides (e.g. budmunchiamines L4 and L5).

Various marine sponges are good sources of bioactive macrocyclic alkaloids. For example, manzamine A, a cytotoxic macrocyclic alkaloid, was isolated from the marine sponge *Haliclona* sp. To date, around 200 different macrocyclic alkaloids have been reported from marine sponges and molluscs. Sponges from the genera *Haliclona* (e.g. manzamines A–D), *Petrosia* (e.g. petrosins A and B), and *Xestospongia* (e.g. xestospongins A–D) are the predominant sources of these macrocyclic alkaloids.

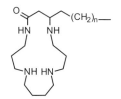

Budmunchiamine L4   n = 11
Budmunchiamine L5   n = 13

Manzamine A
(A cytotoxic macrocyclic alkaloid)

8.2.3 Tests for Alkaloids

Reagent/Test	Composition/ preparation	Result
Meyer's reagent (Potassiomercuric iodide solution)	Freshly prepared by dissolving a mixture of mercuric chloride (1.36 g) and of potassium iodide (5 g) in water (100 ml)	Cream precipitate
Wagner's Reagent (Iodo-potassium iodide)	Prepared by dissolving 2 g of iodine and 6 g of potassium iodide in 100 ml of water	Reddish-brown precipitate
Tannic acid	Tannic acid	Precipitation
Hager's reagent	A saturated solution of picric acid	Yellow precipitate
Dragendorff's reagent (Solution of potassium bismuth iodide)	Bismuth sub-nitrate (1.7 g), glacial acetic acid (20 ml), water (80 ml) and 50% solution of potassium iodide in water (100 ml) are mixed together and stored as stock solution. 10 ml of stock, 20 ml glacial acetic acid and water to make up to 100 ml gives the working solution.	Orange or reddish-brown precipitate (except with caffeine and a few other alkaloids)

Caffeine and other purine derivatives can be detected by the Murexide test. In this test, alkaloids are mixed with a tiny amount of potassium chlorate and a drop of hydrochloric acid, evaporated to dryness and the resulting residue is exposed to ammonia vapour. Purine alkaloids generate a pink colour in this test.

8.3 CARBOHYDRATES

Carbohydrates are the primary fuel for our muscles and the brain. Eating a high carbohydrate diet will ensure maintenance of muscle and liver glycogen (storage forms of carbohydrate), improve performance and delay fatigue. The word *carbohydrate* means 'hydrate of carbon'. Thus, carbohydrates are a group of polyhydroxy aldehydes, ketones and acids or their derivatives, together with linear and cyclic polyols.

Most of these compounds are in the form $C_nH_{2n}O_n$ or $C_n(H_2O)_n$. For example, glucose, $C_6H_{12}O_6$ $C_6(H_2O)_6$. Sometimes, carbohydrates are referred to simply as sugars and their derivatives or saccharides.

β-D-Glucose

Carbohydrates are found abundantly in nature, both in plants and animals, and are essential constituents of all living matter. Photosynthesis is the means by which plants yield sugars from CO_2 and water.

8.3.1 Classification of Carbohydrates

Carbohydrates can be classified in two different ways: general classification is based on the number of sugar units present and the other classification, only for monosaccharides, is based on the functional group present; that is, aldehyde or ketone.

8.3.1.1 General Classification

Generally, carbohydrates are classified into four different categories, *mono-, di-, tri-* and *tetra-saccharides, oligosaccharides* and *polysaccharides.*

8.3.1.1.1 Monosaccharides

These carbohydrates, commonly referred to as 'sugars', contain 3–9 carbon atoms. Monosaccharides (from Greek *monos*, single; *sacchar*, sugar) are the most fundamental units of carbohydrates and cannot be hydrolysed further to simpler units. Most common monosaccharides in nature possess five (*pentose*, $C_5H_{10}O_5$) and six (*hexose*, $C_6H_{12}O_6$) carbon atoms. For example, glucose ($C_6H_{12}O_6$), a six carbon containing sugar, is the most common monosaccharide that is metabolized in our body to provide energy, and fructose is also a hexose found in many fruits. Monosaccharides are usually colourless, water-soluble crystalline solids and some of them have a sweet taste. They are optically active and exist in different isomeric forms.

8.3.1.1.2 Di-, Tri- and Tetra-Saccharides

These carbohydrates are dimers, trimers and tetramers of monosaccharides, and are formed from two, three or four monosaccharide molecules, with the elimination of one, two or three molecules of water, respectively. For example, sucrose is a disaccharide composed of two monosaccharides; glucose and fructose.

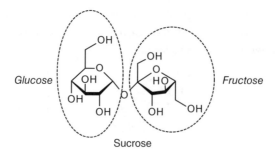

Glucose — Fructose

Sucrose

8.3.1.1.3 Oligosaccharides

The name 'oligosaccharide' (from the Greek *olígos* meaning a few, and *sacchar* for sugar) refers to saccharides containing 2–10 monosaccharides. For example, raffinose, found in beans and pulses, cabbage, brussels sprouts, broccoli, asparagus and other vegetables, is an oligosaccharide composed of three monosaccharide units; for example, galactose, glucose and fructose. Oligosaccharides are useful in cell recognition and cell binding.

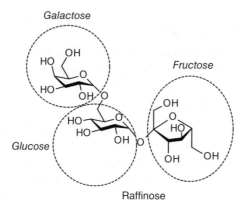

Galactose

Fructose

Glucose

Raffinose

8.3.1.1.4 Polysaccharides

Polysaccharides are composed of a huge number of monosaccharide units, and the number forming the molecule is often approximately known. They are polymeric carbohydrate molecules composed of long chains of monosaccharide units bound together by glycosidic linkages, and on hydrolysis they give the constituent monosaccharides or oligosaccharides. Polysaccharides are usually amorphous and insoluble in water. For example, cellulose and starch are polysaccharides composed of hundreds of glucose units.

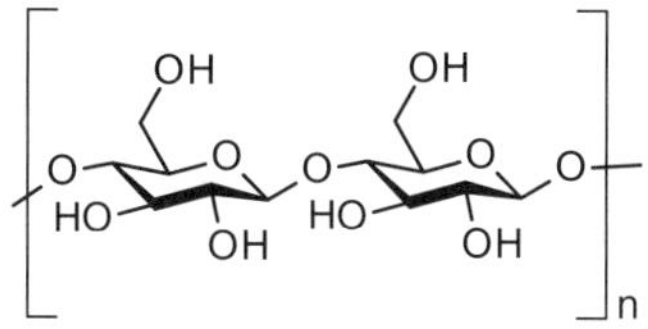

Celluose, composed of several β (1→4) linked D-glucose units

8.3.1.2 Classification of Monosaccharides According to Functional Groups and Carbon Numbers

The two most common functional groups found in monosaccharides (in open chain form) are aldehydes and ketones. When a monosaccharide contains an aldehyde, it is known as *aldose*, for example glucose, and in the case of ketone, it is called *ketose* or *keto sugars*; for example, fructose.

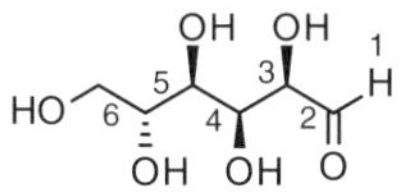

D-Glucose
(An aldose, contains an aldehyde group)

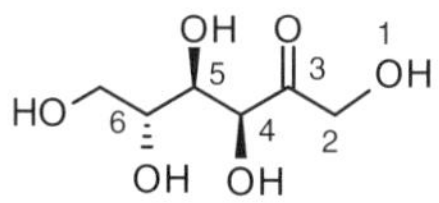

D-Fructose
(A ketose, contains a ketone group)

Depending on the number of carbon atoms present, monosaccharides are classified as *triose, tetrose, pentose* and *hexose*, containing three, four, five and six carbon atoms, respectively. Glucose is a hexose as it contains six carbon atoms. Sometimes, monosaccharides are classified more precisely to denote the functional group as well as the number of carbon atoms. For example, glucose can be classified as an *aldohexose*, as it contains six carbon atoms as well as an aldehyde group.

If any monosaccharide lacks the usual numbers of hydroxyl groups, it is often called a *deoxy sugar*. For example, 2-deoxyribose, which is a component of DNA nucleosides, has one less hydroxyl group than its parent sugar, ribose.

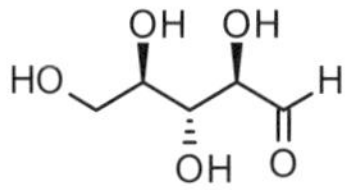

D-Ribose
(An aldopentose, a component of RNA nucleosides)

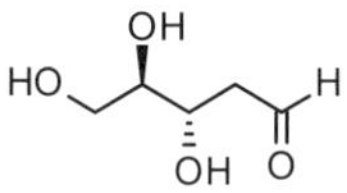

D-2-Deoxyribose
(A deoxyaldopentose. a component of DNA nucleosides)

Hydroxyls, aldehyde and keto groups are not the only functional groups that are present in monosaccharides. Monosaccharides containing carboxylic acid

(—COOH) and amino (—NH$_2$) groups are common structural units in biologically important carbohydrates. For example, 2-amino-2-deoxy-D-glucose, also known as *glucosamine*, is an *amino sugar*, and glucuronic acid is a *sugar acid*.

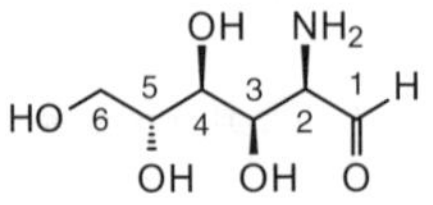

Glucosamine
(An amino sugar)

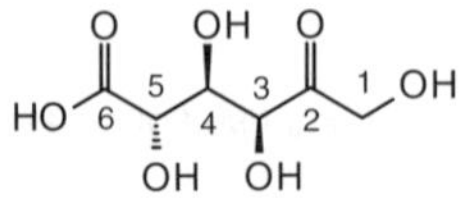

Glucuronic acid
(A sugar acid)

8.3.2 Stereochemistry of Sugars

With monosaccharides, the configuration of the highest numbered chiral carbon is compared with that of D- or L-glyceraldehyde (the simplest aldose), for example, D-sugar has the same configuration as D-glyceraldehyde and L-sugar has same configuration as L-glyceraldehyde. It can be noted that D- and L-notations have no relation to the direction in which a given sugar rotates the plane polarized light, that is, (+) or (−).

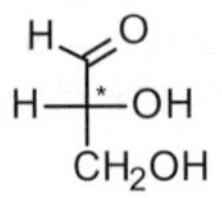

(+)-D-Glyceraldehyde
The hydroxyl group on the chiral carbon is on the right hand side

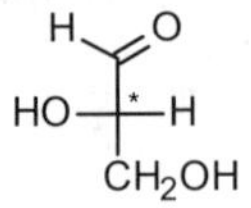

(−)-L-Glyceraldehyde
The hydroxyl group on the chiral carbon is on the left hand side

Glucose, fructose and many other natural monosaccharides have the same configuration as D-glyceraldehyde at the chiral centre farthest from the carbonyl group. In Fischer projections, most natural sugars have the hydroxyl group at the highest numbered chiral carbon pointing to the right. All these sugars are referred to as *D-sugars*; for example, D-glucose.

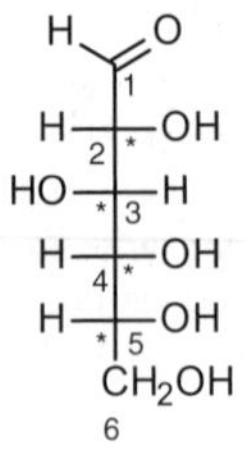

D-Glucose
Hydroxyl group at the highest numbered chiral carbon (C-5) is pointing to the right, i.e similar to D-gluceraldehyde

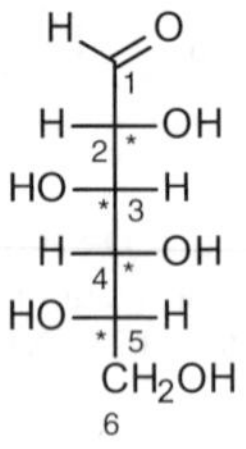

L-Glucose
Hydroxyl group at the highest numbered chiral carbon (C-5) is pointing to the left, i.e similar to L-gluceraldehyde

All *L-sugars* have the configuration as L-glyceraldehyde at the chiral centre farthest from the carbonyl group; for example, L-glucose. In Fischer projections, L-sugars have the hydroxyl group at the highest numbered chiral carbon pointing to the left. Thus, an L-sugar is the mirror image (enantiomer) of the corresponding D-sugar.

8.3.3 Cyclic Structures of Monosaccharides

Monosaccharides exist not only as open chain molecules (acyclic), but also as cyclic compounds.

Cyclization, leading to the formation of a cyclic *hemiacetal* or *hemiketals*, occurs due to an intramolecular nucleophilic addition reaction between a —OH and a C—O group. Many monosaccharides exist in an equilibrium between open chain and cyclic forms.

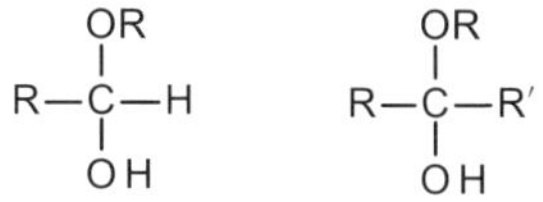

Hemiacetal   Hemiketal

Hemiacetals and hemiketals have the important structural feature —OH and —OR attached to the same carbon as shown before. Through cyclization, sugars attain a *pyranose* and/or a *furanose* form.

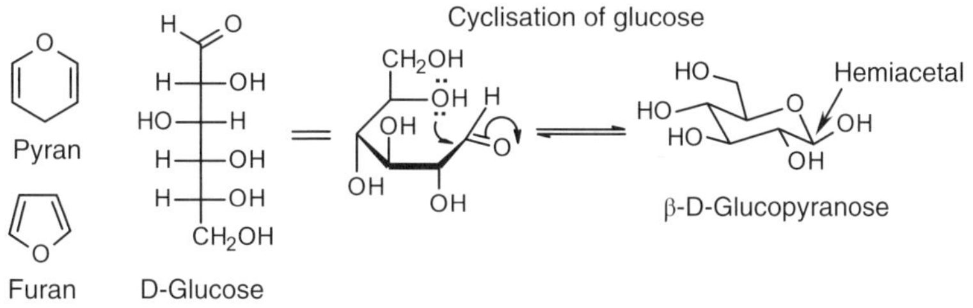

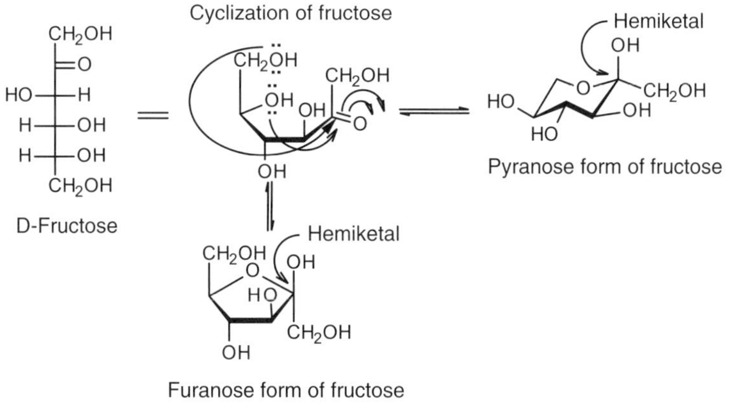

Pyran
Furan   D-Glucose
β-D-Glucopyranose

Cyclisation of glucose

D-Fructose
Cyclization of fructose
Pyranose form of fructose
Furanose form of fructose

Cyclization provides a new chiral centre at C-1 in the cyclic form. This carbon is called the *anomeric carbon*.

At the anomeric carbon, the —OH group can project upwards (β configuration) or downwards (α configuration).

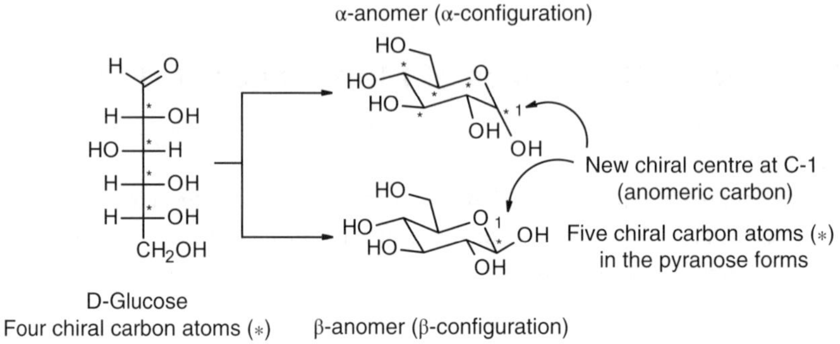

α-anomer (α-configuration)

New chiral centre at C-1 (anomeric carbon)

Five chiral carbon atoms (∗) in the pyranose forms

D-Glucose
Four chiral carbon atoms (∗)   β-anomer (β-configuration)

8.3.3.1 Mutarotation

The term *mutarotation* means the variation of optical rotation with time observed in a solution of sugar when standing. Let's have a look at this phenomenon in a glucose solution. A pure α anomer of glucose has an mp: 146 °C, specific rotation $[\alpha]_D$ +112.2°, and the specific rotation on standing is +52.6°, while a pure β anomer has a mp: 148–155 °C, specific rotation $[\alpha]_D$ +18.7°, and the specific rotation on standing is +52.6°. When a sample of either pure anomer is dissolved in water, its optical rotation slowly changes and ultimately reaches a constant value of +52.6°. Both anomers in solution reach an equilibrium with fixed amounts of α (35%), β (64%) and open chain (~1%) forms.

8.3.4 Acetal and Ketal Formation in Sugars

We have already learnt that a *hemiacetal* or *hemiketal* exists in the cyclic structure of a sugar. For example, the anomeric carbon (C-1) in glucose is a hemiacetal, and that in fructose is hemiketal. When the hydroxyl group in a hemiacetal or hemiketal is replaced by a RO— group, an *acetal* or *ketal* is formed, respectively. R can be an alkyl group or another sugar. The following example shows the acetal formation in glucopyranose.

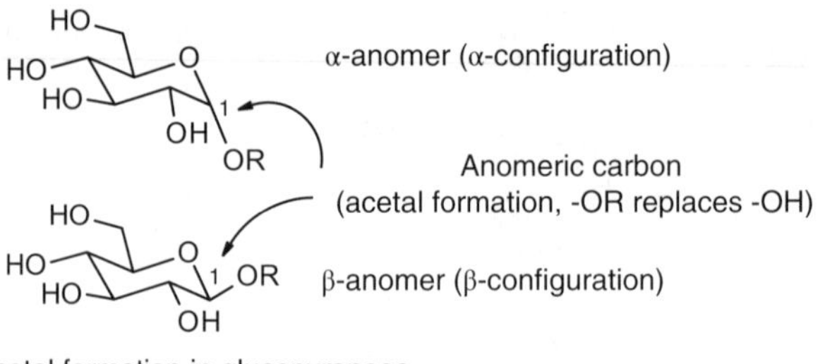

α-anomer (α-configuration)

Anomeric carbon
(acetal formation, -OR replaces -OH)

β-anomer (β-configuration)

Acetal formation in glucopyranose

Acetals and ketals are also called *glycosides*. Acetals and ketals (glycosides) are not in equilibrium with any open chain form. Only hemiacetals and hemiketals can exist in equilibrium with an open chain form. Acetals and ketals do not undergo mutarotation or show any of the reactions specific to the aldehyde or ketone groups. For example, they cannot be oxidized easily to form sugar acids. As an acetal, the carbonyl group is effectively protected.

When glucose is treated with methanol containing hydrogen chloride and prolonged heat is applied, acetals are formed. In this reaction the hemiacetal function is converted to the monomethyl acetal.

D-Glucopyranose

Stereochemistry is not specified

Methyl-β-D-glucopyranoside (Minor product)

Methyl-α-D-glucopyranoside (Major product)

8.3.5 Oxidation, Reduction, Esterification and Etherification of Monosaccharides

For most of the reactions of monosaccharides that involve the aldehyde or ketone functional group, the presence of an open chain form is crucial, as only in this form can these functional groups exist. A sugar solution contains two cyclic anomers and the open chain form in an equilibrium. Once the aldehyde or ketone group of the open chain form is used up in a reaction, the cyclic forms open up to produce a more open chain form to maintain the equilibrium at the reaction.

8.3.5.1 Oxidations of Monosaccharides

Aldoses are easily oxidized to aldonic acid. Aldoses react with *Tollens'* (Ag^+ in aq. NH_3), *Fehling's* (Cu^{2+} in aq. sodium tartrate) and *Benedict's reagents* (Cu^{2+} in aq. sodium citrate), and give characteristic colour changes. All these reactions yield oxidized sugar and a reduced metallic species. These reactions are simple chemical tests for *reducing sugars* (sugars that can reduce an oxidizing agent).

Reaction with Fehling's (and Benedict's) reagent, aldehydes and ketones (e.g. with sugars – aldoses and ketoses) can reduce Fehling's (and Benedict's) reagents, and they themselves are oxidized.

$$Cu^{2+}(Blue) + aldose \text{ or } ketose \rightarrow Cu_2O(red/brown) + oxidized \text{ sugar}$$

Although most sugar molecules are in the cyclic form, the small amounts of open chain molecules are responsible for this reaction.

Therefore, glucose (open chain is an aldose) and fructose (open chain is a ketose) give a positive test and are *reducing sugars*.

When an oxidizing agent, for example nitric acid, is used, a sugar is oxidized at both ends of the chain to the dicarboxylic acid, called *aldaric acid*. For example, galactose is oxidized to galactaric acid by nitric acid.

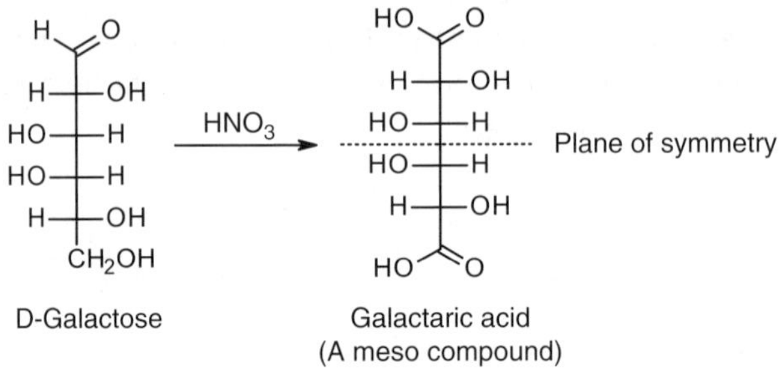

D-Galactose

Galactaric acid
(A meso compound)

8.3.5.2 Reductions of Monosaccharides

8.3.5.2.1 Reduction with Sodium Borohydride

Treatment of a monosaccharide with sodium borohydride ($NaBH_4$) reduces it to a polyalcohol called an *alditol*.

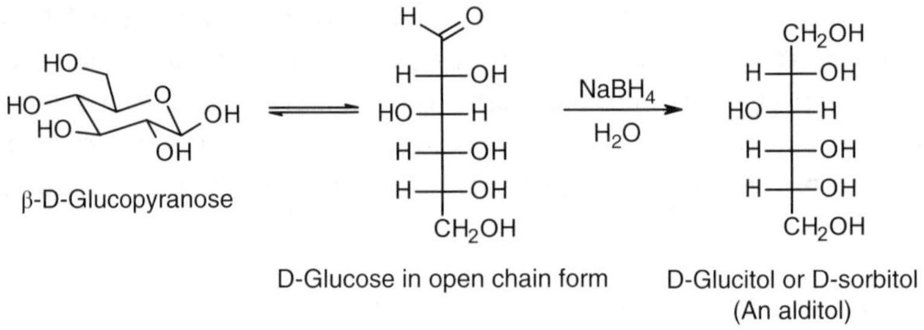

β-D-Glucopyranose

D-Glucose in open chain form

D-Glucitol or D-sorbitol
(An alditol)

The reduction occurs by interception of the open chain form present in the alde-hyde/ketone-hemiacetal/hemiketal equilibrium.

Although only a small amount of the open chain form is present at any given time, that small amount is reduced. Then, more is produced by the opening of the pyranose form, that additional amount is reduced and so on until the entire sample has undergone reaction.

8.3.5.2.2 Reduction with Phenylhydrazine (Osazone Test)

The open chain form of the sugar reacts with phenylhydrazine to afford a pheny-losazone. Three moles of phenylhydrazine are used, but only two moles are taken up at C-1 and C-2.

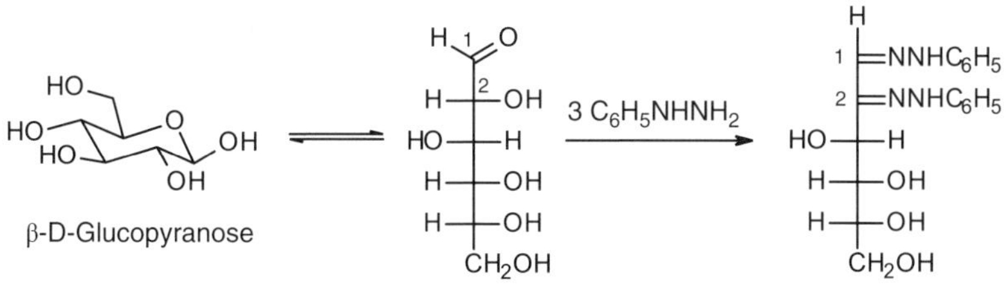

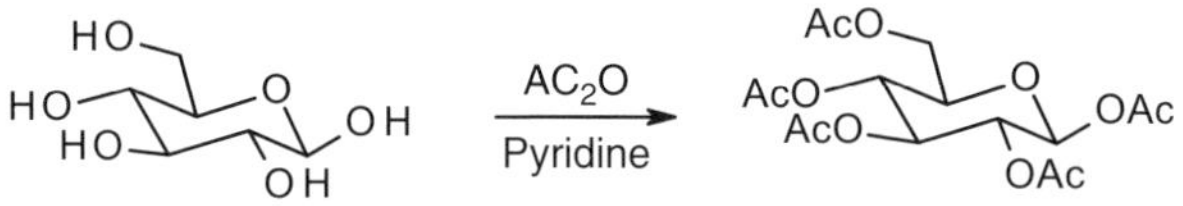

β-D-Glucopyranose ⇌ D-Glucose in open chain form → (3 C₆H₅NHNH₂) Glucosazone (A phenylosazone)

D-Glucose open chain Fischer projection:

H—C(1)=O
H—(2)C—OH
HO—C—H
H—C—OH
H—C—OH
CH₂OH

Glucosazone:

(1) H—C=NNHC₆H₅
(2) C=NNHC₆H₅
HO—C—H
H—C—OH
H—C—OH
CH₂OH

In monosaccharides where structures differ at C-1 and C-2 but are the same in the rest of molecule, we get the same phenylosazone. If we examine the structures of glucose and mannose, the only structural difference we can identify is the orientation of the hydroxyl group at C-2. The rest of the molecules are exactly the same. Therefore, glucose and mannose generate the same phenylosazone. Phenylosazones are highly crystalline solids with characteristic shaped crystals. Shapes are diagnostic of the phenylosazone type.

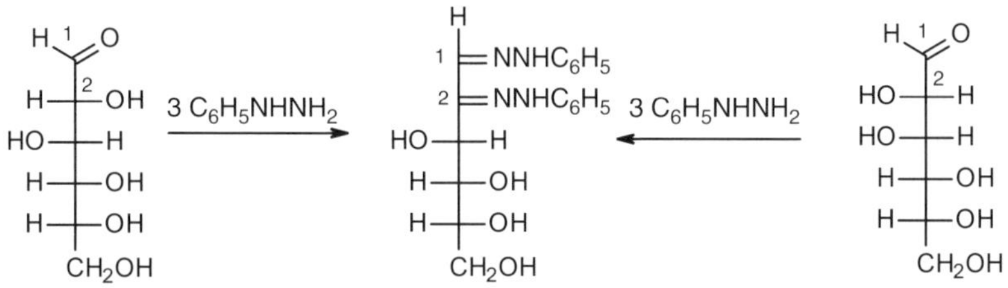

D-Glucose in open chain form → 3 C₆H₅NHNH₂ → A phenylosazone ← 3 C₆H₅NHNH₂ ← D-Mannose in open chain form

8.3.5.3 Esterification of Monosaccharides

Monosaccharides contain a number of alcoholic hydroxyl groups and thus can react with acid anhydrides to obtain corresponding esters. For example, when glucose is treated with acetic anhydride and pyridine it forms a pentaacetate. The ester functions in glucopyranose pentaacetate undergo the typical ester reactions.

β-D-Glucopyranose → (AC₂O / Pyridine) → β-D-glucopyranose

Monosaccharides also form phosphate esters with phosphoric acid. Monosaccharide phosphate esters are important molecules in biological system. For example, in the DNA and RNA nucleotides, phosphate esters of 2-deoxyribose and ribose are present, respectively. Adenosine triphosphate (ATP), the triphosphate ester at C-5 of ribose in adenosine, is found extensively in living systems.

8.3.5.4 Etherification of Monosaccharides

When methyl α-D-glucopyranoside (an acetal) is treated with dimethyl sulphate in the presence of aqueous sodium hydroxide, the methyl ethers of the alcohol functions are formed. The methyl ethers formed from monosaccharides are stable in bases and dilute acids.

Methyl-α-D-Glucopyranoside   Methyl-2,3,4,6-tetra-O-methyl-α-D-Glucopyranoside

8.3.6 Pharmaceutical Uses of Monosaccharides

Pharmaceutically, glucose is probably the most important of all regular monosaccharides. A solution of pure glucose has been recommended for use by subcutaneous injection as a restorative after severe operations or as a nutritive in wasting diseases. It is used to augment the movements of the uterus. Glucose is added to nutritive enemata for rectal alimentation. Its use has also been recommended for rectal injection and by the mouth in delayed chloroform poisoning.

Glucose is used as a pharmaceutical additive. Liquid glucose is used mainly as a pill or tablet additive. For coloured pills, many dispensers prefer a mixture of equal weights of extract of gentian and liquid glucose. Liquid glucose is particularly suitable for the preparation of pills containing ferrous carbonate. It preserves the ferrous salt from oxidation, and will even reduce any ferric salt present. Conversely, it should not be used where such reduction is to be avoided, such as in the preparation of pills containing cupric salts. Apart from the pharmaceutical or medicinal uses, glucose is also used in large quantities in the food and confectionary industries, often in the form of a thick syrup. Fructose, another common monosaccharide found in fruits and honey, is more soluble in water than glucose and is also sweeter than glucose. It is used as a sweetener for diabetic patients and in infusion for parental nutrition.

8.3.7 Disaccharides

Disaccharides contain a glycosidic acetal *bond* between the anomeric carbon of one sugar and an —OH group at any position on the other sugar. A glycosidic bond between C-1 of the first sugar and the —OH at C-4 of the second sugar is particularly common. Such a bond is called the 1,4'-link. For example, maltose, where two glucose units linked between C-1 and C-4 via oxygen. A glycosidic bond to the anomeric carbon can be either α or β.

The most common naturally occurring disaccharides are sucrose (table sugar) and lactose (milk sugar). While sucrose is derived from plants and is prepared commercially from sugar cane and sugar beets, lactose is found in the milk of animals. Other common disaccharides that originate from breaking down of polysaccharides include maltose (obtained from starch) and cellobiose (attained from cellulose).

8.3.7.1 Maltose and Cellobiose

Maltose is a disaccharide, composed of two units of glucose linked (α linkage) between C-1 of one and C-4 of the other via oxygen. Chemically, it can be called 4-*O*-α-D-glucopyranosyl-D-glucopyranose. Cellobiose is also composed of two units of glucose, but the 1,4'-link is β, instead of α. Thus, it can be called 4-*O*-β-D-glucopyranosyl-D-glucopyranose. 'Linkage' always refers to a 'left hand' sugar. For example, in maltose, since the 'linkage' is α, and is in between C-1 of one glucose and C-4 of the other, the 'linkage' is called α 1,4'.

Both maltose and cellobiose exist as α and β anomers and undergo mutarotation. These are reducing sugars. They react with Benedict's and Fehling's reagent, and also react with phenylhydrazine to yield characteristic phenylosazone. If you have a closer look at the following structures of maltose and cellobiose, you will see that the left hand glucose possesses an acetal link (glycosidic link), but the right hand glucose still has the hemiacetal at C-1'. The right hand glucose can exist in an equilibrium of α and β anomers and the open chain form. This is why maltose and cellobiose behave like glucose in chemical reactions.

Maltose is hydrolysed by the enzyme maltase (specific for α-glycosidic linkage) to two units of glucose, but, for the hydrolysis of cellobiose, the enzyme emulsin (specific for β-glycosidic linkage) is necessary. While maltose is the building block of the polysaccharide, starch, cellobiose is the building block of another polysaccharide, cellulose.

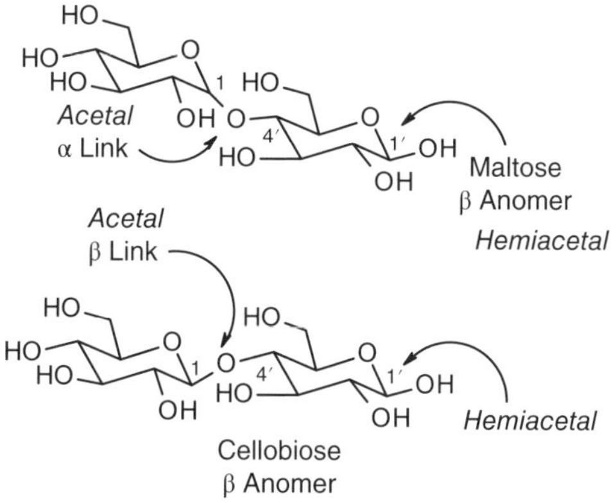

Malt consists of the grain of barley, *Hordeum distichon* (Family: Gramineae), partially germinated and dried. Maltose is the major carbohydrate of malt and malt extracts. Pharmaceutically, the extract of malt is used as a vehicle for the administration of cod-liver oil, and the liquid extract is given with haemoglobin, extract of cascara and various salts.

8.3.7.2 Lactose

Lactose, found in milk and a major component of whey, is a disaccharide that is composed of a unit of glucose and a unit of galactose through a β 1,4'-linkage. Chemically, it can be called 4-O-β-D-galactopyranosyl-D-glucopyranose. Like maltose and cellobiose, lactose is a reducing sugar because of the presence of hemiacetal on the right hand sugar (glucose). Therefore, it also undergoes similar reactions to those of cellobiose and maltose and shows mutarotation.

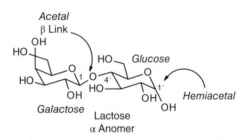

Lactose has a sweetish taste and is used extensively in pharmaceutical industries. It is the second most widely used compound and employed as a diluent, filler or binder in tablets, capsules and other oral product forms, because of its cost-effectiveness, ready availability, neutral taste, low hygroscopicity, compatibility with active ingredients and excipients, chemical and physical stability and water solubility. α-Lactose is used for the production of lactitol, which is present in diabetic products, low calorie sweeteners and slimming products. As lactose is only 30% as sweet as sugar it is used as a sugar supplement and also in food and confectionery.

8.3.7.3 Sucrose

Sucrose is a disaccharide that is composed of a unit of glucose (acetal form) and a unit of fructose (ketal form) linked through C-1 of glucose and C-2 of fructose; that is, a 1,2' link. In sucrose, neither glucose nor fructose can exist in open chain form because of the formation of acetal and ketal as shown here. As a result, sucrose is not a reducing sugar and does now exhibit mutarotation. The specific rotation $[\alpha]_D$ of sucrose is +66°.

A Sucrose molecule

Hydrolysis of sucrose leads to the formation of glucose and fructose with specific rotations $[\alpha]_D$ +52.5° and –92°, respectively, and makes the resulting mixture levorotatory (–). This phenomenon of sucrose is called the *inversion of sucrose* and the resulting mixture is known as an *invert sugar*, which is the main component of honey and is sweeter than sucrose itself. Sucrose is used in pharmaceutical formulations as a masking agent; it is combined with other ingredients to provide palatable dosage forms of many drugs, thus aiding compliance. Sucrose is found in many pharmaceutical dosage forms such as chewable tablets, syrups, lozenges or gums. It is used in the biopharmaceutical industry to stabilize proteins, lipids and carbohydrates throughout the formulation and freeze/thaw lifecycle of therapeutics.

8.3.8 Polysaccharides

A number of monosaccharide units combine to form a *polysaccharide*; for example, starch, cellulose and inulin. Starch and cellulose are the two most important polysaccharides from biological as well as economical viewpoints.

8.3.8.1 Starch

Starch, an essential component of our diet, is a high molecular weight polymer of glucose where the monosaccharide (glucose) units are linked mainly by 1,4'-α-glycoside bonds similar to maltose. Plants are the main source of starch. Starch is obtained from wheat (*Triticum aestivum*), rice (*Oryza sativa*) and maize (*Zea mays*), all from the plant family Gramineae. Potato (*Solanum tuberosum*; family: Solanaceae) and maranta (*Maranta arundinacea*; family: Marantaceae) are also good sources of starch.

Starch consists of two main components: *amylose* (insoluble in cold water) and *amylopectin* (soluble in cold water). Amylose, which accounts for about 20% by weight of starch, has an average molecular weight of over 10^6. It is a polymer of glucopyranose units linked together through α 1,4'-linkages in a linear chain. Hydrolysis of amylose yields maltose. Amylose and iodine form a colour complex, which is blue/black. This is the colour reaction of iodine in starch and a confirmatory test for the presence of starch.

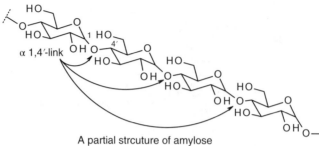

A partial strcuture of amylose

On the other hand, amylopectin accounts for about 80% by weight of starch and consists of hundreds of glucose molecules linked together by 1,4'-α- and also 1,6'-α-glycoside bonds.

Amylopectin contains branches (nonlinear) in approximately one in every 20–25 glucose units. Hydrolysis of amylopectine gives maltose.

A partial strcuture of amylopectine

The pharmaceutical and cosmetic uses of starch include its use as dusting powder, binder, dispersing agent, thickening agent, lubricant, disintegrant, coating agent and diluent because of its nontoxic and non-irritant properties, low cost, ease of modification and versatility.

Products	Composition	Applications
Amylum Iodisatum BPC	Iodized starch	It is administered internally in syphilis and other cachexias, and may be given in milk, water, gruel or arrowroot. Externally, it is used as a dry dressing, being a good substitute for iodoform.

Products	Composition	Applications
Cataplasma Amyli BPC	Starch poultice	Used as a substitute for the domestic bread poultice for application to small superficial ulcerations.
Cataplasma Amyli et Acidi Borici BPC	Starch and boric acid poultice	Starch, 10; boric acid, 6; water, 100. An antiseptic poultice for application to ulcerated wounds.
Glycerinum Amyli BP and Glycerinum Amyli USP	Glycerine of starch	It is a soothing and emollient application for the skin, and is used for chapped hands and chilblains.
Mucilago Amyli BPC	Mucilage of starch	This mucilage is used as a basis for enemata.

Starch soaks up secretions and helps to render injured parts less liable to bacterial infections. As a dusting powder for application to chafings (irritation of skin caused by repetitive friction, usually generated through skin to skin contact of multiple body parts) and excoriations, it is used either alone or mixed with zinc oxide, boric acid and other similar substances. It also forms the basis of violet powder. Boiled with water, it may be employed as an emollient for the skin. Starch is the best antidote for poisoning by iodine. Some examples of commercial preparations of starch are presented in the previous table.

8.3.8.2 Glycogen

Glycogen, a homopolymer of glucose, is the major form of stored carbohydrate in animals and serves the energy storage function. Dietary carbohydrates, which are not needed for immediate energy, are converted by the body to glycogen for long-term storage. It can release glucose units if cellular glucose levels are low. Glycogen 'mops up' excess glucose in cells. Like amylopectin, glycogen contains a complex branching structure with both 1,4′ and 1,6′ links, but it is larger than amylopectin (up to 100 000 glucose units) and far more branched. It has one end glucose unit (where glucose can be added or released) for every 12 units and a branch in every 6 glucose units.

8.3.8.3 Cellulose

Cellulose, the most abundant natural organic polymer, consists of several thousands of D-glucose units linked by 1,4′-β-glycoside bonds as in cellobiose. Cellulose has a linear chain structure.

Different cellulose molecules then can interact to form a large aggregate structure held together by *hydrogen bonds*. On hydrolysis, cellulose provides cellobiose and finally glucose.

Nature uses cellulose mainly as a structural material to provide plants with strength and rigidity. Human digestive enzymes contain α-glucosidase, but not β-glucosidase. Therefore, human digestive enzymes cannot hydrolyse β-glycosidic links between glucose units. In human beings, starch (but not cellulose), is hydrolysed enzymatically to produce glucose. Therefore, cellulose does not have any dietary importance. While there is no food value in cellulose for humans, cellulose and its derivatives are commercially important. Cellulose is used as raw material for the manufacture of cellulose acetate, known commercially as *acetate rayon*, and cellulose nitrate, known as *guncotton*. Commercially important fibres, for example cotton and flax, consist almost completely of cellulose.

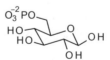

Partial structure of cellulose

Water-soluble, high-viscosity grade cellulose ether compositions are useful for the reduction of serum lipid levels, particularly total serum cholesterol, serum triglycerides and low-density lipoprotein (LDL) levels and/or attenuate the rise of blood glucose levels. The compositions may be in the form of a prehydrated ingestible composition, for example a gelatine, or a comestible, for example a cookie. Cellulose derivatives, such as hydroxyethylcellulose, are used in the formulation of sustained release tablets and suspensions. Natrosol (hydroxyethylcellulose) is a nonionic water-soluble polymer that is extensively used as a thickener, protective colloid, binder, stabilizer and suspending agent, particularly in applications where a non-ionic material is desired. Natrosol is also used in cosmetic preparations as a thickening agent for shampoos, conditioners, liquid soaps and shaving creams.

8.3.9 Miscellaneous Carbohydrates

8.3.9.1 Sugar Phosphates

These sugars are formed by phosphorylation with ATP; for example, glucose 6-phosphate. They are extremely important in carbohydrate metabolism. We already know that nucleotides contain sugar phosphates.

Glucose 6-phosphate

8.3.9.2 Nitrogen Containing Sugars

8.3.9.2.1 Glycosylamines

In these sugars, the anomeric —OH group (of common sugars) is replaced by an amino (—NH$_2$) group. An example is adenosine.

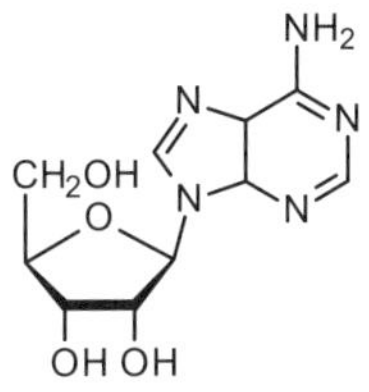

Adenosine
(A nucleoside)

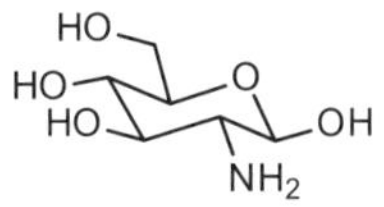

Glucosamine
(An amino sugar)

8.3.9.2.2 Amino Sugars

In amino sugars, a non-anomeric —OH group (of common sugars) is replaced by an amino (—NH$_2$) group. An example is glucosamine, which is found in the exoskeletons of insects and crustaceans, and also isolated from *heparin* (anticoagulant in mast cells in arterial cell walls).

Other amino sugars are found in antibiotics such as *streptomycin* and *gentamycin*.

8.3.9.2.3 Carbohydrate Antibiotics

Antibiotics that contain one or more amino sugars within the molecule are called *carbohydrate antibiotics*. For example, gentamycin is composed of three different units: purpurosamine, 2-deoxystreptamine and garosamine. Other examples include streptomycin and neomycin.

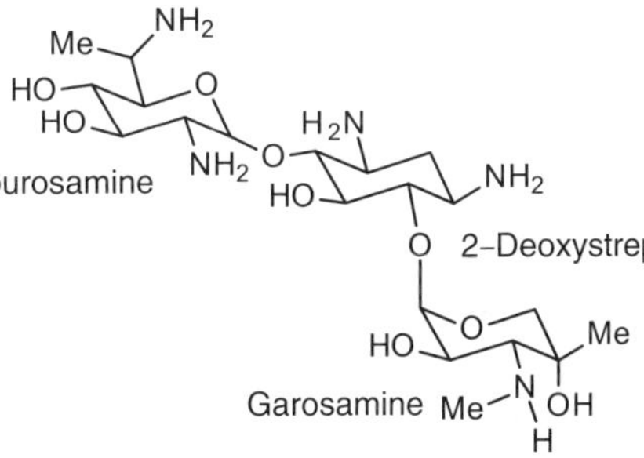

Gentamycin
(A carbohydrate antibiotic)

8.3.9.2.4 Sulphur Containing Carbohydrate

In these sugars, the anomeric —OH group (of common sugars) is replaced by a sulphur-containing group such as lincomycin.

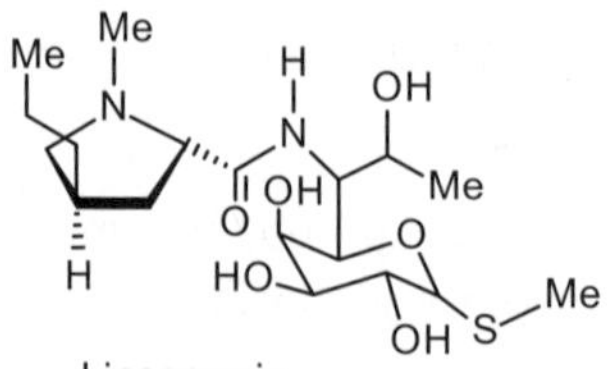

Lincomycin
(An antibiotic)

8.3.9.3 Ascorbic Acid (Vitamin C)

Ascorbic acid, commonly known as vitamin C, is a sugar acid, biosynthesized in plants and also found in the livers of most vertebrates, except human beings. Therefore, human beings need an external supply of this vitamin, mainly from fresh vegetables and fruits. In many pharmaceutical preparations ascorbic acid is used as an antioxidant preservative. Ascorbic acid is highly susceptible to oxidation and oxidized easily to dehydroascorbic acid.

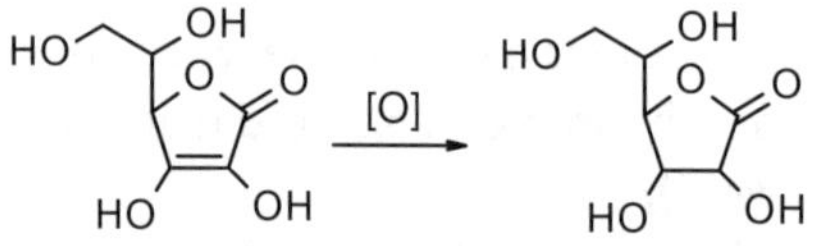

Vitamin C
Ascorbic acid
(A natural antioxidant)

Dehydroascorbic acid

8.3.9.4 Glycoprotein and Glycolipids

Glycoproteins and glycolipids are formed when sugars combine, respectively, with proteins and lipids. Biologically these are important compounds as they are an integral part of cell membranes. Biological membranes are composed of proteins, lipids and carbohydrates. The carbohydrates in the membrane are covalently bonded to proteins (*glycoproteins*) or with lipids (*glycolipids*).

8.3.10 Cell Surface Carbohydrates and Blood Groupings

Small polysaccharide chains, covalently bonded by glycosidic links to hydroxyl groups on proteins (glycoproteins), act as biochemical markers (i.e. *antigenic determinants*) on cell surfaces.

The membrane of red blood cells (RBCs) contains glycoproteins/glycolipids and the type of sugar that combines with these proteins/lipids varies from person to person. This gives rise to different *blood groups* (A, B, AB and O). Human blood group compatibilities are presented in the following table.

	Acceptor blood type			
Donor blood type	A	B	AB	O
A	Compatible	Incompatible	Compatible	Incompatible
B	Incompatible	Compatible	Compatible	Incompatible
AB	Incompatible	Incompatible	Compatible	Incompatible
O	Compatible	Compatible	Compatible	Compatible

8.4 GLYCOSIDES

Compounds that yield one or more sugars on hydrolysis are known as *glycosides*. A glycoside is composed of two moieties: a sugar portion (*glycone*) and non-sugar portion (*aglycone or genin*). For example, the hydrolysis of salicin affords a glucose unit and salicyl alcohol. Glycosides play important roles in living organisms. Many plants store chemicals in the form of inactive glycosides, which can be activated by enzymatic hydrolysis. Many such plant glycosides are used as medications; for example, cardiac glycosides from *Digitalis* species. In animals and humans, poisons are often bound to sugar molecules as part of their elimination from the body.

Salicin
(A glycoside)

Glucose
Sugar
(A glycone)

Salicyl alcohol
(An aglycone)

Glycosides of many different aglycones are extensively found in the plant kingdom. Many of these glycosides are formed from phenols, polyphenols, steroidal and terpenoidal alcohols through glycosidic attachment to sugars. Among the sugars found in natural glycosides, D-glucose is the most prevalent one, but L-rhamnose, D- and L- fructose and L-arabinose also occur quite frequently.

Of the pentoses, L-arabinose is more common than D-xylose and the sugars often occur as oligosaccharides.

The sugar moiety of a glycoside can be joined to the aglycone in various ways, the most common being via an oxygen atom (*O-glycoside*). However, this bridging atom can also be a carbon (*C-glycoside*), a nitrogen (*N-glycoside*) or a sulphur atom (*S-glycoside*). By virtue of the aglycone and/or sugar, glycosides are extremely important pharmaceutically and medicinally. For example, digitoxin is a cardiac glycoside found in the 'fox glove' plant (*Digitalis purpurea*).

8.4.1 Biosynthesis of Glycosides

The biosynthetic pathways are widely variable depending on the type of the agly-cones as well as the glycone units present in the glycosides. The aglycone and the sugar part are biosynthesized separately and then coupled to form a glycoside. The coupling of the sugar and aglycone takes place in the same way irrespective of the structural type of the aglycone. Phosphorylation of a sugar yields a sugar 1-phos-phate, which reacts with a uridine triphosphate (UDP) to form a uridine diphos-phate sugar (UDP-sugar) and inorganic phosphate. This UDP-sugar reacts with the aglycone to form the glycoside and a free UDP.

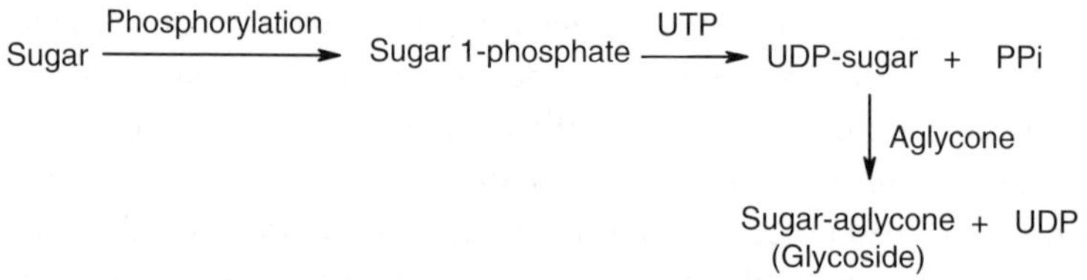

8.4.2 Classification

Glycosides can be classified in various ways; for example, based on the sugar com-ponent, aglycone structure or physical/pharmacological properties.

8.4.2.1 Based on Sugar Component

Glycosides that contain glucose are called *glucosides*. Similarly, when the sugars are fructose or galactose, the glycosides are called *fructoside* or *galactoside*, respectively.

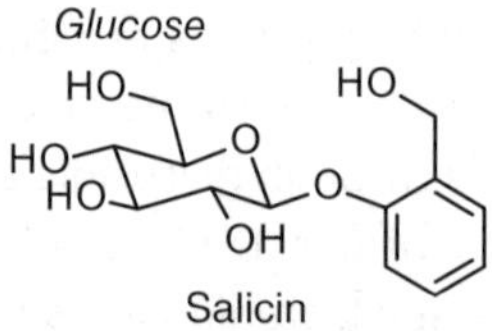

Salicin
(A glucoside)

8.4.2.2 Based on Aglycone

Glycosides can be classified based on the structural types of aglycones present in the glycoside. For example, in anthraquinone, flavonoid, iridoid, lignan or steroid glycosides, the aglycones are anthraquinone, flavonoid, iridoid, lignan or steroid, respectively.

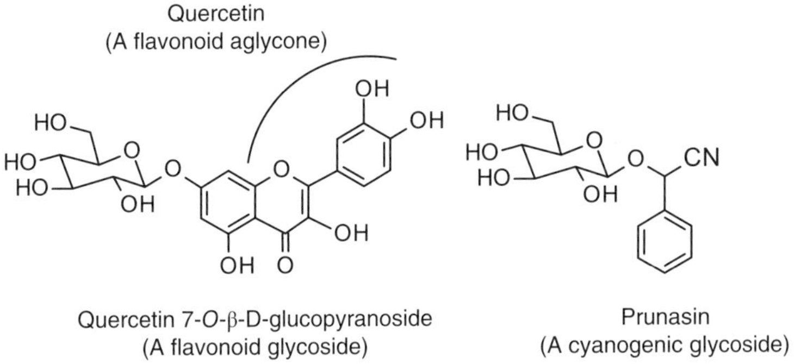

Quercetin
(A flavonoid aglycone)

Quercetin 7-O-β-D-glucopyranoside
(A flavonoid glycoside)

Prunasin
(A cyanogenic glycoside)

8.4.2.3 Based on Properties or Functions

Glycosides that have 'soap-like' properties are called *saponins*. Similarly, glycosides that liberate hydrocyanic acid (HCN) on hydrolysis are known as *cyanogenic glycosides*, and glycosides that have an effect on heart muscles are called *cardiac glycosides*.

8.4.2.3.1 Cyanogenic Glycosides

Amygdalin, prunasin and a number of other related glycosides belong to this class of glycosides that liberate HCN upon hydrolysis. HCN is released from the cyanogenic glycosides when fresh plant material is macerated, as in chewing, which allows enzymes and cyanogenic glycosides to come together releasing HCN. Biosynthetically, the aglycones of cyanogenic glycosides are derived from L-amino acids; for example, amygdalin from L-phenylalanine, linamarin from L-valine and dhurrin from L-tyrosine.

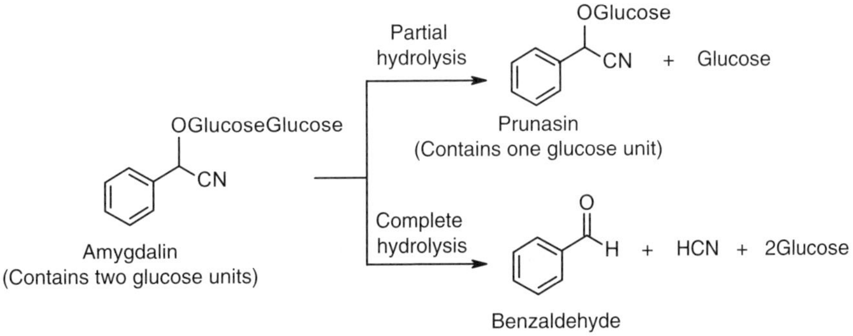

Amygdalin
(Contains two glucose units)

Partial hydrolysis → Prunasin (Contains one glucose unit) + Glucose

Complete hydrolysis → Benzaldehyde + HCN + 2Glucose

Cyanogenic glycosides are common in certain plant families including the Euphorbiaceae, Fabaceae, Linaceae and Rosaceae, and identification of their constituents is a useful tool for informative chemotaxonomic markers. More than 75 different cyanogenic glycosides have been reported from at least 2650 plants from 130 families, including the Asteraceae, Euphorbiaceae, Fabaceae, Linaceae,

Passifloraceae, Poaceae and Rosaceae. However, there are only approximately 25 well-known cyanogenic glycosides and these are generally found in the edible parts of plants, such as apples, apricots, cherries, peaches, plums and quinces, particularly in the seeds of such fruits. The chemicals are also found in almonds, stone fruit, pome fruit, cassava, bamboo shoots, linseed/flaxseed, lima beans, coco yam, chick peas, cashews and kirsch. Cyanogenic glycosides, particularly amygdalin and prunasin, are found in the kernels of apricot, bitter almonds, cherries, plums and peaches. The following are a few other major sources of common cyanogenic glycosides.

Linamarin
(A cyanogenic glucoside)

Common name	Botanical name	Family	Major cyanogenic glycoside present
Almond	*Prunus amygdalus*	Rosaceae	Amygdalin
Cassava	*Manihot utilissima*	Euphorbiaceae	Linamarin (also known as manihotoxin)
Linseed/flax	*Linum usitatissimum*	Linaceae	Linamarin
Butter bean	*Phaseolus lunatus*	Fabaceae	Linamarin
White clover	*Trifolium repens*	Fabaceae	Linamarin
Black cherry	*Prunus serotina*	Rosaceae	Prunasin

8.4.3 Test for Hydrocyanic Acid (HCN)

The liberation of HCN due to complete hydrolysis of cyanogenic glycosides can be determined by a simple colour test using sodium picrate paper (yellow), which turns brick red (sodium isopupurate) in contact with HCN.

8.4.4 Pharmaceutical Uses and Toxicity

The extracts of plants that contain cyanogenic glycosides are used as flavouring agents in many pharmaceutical preparations. Amygdalin has been used in the treatment of cancer (HCN liberated in stomach kills malignant cells) and also as a cough suppressant in various preparations. Cyanogenic glycosides play pivotal roles in the organization of chemical defence systems in plants and in plant-insect interactions.

Excessive ingestion of cyanogenic glycosides can be fatal. Some foodstuffs containing cyanogenic glycosides can cause poisoning (severe gastric irritations and damage) if not properly handled. The toxicity of cyanogenic glycosides and their derivatives is dependent on the release of HCN. Toxicity may result in acute cyanide poisoning and has also been implicated in the aetiology of several chronic diseases. Cyanide toxicity can occur in animal including humans at doses between 0.5 and 3.5 mg HCN per kg of body weight. Symptoms of cyanide toxicity in humans include vomiting, stomach ache, diarrhoea, convulsions and, in severe cases, death. Children are particularly at risk because of their smaller body size. The toxicity of cyanogenic glycosides is associated with their ability to be hydrolysed either spontaneously or in the presence of enzyme to give cyanide as an end product of their hydrolysis.

8.4.5 Anthracene/Anthraquinone Glycosides

The aglycones of anthracene glycosides belong to a structural category of anthracene derivatives. Most of them possess an anthraquinone skeleton and are called *anthraquinone glycosides*; for example, rhein 8-*O*-glucoside and aloin (a C-glucoside). The most common sugars present in these glycosides are glucose and rhamnose.

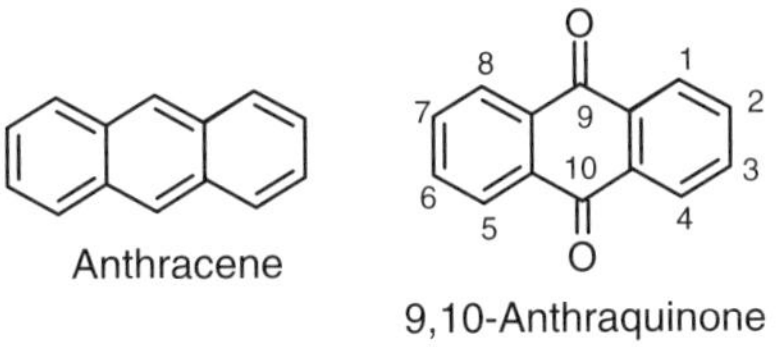

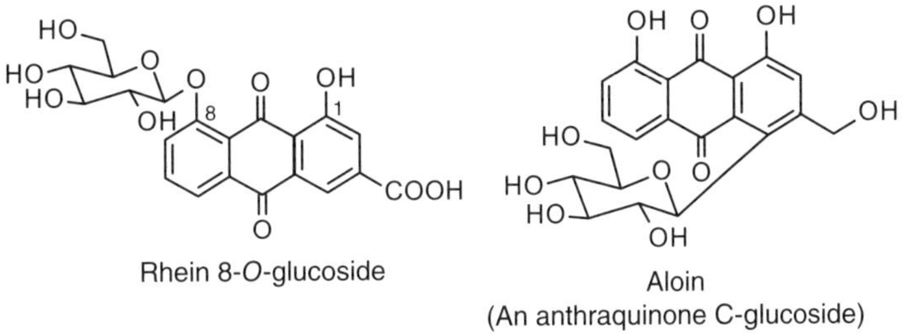

Anthracene

9,10-Anthraquinone

Rhein 8-*O*-glucoside

Aloin
(An anthraquinone C-glucoside)

Anthraquinone glycosides are coloured substances and are the active components in a number of crude drugs, especially with laxative and purgative properties. Anthraquinone aglycone increases peristaltic action of the large intestine.

A number of 'over the counter' laxative preparations contain anthraquinone glycosides. The use of anthraquinone drugs, however, should be restricted to short-term treatment of constipation only, as frequent or long-term use may cause intestinal tumours.

Anthraquinones are found extensively in various plant species, especially from the families Fabaceae, Liliaceae, Polygonaceae, Rhamnaceae and Rubiaceae. They are also biosynthesized in microorganisms; for example, *Penicillium* and *Aspergillus* species. The following structural variations within anthraquinone aglycones are most common in nature.

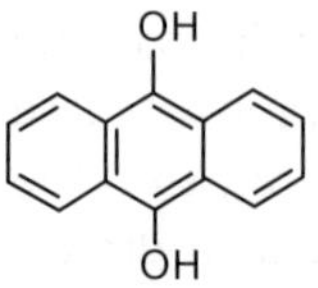

Anthrahydroquinone 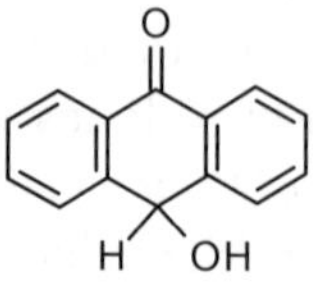   Oxanthrone 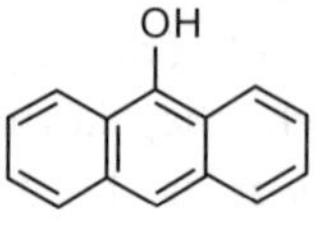   Anthranol 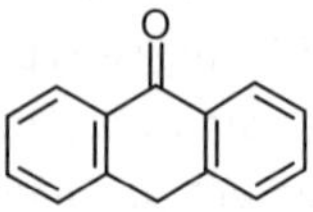   Anthrone

Dimeric anthraquinone and their derivatives are also present as aglycones in anthraquinone glycoside found in the plant kingdom.

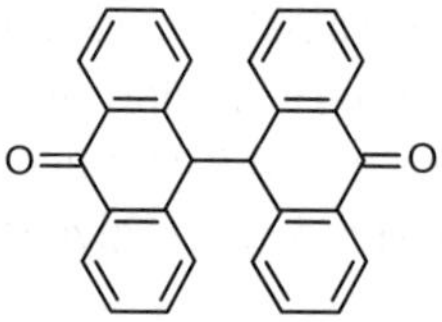

Dianthrone 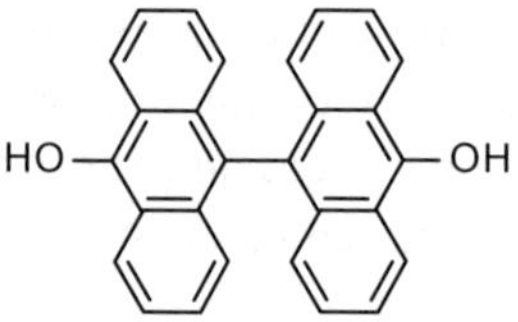   Dianthranol

8.4.5.1 Sennosides

The most important anthraquinone glycosides are sennosides, found in Senna leaves and fruits (*Cassia senna* or *Cassia angustifolia*). These are, in fact, dimeric anthraquinone glycosides. However, monomeric anthraquinone glycosides are also present in this plant.

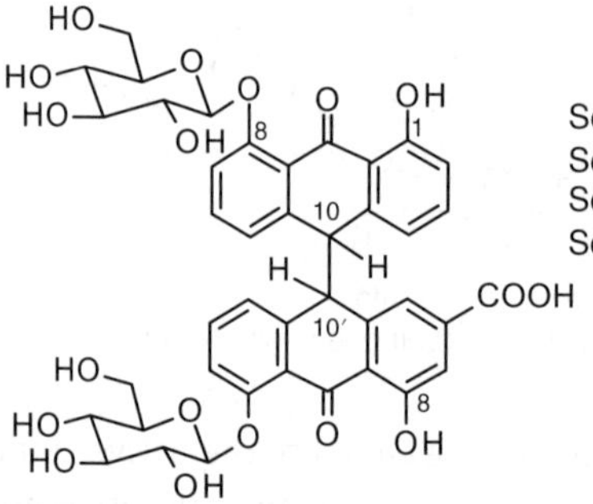

Sennoside A	R = COOH	10,10′-*trans*
Sennoside B	R = COOH	10,10′-*cis*
Sennoside C	R = CH$_2$OH	10,10′-*trans*
Sennoside D	R = CH$_2$OH	10,10′-*cis*

8.4.5.2 Cascarosides

Cascara bark (*Rhamnus purshianus*) contains various anthraquinone *O*-glycosides, but the main components are C-glycosides, which are known as *cascarosides*. Rhubarb (*Rheum palmatum*) also contains several different *O*-glycosides and cascarosides. *Aloe vera* mainly produces anthraquinone C-glycosides; for example, aloin.

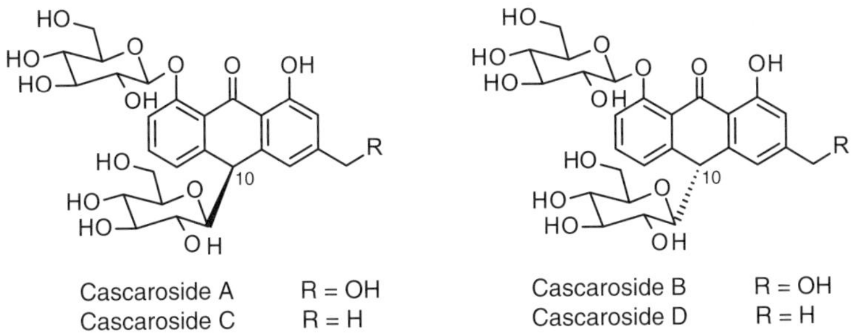

| Cascaroside A | R = OH |
| Cascaroside C | R = H |

| Cascaroside B | R = OH |
| Cascaroside D | R = H |

8.4.5.3 Test for Anthraquinone Glycosides

For free anthraquinones, powdered plant material is mixed with organic solvent, filtered and then an aqueous base, for example NaOH or NH_4OH solution, is added to it. A pink, red or violet colour in the base layer indicates the presence of anthraquinones in the plant sample.

For O-glycosides, the plant samples are boiled with HCl/H_2O to hydrolyse the anthraquinone glycosides to respective aglycones and then this method for free anthraquinones is carried out.

For C-glycosides, the plant samples are hydrolysed using $FeCl_3/HCl$, and then the method for free anthraquinones is carried out.

8.4.5.4 Biosynthesis of Anthraquinone Glycosides

In higher plants anthraquinones are biosynthesized either via acylpolymalonate (as in the plants of the families Polygonaceae and Rhamnaceae) or via shikimic acid pathways (as in the plants of the families Rubiaceae and Gesneriaceae) as presented in the following biosynthetic schemes.

8.4.5.4.1 Acylpolymalonate Pathway

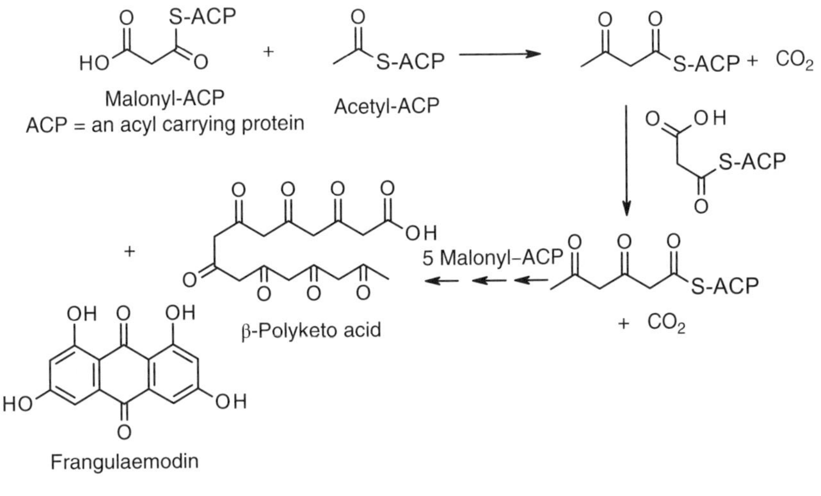

8.4.5.4.2 Shikimic Acid Pathway

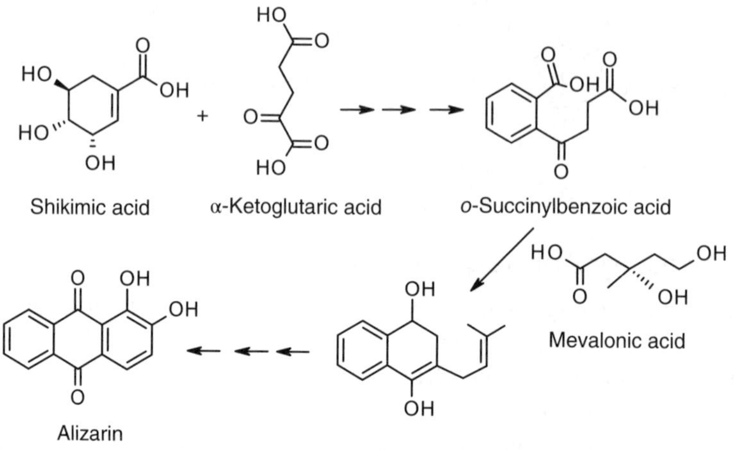

Shikimic acid   α-Ketoglutaric acid   o-Succinylbenzoic acid

Mevalonic acid

Alizarin

8.4.6 Isoprenoid Glycosides

The aglycone of this type of glycoside is biosynthetically derived from isoprene units. There are two major classes of isoprenoid glycosides: saponins and cardiac glycosides.

8.4.6.1 Saponins

Saponin glycosides possess a 'soap-like' behaviour in water; that is, they generate foam. On hydrolysis, an aglycone is produced, which is called *sapogenin*. There are two types of sapogenins: steroidal and triterpenoidal. Usually, the sugar is attached at C-3 in saponins because in most sapogenins there is a hydroxyl group at C-3.

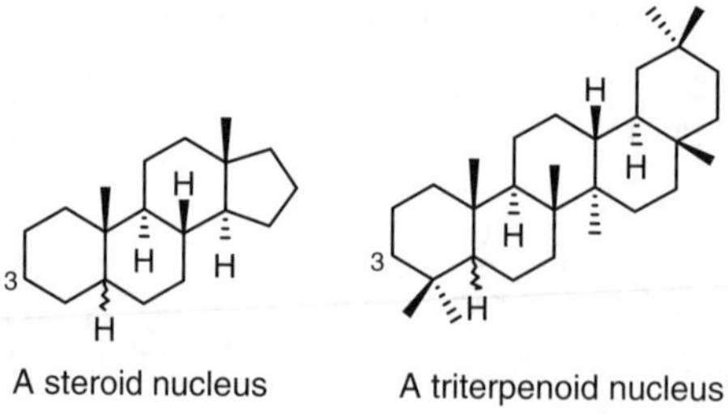

A steroid nucleus   A triterpenoid nucleus

The two major types of steroidal sapogenins are *diosgenin* and *hecogenin*. Steroidal saponins are used in the commercial production of sex hormones for clinical use. For example, progesterone is derived from diosgenin.

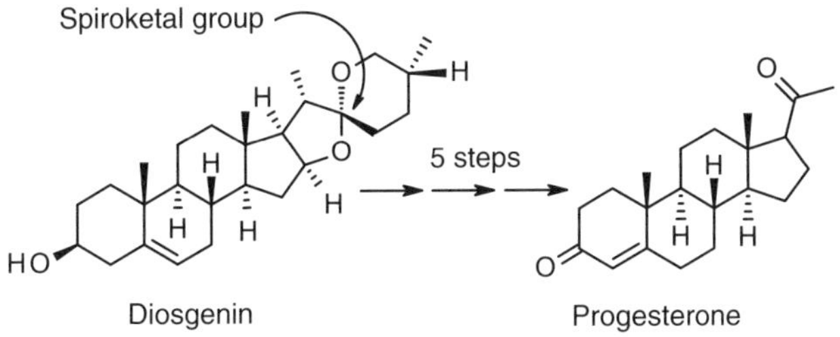

Diosgenin → 5 steps → Progesterone

The most abundant starting material for the synthesis of progesterone is dios-genin isolated from *Dioscorea* species, formerly supplied by Mexico and now China. The spiroketal group attached to the D ring of diosgenin can easily be removed. Other steroidal hormones, for example cortisone and hydrocortisone, can be pre-pared from the starting material hecogenin, which can be isolated from Sisal leaves found extensively in East Africa.

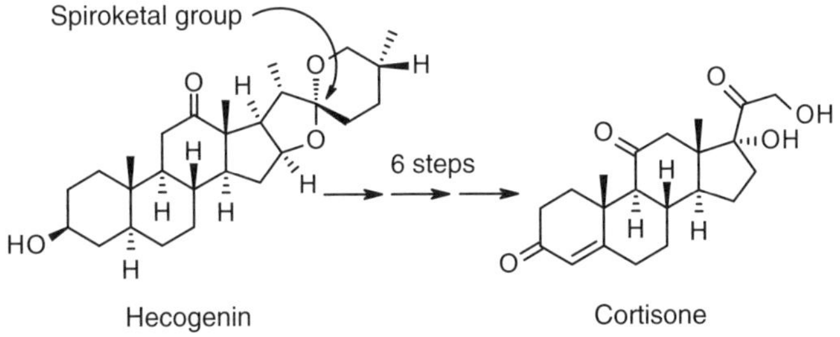

Hecogenin → 6 steps → Cortisone

In *triterpenoidal saponins*, the aglycone is a triterpene. Most aglycones of triter-penoidal saponins are pentacyclic compounds derived from one of the three basic structural classes represented by α-amyrin, β-amyrin and lupeol. However, tetracyclic triterpenoidal aglycones are also found; for example, ginsenosides. These glycosides occur abundantly in many plants; for example, liquorice and ginseng roots contain glycyrrhizinic acid derivatives and ginsenosides, respectively. Most crude drugs con-taining triterpenoid saponins are usually used as expectorants. Three major sources of triterpenoidal glycosides along with their uses are summarized here.

Plants Botanical names (Family)	Main constituents	Uses
Liquorice root *Glycyrrhiza glabra* (Fabaceae)	Glycyrrhizinic acid derivatives	In addition to expectorant action, it is also used as a flavouring agent.
Quillaia bark *Quillaja saponaria* (Rosaceae)	Several complex triter-penoidal saponins, for example, senegin II	Tincture of this plant is used as an emulsi-fying agent

Plants Botanical names (Family)	Main constituents	Uses
Ginseng *Panax ginseng* (Araliaceae)	Ginsenosides	As a tonic, and to promote the feeling of well-being.

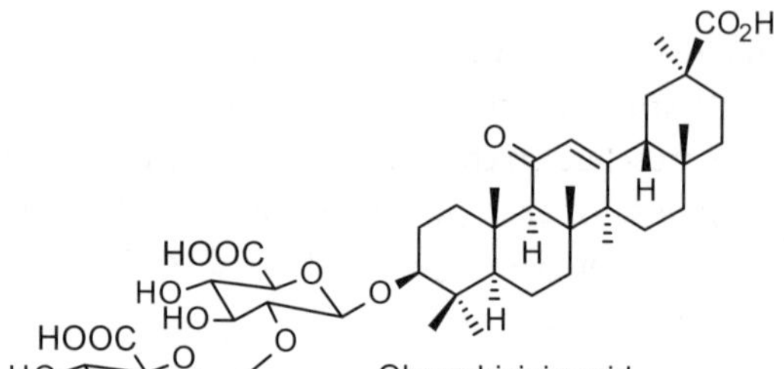

Glycyrrhizinic acid
(A glycoside of liquorice)

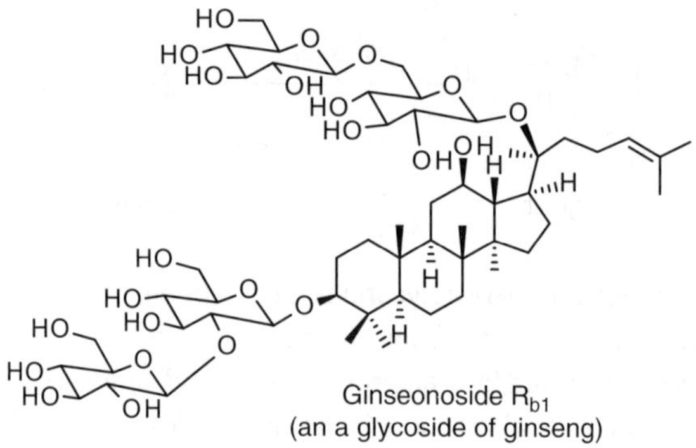

Ginseonoside R_{b1}
(an a glycoside of ginseng)

8.4.6.2 Cardiac Glycosides

Glycosides that exert prominent effect on heart muscle are called cardiac glycosides; for example, digitoxin from *Digitalis purpurea*. Their effect is specifically on myocardial contraction and atrioventricular conduction. The aglycones of cardiac glycosides are steroids with a side chain containing an unsaturated lactone ring, either five-membered γ-lactone (called *cardenolides*) or six-membered δ-lactone (called *bufadienolides*). The sugars present in these glycosides are mainly digitoxose, cymarose, digitalose, rhamnose and sarmentose. Digitoxose, cymarose and sarmentose are 2-deoxysugars.

Cardiac glycosides are found only in a few plant families; for example, Liliaceae, Ranunculaceae, Apocynaceae and Scrophulariaceae are the major sources of these glycosides. Among the cardiac glycosides isolated to date, digitoxin and digoxin, isolated from *Digitalis purpurea* and *Digitalis lanata*, respectively, are the two most important cardiotonics. Digitoxin and digoxin are also found in in *Strophanthus* seeds and squill. Both these cardiac glycosides are cardenolides and the sugar present is the 2-deoxysugar, digitoxose.

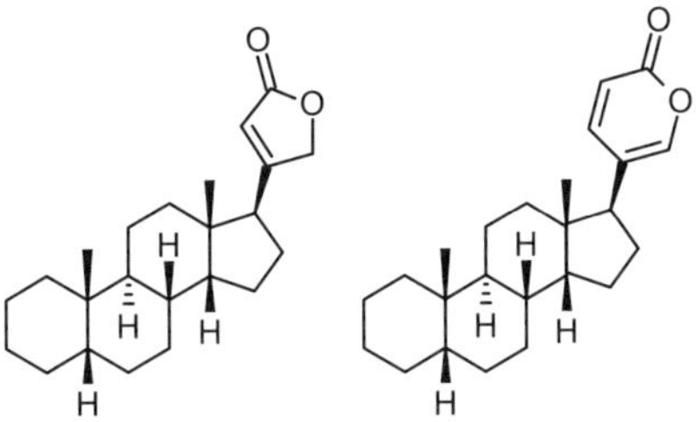

Cardenolide   Bufadienolide

Both sugar and aglycone parts are critical for biological activity. The sugar part possibly is responsible for binding the glycoside to the heart muscle, and the agly-cone moiety has the desired effect on the heart muscle once bound. The lactone ring is essential for the pharmacological action. In addition, the orientation of the 3-OH groups is important. For more prominent activity, this 3-OH group has to be β. In large doses these glycosides lead to cardiac arrest and can be fatal, but in lower doses these glycosides are used in the treatment of congestive heart failure.

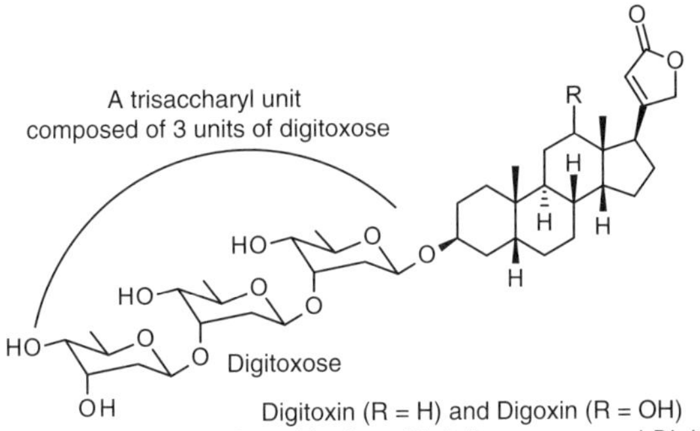

A trisaccharyl unit composed of 3 units of digitoxose

Digitoxose

Digitoxin (R = H) and Digoxin (R = OH)
(Cardiac glycosides from *Digitalis purpurea* and *Digitalis lanata*)

Cardiac glycosides with bufadienolide skeleton, for example proscillaridin A, have been found in plants (e.g. squill, *Drimia maritima*).

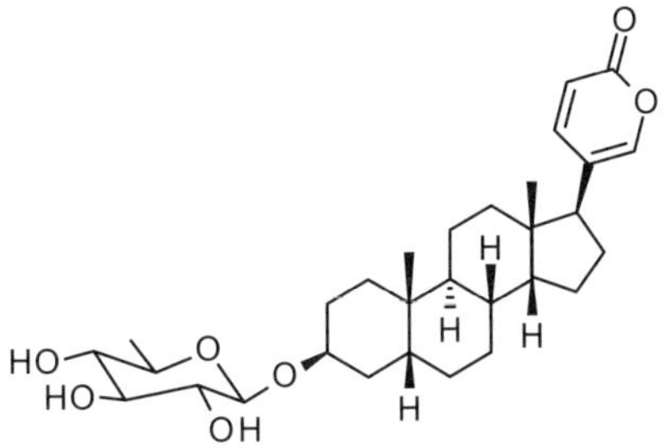

Proscillaridin A
(A bufadienolide cardiac glycoside)

8.4.7 Iridoid and Secoiridoid Glycosides

The iridoids and secoiridoids form a large group of plant constituents that are found usually, but not invariably, as glycosides. For example, harpagoside, an active constituent of *Harpagophytum procumbens*, is an iridoid glycoside. The plant families, for example Lamiaceae (especially genera *Phlomis*, *Stachys* and *Eremostachys*), Gentianaceae, Valerianaceae and Oleaceae are good sources of these glycosides.

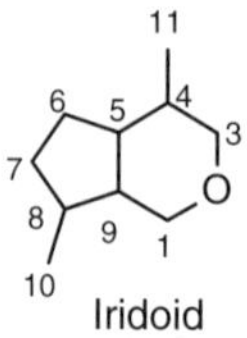

Iridoid

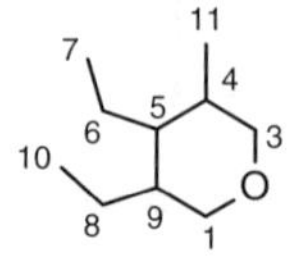

Secoiridoid

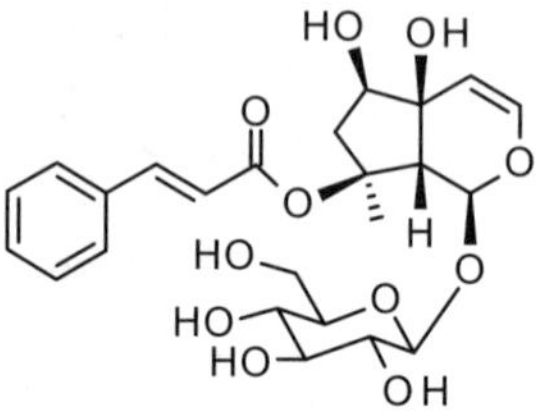

Harpagoside

In most natural iridoids and secoiridoids, there is an additional oxygenation (hydroxy) at C-1 that is generally involved in the glycoside formation.

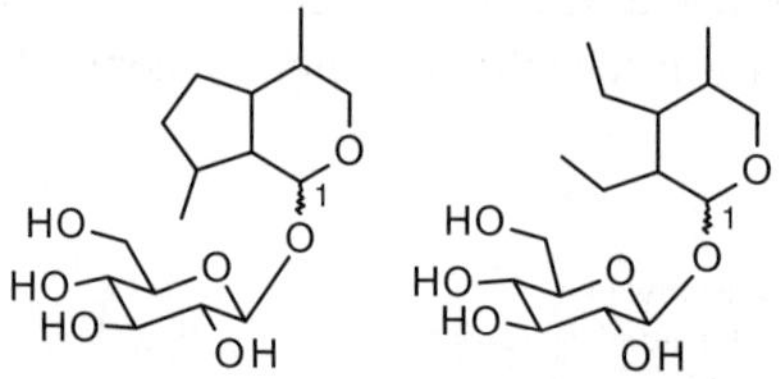

Iridoid and secoiridoid glycosides
Glycosylation is at C-1

It is also extremely common among natural iridoids and secoiridoids to have a double bond between C-3 and C-4 and a carboxylation at C-11. Changes in functionalities at various other carbons in iridoid and secoiridoid skeletons are also found in nature as shown here.

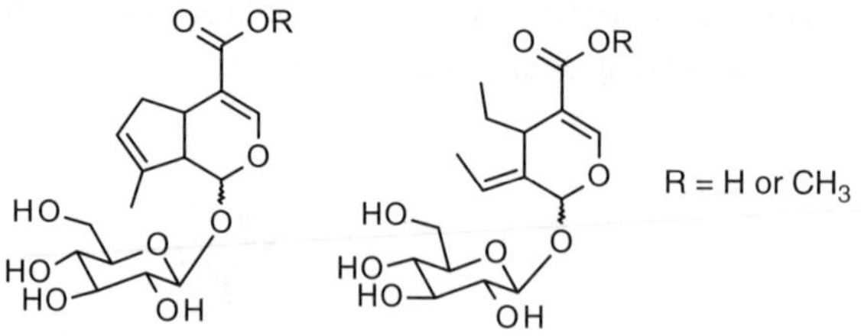

R = H or CH₃

Iridoid and secoiridoid glycosides
with modified functionalities

Some examples of plants that afford iridoid or secoiridoid glycosides and their medicinal uses are summarized here.

8.4.7.1 Devil's Claw (*Harpagophytum procumbens*)

Harpagophytum procumbens of the family Pedaliaceae is native to South Africa, Namibia and Madagascar, and traditionally used in the treatment of osteoarthritis, rheumatoid arthritis, indigestion and low back pain. This plant contains 0.5–3% iridoid glycosides, harpagoside, harpagide and procumbine being the major active iridoid glycosides present.

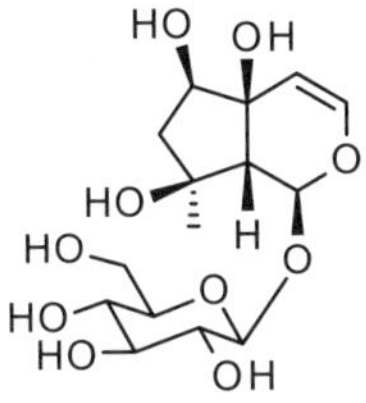

Harpagide 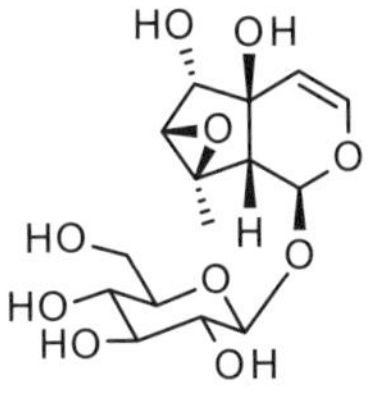   Procumbine

The toxicity of *H. procumbens* is considered to be extremely low. To date, there have been no reported side effects following its use. However, this plant is said to have oxytocic properties and should be avoided in pregnancy. In addition, due to its reflex effect on the digestive system, it should be avoided in patients with gastric or duodenal ulcers.

8.4.7.2 Picrorhiza (*Picrorhiza kurroa*)

Picrorhiza kurroa is a small perennial herb that grows in hilly parts of India, particularly in the Himalayas between 3000 and 5000 m. It belongs to the family Plantaginaceae. The bitter rhizomes of this plant have been used for thousands of years in the *Ayurvedic traditional medicine* to treat indigestion, dyspepsia, constipation, liver dysfunction, bronchial problems and fever.

It is, in combination with various metals, useful in the treatment of acute viral hepatitis. The active constituents of Picrorhiza are a group of iridoid glycosides known as picrosides I–IV and kutkoside.

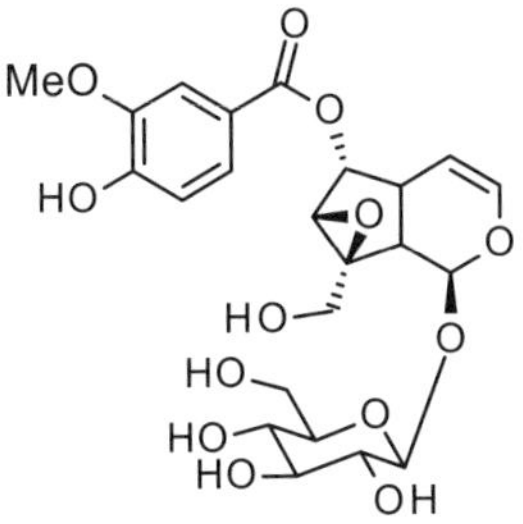

Picroside II 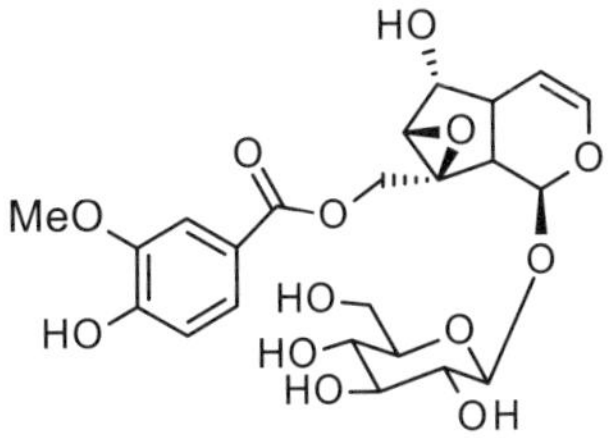   Kutkoside

Picrorhiza has been used widely in India, and no significant adverse reactions have been reported to date.

The oral LD$_{50}$ of *Picrorhiza* iridoid glycosides (known as 'kutkin') is greater than 2600 mg kg^{-1} in rats.

8.4.7.3 Oleuropein

Fraxinus excelsior (ash tree), *Olea europaea* (olive tree) and *Ligustrum obtusifolium* from the family Oleaceae are the major sources of oleuropein. It is a secoiridoid glycoside, which has hypotensive, antioxidant, antiviral and antimicrobial properties. There is no known toxicity or contraindications for this compound.

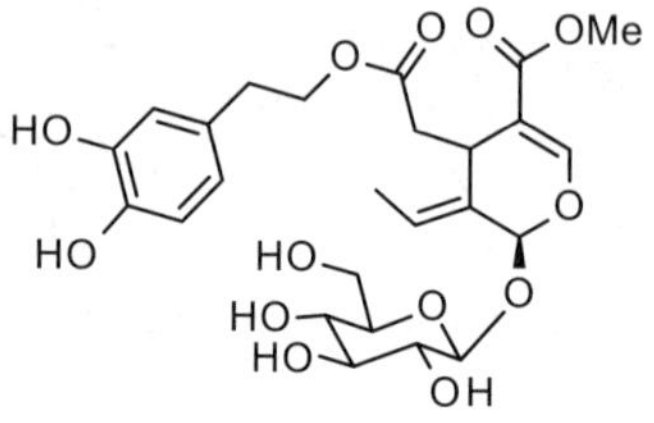

Oleuropein
(A bioactive secoiridoid glycoside)

8.5 TERPENOIDS

Terpenoids, also known as isoprenoids, are a large and structurally diverse group of compounds derived from a combination of two or more *isoprene* units. Isoprene is a five carbon unit, chemically known as 2-methyl-1,3-butadiene. According to the *isoprene rule* proposed by Leopold Ruzicka, terpenoids arise from *head-to-tail* joining of isoprene units. Carbon 1 is called the 'head' and carbon 4 is the 'tail'. For example, myrcene is a simple 10 carbon containing terpenoid formed from the head-to-tail union of two isoprene units as follows.

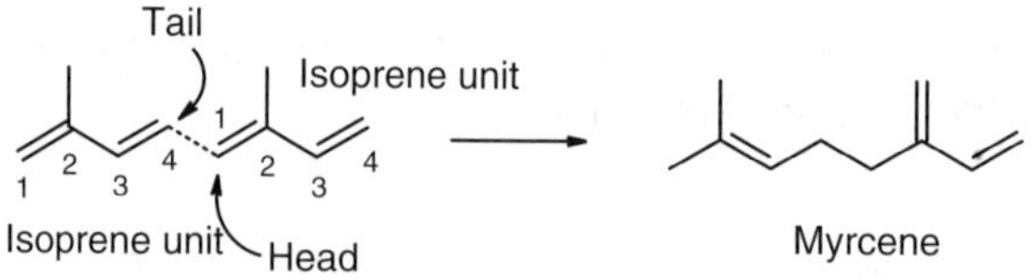

Terpenoids are found in all parts of higher plants and occur in mosses, liverworts, algae and lichens. Terpenoids of insect and microbial origins have also been found. About 60% of known natural products are terpenoids and many of them possess various pharmacological properties.

8.5.1 Classification

Terpenoids are classified according to the number of isoprene units involved in the formation of these compounds.

Type of terpenoids	Number of carbon atoms	Number of isoprene units	Example
Monoterpene	10	2	Limonene
Sesquiterpene	15	3	Artemisinin
Diterpene	20	4	Forskolin
Triterpene	30	6	α-Amyrin
Tetraterpene	40	8	β-Carotene
Polymeric terpenoid	Several	Several	Rubber

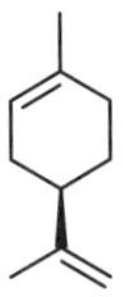

(+)-Limonene
(A monoterpene)

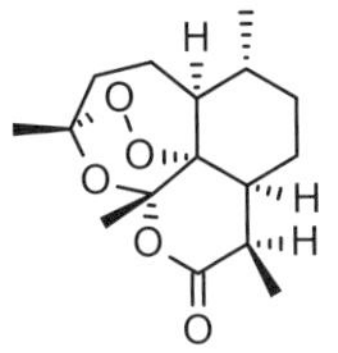

Artemisinin
(An antimalarial sesquiterpene)

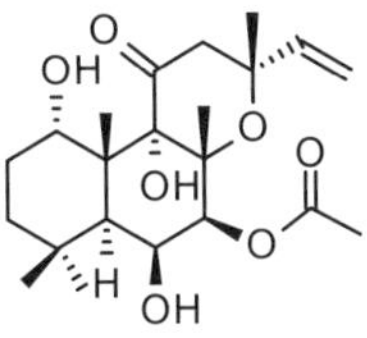

Forskolin
(An antihypertensive diterpene)

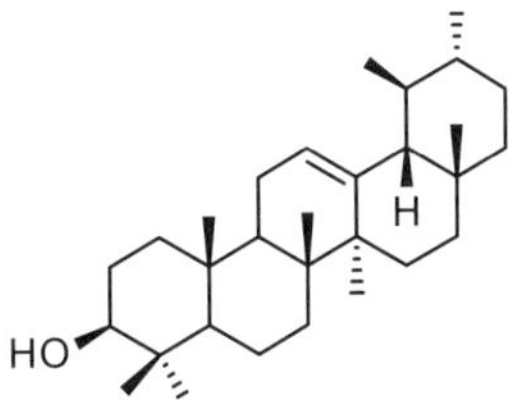

α-Amyrin
(A pentacyclic triterpene)

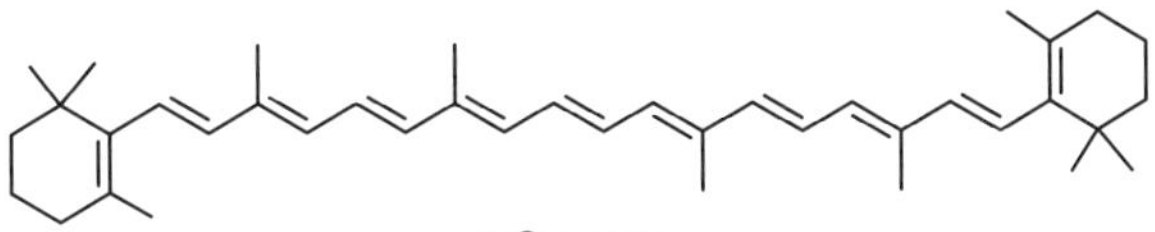

β-Carotene
(A tetraterpene)

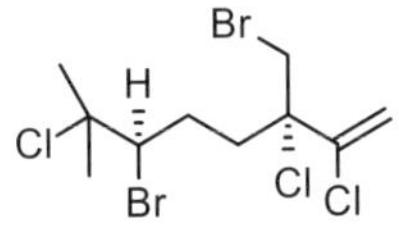

Halomon
(A polyhalogenated monoterpene form
the marine red algae *Portieria hornemannii*)

8.5.2 Biosynthesis of Terpenoids

3R-(+)-Mevalonic acid is the precursor of almost all terpenoids. The enzymes mevalonate kinase and phosphomevalonate kinase catalyse phosphorylation of mevalonic acid to yield 3R-(+)-mevalonic acid 5-diphosphate, which is finally transformed

to isopentenyl diphosphate, also known as isopentenyl pyrophosphate (IPP) by the elimination of a carboxyl and a hydroxyl group mediated by mevalonate 5-diphosphate decarboxylase. IPP is isomerized by isopentenyl isomerase to dimethylallyllylpyrophosphate (DMAPP). A unit of IPP and a unit of DMAPP combine together head-to-tail by dimethylallyl transferase to form geranyl pyrophosphate, which is finally hydrolysed to geraniol, a simple monoterpene. Geranyl pyrophosphate is the precursor of all monoterpenes. Elimination of the pyrophosphate group leads to the formation of acyclic monoterpenes; for example, ocimene and myrcene. Hydrolysis of the phosphate groups leads to the prototypical acyclic monoterpenoid geraniol. Additional rearrangements and oxidations provide compounds such as citral, citronellal, citronellol, linalool and many others. Many monoterpenes found in marine organisms are halogenated; for example, halomon.

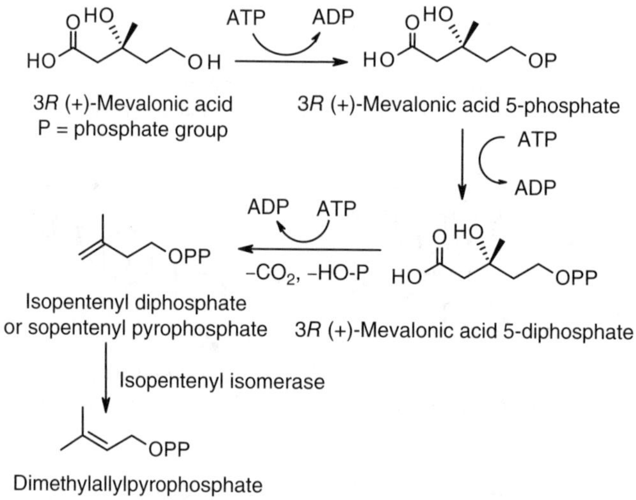

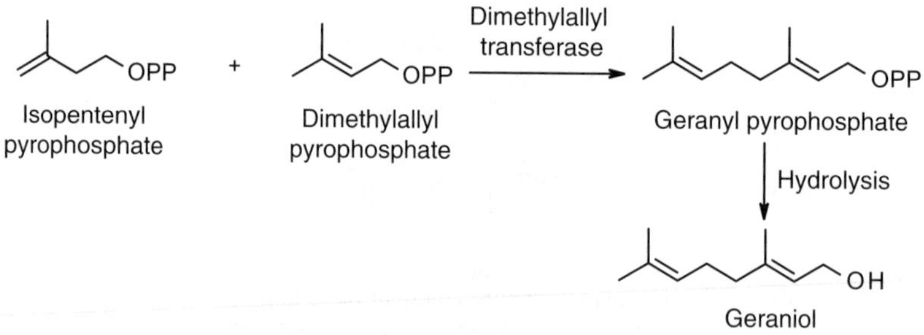

In similar fashions, the core pathway up to C_{25} compounds (five isoprene units) is formed by sequential addition of C_5-moieties derived from IPP to a starter unit derived from DMAPP. Thus, sesquiterpenes are formed form the precursor 2E, 6E-farnesyl pyrophosphate (FPP), and diterpenes from 2E, 6E, 10E-geranylgeranyl pyrophosphate (GGPP). The parents of triterpenes and tetraterpenes are formed by reductive coupling of two FPP or GGPP, respectively. Rubbers and other poly-isoprenoids derive from repeated additions of C_5 units to the starter unit GGPP.

There is also another biosynthetic pathway, discovered in the late 1908s, for certain terpenoids, which is known as the 2-C-methyl-D-erythriotol 4-phosphate/1-deoxy-D-xylulose 5-phosphate pathway (MEP/DOXP pathway) or simply the non-mevalonate pathway. Pyruvate and glyceraldehyde 3-phosphate are converted by DOXP synthase to 1-deoxy-D-xylulose 5-phosphate (DOXP), and by DOXP reductase to 2-C-methyl-D-erythritol 4-phosphate (MEP). The subsequent three reaction steps, catalysed by 4-diphosphocytidyl-2-C-methyl-D-erythritol synthase, 4-diphosphocytidyl-2-C-methyl-D-erythritol kinase and 2-C-methyl-D-erythritol 2,4-cyclodiphosphate synthase, govern the formation of 2-C-methyl-D-erythritol 2,4-cyclopyrophosphate, which is finally converted to (E)-4-hydroxy-3-methyl-but-2-enyl pyrophosphate (HMB-PP) by HMB-PP synthase. HMB-PP reductase converts HMB-PP to IPP and DMAPP, which are the end products of both pathways and are the precursors of isoprene and subsequent terpenoids.

8.5.3 Monoterpenes

Monoterpenes, 10 carbon containing terpenoids, are composed of two isoprene units and found abundantly in plants; for example, (+)-limonene from lemon oil and (–)-linalool from rose oil. Monoterpenes could be linear (acyclic) or could contain rings. Biochemical modifications such as oxidation or rearrangement afford other functionalized monoterpenoids. Many monoterpenes are the constituents of plant volatile oils or essential oils. These compounds are particularly important as flavouring agents in pharmaceutical, confectionary and perfume products. However, a number of monoterpenes show various types of bioactivity and are used in medicinal preparations. For example, camphor is used in liniments against rheumatic pain, menthol is used in ointments and liniments as a remedy against itching, bitter-orange peel is used as an aromatic bitter tonic and as a remedy for poor appetite, and thymol and carvacrol are used in bactericidal preparations.

8.5.3.1 Types of Monoterpenes

Monoterpenes occur in plants in different structural forms, some are cyclic while the others are acyclic. They may also contain various functional groups and depending on their functional groups they can be classified as simple hydrocarbons, alcohols, ketones, aldehydes, acids or phenols. Some examples are cited here.

Geraniol
(An acyclic monoterpene)

(+)-α-Pinene
(A cyclic monoterpene)

(–)-Menthol
(A monoterpene alcohol)

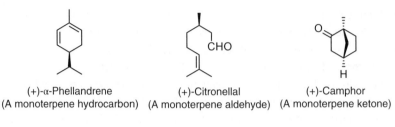

(+)-α-Phellandrene	(+)-Citronellal	(+)-Camphor
(A monoterpene hydrocarbon)	(A monoterpene aldehyde)	(A monoterpene ketone)

(+)-*trans*-Chrysanthemic acid	Carvacrol
(A monoterpene acid)	(A phenolic sesquiterpene)

8.5.3.2 Botanical Sources

A number of plants biosynthesize a variety of monoterpenes. The following table lists just a few of these sources and their major monoterpene components.

Source		Major monoterpenes
Common name	Botanical name (Family)	
Black pepper	*Piper nigrum* (Piperaceae)	α- and β-Pinene, phellandrene
Peppermint leaf	*Mentha piperita* (Lamiaceae)	Menthol and menthone
Oil of rose	*Rosa centifolia* (Rosaceae)	Geraniol, citronellol and linalool
Cardamom	*Elettaria cardamomum* (Zingiberaceae)	α-Terpineol and α-terpinene
Rosemary	*Rosmarinus officinalis* (Lamiaceae)	Borneol, camphene and cineole
Bitter orange	*Citrus aurantium* (Rutaceae)	Geranial and (+)-limonene
Camphor	*Cinnamomum camphora* (Lauraceae)	(+)-Camphor
Caraway	*Carum carvi* (Apiaceae)	(+)-Carvone and (+)-limonene
Thyme	*Thymus vulgaris* (Lamiaceae)	Carvacrol and thymol

8.5.4 Sesquiterpenes

Sesquiterpenes, 15-carbon-containing terpenoids, are composed of three isoprene units and found abundantly in plants; for example, artemisinin from *A. annua* and (–)-α-bisabolol from *Matricaria recutita* (German chamomile). Sesquiterpenes, found naturally in plants and insects as semiochemicals (defensive agents or pheromones), can be either cyclic or acyclic. Addition of IPP to GPP provides 2*E*, 6*E*-FPP, the precursor for all sesquiterpenes. Farnesylpyrophosphate can cyclize by various cyclase enzymes in various ways leading to the production of a variety of sesquiterpenes. Some of these sesquiterpenes are medicinally important bioactive compounds. For example, (–)-α-bisabolol and its derivatives are potent anti-inflammatory and spasmolytic agents, and artemisinin is an antimalarial drug.

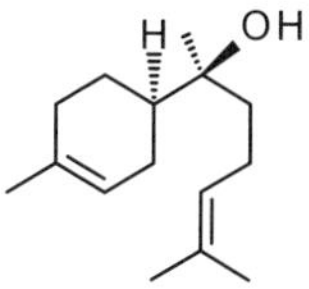

(–)-α-Bisabolol

8.5.4.1 Structural Types of Sesquiterpenes

Sesquiterpenes can be of various structural types, some of which are presented with specific examples in the following table.

Major structural types	Specific examples	
Acoranes	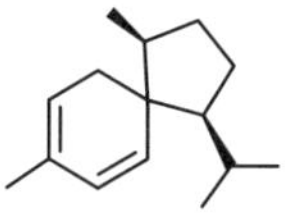 3,5-Acoradiene	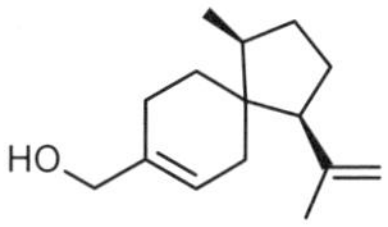 3,11-Acoradien-15-ol
Africananes	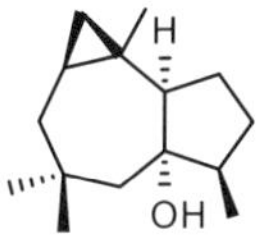 2-Africananol	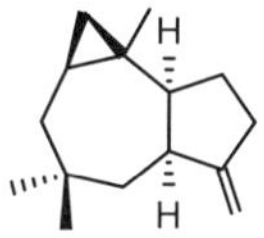 3(15)-Africanene
Alliacanes	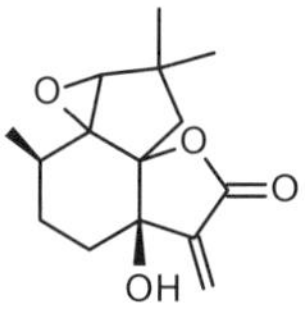 Alliacol A	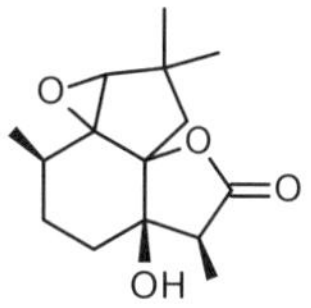 Alliacolide
Aristolanes	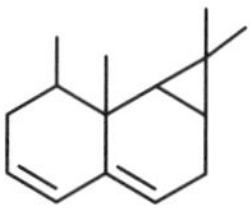 1,9-Aristoladiene	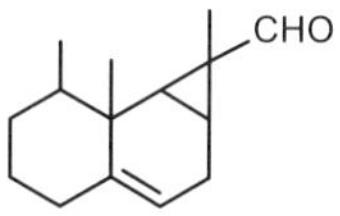 9-Aristolen-12-al
Aromadendranes	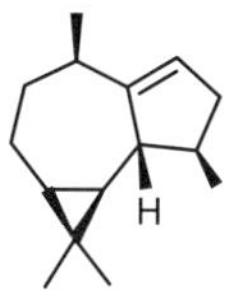 1-Aromadendrene	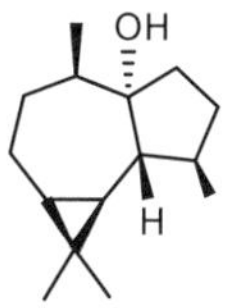 1-Aromadendranol

Major structural types	Specific examples

Bisabolanes

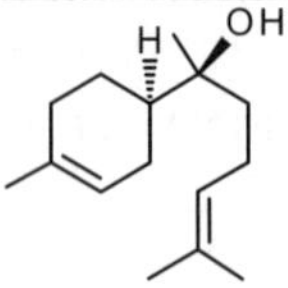

(−)-α-Bisabolol   (−)-β-Bisabolene

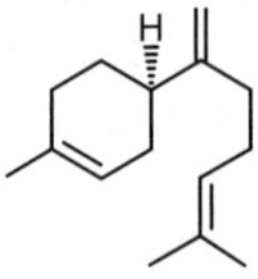

Botrydials

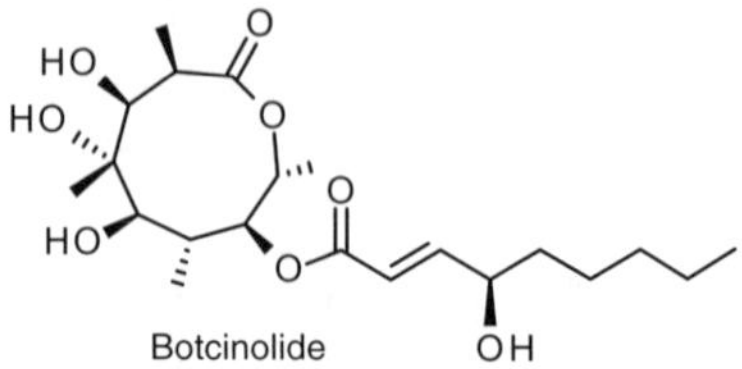

Botcinolide

Cacalols

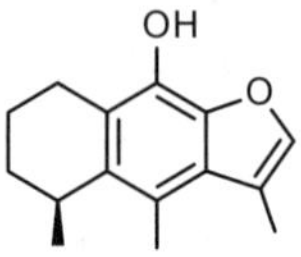

Cacalol   Cacalolide

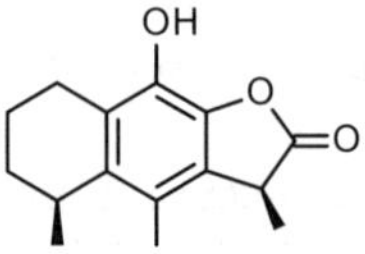

Cadinanes

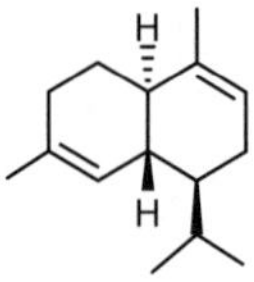

α-Cadinene   Artemisinic acid

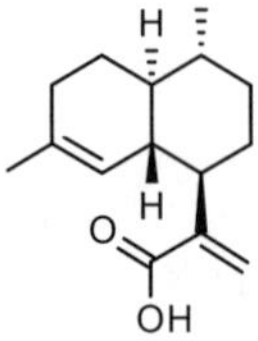

Campherenanes, α-santalanes and β-santalanes

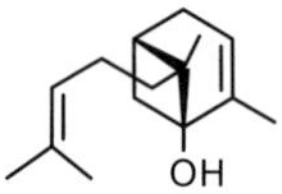

α-Bergamotene   α-Santalene

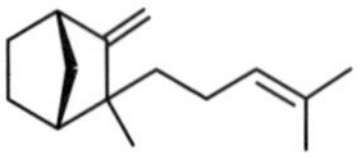

Carabranes

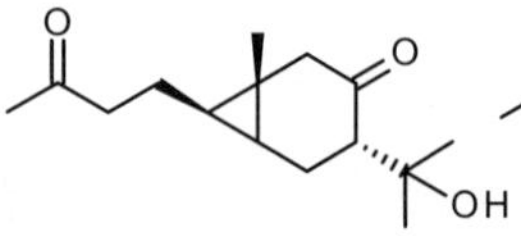

Curcurabranol A   Curcumenone

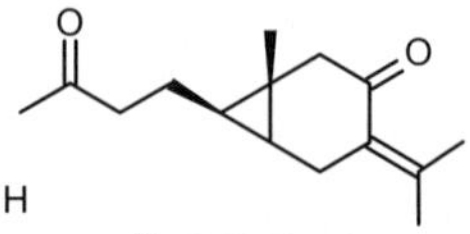

Caryophyllanes

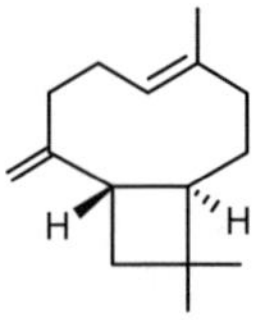

β-Caryophyllene
(A bioactive component of
Cinnamomum zeylanicum)

Major structural types	Specific examples

Cedranes and isocedranes

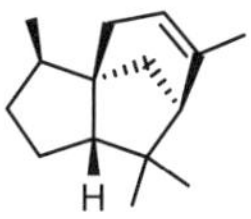

α-Cedrene

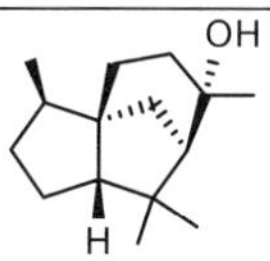

Cedrol

Chamigranes

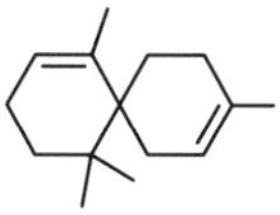

2,7-Chamigradiene

Majusculone

Copaanes

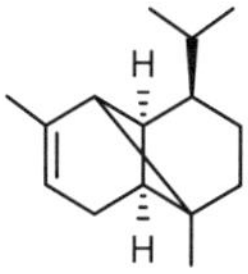

3-Copaene

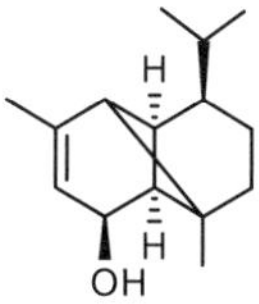

3-Copaen-2-ol

Cuparanes and cyclolauranes

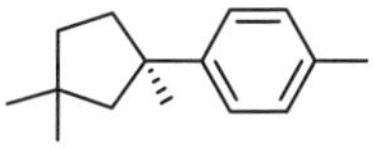

Cuparene

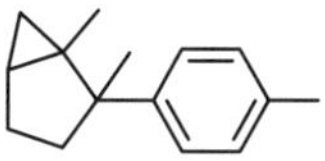

Cyclolaurene

Cyclobutanes and cyclopentanes

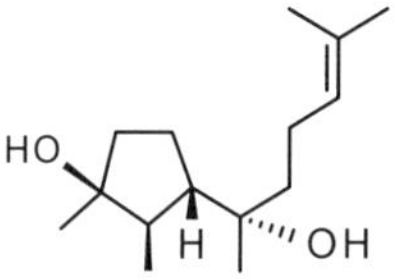

Cyclonerodiol
(A fungal metabolite)

Cyclobisabolanes

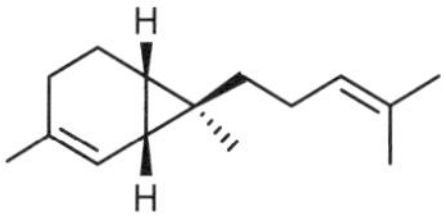

Sesquicaren

Cycloeudesmanes

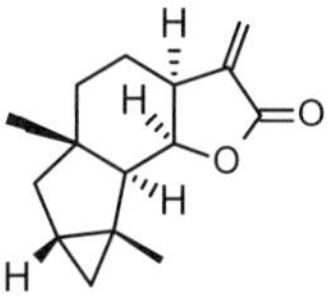

Brothenolide

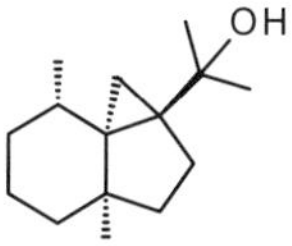

Cycloeudesmol

Cyclofarnesanes

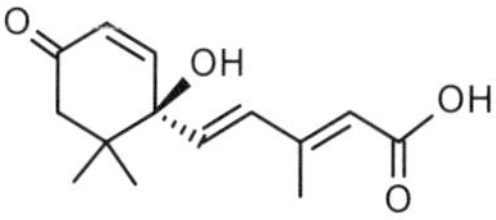

Abscisic acid
(A plant-growth regulatory agent)

Major structural types	Specific examples

Daucanes and isodaucanes

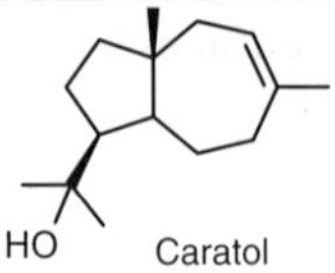

HO   Caratol

Drimanes

CHO
8-Drimen-11-al

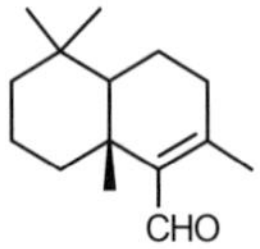

MeO

O   OH

O   OMe

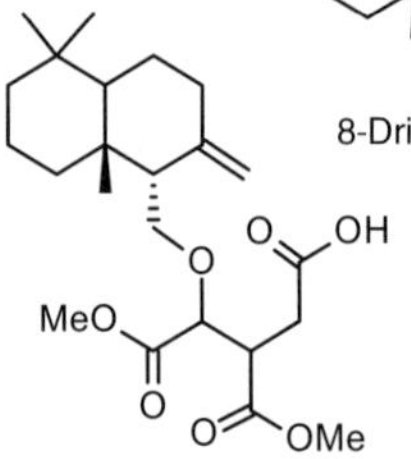

Cryptoporic acid A   Marasmene

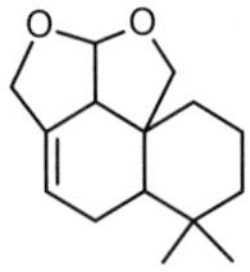

Elemanes

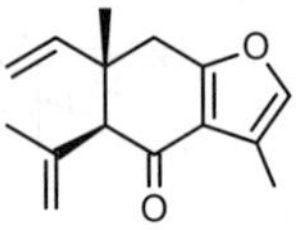

Curzerenone   β-Elemene
(A component of turmaric)   (A component of turmaric)

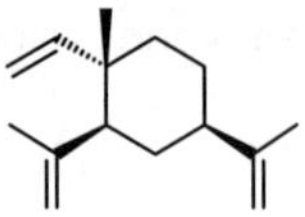

Emmotins

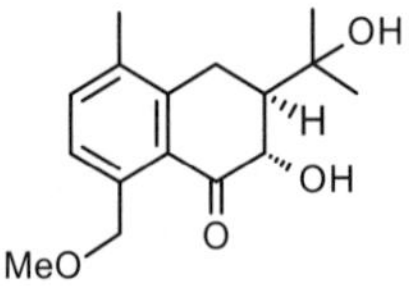

OH   OH

H   H
OH   OH

MeO

Emmotin A   Emmotin F

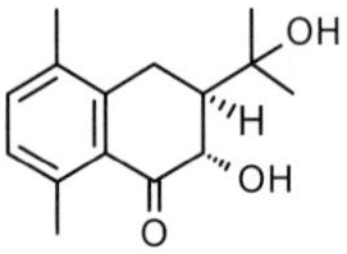

Eremophilanes

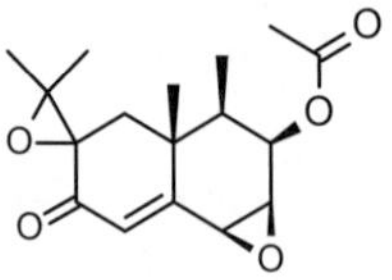

O   HO
O   OAc
O   O
O   O

Eremofortin A   Eremofortin C

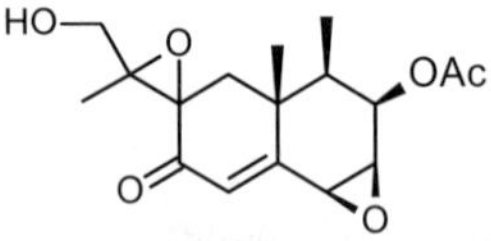

Eudesmanes

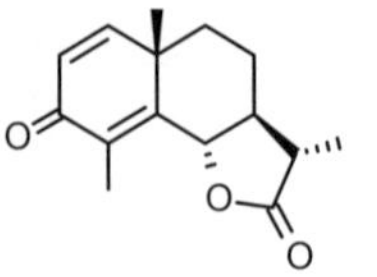

O
O
O

α-Santonin   β-Eudesmol
(An anthelmintic component of
Artemisia species)

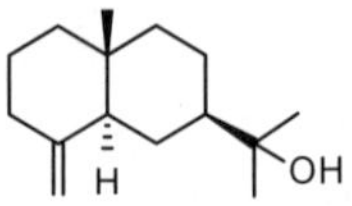

Major structural types	Specific examples

Furanoid farnesanes

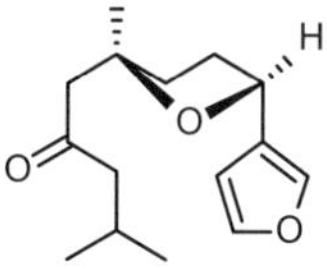

Ipomeamarone
(A phytoalexin)

Germacranes

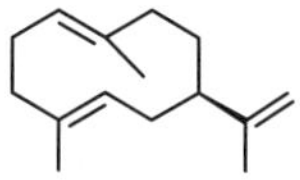

Germacrene A   Germacrene D   Parthenolide

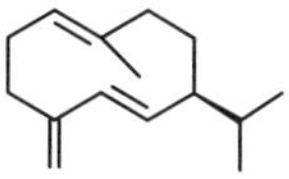

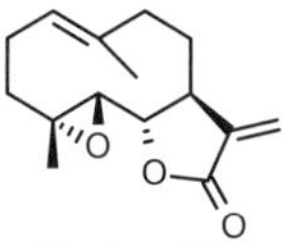

Gymnomitranes

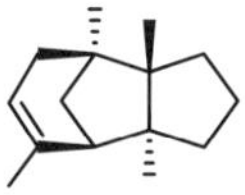

3-Gymnomitrene   3-Gymnomitren-15-al

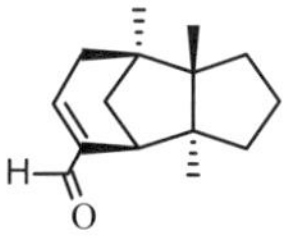

Guaianes

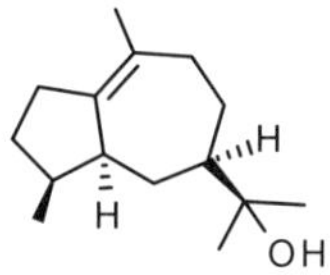

Bulnesol   Matricin

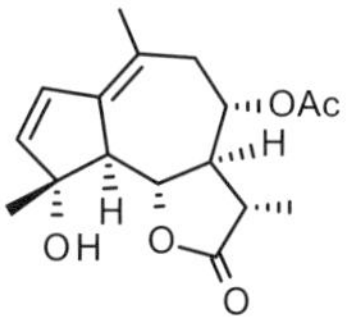

Hirsutanes and rear-ranged hisutanes

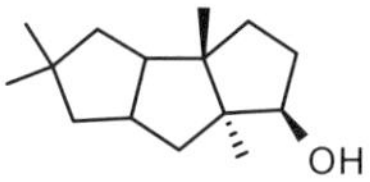

Ceratopicanol   5α, 7β, 9α-Hirsutanetriol

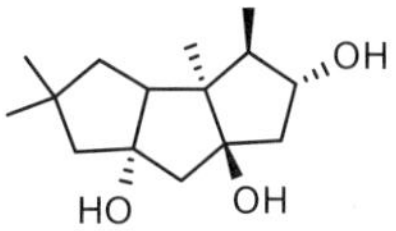

Humulanes

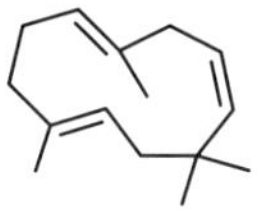

Humulene
(A sesquiterpene from hops)

Illudalanes

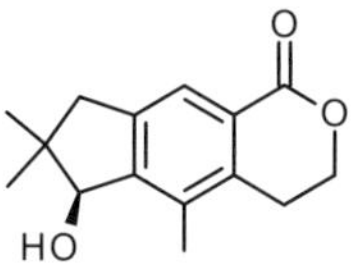

Calomelanolactone   Candicansol

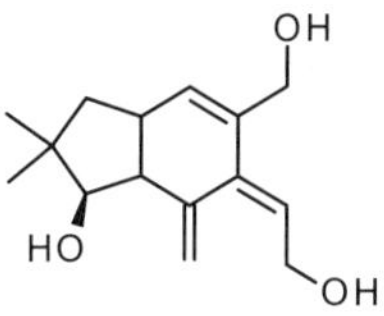

Major structural types	Specific examples

Illudanes and protoilludanes

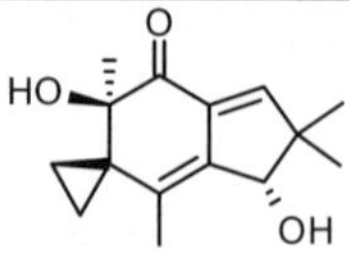

Illudin M
(A component of luminescent mushrooms)

Isolactaranes, merulanes, lactaranes and marasmanes

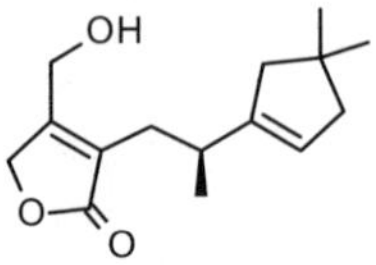

Blennin C   Lacterorufin C

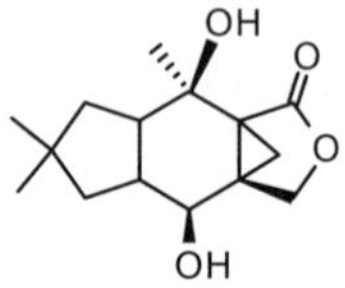

Lauranes

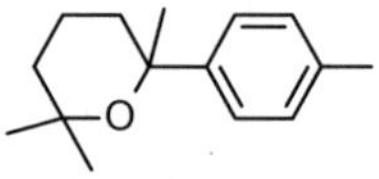

Curcumene ether   Aplysin

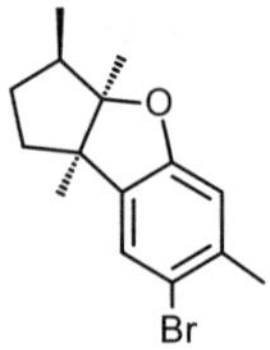

Lepidozanes and bicyclo-germacranes

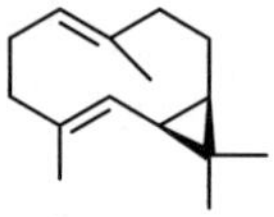

Bicyclogermacrene

Lippifolianes and himachalanes

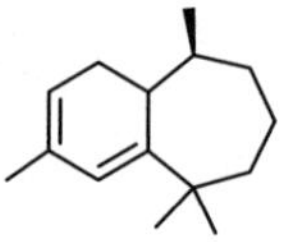

] 1,3-Himachaladiene   2-Himachalen-6-ol

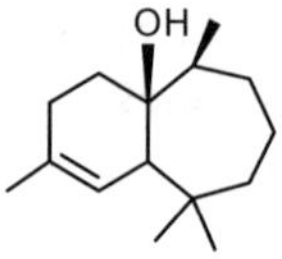

Longifolanes

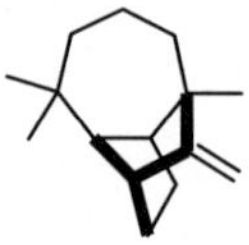

(+)-Longifolene

Longipinanes

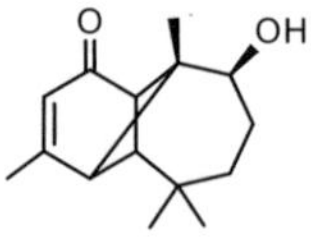

8β-Hydroxy-3-longipinen-5-one   3-Longipinene

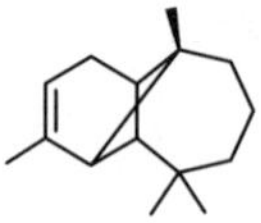

Major structural types	Specific examples

Nardosinanes

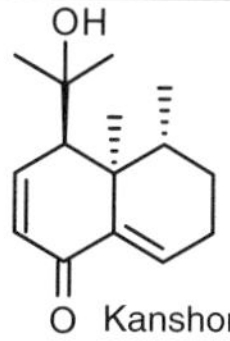

O Kanshone A   Lemnacarnol

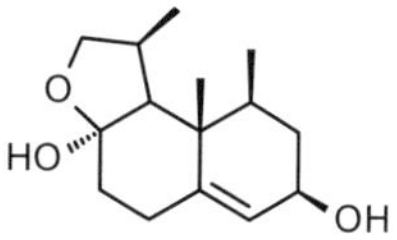

Oplopanes

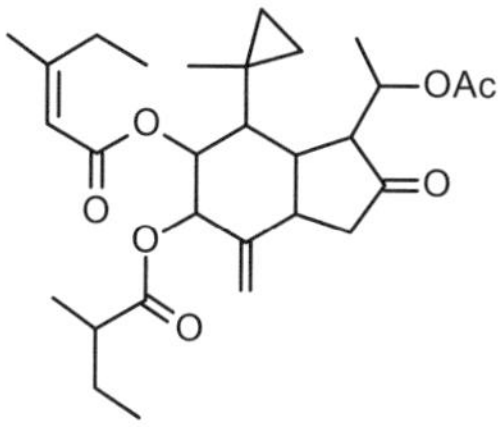

Abrotanifolone   10-Hydroxy-4-oplopanone

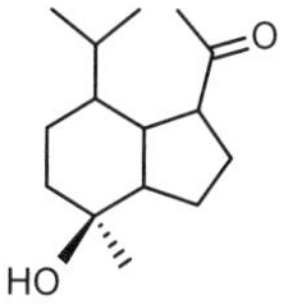

Oppositanes

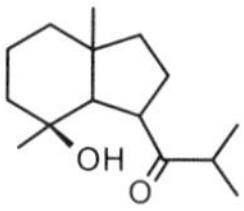

4-Hydroxy-7-oppositanone   1β,4β,11-Oppositanetriol

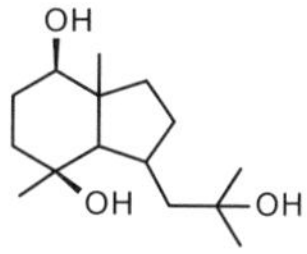

Patchoulanes and rearranged patchoulanes

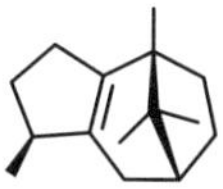

β-Patchoulene

Pentalenanes

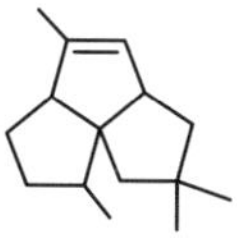

Pentalenene   Pentalenic acid

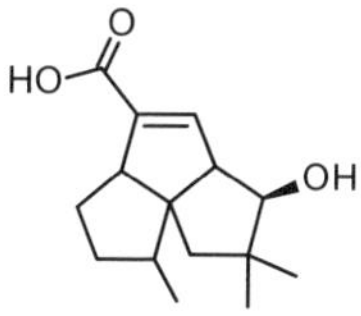

Picrotoxanes

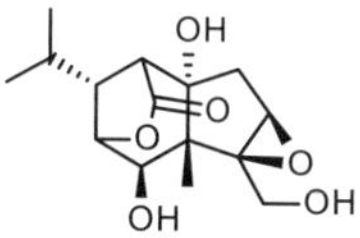

Amoenin   Nobiline

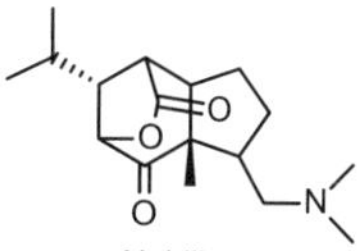

Pinguisanes

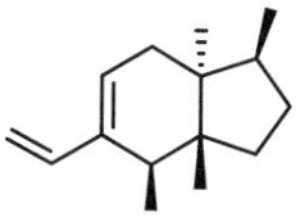

5,10-Pinguisadiene   Pinguisanin

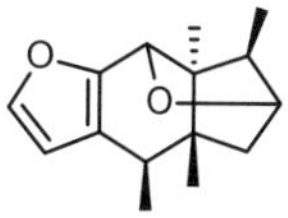

Major structural types	Specific examples

Precapnellanes and capnellanes

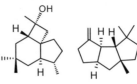

Viridianol   9,12-Capnellene

Simple farnesanes (acyclic)

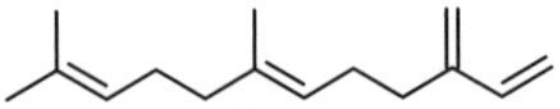

trans-β-Farnesene
(A potent aphid-repellant found in
hops and sweet potatoes)

Spirovetivanes

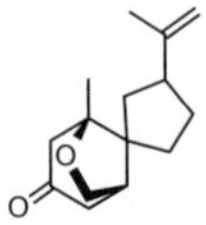

Cyclodehydroisolubimin   1(10),7(11)-Spirovetivadien-2-one

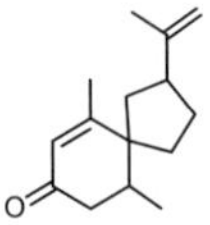

Sterpuranes

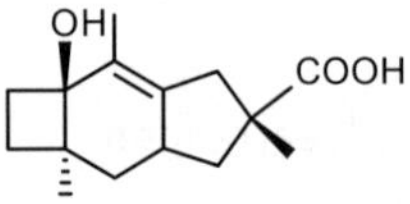

6-Hydroxy-6-sterpuren-12-oic acid

Trichothecanes

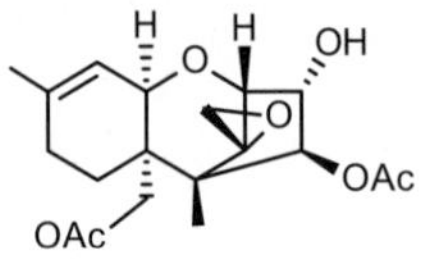

Diacetoxyscirpenol
(A fungal toxin)

Valerenanes

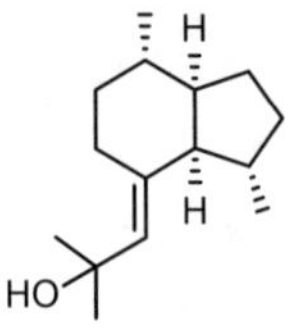

6-Valerenen-11-ol   Valerenic acid

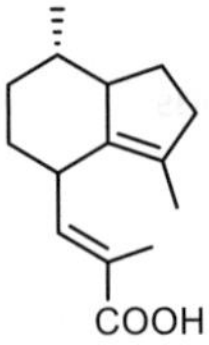

Xanthanes

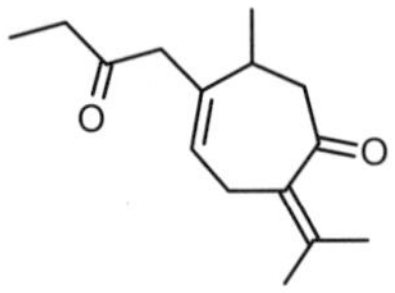

Curmadione
(A component of turmaric)

8.5.4.2 Botanical Sources

Plants produce a variety of sesquiterpenes. The following table lists just a few of these sources and their major sesquiterpene components.

Source		Major sesquiterpenes
Common name	**Botanical name (Family)**	
German chamomile	*Matricaria recutita* (Asteraceae)	(−)-α-Bisabolol and its derivatives
Feverfew	*Tanacetum parthenium* (Asteraceae)	Farnesene, germacrene D, parthenolide
Qinghao	*Artemisia annua* (Asteraceae)	Artemisinin and its derivatives
Holy thistle	*Cnicus benedictus* (Asteraceae)	Cnicin
Cinnamon	*Cinnamomum zeylanicum* (Lauraceae)	β-Caryophyllene
Cloves	*Syzygium aromaticum* (Myrtaceae)	β-Caryophyllene
Hop	*Humulus lupulus* (Cannabaceae)	Humulene
Wormseed	*Artemisia cinia* (Asteraceae)	α-Santonin
Valerian	*Valeriana officinalis* (Valerianaceae)	Valeranone
Juniper berries	*Juniperus communis* (Cupressaceae)	α-Cadinene
Curcuma or Turmeric	*Cucuma longa* (Zingiberaceae)	Curcumenone, curcumabranol A, curcumabranol B, β-elemene, curzerenone

8.5.5 Diterpenes

The *diterpenoids* constitute a large group of 20-carbon-containing compounds derived from 2E,6E,10E-GGPP or its allylic geranyl linaloyl isomer through condensation of IPP with 2E,6E-FPP. They are found in higher plants, fungi, insects and marine organisms. Diterpenes form the basis for biologically important compounds such as retinol, retinal and phytol. They are known to be antimicrobial and anti-inflammatory. One of the simplest and most significant of the diterpenes is phytol, a reduced form of geranylgeraniol, which constitutes the lipophilic side-chain of the chlorophylls. Phytol also forms a part of vitamin E (tocopherols) and vitamin K molecules. Vitamin A is also a 20-carbon-containing compound, and can be regarded as a diterpene. However, vitamin A is formed from a cleavage of a tetraterpene

Among the medicinally important diterpenes, paclitaxel, isolated from *T. brevifolia* (family: Taxaceae), is one of the most successful anticancer drugs of modern time. The anticancer drug taxol, used in therapy against ovarian, breast and lung cancer, along with its synthetic water-soluble analogue taxotere is an example of unusual structure discovered from nature and used as a medicine. Cembrene is a type of macrocyclic diterpene consisting of four isoprene units bonded 'head to tail'. It was first found in the oleoresin secreted from the trunk of a pine tree. All cembranoids feature a 14-carbon-macrocyclic skeleton. Cembranoids are primarily distributed in plants of the *Nicotiana* and *Pinus* genera, as well as in some marine organisms (soft coral). About 90 naturally occurring cembranoids have been isolated from tobacco. Abietic acid is an irritant compound present in pine wood and resin. Tanshinones are abietane diterpenes, isolated mainly from *Salvia miltiorrhiza* (Lamiaceae), a plant largely used in TCM for the treatment of cardiovascular and inflammatory diseases. Among them, tanshinone I is an apoptose inductor and displays several anticancerous biological properties. Cannabinoids are a group of diterpenes present in Cannabis (*Cannabis sativa*). All these substances are structurally related to tetrahydrocannabinol (THC) and are able to bind to specific cannabinoid receptors.

Phytol
(A diterpene)

α-Tocopherol
(A member of the vitamin E group)

Vitamin K$_1$
(Contains a diterpenoidal part)

Forskolin
(An antihypertensive agent)

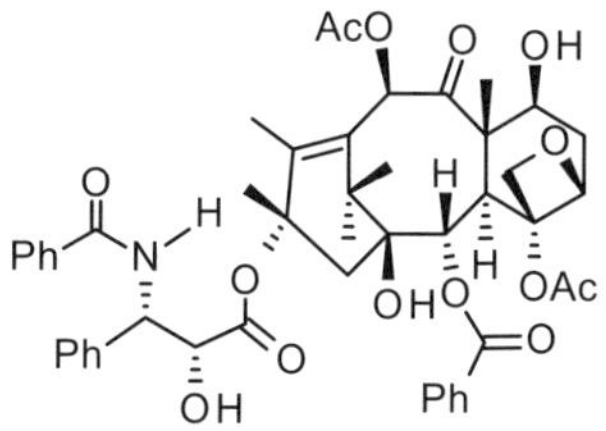

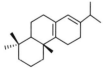

AcO O OH

Ph

Paclitaxel or Taxol
(An anticancer drug)

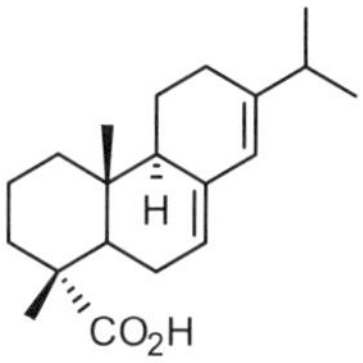

Abietic acid 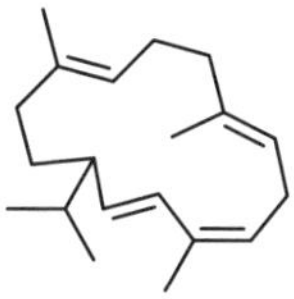   Cambrene 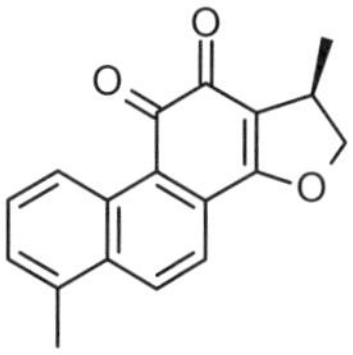   Tashinone I

CO₂H

Promising diterpenes are the ginkgolides, which possess antagonistic activity towards platelet-activating factor, increasing in conditions of shock, burns, ulceration and inflammation. Also used in therapy is the diterpene resiniferatoxin, an ultrapotent vanilloid, isolated from the *Euphorbia resinifera* latex, in clinical trials for bladder hyperreflexia and diabetic neuropathy.

8.5.5.1 Major Structural Types of Diterpenes

While there are a number of acyclic diterpenes like phytol, cyclization of these acyclic diterpenes, driven by various enzymes, leads to the formation of several cyclic diterpenes containing 2–4 rings; for example, cembrene A has one ring, labdanes have two rings, abietanes have three and stemacrenes have four. A number of other biogenetic reactions, for example oxidation, also bring in variations among these cyclic diterpenes. Some of the major structural types encountered in diterpenes are shown here.

Abietanes   **Amphilectanes**

8,13-Abietadiene

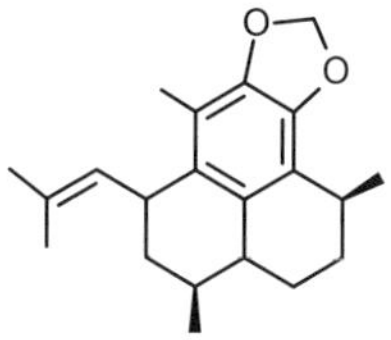

Helioporin E

Beyeranes

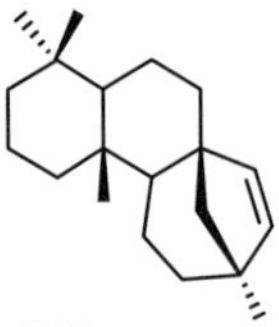

15-Beyerene

Briaranes

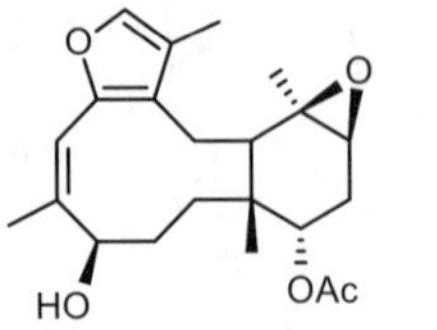

Verecynarmin G

Cassanes

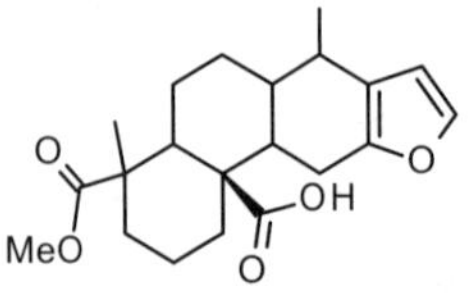

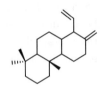

Caesaljapin

Cembranes

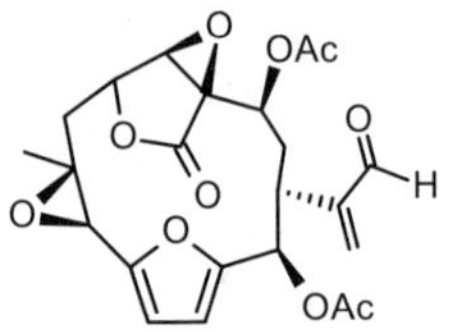

Bippinatin B

Cleistanthanes

13(17),15-Cleistanthadiene

Cyathanes

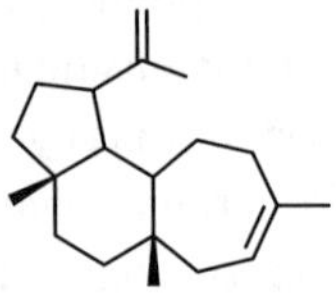

12,18-Cyathadiene

Daphnanes

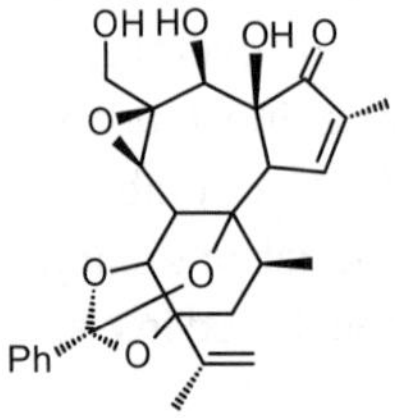

Daphnetoxin

Dolabellanes

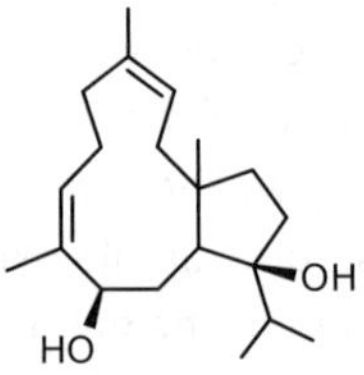

3,7-Dolabelladiene-9,12-diol

Dolastanes

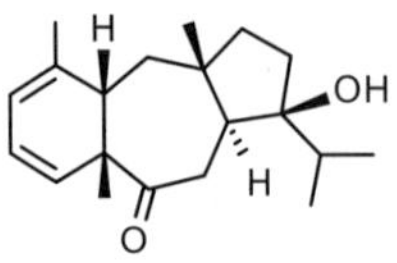

9-Hydroxy-1,3-dolastadien-6-one

Eunicellanes and asbestinanes

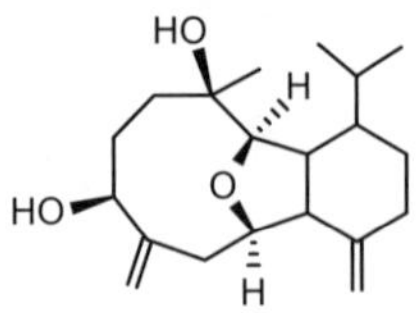

6,13-Epoxy-4(18),8(19)-eunicelladiene-9,12-diol

Fusicoccanes

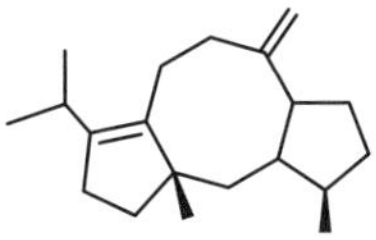

7(17),10(14)-Fusicoccadiene

Gibberellins

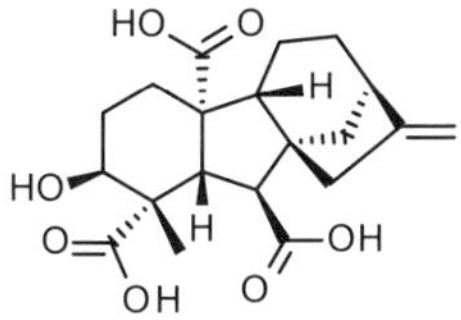

Gibberellin A$_{13}$

Isocopalanes

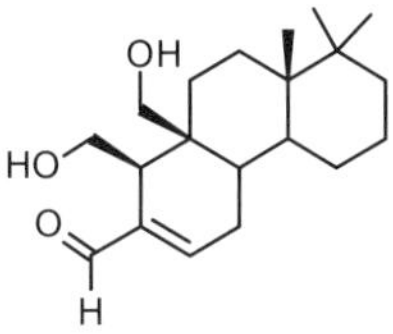

15,17-Dihydroxy-12-isocopalen-16-al

Jatrophanes

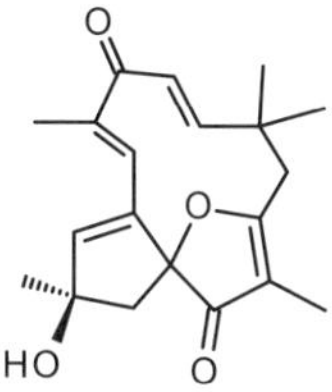

2β-Hydroxyjatrophone

Kauranes and phyllocladanes

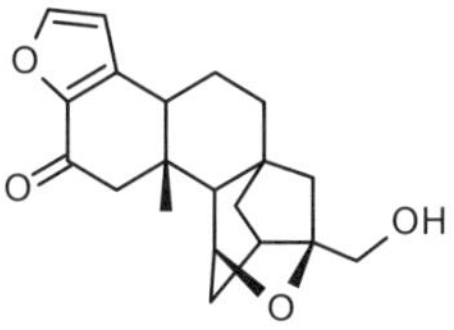

Bengalensol

Labdanes

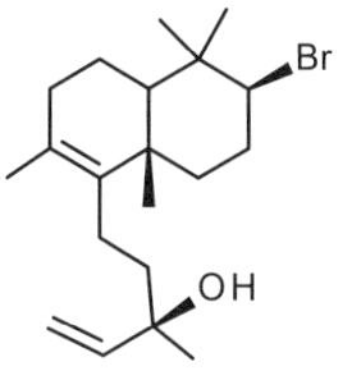

3-Bromo-7,14-labdadien-13-ol

Lathyranes

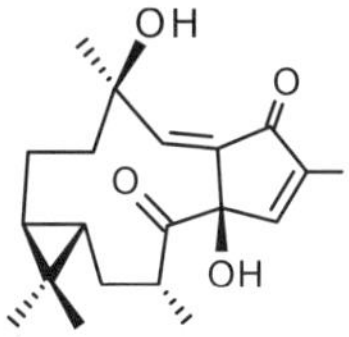

Curculathyrane A

Lobanes

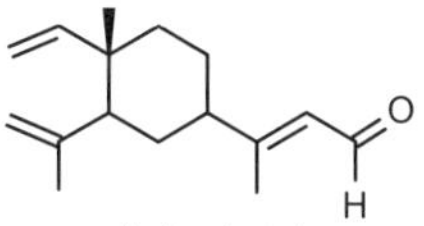

Lobophytal

Pachydictyanes

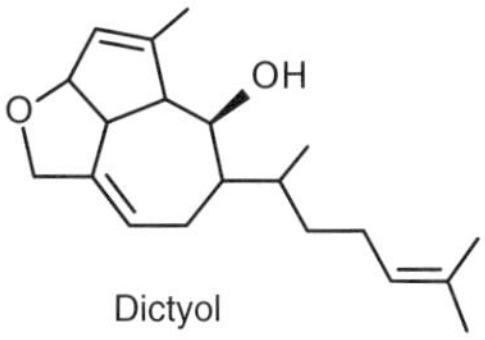

Dictyol

Phytanes

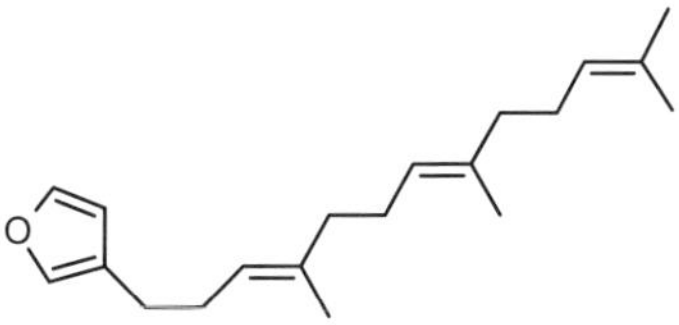

Ambliofuran

Pimaranes

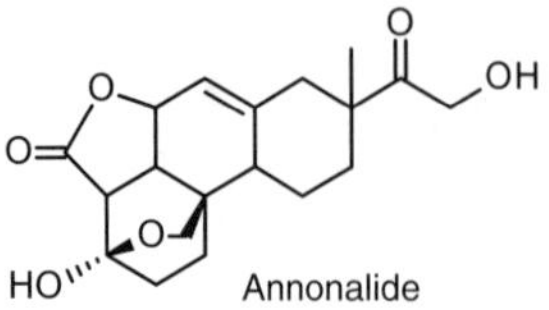

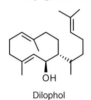

Annonalide

Podocarpanes

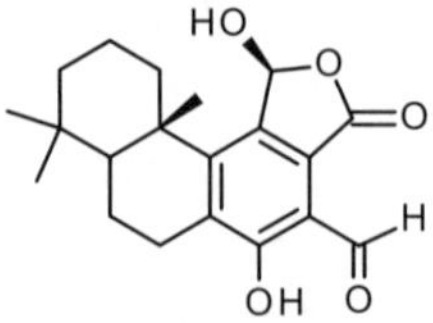

HO

Betolide

Prenylgermacranes

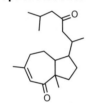

OH

Dilophol

Serrulatanes and bifloranes

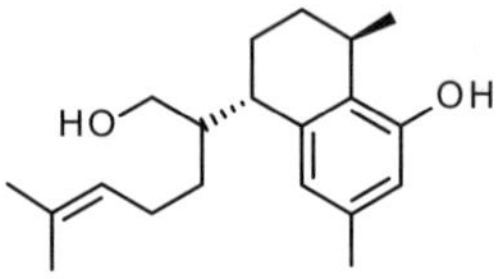

HO

OH

14-Serrulatene-8,18-diol

Sphenolobanes

3-Spenolobene-5,16-dione

Taxanes

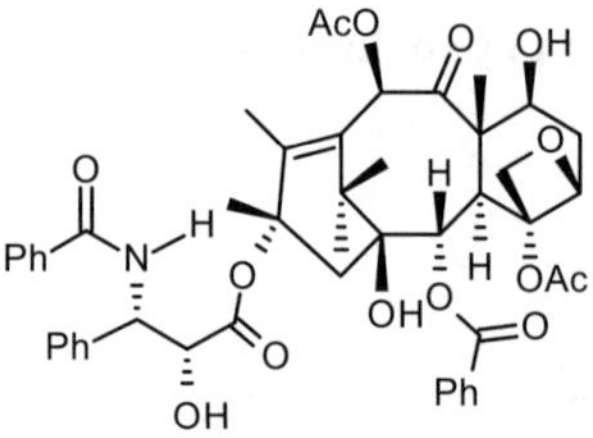

AcO   O   OH

O
Ph   N   H

Ph   O

OH   O
Ph

OH   OAc

OH

Paclitaxel or Taxol

Tiglianes and ingenanes

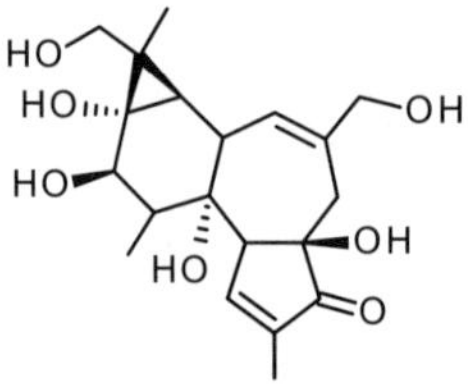

HO
HO,,,,
OH
HO,
OH
HO
O

4β, 9α, 12β, 13α, 16α, 20β-Hexa-
hydroxy-1,6-tigliadien-3-one

Verrucosanes

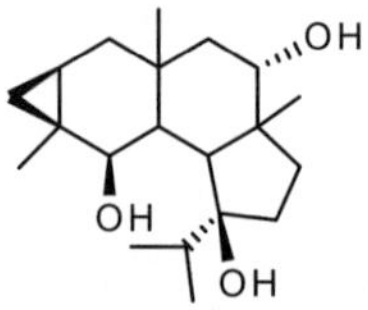

OH

OH
OH

2β, 9α, 13β-Verrucosanetriol

Xenicanes and xeniaphyllanes

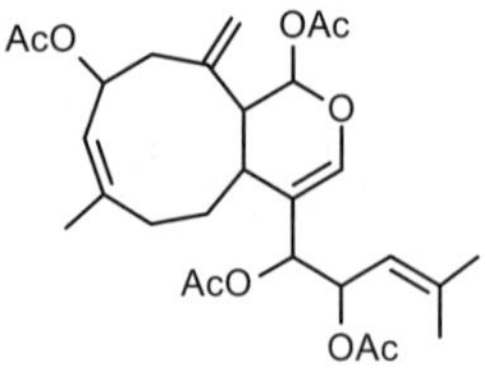

AcO   OAc
O

AcO
OAc

Xenicin

8.5.5.2 Botanical Sources

Diterpenes are found in nature, mainly in plants, but also in other natural sources; for example, microorganisms, marine organisms and insects. There are over 2500 diterpenes reported to date, and almost all plant families biosynthesize various diterpenes. Diterpenes gibberellins are plant hormones and phytol occurs as a side chain on chlorophyll; both these diterpenes are widely distributed in the plant kingdom. Manool, sclareol and labdane derivatives are common in resins of the Pinaceae. Highly oxygenated derivatives occur in the families, Asteraceae, Lamiaceae and Verbenaceae, and in the genus *Croton* of the Euphorbiaceae. *Ent*-Kaurene diterpenes, enmeins are found in the Lamiaceae, gibberellic acids (found in all plants and in some fungi) and grayanotoxins, extremely poisonous compounds in the family Ericaceae. These last compounds are also common in *Rhododendron* species. This precursor is also the source of stevioside, a sweetener from the Asteraceae plant *Stevia rebaudiana*. The following table presents just a few of these sources and their major diterpenoidal components.

Source		Major diterpenes
Common name	**Botanical name (Family)**	
Yew tree	*Taxus brevifolia* (Taxaceae)	Paclitaxol
Sawada fungus	*Gibberella fujikuroi*	Gibberellins
Coleus	*Coleus forskohlii* (Lamiaceae)	Forskolin
Stevia	*Stevia rebaudiana* (Asteraceae)	Stevioside
Ginkgo	*Ginkgo biloba* (Ginkgoaceae)	Ginkgolides

8.5.6 Triterpenes

The *triterpenoids*, encompass a large and diverse group of naturally occurring 30-carbon-atom-containing compounds derived from squalene or, in the case of 3β-hydroxytriterpenoids, the 3S-isomer of squalene 2,3-epoxide. Two molecules of farnesyl pyrophosphate are joined tail-to-tail to yield squalene. The conformation that *all-trans*-squalene 2,3-epoxide adopts, when the initial cyclization takes place, determines the stereochemistry of the ring junctions in the resulting triterpenoids. The initially formed cation intermediate may undergo a series of 1,2-hydride and methyl migrations, commonly called backbone rearrangements, to provide a variety of skeletal types.

A number of triterpenoids are bioactive compounds and used in medicine. For example, fusidic acid is an antimicrobial fungal metabolite, isolated from *Fusidium coccineum*, and cytotoxic dimeric triterpenoids, crellastatins, isolated from marine sponge *Crella* species.

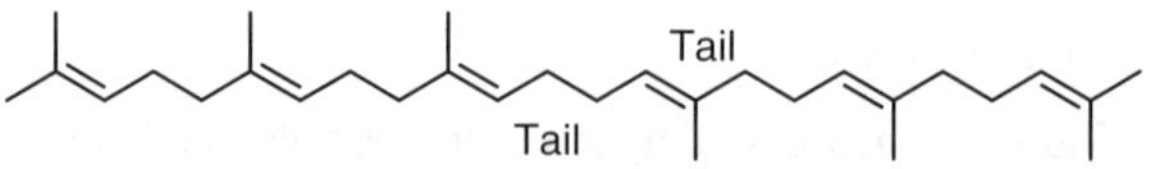

Squalene
(Formed from tail-to-tail combination of 2 FPP)

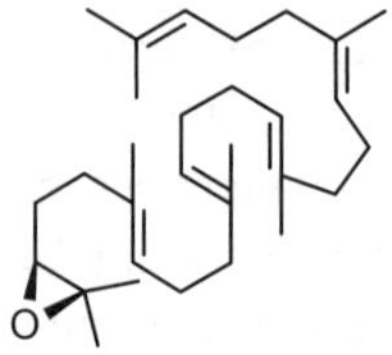

Squalene 2,3-epoxide

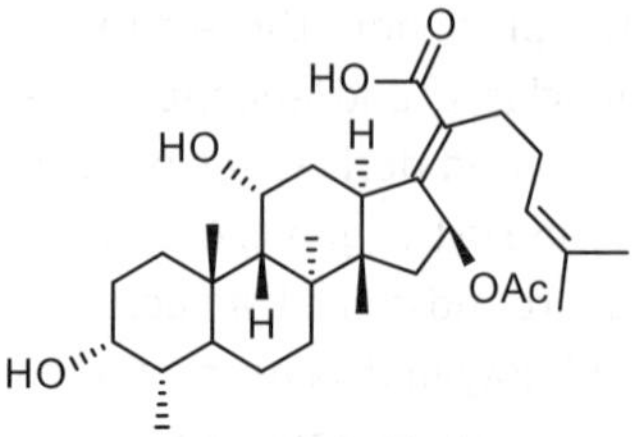

Fusidic acid
(An antimicrobial agent)

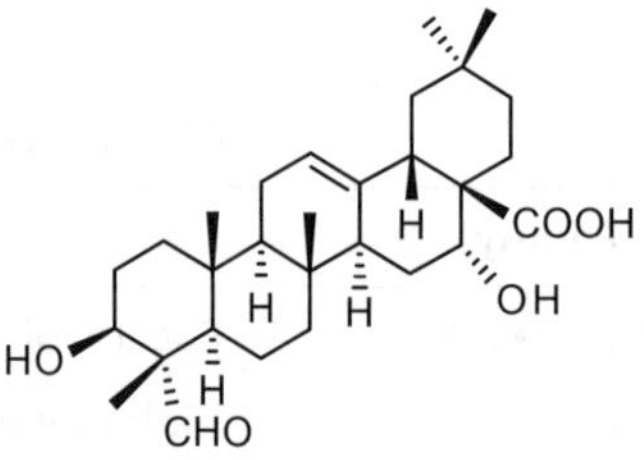

Quillaic acid
(A triterpene from Quillaja)

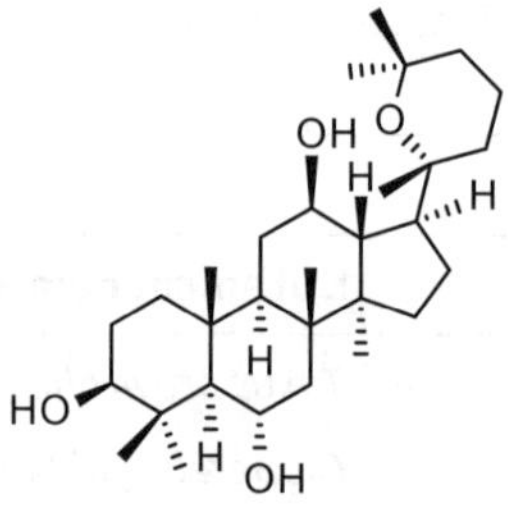

Panaxatriol
(A triterpene from Panax ginseng)

8.5.6.1 Major Structural Types

While squalene, the parent of all triterpenoids, is a linear acyclic compound, most triterpenoids exist in cyclic forms, penta- and tetra-cyclic triterpenes being the major types. Within these cyclic triterpenoids distinct structural variations lead to several structural classes of triterpenoids. Some of the major structure types of triterpenoids are shown here.

Tetracyclic triterpenes

Apotirucallanes

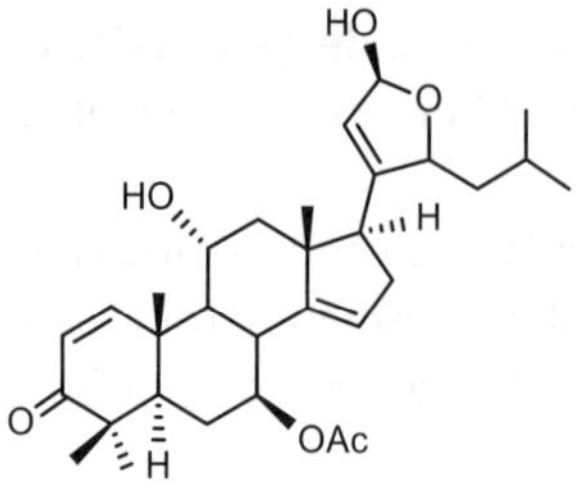

Azadirachtol

Cucurbitanes

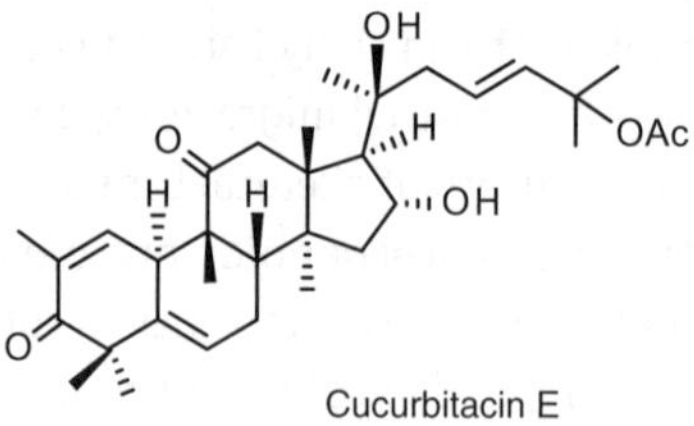

Cucurbitacin E
(Occurs in the family Cucurbitaceae)

Cycloartanes

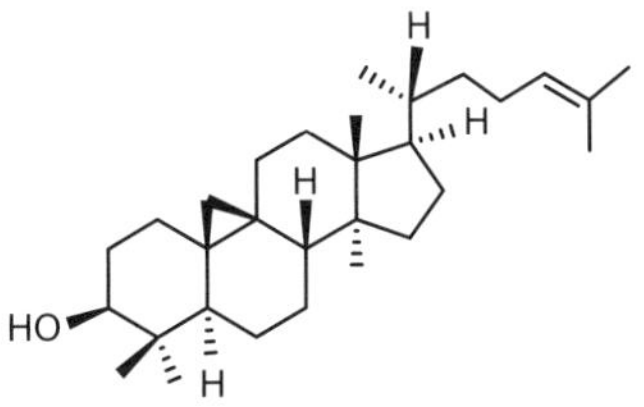

Cycloartenol
(Precursor of phytosterols)

Dammaranes

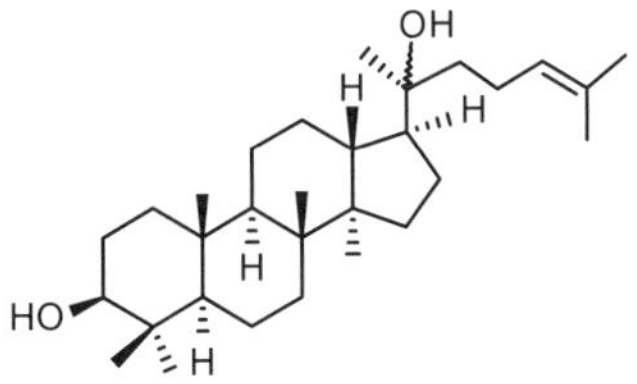

Dammarenediols

Euphanes

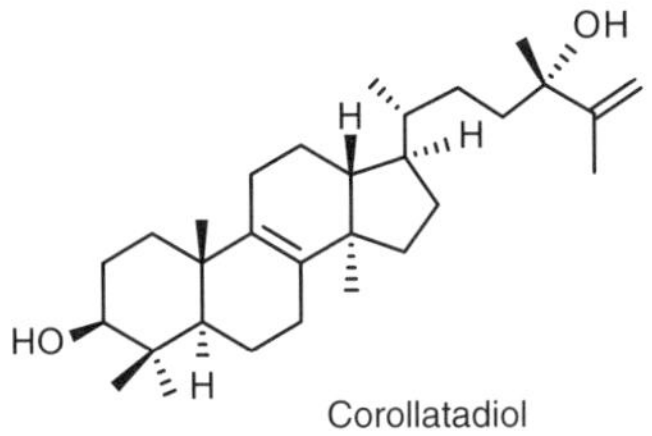

Corollatadiol

Lanostanes

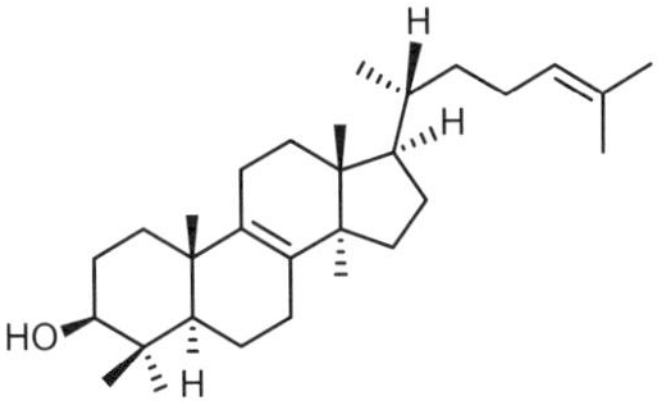

Lanosterol

Prostostanes and fusidanes

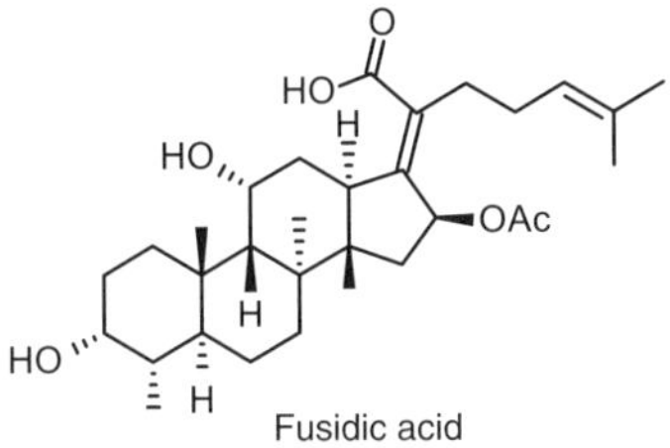

Fusidic acid
(An antibiotic)

Pentacyclic triterpenes

Friedelanes

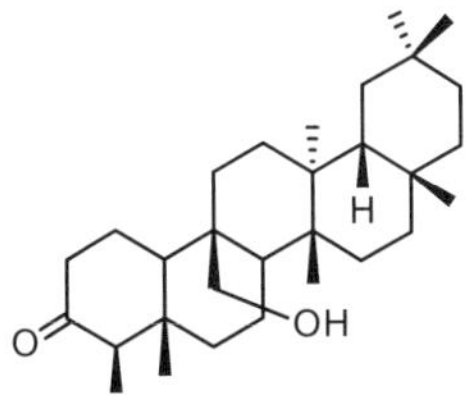

25-Hydroxy-3-friedelanone

Hopanes

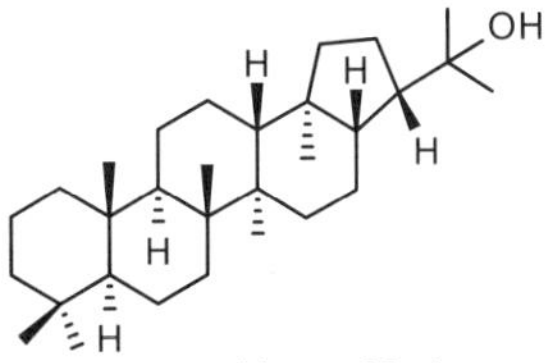

Hopan-22-ol

Lupanes

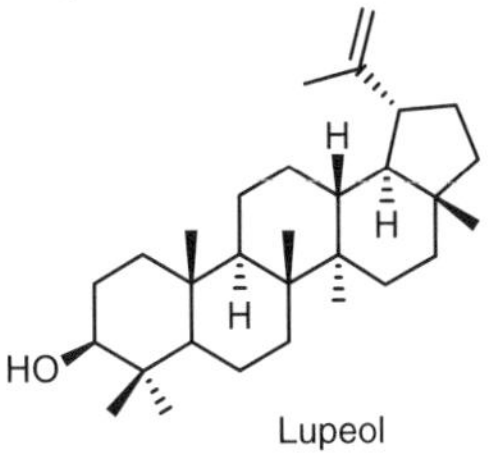

Lupeol

Oleananes

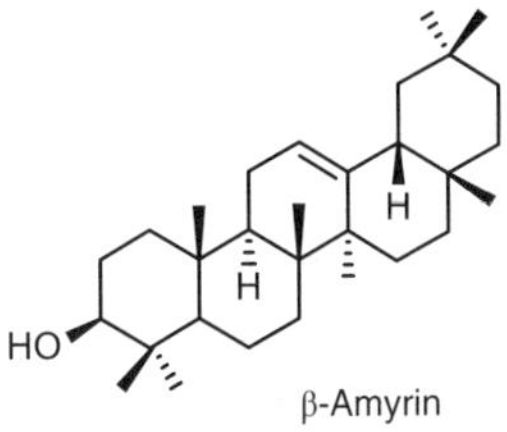

β-Amyrin

Serratanes

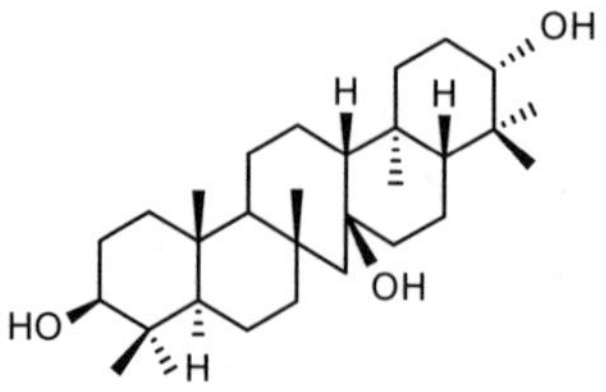

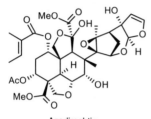

3,14,21-Serratanetriol

Modified triterpenes

Limonoids

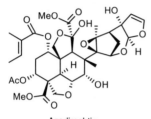

Azadirachtin
(A limonoid from Neem, *Azadirachta indica*)

Quassinoids

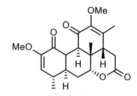

Quassin
(A quassinoid from *Quassia amara*)

Steroids

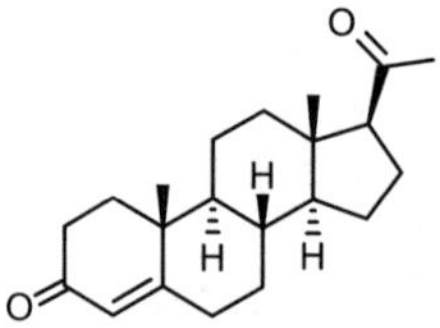

Progesterone
(A female sex hormone)

8.5.6.2 Botanical Sources

Plants are the main sources of natural triterpenes. However, they are also found in other natural sources; for example, fungus. The following table presents just a few of these sources and their major triterpenoidal components.

Source		Major diterpenes
Common name	**Botanical name (Family)**	
Fusidium	*Fusidium coccineum*	Fusidic acid
Ganoderma	*Ganoderma lucidum*	Lanosterol
Dammar resin	*Balanocarpus heimii* (Dipterocarpaceae)	Dammarenediols
Ginseng	*Panax ginseng* (Araliaceae)	Dammarenediols
Lupin	*Lupinus luteus* (Fabaceae)	Lupeol
Quillaia	*Quillaja saponaria* (Rosaceae)	Quillaic acid

8.5.7 Tetraterpenes

The tetraterpenes arise by tail-to-tail coupling of two GGPP molecules. Tetra-terpenes are represented by the carotenoids and their analogues; for example, β-carotene, an orange colour pigment of carrots (*Daucus carota*, family: Apiaceae); lycopene, a characteristic pigment in ripe tomato fruit (*Lycopersicon esculentum*, family: Solanaceae) and capsanthin, the brilliant red pigment of peppers (*Capsicum annuum*, family: Solanaceae).

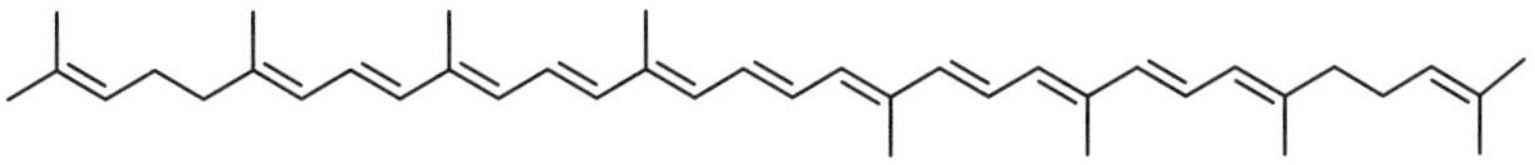

Lycopene

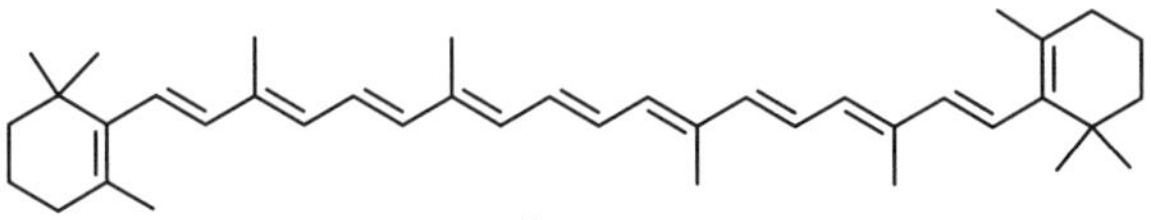

β-Carotene

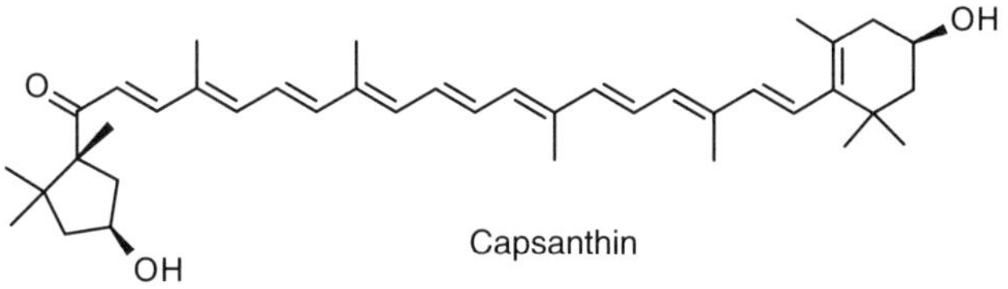

Capsanthin

Carotenoids are found abundantly in plants and have been used as colour-ing agents for foods, drinks, confectionary and drugs. The vitamin A group of compounds are important metabolites of carotenoids; for example, vitamin A$_1$ (retinol).

8.5.7.1 Chemistry of Vision: Role of Vitamin A

β-Carotene is converted to vitamin A$_1$ (retinol) in our liver. Vitamin A$_1$ is a fat-soluble vitamin found in animal products; for example, eggs, dairy products, liver and kid-neys. It is oxidized to an aldehyde called all-*trans*-retinal, and then isomerized to give 11-*cis*-retinal, which is the light-sensitive pigment present in the visual sys-tems of all living beings.

Rod and *cone* cells are the light sensitive receptor cells in the retina of the human eye. About three million rod cells are responsible for our vision in dim light, whereas the hundred million cone cells are responsible for our vision in bright light and for the perception of bright colours. In the rod cells, 11-*cis*-retinal is converted to *rhodopsin.*

When light strikes the rod cells, isomerization of the C-11/C-12 double bond takes place and *trans*-rhodopsin (metarhodopsin II) is formed. This *cis-trans* isomerization is accompanied by an alteration in molecular geometry, which generates a nerve impulse to be sent to the brain resulting in the perception of *vision*. Metarhodopsin II is recycled back to rhodopsin by a multi-step sequence that involves the cleavage to all-*trans*-retinal and *cis-trans* isomerization back to 11-*cis*-retinal. A deficiency of vitamin A leads to vision defects; for example, night blindness. Vitamin A is quite unstable and sensitive to oxidation and light. Excessive intake of the vitamin, however, can lead to adverse effects; for example, pathological changes in the skin, hair loss, blurred vision and headaches.

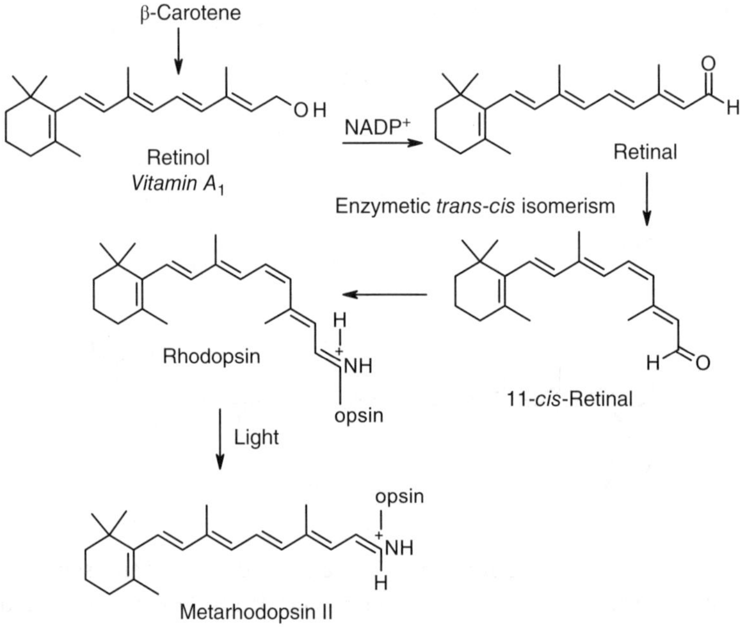

8.6 STEROIDS

You have surely come across the items of news that appear quite frequently in the media, related to world-class athletes and sports personalities abusing *anabolic steroids*, for example nandrolone, to enhance performance and also to improve physical appearance. What are these substances? Well, all these drugs and many other important drugs belong to the class of compounds called steroids. *Steroids* are chemical messengers, also known as hormones; for example, the sex hormones estradiol and testosterone. They are synthesized in glands and delivered by the bloodstream to target tissues to stimulate or inhibit some process. Steroids are non-polar and therefore lipids. Their non-polar character allows them to cross cell membranes so they can leave the cells in which they are synthesized and enter their target cells.

8.6.1 Structures of Steroids

Structurally, a steroid is a lipid characterized by a carbon skeleton with four fused rings. The steroid core structure is composed of 17 carbon atoms. All steroids are derived from the acetyl CoA biosynthetic pathway. Hundreds of distinct steroids have been identified in plants, animals and fungi, and most of them have interesting biological activity. They have a common basic ring structure; three-fused cyclo-hexane rings together form the phenanthrene part, fused to a cyclopentane ring system known as *cyclopentaphenanthrene*. The four rings are lettered A, B, C and D, and the carbon atoms are numbered beginning in the A ring as shown in gonane. These fused rings can be *trans*-fused or *cis*-fused. In steroids, the B, C and D rings always are *trans*-fused. In most naturally occurring steroids, rings A and B also are *trans*-fused. Steroids vary by the functional groups attached to this four-ring core and by the oxidation state of the rings.

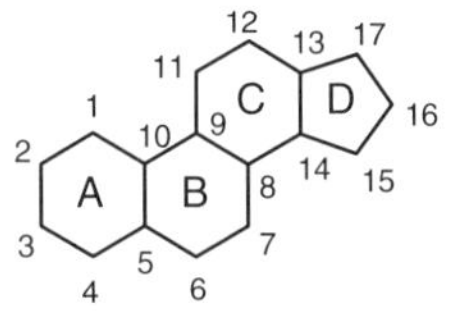

Gonane

Many steroids have methyl groups at C-10 and C-13 positions. These are called angular methyl groups. Steroids also have a side chain attached to C-17 and a related series of steroids are named after their fundamental ring systems, which are shown in the following structures.

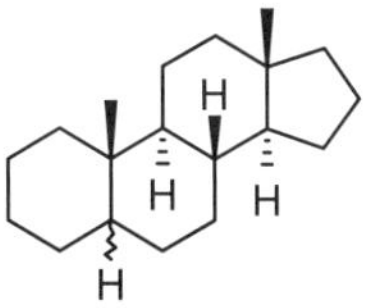

Androstane	Pregnane	Cholane
(A C_{19} steroid)	(A C_{21} steroid)	(A C_{24} steroid)

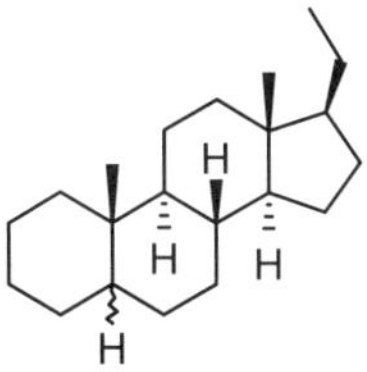

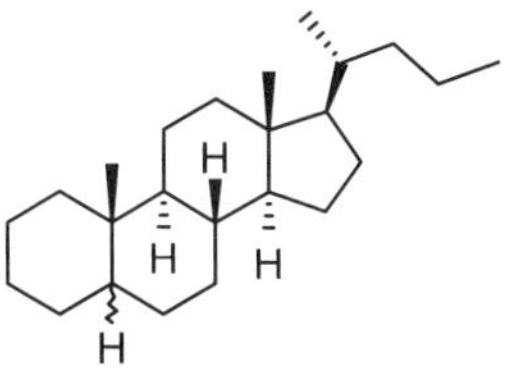

Many steroids have an alcoholic hydroxyl attached to the ring system, normally at C-3 of the ring A, and are known as *sterols*, which normally derive from choles-tane. The most common sterol is cholesterol occurs in most animal tissues. There are many different steroid hormones and cholesterol is the precursor for all of them. Cholesterol is also the precursor of vitamin D.

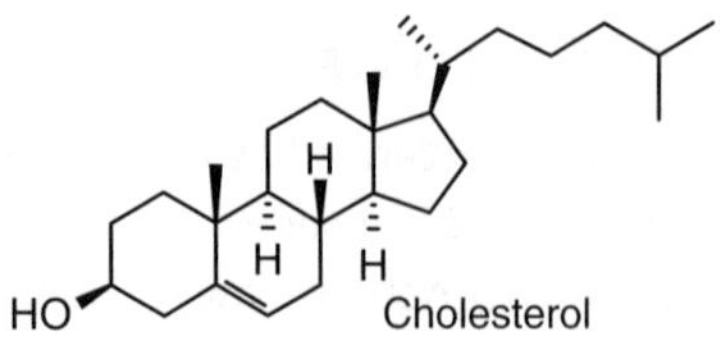

Cholesterol
(A C$_{27}$ sterol)

8.6.2 Stereochemistry of Steroids

Sir Derek H. R. Barton of Great Britain received the Nobel Prize in 1969 for recognizing that functional groups could vary in reactivity depending on whether they occupied an *axial* or an *equatorial* position on a ring. The steroid skeleton shows a specific stereochemistry. All three of the six-membered rings can adopt strain-free chair conformations as shown next. Unlike simple cyclohexane rings, which can undergo chair-chair inter-conversions, steroids being the large rigid molecules cannot undergo ring-flips. Steroids can have either *cis* or *trans* fusion of the A and B rings, both kind of steroids are relatively long, flat molecules but the A, B *trans*-fused steroids are by far the more common, though *cis*-fused steroids are found in bile. Furthermore, the presence of two angular methyl groups at C-10 and C-13 positions is characteristic in cholesterol. Substituents on the steroid ring system may be either *axial* or *equatorial* and, as usual, *equatorial* substitution is more favourable than *axial* substitution for *steric* reasons. Thus, the hydroxyl group at C-3 of cholesterol has the more stable *equatorial* orientation.

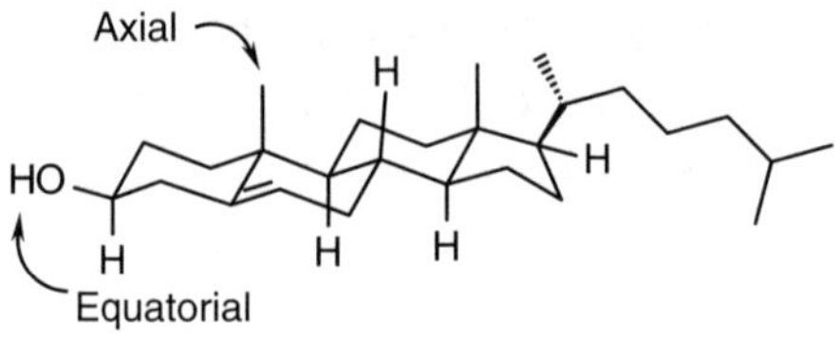

3D structure of cholesterol

In most steroids, the B–C and C–D rings are fused usually in a *trans* manner. The lower side of the steroid nucleus is denoted α, the upper side of the steroid is denoted β, usually drawn as projected beneath the plane of the paper, which is shown as broken lines and above the plane of the paper, which is drawn as solid lines. Thus, substituents attached to the steroid skeleton are also characterized as α and β. Cholesterol has eight *chiral* centres, therefore 256 stereoisomers are theoretically possible, but only one exists in nature! Stereogenic centres in steroid side chains are denoted preferentially with the *R* and *S* nomenclature.

8.6.3 Physical Properties of Steroids

The main feature, as in all lipids, is the presence of a large number of carbon-hydrogen that makes steroids non-polar or lipophilic. The solubility of steroids in non-polar organic solvents, for example ether, chloroform, acetone and benzene, and general insolubility in water results from their significant hydrocarbon components. However, with the increase in number of hydroxyl or other polar functional groups on the steroid skeleton, the solubility in polar solvents increases.

8.6.4 Types of Steroid

On the basis of the physiological and chemical functions, steroids can be classified as follows.

8.6.4.1 Anabolic Steroids or Anabolic Androgenic Steroids

There is a class of natural and synthetic steroids that interact with androgen receptors to promote cell growth and cell division, resulting in growth of several types of tissues, especially muscle and bone. *Anabolic steroids* are actually steroidal androgens that include natural androgens like testosterone as well as synthetic androgens that are structurally related and have similar effects to testosterone. There are natural and synthetic anabolic steroids. For example, testosterone, nandrolone and methandrostenolone.

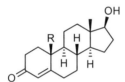

Testosterone R = Me
Nandrolone R = H
Natural anabolic steroids

Methandrostenolone
(A synthetic anabolic steroid)

8.6.4.2 Corticosteroids (Glucocorticoid and Mineralocorticoids)

These are a group of steroid hormones characterized by their ability to bind with the *cortisol receptor* and trigger similar effects. Glucocorticoids regulate many aspects of metabolism and immune functions, and are often prescribed as a remedy for inflammatory conditions like asthma and arthritis. For example, cortisol.

Cortisol
(Produced in adrenal cortex)

Mineralocorticoids are corticosteroids that help maintain blood volume and control renal excretion of electrolytes. For example, aldosterone.

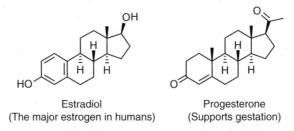

Aldosterone
(Produced in adrenal cortex)

8.6.4.3 Sex Steroids or Gonadal Steroids

Sex hormones are the steroids that interact with vertebrate androgen or estrogen receptors to produce sex differences (primary and secondary sex characters) and support reproduction. They include androgens, estrogen and progestogens. Examples include testosterone, estradiol and progesterone.

Estradiol
(The major estrogen in humans)

Progesterone
(Supports gestation)

8.6.4.4 Phytosterols or Plant Sterols

These are steroid alcohols that occur naturally in plants. For example, β-sitosterol.

β-Sitosterol
(A plant sterol)

8.6.4.4.1 Ergosterols

This group of sterols occur in fungi and include some Vitamin D supplements.

Ergosterol
(A precursor of vitamin D_2)

However, broadly, steroids can be classified only into two main classes: *sex or reproductive hormones* and *adrenocorticoid or adrenocortical hormones*.

8.6.5 Biosynthesis of Steroids

All steroids are biosynthesized in cells from the sterols lanosterol (animals and fungi) or cycloartenol (plants). Lanosterol and cycloartenol are derived from the cyclization of the triterpene squalene as shown next. Steroid biosynthesis is an anabolic pathway that provides steroids from simple precursors. A unique biosynthetic pathway is followed in animals, making the pathway a common target for antibiotics and other anti-infection drugs. Steroid metabolism in humans is also the target of cholesterol-lowering drugs, such as statins. In animals including humans, the biosynthesis of steroids follows the mevalonate pathway, which uses acetyl-CoA as building blocks for DMAPP and IPP. In the subsequent steps, DMAPP and IPP join to form GPP, which leads to the formation of the steroid lanosterol. Further modifications of lanosterol into other steroids are known as steroidogenesis transformations.

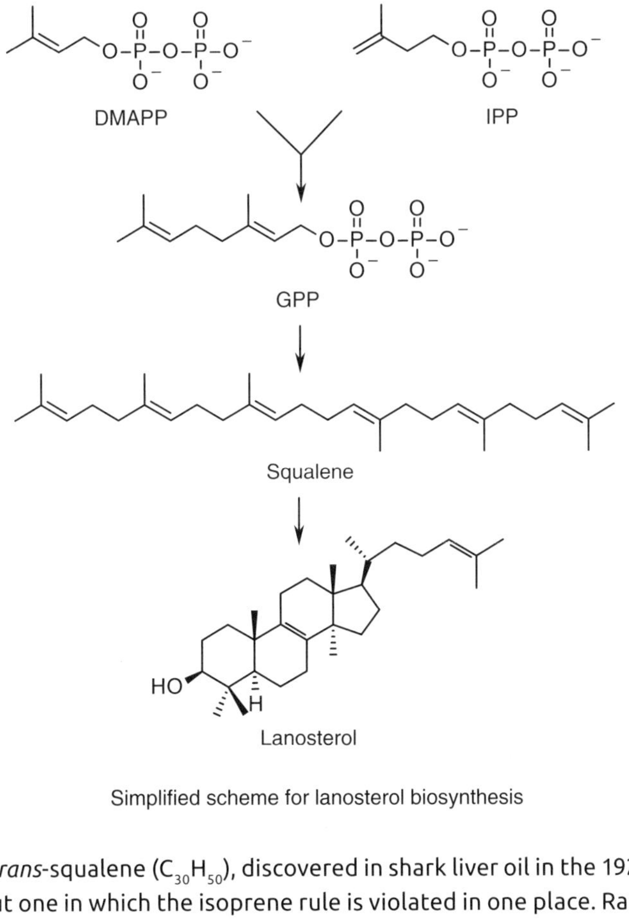

DMAPP

IPP

GPP

Squalene

Lanosterol

Simplified scheme for lanosterol biosynthesis

The all-*trans*-squalene ($C_{30}H_{50}$), discovered in shark liver oil in the 1920s, is a triterpene, but one in which the isoprene rule is violated in one place. Rather than a head-to-tail arrangement of six units of isoprene, there appear to be farnesyl units that have been connected tail-to-tail. Cholesterol is biosynthesized from squalene, which is first converted to lanosterol. The conversion of squalene to the steroid skeleton is an oxirane, squalene-2,3-oxide, which is transformed by enzymes into lanosterol, a steroid alcohol naturally found in wool fat. The whole process is highly stereoselective. The first step in the conversion of squalene to lanosterol is epoxidation of the 2,3-double bond of squalene. Acid-catalysed ring opening of the epoxide initiates a series of cyclization resulting in the formation of protesterol cation. Elimination of a C-9 proton leads to the 1,2 -hydride and 1,2-methyl shifts, resulting in the formation of lanosterol, which in turn is converted to cholesterol by enzymes in a series of 19 steps.

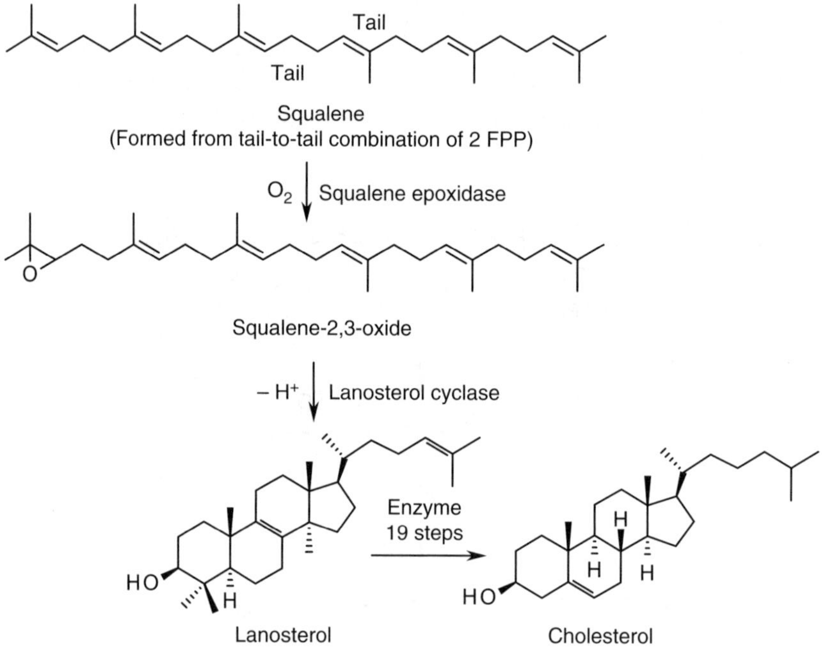

Squalene
(Formed from tail-to-tail combination of 2 FPP)

O_2 | Squalene epoxidase

Squalene-2,3-oxide

$- H^+$ | Lanosterol cyclase

Enzyme
19 steps

Lanosterol

Cholesterol

8.6.6 Synthetic Steroids

Several steroids have been synthesized in an effort to investigate their physiological effects. Prednisone is an example of a synthetic drug. The oral contraceptives and anabolic steroids are the best-known of all steroids. Microbial catabolism of phytosterol side chains gives C-19 steroids, C-22 steroids and 17-ketosteroids (i.e. precursors to adrenocortical hormones and contraceptives). The semisynthesis of steroids often begins from precursors like cholesterol, phytosterols or sapogenins. For example, Syntex, a company involved in the Mexican barbasco trade, used *Dioscorea mexicana* to produce the sapogenin diosgenin in the early days of the synthetic steroid pharmaceutical industry. Some steroidal hormones are economically obtained only by total synthesis from petrochemicals; for example, 13-alkyl steroids. For example, the pharmaceutical norgestrel begins from methoxy-1-tetralone, a petrochemical derived from phenol.

It was discovered in 1930s that the injection of progesterone could be effective as a contraceptive. The pill is an oral contraceptive containing synthetic derivatives of the female sex hormones, progesterone and estrogen. These synthetic hormones prevent ovulation and thus prevent pregnancy. The two most important birth-control pills are norethindrone and ethynylestradiol. Many synthetic steroids have been found to be much more potent than natural steroids. For example, the contraceptive drug, norethindrone is better than progesterone in arresting (terminating) ovulation.

Norethindrone
(A synthetic progestin)

Ethynylestradiol
(A synthetic estrogen)

Steroids that aid in muscle development are called *anabolic steroids*. They are synthetic derivatives of testosterone, thus they have the same muscle-building effect as testosterone.

Androstenedione

Stanozolol

Dianabol
Methandrostenolone

There are more than 100 different anabolic steroids, which vary in structure, duration of action, relative effects and toxicities. Androstenedione, stanozolol and dianabol are anabolic steroids. They are used to treat people suffering from traumas accompanied by muscle deterioration. The use of anabolic steroid can lead to a number of dangerous side effects including lowered levels of high-density lipoprotein cholesterol, which benefits the heart, and elevated levels of harmful LDL, stimulation of prostate tumours, clotting disorders and liver problems.

8.6.7 Functions of Steroids

Steroids and their metabolites, for example steroid hormones, often are signalling molecules, and steroids and phospholipids are components of cell membranes.

Steroids such as cholesterol decrease membrane fluidity. Steroids play critical roles in a number of disorders, including malignancies like prostate cancer, where steroid production inside and outside the tumour promotes cancer cell aggressiveness. However, the most important function of steroids in most living systems is as hormones.

Steroid hormones produce their physiological effects by binding to steroid hormone receptor proteins. The binding of steroids to their receptors causes changes in gene transcription and cell function. From biological and physiological viewpoints, probably the most important steroids are cholesterol, the steroid hormones and their precursors and metabolites. Cholesterol, a common component of animal cell membranes, is an important steroid alcohol. Cholesterol is formed in brain tissue, nerve tissue and in the blood stream. A number of vertebrate hormones, which govern a number of physiological functions from growth to reproduction, are biosynthesized from cholesterol. However, cholesterol is the major compound found in gallstones and bile salts, and it also contributes to the formation of deposits on the inner walls of blood vessels. These deposits harden and obstruct the flow of blood. This condition results in various heart diseases, strokes and high blood pressure, and a high level of cholesterol can be life-threatening. A significant body of research is currently underway to determine if a correlation exists between cholesterol levels in the blood and diet. Not only does cholesterol come from the diet, but also cholesterol is synthesized in the body from carbohydrates and proteins as well as fat. Therefore, the elimination of cholesterol rich food items from the diet does not necessarily lower blood cholesterol levels. Some studies have found that if certain unsaturated fats and oils are substituted for saturated fats, the blood cholesterol level decreases.

Sex hormones control tissue growth and reproduction. Male sex hormones are testosterone and 5α-dihydrotestosterone, also known as androgens, which are secreted by the testes. The primary male hormone, testosterone, is responsible for the development of secondary sex characteristics during puberty. They also promote muscle growth. The two most important female sex hormones are oestradiol and estrone, also known as *estrogens*.

They are responsible for the development of female secondary sex characteristics.

Testosterone

5α-Dihydrotestosterone

Estrogen is biosynthesized from testosterone by making the first ring aromatic, which results in more double bonds, the loss of a methyl group and formation of an alcohol group. Estrogen, along with progesterone regulates changes occurring in the uterus and ovaries known as the *menstrual cycle*. Progesterone is a member of the class called *progestins*. It is also the precursor of sex hormones and adrenal cortex steroids. Progesterone is an essential component for the maintenance of pregnancy. It also prevents ovulation during pregnancy. Many of the steroid hormones are ketones, including testosterone and progesterone.

The male and female hormones have only slight differences in structures, but yet have very different physiological effects. For example, the only difference between testosterone and progesterone is the substituent at C-17.

Estrone

Adrenocorticoid hormones are biosynthesized in the adrenal glands. They regulate a variety of metabolic processes. The most important mineralocorticoid is aldosterone, an aldehyde as well as a ketone, which regulates the reabsorption of sodium and chloride ions in the kidney and increases the loss of potassium ions. Aldosterone is secreted when blood sodium ion levels are too low to cause the kidney to retain sodium ions. If sodium levels are elevated, aldosterone is not secreted, so some sodium will be lost in the urine and water. Aldosterone also controls swelling in the tissues.

Cortisone

Prednisone

Cortisol or hydrocortisone, the most important glucocorticoid, has the function of increasing glucose and glycogen concentrations in the body. These reactions are completed in the liver by taking fatty acids from lipid storage cells and amino acids from body proteins to make glucose and glycogen. Cortisol and its ketone derivative, cortisone, are potent anti-inflammatory agents. Cortisone or similar synthetic derivatives such as prednisolone, the active metabolite of prednisone, are used to treat inflammatory diseases, rheumatoid arthritis and bronchial asthma. There are many side effects with the use of cortisone drugs, so their use must be monitored carefully. Prednisolone is designed to be a substitute for cortisone, which has much higher side effects than prednisolone.

Phytosterols found in plants have many applications as food additives, in medicine and in cosmetics. The best-known, and scientifically proven, benefit of phytosterols is their ability to help lower cholesterol and to protect against the development of certain cancers. A lesser known benefit of phytosterols involves skin care, where phytosterols not only stop the slow-down of collagen production as a result of long exposure to the sun, they actually initiate new collagen production. Ergosterol is a component of fungal cell membranes, serving the same function that cholesterol serves in animal cells. The presence of ergosterol in fungal cell membranes coupled with its absence in animal cell membranes makes it a useful target for antifungal drugs. Ergosterol is also present in the cell membranes of some protists, such as trypanosomes. This explains the use of some antifungal agents against West African sleeping sickness.

8.7 PHENOLICS

Phenolics are a large group of structurally diverse naturally occurring compounds that possess at least one phenolic moiety in their structures. For example, umbelliferone, a coumarin, has a phenolic hydroxyl functionality at C-7, quercetin is a flavonoid, which has four phenolic hydroxyls at C-5, C-7, C-3′ and C-4′. Although phenolic groups of compounds can be of various structural types, in this section we will mainly focus our discussion on phenylpropanoids, coumarins, flavonoids and isoflavonoids, lignans and tannins.

Umbelliferone
(The most common natural coumarin)

Quercetin
(A natural antioxidant)

Etoposide R = Me
Teniposide R = thiophenyl
(Anticancer lignans)

Most of these compounds, for example quercetin, possess various degrees of antioxidant or free radical scavenging properties. Many phenolic compounds have medicinal properties and long been used as drugs. For example, etoposide and teniposide, two lignans, are anticancer drugs.

8.7.1 Phenylpropanoids

Phenylpropanes, deriving directly from phenylalanine, are aromatic compounds with a propyl side chain attached to the benzene ring. Naturally occurring phenylpropanoids often contain oxygenated substituents, for example OH, OMe or methylenedioxy, on the benzene ring. Phenylpropanoids with hydroxyl substituent(s) on the benzene ring belong to the group phenolics; for example, caffeic acid and coumaric acid.

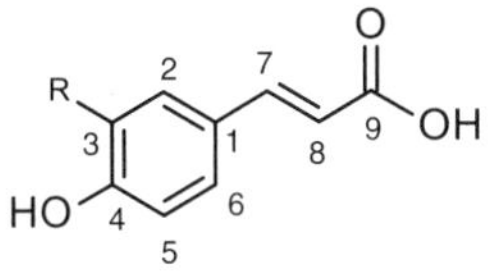

Coumaric acid (or 4-hydroxycinnamic acid): R = H
Caffeic acid (or 3,4-dihydroxycinnamic acid): R = OH

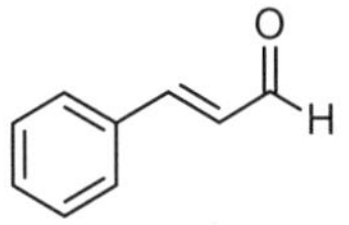

Cinnamaldehyde

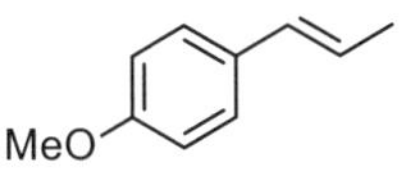

Anethole

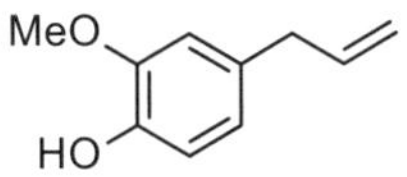

Eugenol

Phenylpropanoids are found widespread in higher plants, especially in the plants that produce essential oils; for example, plants in the families, Apiaceae, Lamiaceae, Lauraceae, Myrtaceae and Rutaceae. As examples, Tolu balsam (*Myroxylon balsamum*, family: Fabaceae) yields a high concentration of cinnamic acid esters; cinnamon (*Cinnamomum verum*, family: Lauraceae) provides cinnamaldehyde; fennel (*Foeniculum vulgare*, family: Apiaceae) is a good source of eugenol and star anise (*Illicium verum*, family: Illiaceae) yields high amounts of anethole. The biosynthesis of phenylpropanoids follows the shikimic acid pathway and the immediate precursor of cinnamic acid is phenylalanine. Other phenylpropanoids, and a number of other phenolics, for example coumarins, flavonoids and lignans, originate from the phenylpropanoid, cinnamic acid.

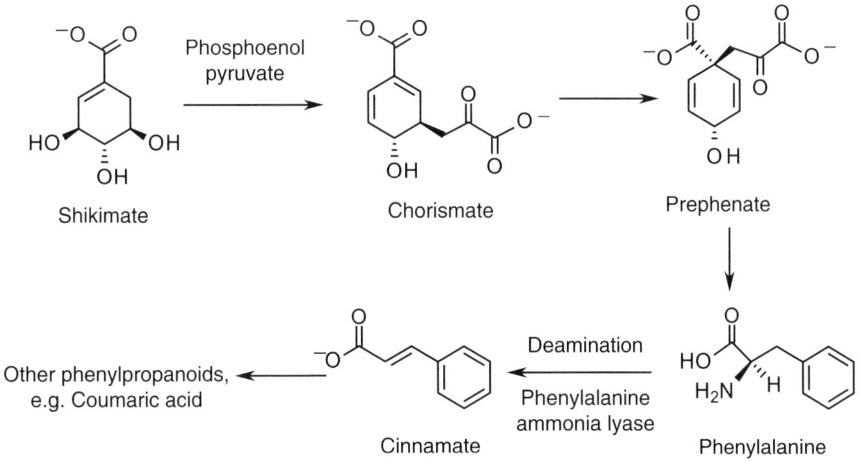

Shikimate

Phosphoenol pyruvate

Chorismate

Prephenate

Other phenylpropanoids, e.g. Coumaric acid

Cinnamate

Deamination

Phenylalanine ammonia lyase

Phenylalanine

Several phenylpropanoids possess important pharmacological and medicinal properties; for example, analgesic, antioxidant, antibacterial, anti-inflammatory, anticancer, antiviral, antiproliferative, hypotensive and wound healing activities. Certain phenylpropanoids are used in sunscreen. Cinnamaldehyde and eugenol, found in the *Cinnamomum* species, are well-known for their antibacterial, anti-inflammatory and antiproliferative properties. Coumaric acid has antioxidant, anti-inflammatory properties and may reduce the risk of stomach cancer by reducing the formation of carcinogenic nitrosamines.

Similarly, caffeic acid, found abundantly in the bark of *Eucalyptus globulus*, freshwater fern *Salvinia molesta*, the mushroom *Phellinus linteus* and many other flowering plants, possesses immunomodulatory, anticancer, anti-inflammatory, antimicrobial, antiproliferative, antioxidant, antitumour and fungicidal properties. Caffeic acid is also present in various food sources such as coffee beans, apples, artichoke, berries and pears, and is used in supplements for boosting athletic performance, exercise-related fatigue, weight loss, cancer, HIV/AIDS, herpes and other conditions. However, the problem of low bioavailability, fast metabolism and potential toxicity/sensitization are some of the limiting factors for the development of phenylpropanoid-based drugs.

8.7.2 Coumarins

Coumarins (2*H*-1-benzopyran-2-one or 2*H*-chromen-2-one) are the largest class of *1-benzopyran* derivatives, found mainly in higher plants. The first member of this class, coumarin, was isolated from Tonka beans (*Dipteryx odorata*) by the German chemist A. Vogel in 1820. The French word for tonka bean is *coumarou*, which influenced the naming of this compound as coumarin. Coumarins are actually the modified forms of phenylpropanoids. Most natural coumarins are oxygenated at C-7; for example, umbelliferone (7-hydroxycoumarin).

Umbelliferone R = H
Scopoletin R = OMe

Imperatorin
(An *O*-prenylated furanocoumarin)

Osthol
(An *C*-prenylated furanocoumarin)

Umbelliferone is considered as the structural and biogenetic parent of the more highly oxygenated coumarins; for example, scopoletin. C- and *O*-prenylations are common in a large number of coumarins; for example, imperatorin and osthol. The prenyl groups found in coumarins exhibit the greatest number of biogenetic modifications including cyclization to dihydropyrans, pyrans, dihydrofurans and furans.

Coumarins occur abundantly in various plant families; for example, Apiaceae, Asteraceae, Fabaceae, Lamiaceae, Moraceae, Poaceae, Rutaceae and Solanaceae.

However, Apiaceae (*alt.* Umbelliferae) and Rutaceae are the two most important coumarin-producing plant families.

Many coumarins are used in sunscreen preparations for protection against sunlight, because these compounds absorb short-wave UV radiation (280–315 nm), which is harmful to human skin, but transmits long-wave UV radiation (315–400 nm) that provides a brown sun-tan. Dicoumarol, a dimeric coumarin, occurs in mouldy sweet clover, *Melilotus officinalis* (family: Fabaceae), has a prominent anticoagulant property and has been used in medicine as an anti-blood-clotting agent for the prevention of thrombosis. Psoralen, a linear furanocoumarin isolated from *Psoralea corylifolia* (family: Fabaceae) and also found in the families, Apiaceae, Fabaceae, Moraceae and Rutaceae, has long been used in the treatment of vertigo.

A number of coumarins also possess antifungal and antibacterial properties. Coumarins have shown some evidence of biological activity and have limited approval for few medical uses as pharmaceuticals, such as in the treatment of lymphedema and their ability to increase plasma antithrombin levels. Both coumarin and indandione derivatives produce a uricosuric effect by interfering with the renal tubular reabsorption of urate.

Various coumarins are known to possess anticancer, anti-infective and anti-inflammatory properties, and are used to treat oedema, such as in the treatment of lymphedema, elephantiasis and other high protein oedema conditions. Coumarin itself has been used as a flavouring agent in food, but because of several reported cases of drug-related hepatitis after taking high doses of coumarin or dicoumarols as an anticoagulant, coumarin has been withdrawn from the market, except in brands containing low doses of coumarin and troxerutin.

8.7.2.1 Biosynthesis of Coumarins

The pathway of coumarin biosynthesis was largely investigated in the 1960s and 1970s, employing tracer feeding experiments. Radiolabelled cinnamic acid was incorporated into coumarin and 7-hydroxycoumarins. Other tracer experiments conducted with *Lavandula officinalis*, a plant that biosynthesizes coumarin as well as 7-hydroxylated coumarins, revealed that in the latter instance *para*-hydroxylation preceded the *ortho*-hydroxylation required for lactonization. This supported the hypothesis that umbelliferone is derived from *cis-p*-coumaric acid, whereas coumarin originates from *cis*-cinnamic acid. Different enzymes for the ortho-hydroxylation/lactonization of coumarin versus umbelliferone could be involved.

Briefly, the biosynthesis of coumarins begins with *trans*-cinnamic acid, which is oxidized to *ortho*-coumaric acid (2-hydroxy cinnamic acid) followed by formation of the glucoside. This glucoside isomerizes to the corresponding *cis*-compound, which finally through ring closure forms coumarin. However, as most natural coumarins contains an oxygenation at C-7, the biosynthesis proceeds through 4-hydroxylation of cinnamic acid. It should be noted that the *ortho*-hydroxylation on the cinnamic acid is the key step in the biosynthesis of coumarins.

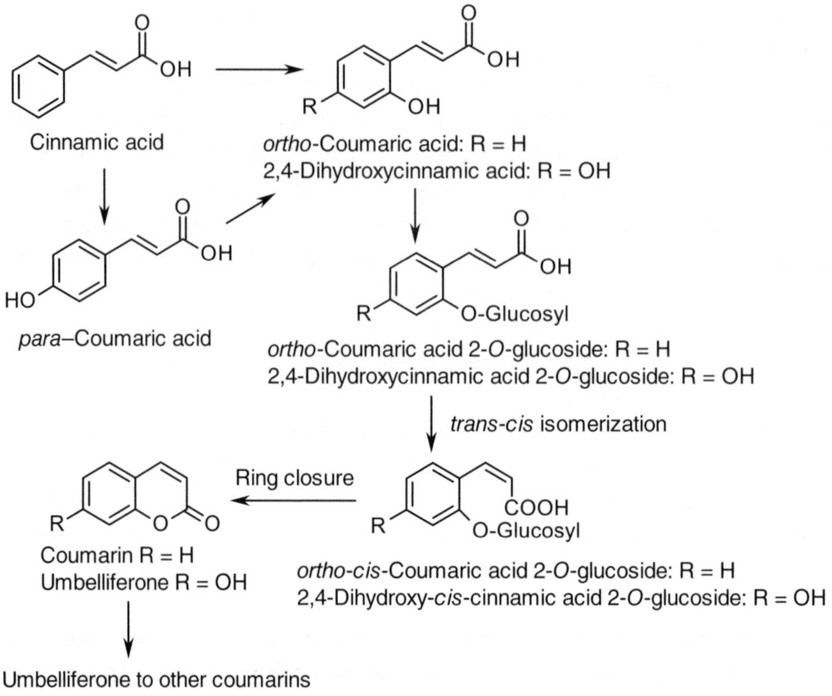

Cinnamic acid

ortho-Coumaric acid: R = H
2,4-Dihydroxycinnamic acid: R = OH

para–Coumaric acid

ortho-Coumaric acid 2-*O*-glucoside: R = H
2,4-Dihydroxycinnamic acid 2-*O*-glucoside: R = OH

trans-cis isomerization

Ring closure

Coumarin R = H
Umbelliferone R = OH

ortho-cis-Coumaric acid 2-*O*-glucoside: R = H
2,4-Dihydroxy-*cis*-cinnamic acid 2-*O*-glucoside: R = OH

Umbelliferone to other coumarins

8.7.2.2 Structural Types of Coumarins

Different types of coumarins with specific examples are shown here.

Simple coumarins

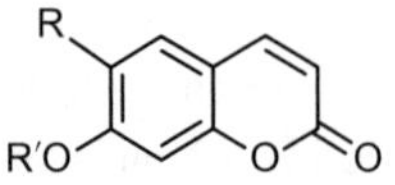

Umbelliferone R = R′ = H
Aesculetin R = OH, R′ = H
Scopoletin R = OMe, R′ = H
Scopolin R = OMe, R′ = glucosyl

Simple prenylated coumarins

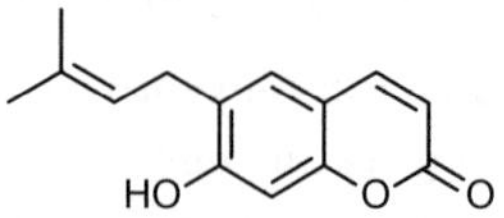

Demethylsuberosin

Linear furanocoumarins

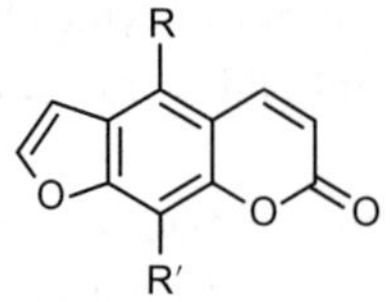

Psoralen R = R′ = H
Bergapten R = OMe, R′ = H
Xanthotoxin R = H, R′ = OMe
Isopimpinellin R = R′ = OMe

Angular furanocoumarins

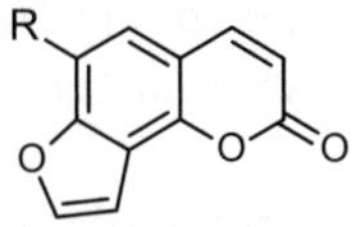

Angelicin, R = H
(A typical coumarin of *Angelica* species)
Sphondin, R = OMe

Linear dihydrofuranocoumarins

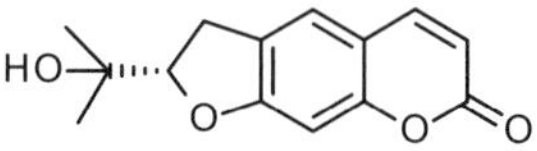

Marmesin

Angular dihydrofuranocoumarins

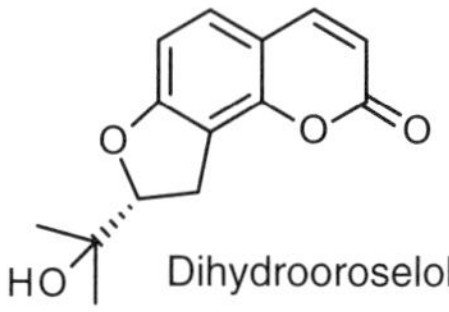

Dihydrooroselol

Linear pyranocoumarins

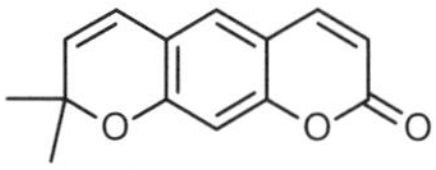

Xanthyletin

Angular pyranocoumarins

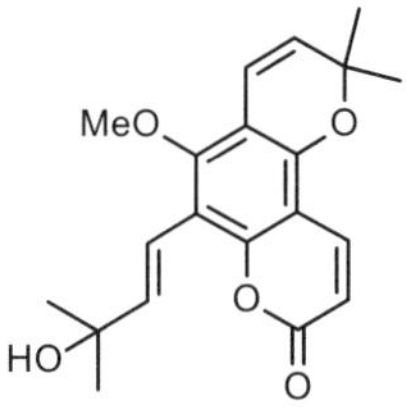

Avicennol

Linear dihydropyranocoumarins

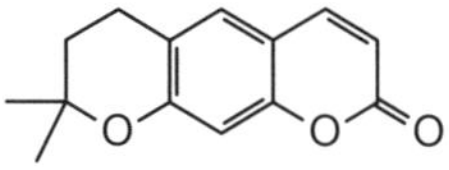

1′,2′-Dihydroxanthyletin

Angular hydropyranocoumarins

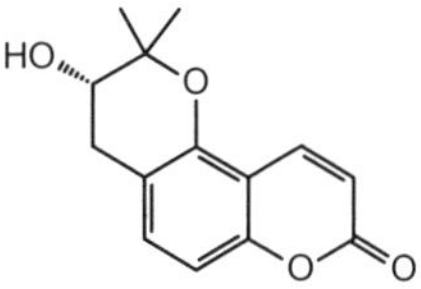

Libanotin A

Sesquiterpeneyl coumarins

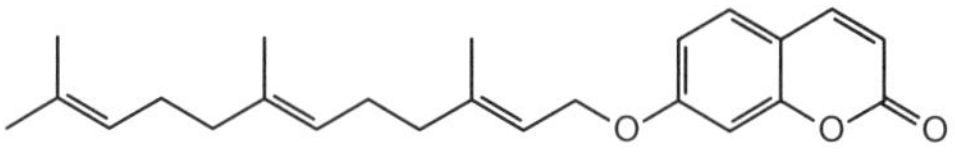

Umbelliprenin

Dimeric coumarins

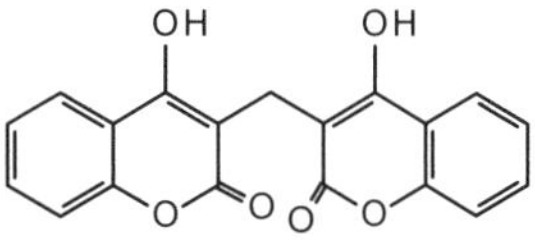

Dicoumarol

8.7.3 Flavonoids and Isoflavonoids

Flavonoids, the derivatives of 1,3-diphenylpropane (benzo-gamma-pyrone derivatives), form a large group of natural products, which are widespread in higher plants but also found in some lower plants including algae. Most flavonoids are yellow compounds and contribute to the yellow colour of the flowers and fruits where they are usually present as glycosides.

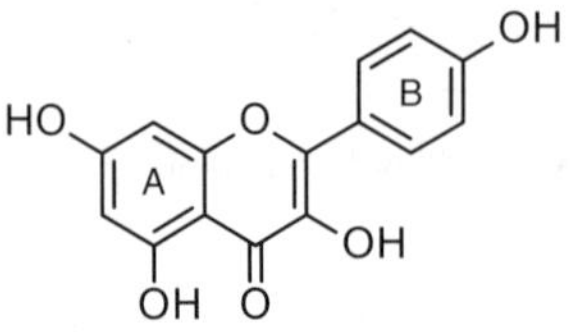

Kaempferol 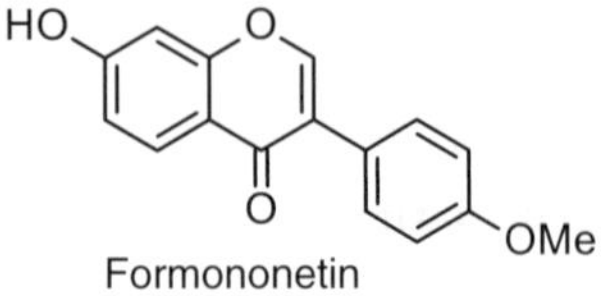   Formononetin

Most flavonoids occur as glycosides and within any one class may be character-ized as mono glycosidic, diglycosidic and so on. There are well over 2000 glycosides of the flavones and flavonols isolated to date. Both O- and C-glycosides are common in plant flavonoids; for example, rutin is an O-glycoside, whereas isovitexin is a C-gly-coside. Sulphated conjugates are also common in the flavone and flavonol series, where the sulphate conjugation may be on a phenolic hydroxyl and/or on an aliphatic hydroxyl of a glycoside moiety.

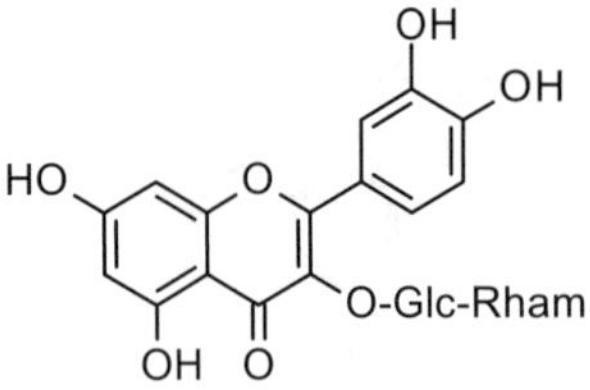

Rutin 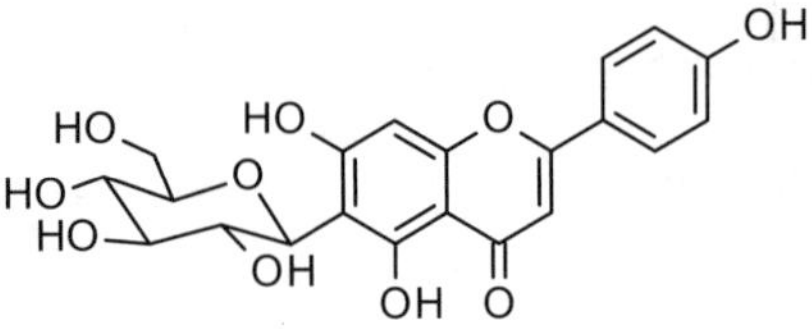   Isovitexin

Most flavonoids are potent antioxidants, which is probably the most important property of flavonoids from a therapeutic point of view. The antioxidant activity of flavonoids are the result of a high propensity to electron transfer, ferrous ions chelating activity and direct scavenging of reactive oxygen species. Several flavo-noids possess anticancer, anti-ageing, anti-inflammatory (because of their ability to diminish production of proinflammatory substances or mediators, e.g. prosta-glandins, leukotrienes, reactive oxygen species and nitric oxide), antihepatotoxic (hepatoprotective), antitumour, antimicrobial, antithrombotic (because of their ability to scavenge superoxide anions) and antiviral properties. Many traditional medicines and medicinal plants contain flavonoids as the bioactive compounds. The antioxidant properties of flavonoids present in fresh fruits and vegetables are thought to contribute to their preventative effect against cancer and heart diseases. Rutin, a flavonoid glycoside found in many plants, for example *Sophora japonica* (Fabaceae), buckwheat (*Fagopyrum esculentum*, family: Polygonaceae) and rue (*Ruta graveolens*, family: Rutaceae), is probably the most studied of all flavonoids and is included in various multivitamin preparations. Another flavonoid glycoside, hesperidin from *Citrus* peels, is also included in a number of dietary supplements and is claimed to have a beneficial effect in the treatment of capillary bleeding.

Hesperidin

8.7.3.1 Biosynthesis of Flavonoids

Structurally, flavonoids are derivatives of 1,3-diphenylpropane; for example, kaempferol. One of the phenyl groups, ring B, originates from the shikimic acid pathway, while the other ring, ring A, is from the acetate pathway through ring closure of a polyketide. One hydroxyl group in ring A is always situated in the *ortho* position to the side chain and is involved in the formation of the third six-membered ring or a five-membered ring (only found in *aurones*). The 2-phenyl side-chain of flavonoid skeleton isomerizes to the 3-position giving rise to iso-flavones; for example, formononetin. Flavonoids are biosynthesized by the phenylpropanoid metabolic pathway in which the amino acid, phenylalanine, is utilized to generate *para*-coumaroyl CoA. This combines with malonyl-CoA to yield the true backbone of flavonoids, a group of compounds known as chalcones; for example, naringenin chalcone. Conjugate ring-closure of chalcones gives the familiar form of flavonoids; for example, naringenin. The metabolic pathway continues through a series of enzymatic modifications to afford flavanones → dihydroflavonols → anthocyanins. Along this pathway, many products can be formed, including the flavonols, flavan-3-ols, proanthocyanidins (tannins) and several other polyphenolics. The biosynthesis of flavonoids can be summarized as follows.

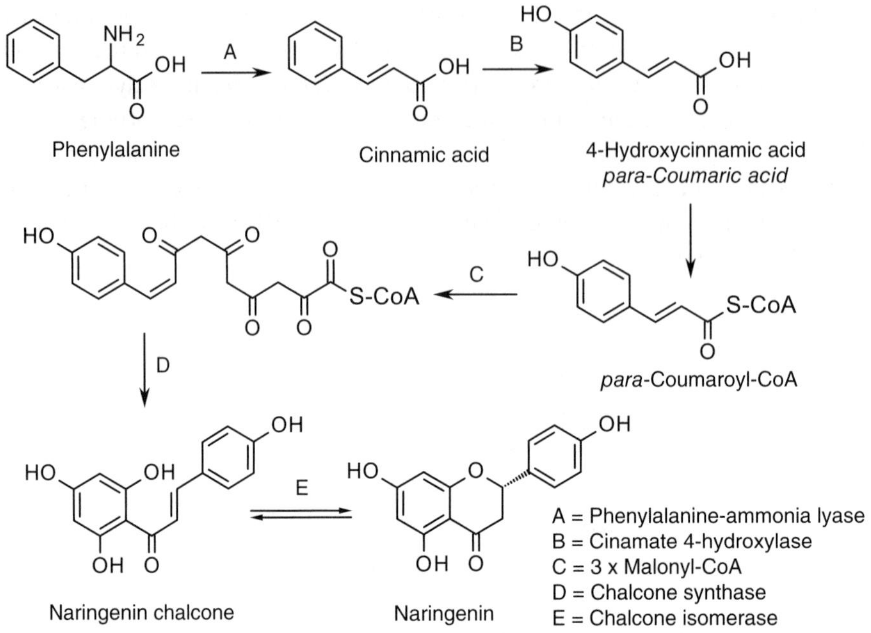

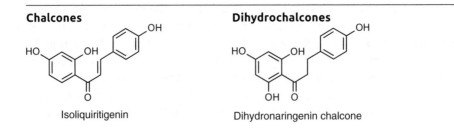

Phenylalanine

Cinnamic acid

4-Hydroxycinnamic acid
para-Coumaric acid

S-CoA

S-CoA

para-Coumaroyl-CoA

Naringenin chalcone

Naringenin

A = Phenylalanine-ammonia lyase
B = Cinamate 4-hydroxylase
C = 3 x Malonyl-CoA
D = Chalcone synthase
E = Chalcone isomerase

8.7.3.2 Classification of Flavonoids

Flavonoids can be classified according to their biosynthetic origins. Some flavonoids are both intermediates in the biosynthesis as well as the end-products; for example, chalcones, flavanones, flavanon-3-ols and flavan-3,4-diols. Other classes are only known as the end-products of biosynthesis; for example, anthocyanins, flavones and flavonols. Two further classes of flavonoids are those in which the 2-phenyl side-chain of flavonoid isomerizes to the 3-position (giving rise to isoflavones and related isoflavonoids) and then to the 4-position (giving rise to the neoflavonoids). The major classes of flavonoids, with specific examples, are summarized here.

Chalcones

Isoliquiritigenin

Dihydrochalcones

Dihydronaringenin chalcone

Flavanones

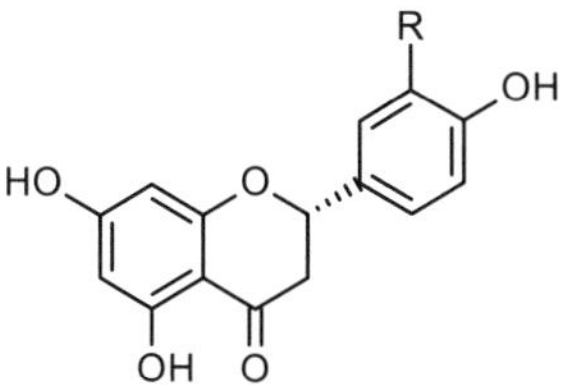

Naringenin R = H
Eriodictyol R = OH

Flavones

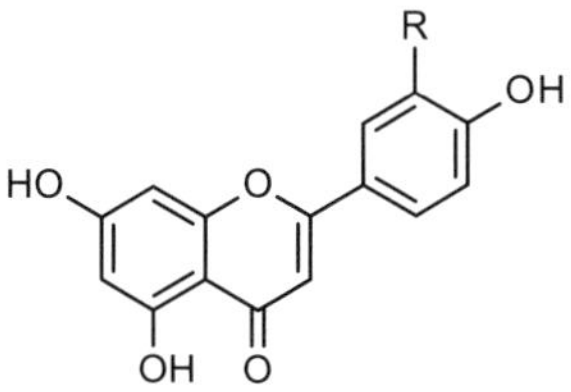

Apigenin R = H
Luteolin R = OH

Flavanon-3-ol

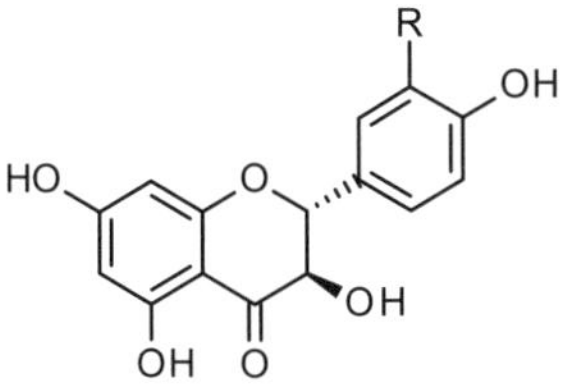

Dihydrokaempferol R = H
Dihydroquercetin R = OH

Flavonol

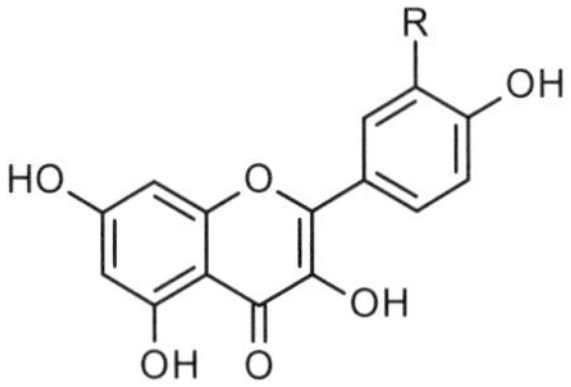

Kaempferol R = H
Quercetin R = OH

Flavan-3,4-diol

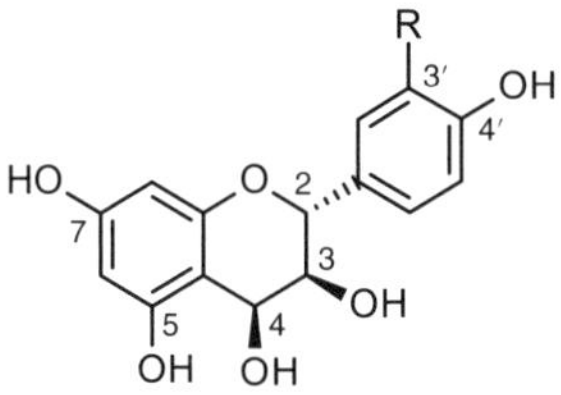

Leucopelargonidin R = H
Leucocyanidin R = OH

Flavan-3-ol

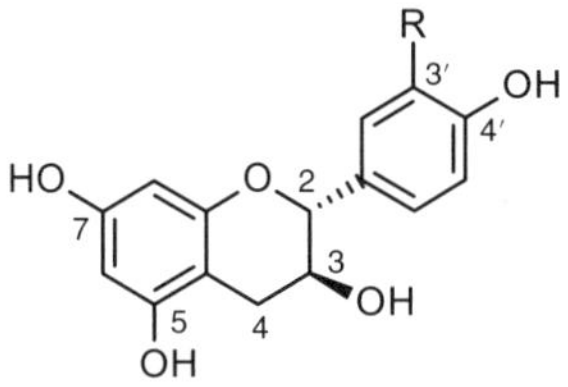

Afzalechin R = H
(+)-Catechin R = OH

Flavan

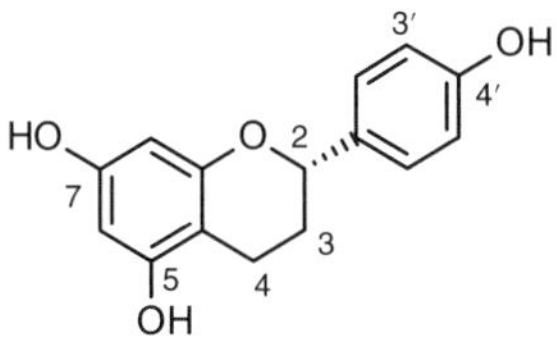

3-Deoxyafzalechin

Anthocyanidin

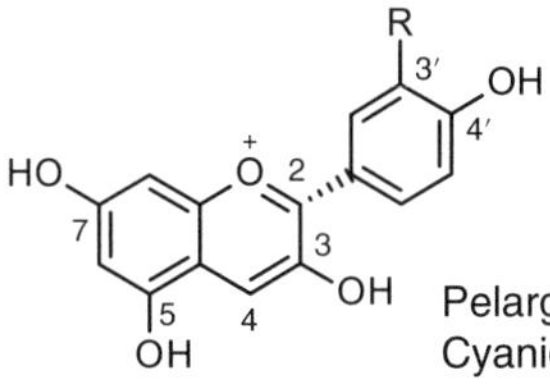

Pelargonidin R = H
Cyanidin R = OH

Flavonoid *O*-glycoside

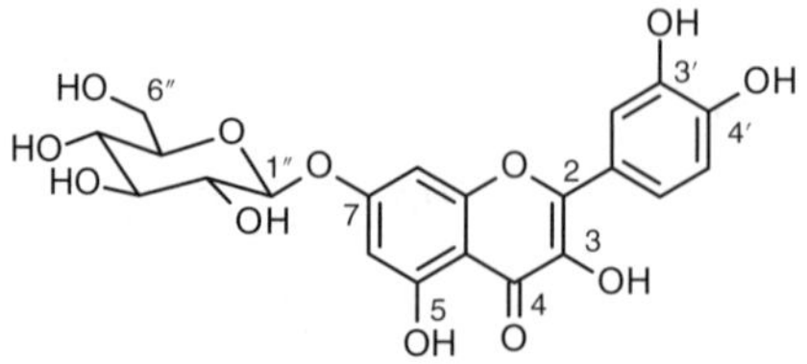

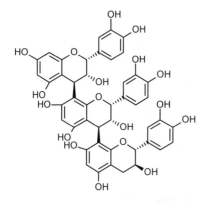

Quercetin 7-O-β-D-glucopyranoside

Flavonoid C-glycoside

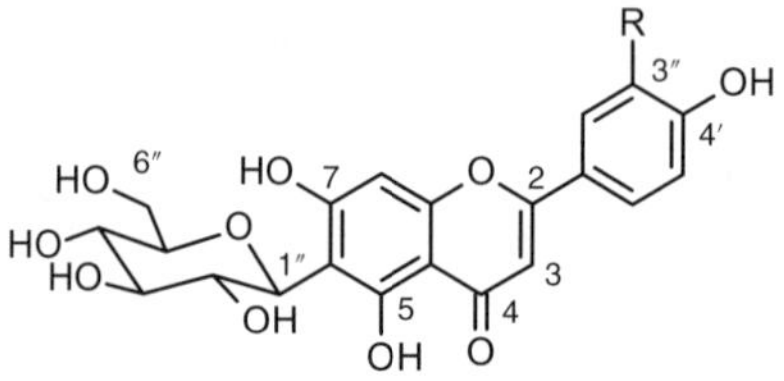

Isovitexin R = H
Isoorientin R = OH

Proanthocyanidin

Epicatechin trimer
Condensed tannin

Isoflavonoid

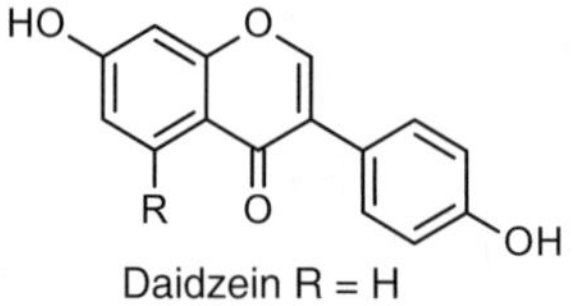

Daidzein R = H
Genistein R = OH

Biflavonoid

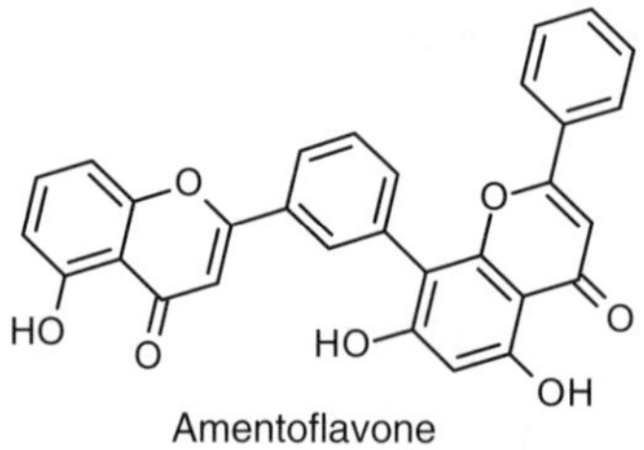

Amentoflavone

8.7.4 Lignans

Lignans are a large group of plant phenolics, biosynthesized from the union of two phenylpropane molecules; for example, both matairesinol (*Centaurea* species, family: Asteraceae) and podophyllotoxin (*Podophyllum peltatum*, family: Berberidaceae) are formed from the phenylpropane, coniferyl alcohol. Lignans are essentially cinnamoyl alcohol dimers. Several lignans are formed though further cyclization and other structural modifications; for example, dibenzylbutyrolactone and epoxy lignan.

Natural lignans are optically active, although a few *meso*-compounds exist in nature. Like any other optically active compounds, important physiological or pharmacological properties of lignans are generally associated with a particular absolute configuration; for example, the antitumour agent podophyllotoxin.

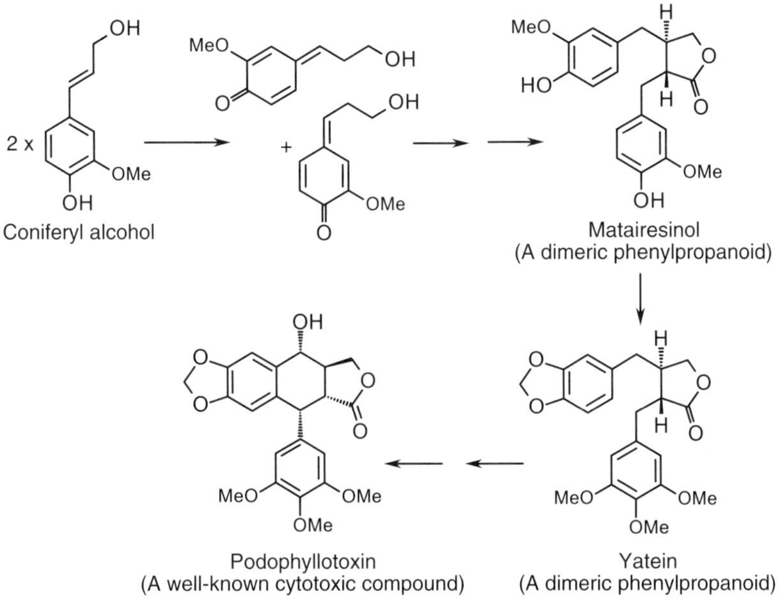

Coniferyl alcohol

Matairesinol
(A dimeric phenylpropanoid)

Podophyllotoxin
(A well-known cytotoxic compound)

Yatein
(A dimeric phenylpropanoid)

Lignans, including neolignans, are quite widespread in the plant kingdom, particularly from the families such as Asteraceae, Berberidaceae, Magnoliaceae, Pinaceae, Piperacae, Phytolaccaceae and Rutaceae, and are well-known for producing a variety of lignans. Among lignin-rich food sources, flax seeds and sesame seeds contain the highest amounts of various lignans; for example, hydroxyresinol, lariciresinol, matairesinol, pinoresinol, secolariciresinol, sesamin and syringaresinol. The main lignan precursor found in the flax seeds is secoisolariciresinol diglucoside. Other sources of lignans include cereals like barley, rye, oat, rye and wheat, soybeans and cruciferous vegetables (e.g. broccoli and cabbage) and some fruits, particularly apricots and strawberries. Secoisolariciresinol and matairesinol were the first plant lignans identified in foods.

Among the classes of identified natural products, lignans are one of the largest groups (more than 7000 naturally occurring lignans identified to date) and have been studied extensively for their diverse structures, biological activities and relevant medicinal values. Lignans possess interesting biological, pharmacological or medicinal activities, including inhibition of carcinogenesis and induction of differentiation in leukaemia or teratocarcinoma cells. In addition, lignans are known to possess antiasthma, antihypertensive, anti-inflammation, antimicrobial, antioxidant, hepatoprotective, hypocholesterolemic, insecticidal and pro-estrogen properties.

8.7.4.1 Structural Types of Lignans

Major structural types encountered in natural lignans are shown next. *Neolignans* are also included as the range of lignoids and their plant sources has widened so that the distinction between lignans and neolignans has become less important. Neolignans are also dimers of cinnamyl units but their structures are generated by coupling of mesomeric radicals other than the β-β link typical of the lignans.

Simple dibenzylbutane lignans

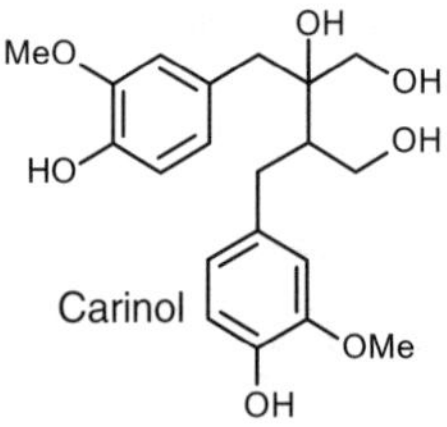

Carinol

Dibenzylbutyrolactone lignans

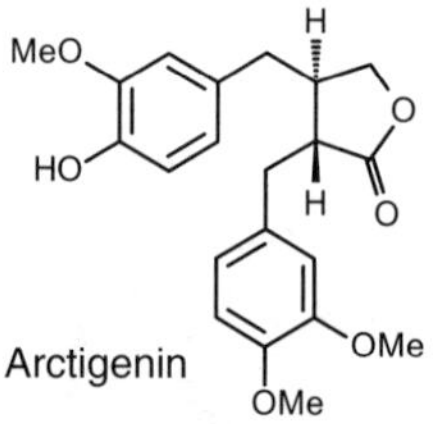

Arctigenin

Epoxy and diepoxy lignans

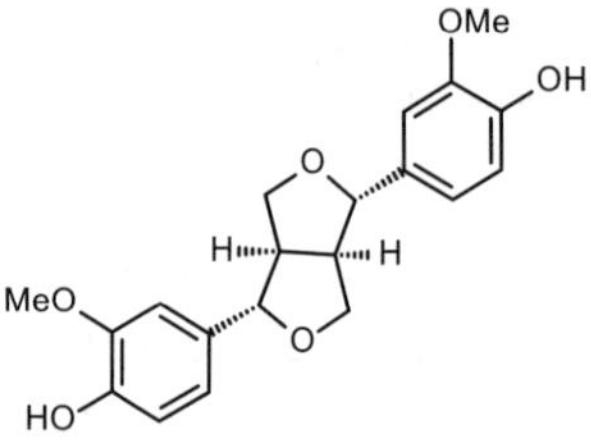

Pinoresinol

Simple aryltetralin lignans (2,7′-cyclolignans)

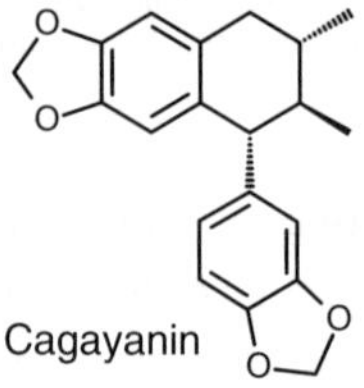

Cagayanin

Dibenzocyclooctadiene lignans (2,2′-cyclolignans)

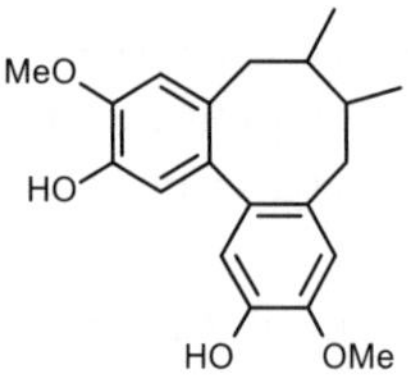

A 2,2′-cyclolignan

Neolignans

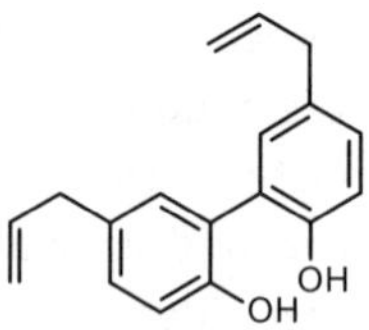

Magnolol
(A bioactive neolignan of *Magnolia* species)

8.7.5 Tannins

Plant polyphenols, also known as vegetable *tannins*, are heterogenous group of natural products widely distributed in the plant kingdom. Tannins are often present in unripe fruits, but disappear during ripening. It is believed that tannins may provide plants with protection against microbial attacks. Tannins naturally occur in black tea and their characteristics emerge when the tea is brewed a few minutes longer than recommended. After brewing the tea, take a sip and you will immediately notice a slight bitterness in the middle of your tongue and a dryness in the front of your mouth – this is a tannin!

Tannins are amorphous substances, which produce colloidal acidic aqueous solutions with astringent taste. Tannins form insoluble and indigestible compounds with proteins, and this is the basis of their extensive use in the leather industry (tanning process) and for the treatment of diarrhoea, bleeding-gum and skin injuries. Tannins are metal ion chelators, protein precipitating agents and biological antioxidants. The detoxification of snake venoms and bacterial toxins by persimmon tannin is based on the strong binding ability of this tannin with proteins. Fruits and tannin-rich extracts of fruits have shown anthelmintic, antidiabetic, anti-*Helicobacter pylori*, anti-inflammatory, antimicrobial, antioxidant, antiulcer and immune-regulating activities. Some tannins also have been reported to inhibit the growth of cancer cells and reduce DNA damage.

8.7.5.1 Classification of Tannins

Tannins can be classified into two major classes: *hydrolysable tannins* and *condensed tannins*, also known as *non-hydrolysable tannins. Phlorotannins* are another group of tannins, generally found in brown algae (e.g. *Ascophyllum nodosum*) and the structure is based on phloroglucinol units. To a much lesser extent, there are oligostilbenoids (oligo- or polystilbenes) are oligomeric forms of stilbenoids that constitute a small class of tannins.

Phloroglucinol
(The building block for phlorotannins)

On treatment with acids or enzymes, while hydrolysable tannins are split into simpler molecules, condensed tannins provide complex water-insoluble products. Hydrolysable tannins are subdivided into *gallotannins* and *ellagitannins.* Gallotannins, on hydrolysis, yield sugar and gallic acid, whereas hydrolysis of ellagitannins results in sugar, gallic acid and ellagic acid. Pentagalloylglucose, which has long been used in the tanning industry, is an example of gallotannin. In fact, hydrolysable tannins are biosynthesized from pentagalloyl glucose, which then undergoes a series of hydrolytic and oxidative couplings to give simple tannins; for example, strictinin and more complex structures where phenolic groups have undergone esterification or ether formation with additional gallic acid moieties.

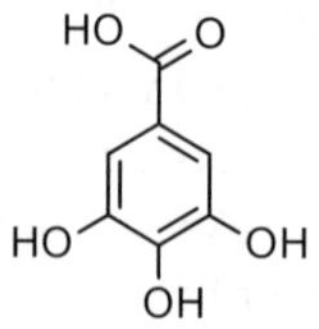

Gallic acid

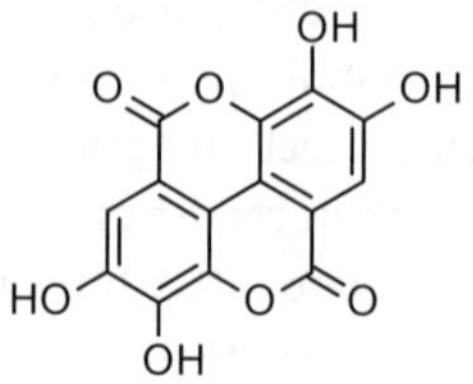

Ellagic acid

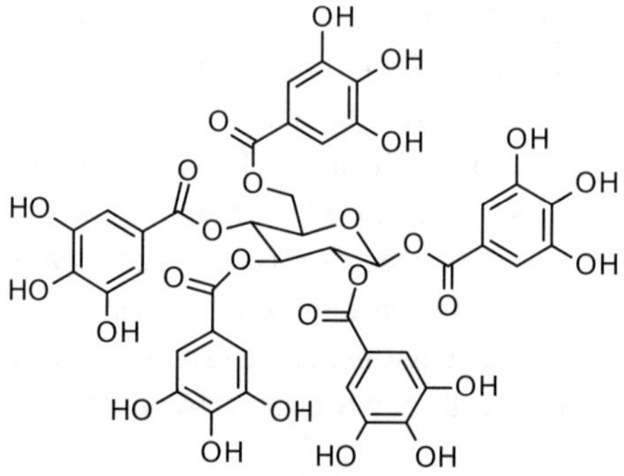

Pentagalloylglucose

Condensed tannins are complex polymers where the building blocks are usually catechins and flavonoids, esterified with gallic acid. An example is epicatechin trimer. In most cases, condensed tannins are carbon–carbon linked polymers of flavonoids.

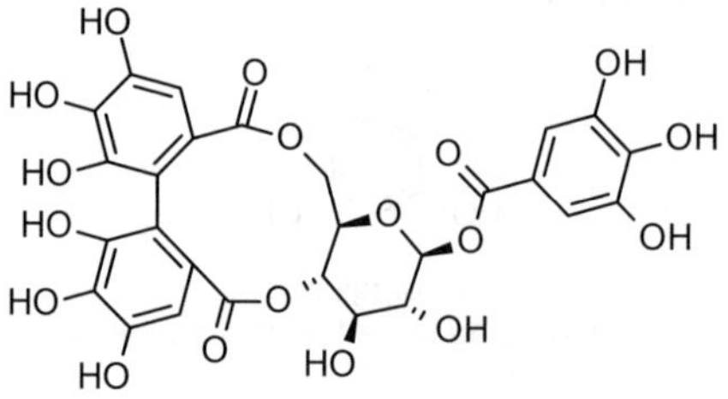

Strictinin
(Derived from pentagalloyl glucose)

8.7.5.2 Occurrence

The best known families of all species tested that contain tannin are: Aceraceae, Actinidiaceae, Anacardiaceae, Bixaceae, Burseraceae, Combretaceae, Dipterocarpaceae,

Ericaceae, Grossulariaceae and Myricaceae for dicot, and Najadaceae and Typhaceae in monocot. Other plant families that are well known for producing tannins include Asteraceae, Fagaceae, Mimosaceae, Solanaceae and Asteraceae. Some families like Boraginaceae, Cucurbitaceae and Papaveraceae contain no tannin-rich species. Various fruits and vegetables in our regular diets are rich in tannins; for example, pomegranates, strawberries, cranberries, blueberries, nuts (e.g. hazelnuts, walnuts and pecans), herbs and spices (e.g. cloves, tarragon, cumin, thyme, vanilla and cinnamon), legumes (e.g. chickpeas), chocolate and various wines and beer.

8.7.5.3 Test for Tannins

There are three groups of tests for tannins: precipitation of proteins or alkaloids, reaction with phenolic rings and de-polymerization. Two of those tests are outlined next.

8.7.5.3.1 Goldbeater's Skin Test

In this test, goldbeater's skin or ox skin is dipped in hydrochloric acid, rinsed in water, soaked in the tannin solution for 5 minutes, washed in water and then treated with 1% $FeSO_4$ solution. It gives a blue black colour indicating the presence of tannins.

8.7.5.3.2 Ferric Chloride (FeCl₃) Test

This test is actually for phenolic compounds. Any compound possessing a phenolic hydroxyl functionality will show a positive result in this test. Powdered plant leaves of the test plant (1.0 g) are weighed, distilled water (10 ml) is added and the mixture is boiled for 5 minutes. Two drops of 5% $FeCl_3$ are then added. Production of a greenish precipitate is an indication of the presence of tannins.

Index

Chemistry for Pharmacy Students: General, Organic and Natural Product Chemistry,
Second Edition. Lutfun Nahar and Satyajit Sarker.
© 2019 John Wiley & Sons Ltd. Published 2019 by John Wiley & Sons Ltd.